Cambridge
International AS and A Level

Biology

C J Clegg

D1347925

HODDER
EDUCATION
AN HACHETTE UK COMPANY

The cover image is of an orang-utan, mother and infant. The name of these primates means 'people of the forest', and they share over 96 per cent of our own genetic make-up. Today, they live in the forests of Indonesia and Malaysia only. The future of orang-utan populations is under threat from deforestation.

Hachette UK's policy is to use papers that are natural, renewable and recyclable products and made from wood grown in sustainable forests. The logging and manufacturing processes are expected to conform to the environmental regulations of the country of origin.

Orders: please contact Bookpoint Ltd, 130 Milton Park, Abingdon, Oxon OX14 4SB. Telephone: (44) 01235 827720. Fax: (44) 01235 400454. Lines are open 9.00–5.00, Monday to Saturday, with a 24-hour message answering service. Visit our website at www.hoddereducation.com

© C J Clegg 2014
First published in 2014 by
Hodder Education, a Hachette UK company
338 Euston Road
London NW1 3BH

Impression number   5 4 3 2 1
Year   2018 2017 2016 2015 2014

Cover photo by Eric Gevaert
Typeset in ITC Garamond by Aptara, Inc.
Printed in Italy
A catalogue record for this title is available from the British Library
ISBN 978 1444 17534 9

Contents

Contents

Student's CD contents

Appendix 1: Background chemistry for biologists
Appendix 2: Investigations, data handling and statistics
Appendix 3: Preparing for your exam

Also, for each topic:

- An interactive test
- A list of key terms
- A topic summary
- Additional work on data handling and practical skills
- Suggested websites and further reading
- A revision checklist
- Answers to all the examination-style questions

Introduction

Cambridge International AS and A Level Biology is an excellent introduction to the subject and a sound foundation for studies beyond A Level, in further and higher education, for professional courses and for productive employment in the future. Successful study of this programme gives lifelong skills, including:

- confidence in a technological world and informed interest in scientific matters
- understanding of how scientific theories and methods have developed
- awareness of the applications of biology in everyday life
- ability to communicate effectively
- concern for accuracy and precision
- awareness of the importance of objectivity, integrity, enquiry, initiative and inventiveness
- understanding of the usefulness and limitations of scientific methods and their applications
- appreciation that biology is affected by social, economic, technological, ethical and cultural factors
- knowledge that biological science overcomes national boundaries
- awareness of the importance of IT
- understanding of the importance of safe practice
- an interest and care for the local and global environment and their conservation.

This book is designed to serve students as they strive for these goals.

The structure of the book

The Cambridge International Examinations AS and A Level Biology syllabus is presented in sections. The contents of this book follows the syllabus sequence, with each section the subject of a separate topic.

Topics 1 to 11 cover Sections 1 to 11 of the AS Level syllabus and are for all students. AS students are assessed only on these.

Topics 12 to 19 cover Sections 12 to 19, the additional sections of the syllabus for A Level students only.

In addition, there are the answers to the self-assessment questions.

Cambridge International AS and A Level Biology has many special features.

- Each topic begins with the syllabus learning outcomes which identify essential objectives.
- The text is written in straightforward language, uncluttered by phrases or idioms that might confuse students for whom English is a second language.
- Photographs, electron micrographs and full-colour illustrations are linked to support the relevant text, with annotations included to elaborate the context, function or applications.
- Explanations of structure are linked to function. The habitat and environment of organisms are identified where appropriate. Application of biology to industries and the economic, environmental and ethical consequences of developments are highlighted, where appropriate.
- Processes of science (scientific methods) and something of the history of developments are introduced selectively to aid appreciation of the possibilities and limitations of science.
- Questions are included to assist comprehension and recall. Answers to these are given at the back of the book. At the end of each topic, examination-style questions are given. Answers to these are given on the CD.

Introduction

A new feature of the syllabus is Key concepts. These are the essential ideas, theories, principles or mental tools that help learners to develop a deep understanding of their subject, and make links between different topics. An icon indicates where each Key concept is covered:

 Cells as the units of life
A cell is the basic unit of life and all organisms are composed of one or more cells. There are two fundamental types of cell: prokaryotic and eukaryotic.

 Biochemical processes
Cells are dynamic: biochemistry and molecular biology help to explain how and why cells function as they do.

 DNA, the molecule of heredity
Cells contain the molecule of heredity, DNA. Heredity is based on the inheritance of genes.

 Natural selection
Natural selection is the major mechanism to explain the theory of evolution.

 Organisms in their environment
All organisms interact with their biotic and abiotic environment.

 Observation and experiment
The different fields of biology are intertwined and cannot be studied in isolation: observation and enquiry, experimentation and fieldwork are fundamental to biology.

Author's acknowledgements

I am indebted to the experienced international teachers and the students who I have been privileged to meet in Asia and in the UK in the process of preparing this material. I am especially indebted to Christine Lea, an experienced teacher and examiner of Biology who has guided me topic by topic on the special needs of the students for whom this book is designed.

Finally, I am indebted to the publishing team of project editor, Lydia Young, editor Joanna Silman and designer Melissa Brunelli at Hodder Education, and to freelance editor Penny Nicholson whose skill and patience have brought together text and illustration as I have wished. I am most grateful to them.

Dr Chris Clegg
Salisbury, Wiltshire, UK
June, 2014

1 Cell structure

All organisms are composed of cells. Knowledge of their structure and function underpins much of biology. The fundamental differences between eukaryotic and prokaryotic cells are explored and provide useful biological background for the topic on Infectious disease. Viruses are introduced as non-cellular structures, which gives candidates the opportunity to consider whether cells are a fundamental property of life.

The use of light microscopes is a fundamental skill that is developed in this topic and applied throughout several other sections of the syllabus. Throughout the course, photomicrographs and electron micrographs from transmission and scanning electron microscopes should be studied.

1.1 The microscope in cell studies

An understanding of the principles of microscopy shows why light and electron microscopes have been essential in improving our knowledge of cells.

By the end of this section you should be able to:

a) compare the structure of typical animal and plant cells by making temporary preparations of live material and using photomicrographs

b) calculate the linear magnifications of drawings, photomicrographs and electron micrographs

c) use an eyepiece graticule and stage micrometer scale to measure cells and be familiar with units (millimetre, micrometre, nanometre) used in cell studies

d) explain and distinguish between resolution and magnification, with reference to light microscopy and electron microscopy

e) calculate actual sizes of specimens from drawings, photomicrographs and electron micrographs

Introducing cells

The cell is the basic unit of living matter – the smallest part of an organism which we can say is alive. It is cells that carry out the essential processes of life. We think of them as self-contained units of structure and function. Some organisms are made of a single cell and are known as **unicellular**. Examples of unicellular organisms are introduced in Figure 1.1. In fact, there are vast numbers of different unicellular organisms in the living world, many with a very long evolutionary history.

Other organisms are made of many cells and are known as **multicellular** organisms. Examples of multicellular organisms are the mammals and flowering plants. Much of the biology in this book is about multicellular organisms, including humans, and the processes that go on in these organisms. But remember, single-celled organisms carry out all the essential functions of life too, only these occur within the single cell.

Question

1 State the essential processes characteristic of living things.

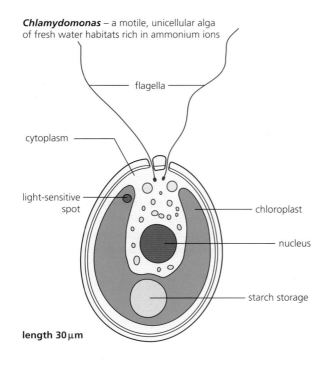

Chlamydomonas – a motile, unicellular alga of fresh water habitats rich in ammonium ions

flagella

cytoplasm

light-sensitive spot

chloroplast

nucleus

starch storage

length 30 µm

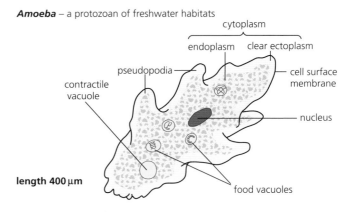

Amoeba – a protozoan of freshwater habitats

cytoplasm

endoplasm   clear ectoplasm

pseudopodia

contractile vacuole

cell surface membrane

nucleus

food vacuoles

length 400 µm

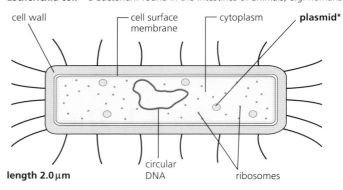

Escherichia coli – a bacterium found in the intestines of animals, e.g. humans

cell wall

cell surface membrane

cytoplasm

plasmid*

circular DNA

ribosomes

length 2.0 µm

Figure 1.1 Introducing unicellular organisation
*Plasmids are illustrated in Figure 1.24 (page 24) and in Figure 19.5 (page 459).

Cell size

Cells are extremely small – most are only visible as distinct structures when we use a **microscope** (although a few types of cells are just large enough to be seen by the naked eye).

Observations of cells were first reported over 300 years ago, following the early development of microscopes (see Figure 1.2). Today we use a **compound light microscope** to investigate cell structure – perhaps you are already familiar with the light microscope as a piece of laboratory equipment. You may have used one to view living cells, such as the single-celled animal, *Amoeba*, shown in Figure 1.1.

Since cells are so small, we need suitable units to measure them. The **metre** (symbol **m**) is the standard unit of length used in science. This is an internationally agreed unit, or **SI unit**. Look at Table 1.1 below. This shows the subdivisions of the metre that we use to measure cells and their contents. These units are listed in descending order of size. You will see that each subdivision is $\frac{1}{1000}$ of the unit above it. The smallest units are probably quite new to you; they may take some getting used to.

The dimensions of cells are expressed in the unit called a **micrometer** or micron (**µm**). Notice this unit is one thousandth (10^{-3}) of a millimetre. This gives us a clear idea about how small cells are when compared to the millimetre, which you can see on a standard ruler.

Bacteria are really small, typically 0.5–10 µm in size, whereas the cells of plants and animals are often in the range 50–150 µm, or larger. In fact, the lengths of the unicellular organisms shown in Figure 1.1 are approximately:

Chlamydamonas	30 µm
Amoeba	400 µm (but its shape and therefore length varies greatly)
Escherichia coli	2 µm

Table 1.1 Units of length used in microscopy

1 metre (**m**) = 1000 millimetres (**mm**)
1 mm = 1000 micrometres (**µm**) (or microns)
1 µm = 1000 nanometres (**nm**)

Question

2 Calculate
  a how many cells of 100 µm diameter will fit side by side along a millimetre
  b the magnification of the image of *Escherichia coli* in Figure 1.1.

Hooke's microscope, and a drawing of the cells he observed

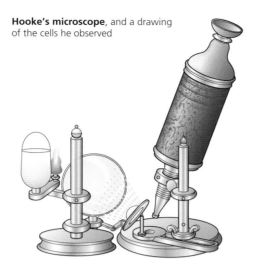

Robert Hooke (1662), an expert mechanic and one of the founders of the Royal Society in London, was fascinated by microscopy. He devised a compound microscope, and used it to observe the structure of cork. He described and drew cork cells, and also measured them. He was the first to use the term 'cells'.

Anthony van Leeuwenhoek (1680) was born in Delft. Despite no formal training in science, he developed a hobby of making lenses, which he mounted in metal plates to form simple microscopes. Magnifications of × 240 were achieved, and he observed blood cells, sperms, protozoa with cilia, and even bacteria (among many other types of cells). His results were reported to the Royal Society, and he was elected a fellow.

Robert Brown (1831), a Scottish botanist, observed and named the cell nucleus. He also observed the random movements of tiny particles (pollen grains, in his case) when suspended in water (Brownian motion).

Matthias Schleiden (1838) and Theodor Schwann (1839), German biologists, established cells as the natural unit of form and function in living things: 'Cells are organisms, and entire animals and plants are aggregates of these organisms arranged to definite laws.'

Rudolf Virchow (1856), a German pathologist, established the idea that cells arise only by division of existing cells.

Louis Pasteur (1862), a brilliant French microbiologist, established that life does not spontaneously generate. The bacteria that 'appear' in broth are microbes freely circulating in the air, which contaminate exposed matter.

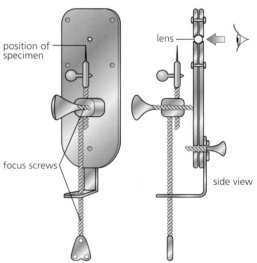

position of specimen

focus screws

lens

side view

Leeuwenhoek's microscope

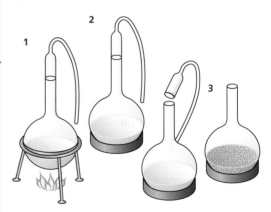

Pasteur's experiment, in which broth was sterilised (**1**), and then either exposed to air (**3**) or protected from air-borne spores in a swan-necked flask (**2**). Only the broth in **3** became contaminated with bacteria.

Figure 1.2 Early steps in the development of the cell theory

Cell theory

Many biologists helped to develop the idea that living things are made of cells. This idea has become known as the **cell theory**. This concept evolved gradually during the nineteenth century, following a steadily accelerating pace in the development of **microscopy** and **biochemistry**. You can see a summary of these developments in Figure 1.2.

Today we recognise that the statement that cells are the unit of structure and function in living things really contains three basic ideas.

- Cells are the building blocks of structure in living things.
- Cells are the smallest unit of life.
- Cells are made from other cells (pre-existing cells) by division.

To this we can now confidently add two concepts.

- Cells contain a blueprint (information) for their growth, development and behaviour.
- Cells are the site of all the chemical reactions of life (metabolism).

We will return to these points later.

Introducing animal and plant cells

No 'typical' cell exists – there is great variety among cells. However, we shall see that most cells have features in common. Using a compound microscope, the *initial* appearance of a cell is of a sac of fluid material, bound by a membrane and containing a nucleus. Look at the cells in Figure 1.3.

We see that a cell consists of a nucleus surrounded by cytoplasm, contained within the cell surface membrane. The **nucleus** is the structure that controls and directs the activities of the cell. The **cytoplasm** is the site of the chemical reactions of life, which we call 'metabolism'. The **cell surface membrane**, sometimes called the plasma membrane, is the barrier controlling entry to and exit from the cytoplasm.

Animal and plant cells have these three structures in common. In addition, there are many tiny structures in the cytoplasm, called **organelles**. Most of these organelles are found in both animal and plant cells. An organelle is a discrete structure within a cell, having a specific function. Organelles are all too small to be seen at this magnification. We have learnt about the structure of organelles using the **electron microscope** (see page 12).

There are also some important basic differences between plant and animal cells. For example, there is a tough, slightly elastic **cell wall**, made largely of cellulose, present around plant cells. Cell walls are absent from animal cells.

A **vacuole** is a fluid filled space within the cytoplasm, surrounded by a single membrane. Plant cells frequently have a large permanent vacuole. By contrast, animal cells may have small vacuoles, and these are often found to be temporary structures.

Green plant cells contain organelles called **chloroplasts** in their cytoplasm. These are not found in animal cells. Chloroplasts are where green plant cells manufacture food molecules by a process known as **photosynthesis**.

The **centrosome**, an organelle that lies close to the nucleus in animal cells, is not present in plants. This tiny organelle is involved in nuclear division in animal cells (see page 107).

Finally, the way organisms store energy-rich reserves differs, too. Animal cells may store **glycogen** (see page 39); plants cells normally store **starch** (see page 38).

Question

3 Draw up a table to highlight the differences between plant and animal cells.

Cells become specialised

Newly formed cells grow and enlarge. A growing cell normally divides into two cells. However, cell division in multicellular organisms is very often restricted to cells which have not specialised. In multicellular organisms the majority of cells become specialised in their structure and in the functions they carry out. As a result, many fully specialised cells are no longer able to divide.

Another outcome of specialisation is that cells show great variety in shape and structure. This variety reflects how these cells have adapted to different environments and to different tasks within multicellular organisms. We will return to these points later.

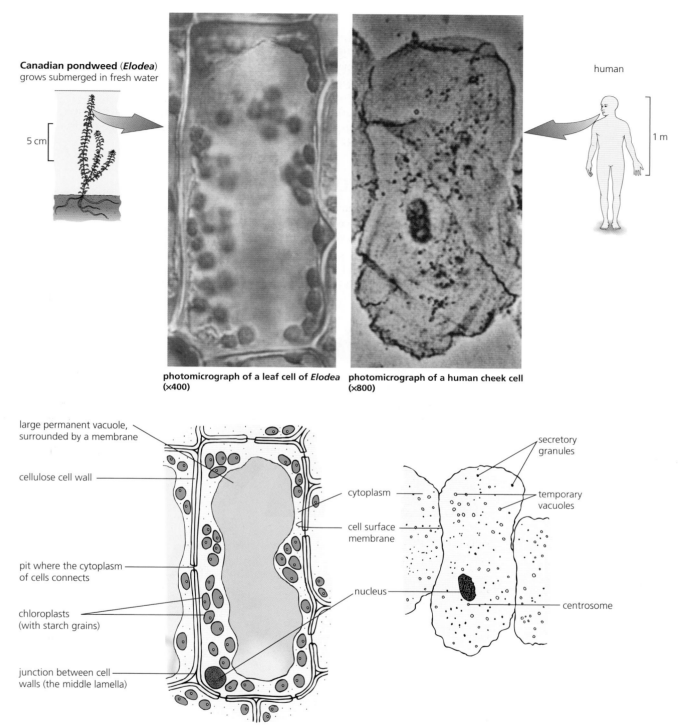

Canadian pondweed (*Elodea*) grows submerged in fresh water

5 cm

human

1 m

photomicrograph of a leaf cell of *Elodea* (×400)

photomicrograph of a human cheek cell (×800)

large permanent vacuole, surrounded by a membrane

cellulose cell wall

pit where the cytoplasm of cells connects

chloroplasts (with starch grains)

junction between cell walls (the middle lamella)

secretory granules

cytoplasm

temporary vacuoles

cell surface membrane

nucleus

centrosome

Figure 1.3 Plant and animal cells from multicellular organisms

Microscopy

A microscope is used to produce a magnified image of an object or specimen. Today, cells can be observed by two fundamentally different types of microscopy:
- the compound light microscope, using visible light;
- the electron microscope, using a beam of electrons.

In this course you will be using the light microscope frequently, and we start here, using it to observe temporary preparations of living cells. Later, we will introduce the electron microscope and the changes this has brought to the study of cell structure.

Light microscopy

We use microscopes to magnify the cells of biological specimens in order to see them at all. Figure 1.4 shows two types of light microscope.

In the simple microscope (the **hand lens**), a single biconvex lens is held in a supporting frame so that the instrument can be held close to the eye. Today a hand lens is used to observe external structure. However, some of the earliest detailed observations of living cells were made with single-lens instruments (see Figure 1.2).

Figure 1.4 Light microscopy

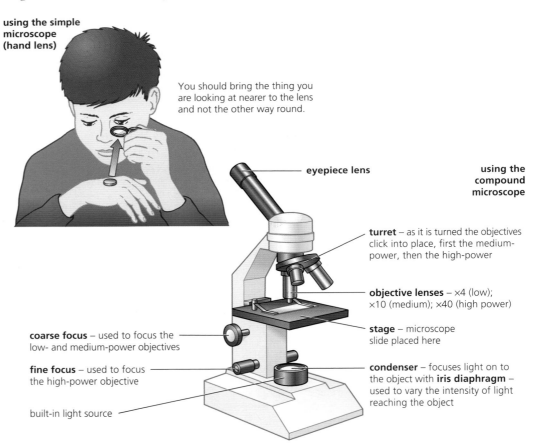

using the simple microscope (hand lens)

You should bring the thing you are looking at nearer to the lens and not the other way round.

using the compound microscope

eyepiece lens

turret – as it is turned the objectives click into place, first the medium-power, then the high-power

objective lenses – ×4 (low); ×10 (medium); ×40 (high power)

stage – microscope slide placed here

condenser – focuses light on to the object with **iris diaphragm** – used to vary the intensity of light reaching the object

coarse focus – used to focus the low- and medium-power objectives

fine focus – used to focus the high-power objective

built-in light source

In the **compound microscope**, light rays are focused by the **condenser** on to a specimen on a microscope slide on the stage of the microscope. Light transmitted through the specimen is then focused by two sets of lenses (hence the name 'compound' microscope). The **objective lens** forms an image (in the microscope tube) which is then further magnified by the **eyepiece lens**, producing a greatly enlarged image.

Cells and tissues examined with a compound microscope must be sufficiently transparent for light rays to pass through. When bulky tissues and parts of organs are to be examined, thin sections are cut. Thin sections are largely colourless.

Examining the structure of living cells

Living cells are not only tiny but also transparent. In light microscopy it is common practice to add dyes or stains to introduce sufficient contrast and so differentiate structure. Dyes and stains that are taken up by living cells are especially useful.

1 Observing the nucleus and cytoplasm in onion epidermis cells.

A single layer of cells, known as the epidermis, covers the surface of a leaf. In the circular leaf bases that make up an onion bulb, the epidermis is easily freed from the cells below, and can be lifted away from a small piece of the leaf with fine forceps. Place this tiny sheet of tissue on a microscope slide in a drop of water and add a cover slip. Irrigate this temporary mount with iodine (I_2/KI) solution (Figure 1.5). In a few minutes the iodine will penetrate the cells, staining the contents yellow. The nucleus takes up the stain more strongly than the cytoplasm, whilst the vacuole and the cell walls are not stained at all.

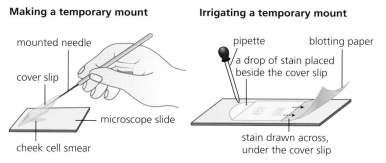

Figure 1.5 Preparing living cells for light microscopy

2 Observing chloroplasts in moss leaf cells.

A leaf of a moss plant is typically mostly only one cell thick. Remove a leaf from a moss plant, mount it in water on a microscope slide and add a cover slip. Then examine individual cells under medium and high power magnification. No stain or dye is used in this investigation.

What structures in these plant cells are visible?

3 Observing nucleus, cytoplasm and cell membrane in human cheek cells.

Take a smear from the inside lining of your cheek, using a fresh, unused cotton bud you remove from the pack. Touch the materials removed by the 'bud' onto the centre of a microscope slide, and then immediately submerge your cotton bud in **1%** sodium hypochlorite solution (or in absolute alcohol). Handle the microscope slide yourself, and at the end of the observation immerse the slide in 1% sodium hypochlorite solution (or in absolute alcohol). To observe the structure of human cheek cells, irrigate the slide with a drop of methylene blue stain (Figure 1.5), and examine some of the individual cells with medium and high power magnification.

How does the structure of these cells differ from plant cells?

4 Examining cells seen in prepared slides and in photomicrographs.

The structures of cells can also be observed in prepared slides and in photomicrographs made from prepared slides. You might choose to examine the cells in mammalian blood smears and a cross-section of a flowering plant leaf, for example. Alternatively (or in addition) you can examine photomicrographs of these (Figure 8.2b on page 153 and Figure 7.2 on page 130).

Recording observations

What you see with a compound microscope may be recorded by **drawings** of various types. For a clear, simple drawing:

- use a sharp HB pencil and a clean eraser
- use unlined paper and a separate sheet for each specimen you record
- draw clear, sharp outlines and avoiding shading or colouring
- use most of the available space to show all the features observed in the specimen
- label each sheet or drawing with the species, conditions (living or stained), transverse section (TS) or longitudinal section (LS), and so forth
- label your drawing fully, with labels positioned clear of the structures shown, remembering that label lines should not cross
- annotate (add notes about function, role or development), if appropriate
- include a statement of the magnification under which the specimen has been observed (for example, see pages 9–10).

view (phase contrast) of the layer of the cells (epithelium) lining the stomach wall

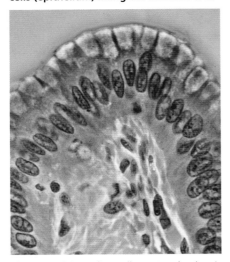

The lining of the stomach consists of columnar epithelium. All cells secrete mucus copiously.

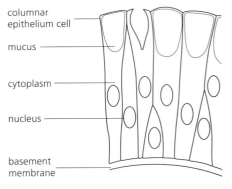

columnar epithelium cell

mucus

cytoplasm

nucleus

basement membrane

Figure 1.6 Recording cell structure by drawing

Extension

Alternatively, images of cells and tissues viewed may be further magnified, displayed or projected (and saved for printing out) by the technique of **digital microscopy**. A digital microscope is used, or alternatively an appropriate video camera is connected by a microscope coupler or eyepiece adaptor that replaces the standard microscope eyepiece. Images are displayed via video recorder, TV monitor or computer.

Question

4 What are the difficulties in trying to describe a typical plant or animal cell?

Measuring microscopic objects

The size of a cell can be measured under the microscope. A transparent scale called a **graticule** is mounted in the eyepiece at a point called the focal plane. There is a ledge for it to rest on. In this position, when the object under observation is in focus, so too is the scale. The size (e.g. length, diameter) of the object may then be recorded in arbitrary units.

Next, the graticule scale is calibrated using a **stage micrometer**. This is a tiny, transparent ruler which is placed on the microscope stage in place of the slide and then observed. With the graticule and stage micrometer scales superimposed, the actual dimensions of the object can be estimated in microns. Figure 1.7 shows how this is done. Once the size of a cell has been measured, a **scale bar** line may be added to a micrograph or drawing to record the actual size of the structure, as illustrated in the photomicrograph in Figure 1.8. Alternatively, the magnification can be recorded.

Figure 1.7 Measuring the size of cells

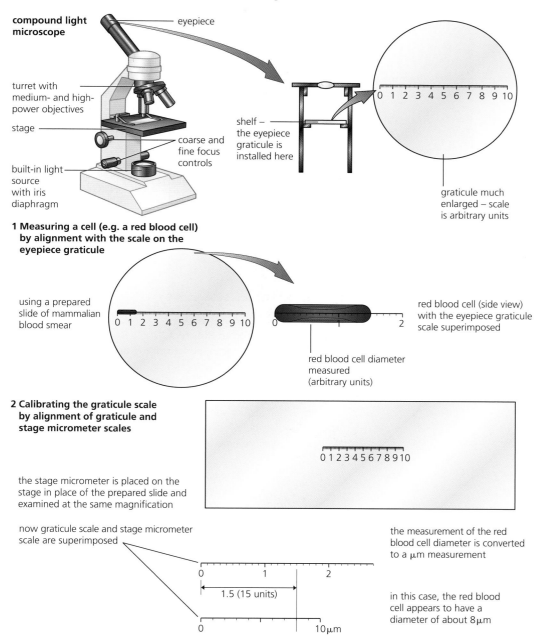

compound light microscope — eyepiece

turret with medium- and high-power objectives

stage

coarse and fine focus controls

built-in light source with iris diaphragm

shelf – the eyepiece graticule is installed here

graticule much enlarged – scale is arbitrary units

1 Measuring a cell (e.g. a red blood cell) by alignment with the scale on the eyepiece graticule

using a prepared slide of mammalian blood smear

red blood cell (side view) with the eyepiece graticule scale superimposed

red blood cell diameter measured (arbitrary units)

2 Calibrating the graticule scale by alignment of graticule and stage micrometer scales

the stage micrometer is placed on the stage in place of the prepared slide and examined at the same magnification

now graticule scale and stage micrometer scale are superimposed

1.5 (15 units)

10 μm

the measurement of the red blood cell diameter is converted to a μm measurement

in this case, the red blood cell appears to have a diameter of about 8 μm

Calculating the linear magnification of drawings, photomicrographs (and electron micrographs)

Since cells and the structures they contain are small, the images of cells and cell structures (photographs or drawings) that we make are always highly magnified so that the details of structure can be observed and recorded. Because of this, these images typically show a scale bar to indicate the actual size of the cell or structure. Look at Figure 1.8 on the next page – this is a case in point.

From the scale bar we can calculate both the size of the image and the magnification of the image.

**photomicrograph of *Amoeba proteus* (living specimen) –
phase contrast microscopy**

interpretive drawing

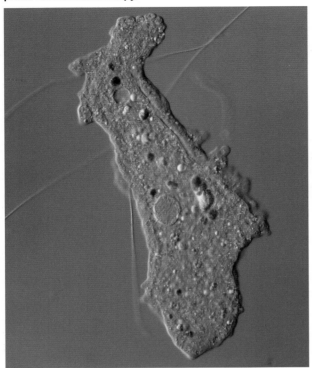

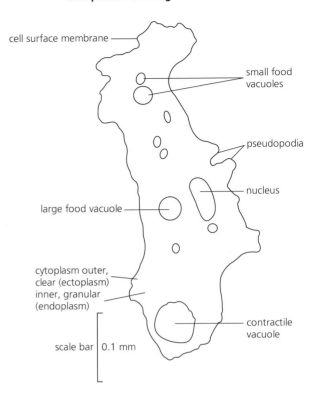

cell surface membrane

small food vacuoles

pseudopodia

nucleus

large food vacuole

cytoplasm outer,
clear (ectoplasm)
inner, granular
(endoplasm)

contractile vacuole

scale bar | 0.1 mm

Figure 1.8 Recording size by means of scale bars

1 Size of the Amoeba in Figure 1.7.

Use a ruler to measure the length of the image of cell. This is 95 mm.

Use a ruler to measure the length of the scale bar. This is 19 mm.

(Note that the scale represents an actual length of 0.1 mm.)

It is very important in these calculations to make sure that the units for the size of the image and the actual size of the specimen are the same – either millimetres (mm) or micrometres (μm). Millimetres can be converted to micrometres by multiplying by one thousand. Micrometres can be converted to millimetres by dividing by one thousand.

So:

The length of the image of the cell is $95 \times 1000 \, \mu m$ $= 95\,000 \, \mu m$
The length of the scale bar is $19 \times 1000 \, \mu m$ $= 19\,000 \, \mu m$
The scale represents an actual length of $0.1 \times 1000 \, \mu m = 100 \, \mu m$

We use the ratio of these values to work out the actual length of the *Amoeba*.

$$\frac{100 \, \mu}{19\,000 \, \mu m} = \frac{\text{actual length of the cell}}{95\,000 \, \mu m}$$

$$\text{actual length} = 100 \, \mu m \times \frac{95\,000 \, \mu m}{19\,000 \, \mu m}$$

$$= 500 \, \mu m$$

(Note that *Amoeba* moves about by streaming movements of its cytoplasm. In this image the cell is extended and its length seems large, perhaps. It is equally likely to be photographed in a more spherical shape, of diameter one tenth of its length here.)

2 Magnification of the Amoeba.

We use the formula: $\quad \text{magnification} = \dfrac{\text{measured size of the cell}}{\text{actual length of the cell}}$

So here: $\quad \text{magnification} = \dfrac{95\,000 \, \mu m}{500 \, \mu m} = \times 190$

Question

5 Using the magnification given for the photomicrograph of the cell of *Elodea* in Figure 1.3 (page 5), calculate the actual length of the cell.

Magnification: the size of an image of an object compared to the actual size. It is calculated using the formula $M = I \div A$ (M is magnification, I is the size of the image and A is the actual size of the object, using the **same units** for both sizes). This formula can be rearranged to give the actual size of an object where the size of the image and magnification are known: $A = I \div M$.

Question

6 Calculate the magnification obtained with a ×6 eyepiece and a ×10 objective lens.

Resolution: the ability of a microscope to distinguish two objects as separate from one another. The smaller and closer together the objects that can be distinguished, the higher the resolution. Resolution is determined by the wavelength of the radiation used to view the specimen. If the parts of the specimen are smaller than the wavelength of the radiation, then the waves are not stopped by them and they are not seen. Light microscopes have limited resolution compared to electron microscopes because light has a much longer wavelength than the beam of electrons in an electron microscope.

3 Finally, given the magnification of an image, we can calculate its real size.

Look at the images of the human cheek cell in Figure 1.3 (page 5).

Measure the observed length of the cell in mm. It is 90 mm.

Convert this length to µm: 90 mm = 90 × 1000 µm = 90 000 µm

Use the equation: actual size (A) = $\dfrac{\text{image size (I)}}{\text{magnification (M)}} = \dfrac{90\,000}{\times 800} = 112.5\ \mu m$

This size is greater than many human cheek cells. It suggests these epithelial cells are squashed flat in the position they have in the skin, perhaps.

The magnification and resolution of an image

Magnification is the number of times larger an image is than the specimen. The magnification obtained with a compound microscope depends on which of the lenses you use. For example, using a ×10 eyepiece and a ×10 objective lens (medium power), the image is magnified ×100 (10 × 10). When you switch to the ×40 objective lens (high power) with the same eyepiece lens, then the magnification becomes ×400 (10 × 40). These are the most likely orders of magnification used in your laboratory work.

Actually, there is no limit to magnification. For example, if a magnified image is photographed, then further enlargement can be made photographically. This is what may happen with photomicrographs shown in books and articles.

Magnification is given by the formula: $\dfrac{\text{size of image}}{\text{size of speciman}}$

So, if a particular plant cell with a diameter of 150 µm is photographed with a microscope and the image is enlarged photographically, so that in a print of the cell the diameter is 15 cm (150 000 µm), the magnification is: $\dfrac{150\,000}{150} = 1000$.

If a further enlargement is made, to show the same cell at a diameter of 30 cm (300 000 µm), then the magnification is: $\dfrac{300\,000}{150} = 2000$.

In this particular case the image size has been doubled, but the **detail will be no greater**. You will not be able to see, for example, details of cell surface membrane structure however much the image is enlarged. This is because the layers making up a cell's membrane are too thin to be seen as separate structures using the light microscope.

The **resolving power** (**resolution**) of a microscope is its ability to separate small objects which are very close together. If two separate objects cannot be resolved they will be seen as one object. Merely enlarging them will not separate them.

The resolution achieved by a light microscope is determined by the wavelength of light. Visible light has a wavelength in the range 400–700 nm. (By 'visible' we mean our eyes and brain can distinguish light of wavelength of 400 nm (violet light) from light of wavelength of 700 nm, which is red light.)

In a microscope, the limit of resolution is approximately half the wavelength of light used to view the object. So, any structure in a cell that is smaller than half the wavelength of light cannot be distinguished from nearby structures. For the light microscope the limit of resolution is about 200 nm (0.2 µm). This means two objects less than 0.2 µm apart may be seen as one object. To improve on this level of resolution the **electron microscope** is required (Figure 1.9).

chloroplast enlarged (× 6000) a) from a transmission electron micrograph

b) from a photomicrograph obtained by light microscopy

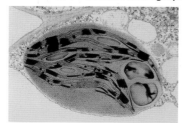 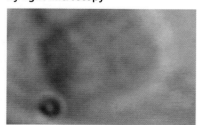

Figure 1.9 Magnification without resolution

Electron microscopy – the discovery of cell ultrastructure

The electron microscope uses electrons to make a magnified image in much the same way as the light microscope uses light. However, because an electron beam has a much shorter wavelength its resolving power is much greater. For the electron microscope used with biological materials, the limit of resolution is about 5 nm. (The size of a nanometre is given in Table 1.1 on page 2.)

Only with the electron microscope can the detailed structure of the cell organelles be observed. This is why the electron microscope is used to resolve the fine detail of the contents of cells – the organelles and cell membranes. The fine detail of cell structure is called **cell ultrastructure**. It is difficult to exaggerate the importance of electron microscopy in providing our detailed knowledge of cells.

Figure 1.10 shows an electron microscope. In an electron microscope, the electron beam is generated by an **electron gun**, and focusing is by **electromagnets**, rather than glass lenses. We cannot see electrons, so the electron beam is focused onto a **fluorescent screen** for viewing or onto a **photographic plate** for permanent recording.

Question

7 Distinguish between 'resolution' and 'magnification'.

a)

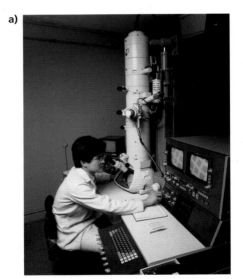

Electrons are negatively charged and are easily focused using electromagnets.

b)

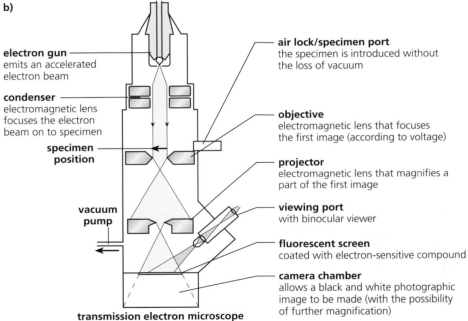

electron gun
emits an accelerated electron beam

condenser
electromagnetic lens focuses the electron beam on to specimen

specimen position

vacuum pump

air lock/specimen port
the specimen is introduced without the loss of vacuum

objective
electromagnetic lens that focuses the first image (according to voltage)

projector
electromagnetic lens that magnifies a part of the first image

viewing port
with binocular viewer

fluorescent screen
coated with electron-sensitive compound

camera chamber
allows a black and white photographic image to be made (with the possibility of further magnification)

transmission electron microscope

Figure 1.10 Using the transmission electron microscope

Limitations of the electron microscope – and how these are overcome

The electron microscope has revolutionised the study of cells. It has also changed the way cells can be observed. The electron beam travels at very high speed but at very low energy. This has practical consequences for the way biological tissue is observed at these very high magnifications and resolution.

Next we look into these outcomes and the way the difficulties are overcome.

1 Electrons cannot penetrate materials as well as light does.

Specimens must be extremely thin for the electron beam to penetrate and for some of the electrons to pass through. Biological specimens are sliced into very thin sections using a special machine called a microtome. Then the membranes and any other tiny structures present in these sections must be stained with heavy metal ions (such as lead or osmium) to make them absorb electrons at all. (We say they become electron-opaque.) Only then will these structures stand out as dark areas in the image.

Question

8 Given the magnification of the TEM of a liver cell in Figure 1.11, calculate
  a the length of the cell
  b the diameter of the nucleus.

2 Air inside the microscope would deflect the electrons and destroy the beam.

The interior of the microscope must be under a vacuum. Because of the vacuum, no *living* specimens can survive inside the electron microscope when in use. Water in cells would boil away in a vacuum.

As a result, before observations are possible, a specimen must have all the water removed. Sections are completely dehydrated. This has to be done whilst keeping the specimen as 'life-like' in structure as is possible. This is a challenge, given that cells are 80–90 per cent water. It is after the removal of water that the sections have the electron-dense stains added.

The images produced when this type of section is observed by the electron microscope are called transmission electron micrographs (TEM) (Figure 1.11).

TEM of liver cells (×15 000)

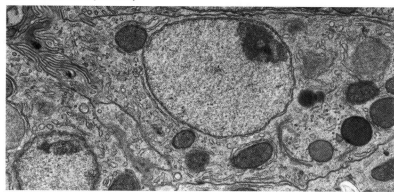

interpretive drawing

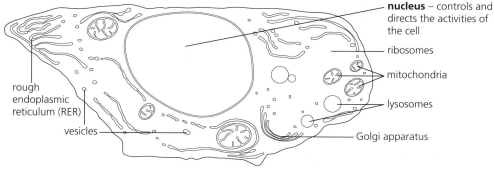

nucleus – controls and directs the activities of the cell

ribosomes

mitochondria

lysosomes

Golgi apparatus

rough endoplasmic reticulum (RER)

vesicles

Figure 1.11 Transmission electron micrograph of a liver cell, with interpretive drawing

In an alternative method of preparation, biological material is *instantly* frozen solid in liquid nitrogen. At atmospheric pressure this liquid is at −196°C. At this temperature living materials do not change shape as the water present in them solidifies instantly.

This solidified tissue is then broken up in a vacuum and the exposed surfaces are allowed to lose some of their ice. Actually, the surface is described as 'etched'.

Finally, a carbon replica (a form of 'mask') of this exposed surface is made and actually coated with heavy metal to strengthen it. The mask of the surface is then examined in the electron microscope. The resulting electron micrograph is described as being produced by **freeze etching**.

A comparison of a cell nucleus prepared as a thin section and by freeze etching is shown in Figure 1.13. The picture we get of nucleus structure is consistent. It explains why we can be confident that our views of cell structure obtained by electron microscopy are realistic.

An alternative form of electron microscopy is **scanning electron microscopy**. In this, a narrow electron beam is scanned back and forth across the surface of the whole specimen. Electrons that are reflected or emitted from this surface are detected and converted into a three-dimensional image. Larger specimens can be viewed by scanning electron microscopy rather than by transmission electron microscopy, but the resolution is not as great (see Figure 1.12).

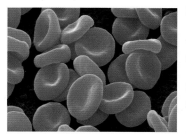

streptococcus pyogenes
(0.7 μm in diameter)

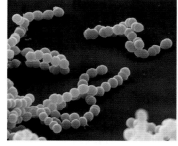

red blood cells
(5.7 μm in diameter)

Figure 1.12 Scanning electron micrographs

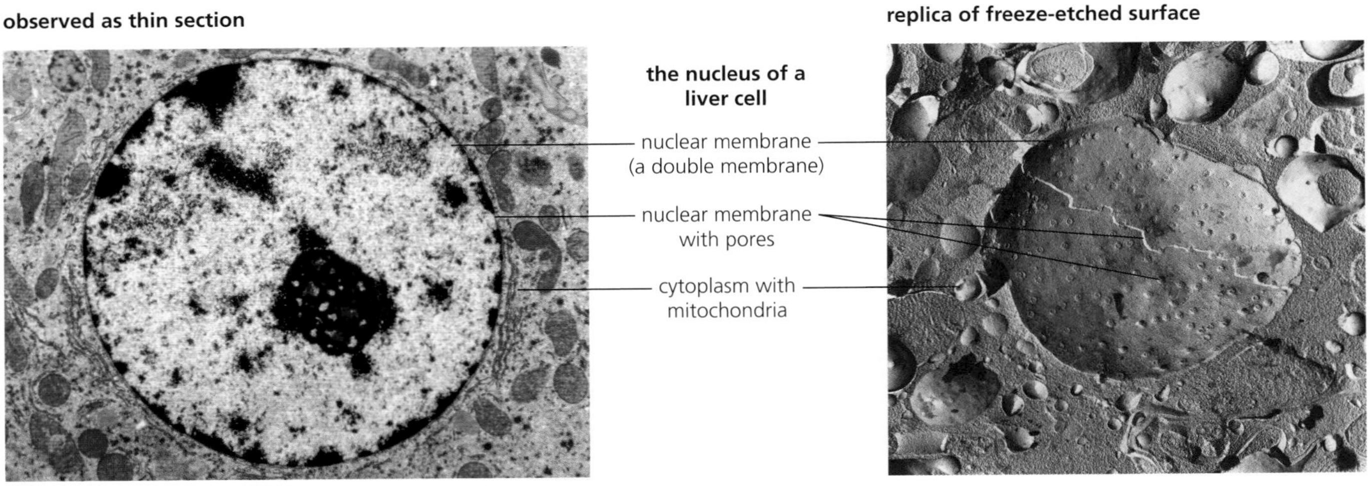

observed as thin section

replica of freeze-etched surface

the nucleus of a
liver cell

nuclear membrane
(a double membrane)

nuclear membrane
with pores

cytoplasm with
mitochondria

Figure 1.13 Transmission electron micrographs from thin-sectioned and freeze-etched material

1.2 Cells as the basic units of living organisms

The cell is the basic unit of all living organisms. The interrelationships between these cell structures show how cells function to transfer energy, produce biological molecules including proteins and exchange substances with their surroundings. Prokaryotic cells and eukaryotic cells share some features, but the differences between them illustrate the divide between these two cell types.

By the end of this section you should be able to:

a) describe and interpret electron micrographs and drawings of typical animal and plant cells as seen with the electron microscope
b) recognise the following cell structures and outline their functions:
   cell surface membrane; nucleus, nuclear envelope and nucleolus; rough endoplasmic reticulum; smooth endoplasmic reticulum; Golgi body (Golgi apparatus or Golgi complex); mitochondria (including small circular DNA); ribosomes (80S in the cytoplasm and 70S in chloroplasts and mitochondria); lysosomes; centrioles and microtubules; chloroplasts (including small circular DNA); cell wall; plasmodesmata; large permanent vacuole and tonoplast of plant cells
c) state that ATP is produced in mitochondria and chloroplasts and outline the role of ATP in cells
d) outline key structural features of typical prokaryotic cells as seen in a typical bacterium (including: unicellular, 1–5 μm diameter, peptidoglycan cell walls, lack of organelles surrounded by double membranes, naked circular DNA, 70S ribosomes)
e) compare and contrast the structure of typical prokaryotic cells with typical eukaryotic cells
f) outline the key features of viruses as non-cellular structures

Studying cell structure by electron microscopy

Look back at Figure 1.11 (on page 13). Here the organelles of a liver cell are shown in a transmission electron micrograph (TEM) and identified in an interpretive diagram. You can see immediately that many organelles are made of membranes – but not all of them.

In the living cell there is a fluid around the organelles. The **cytosol** is the aqueous (watery) part of the cytoplasm in which the organelles are suspended. The chemicals in the cytosol are substances formed and used in the chemical reactions of life. All the reactions of life are known collectively as **metabolism**, and the chemicals are known as metabolites.

Cytosol and organelles are contained within the **cell surface membrane**. This membrane is clearly a barrier of sorts. It must be crossed by all the metabolites that move between the cytosol and the environment of the cell.

Cytosol, organelles and the cell surface membrane make up a cell – a unit of structure and function which is remarkably able to survive, prosper and replicate itself. The molecules present in cells and how the chemical reactions of life are regulated are the subject of Topic 2. The structure of the cell membrane and how molecules enter and leave cells is the subject of Topic 3.

The structure and function of the **organelles** is what we consider next. Our understanding of organelles has been built up by examining TEMs of very many different cells. The outcome, a detailed picture of the ultrastructure of animal and plant cells, is represented diagrammatically in a generalised cell in Figure 1.14.

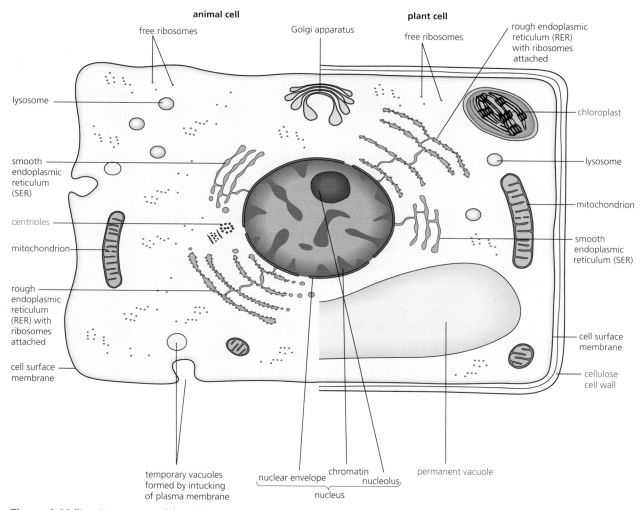

Figure 1.14 The ultrastructure of the eukaryotic animal and plant cell

Organelle structure and function

Question

9 Outline how the electron microscope has increased our knowledge of cell structure.

The electron microscope has enabled us to see and understand the structure of the organelles of cells. However, looking at detailed structure does not tell us what the individual organelles do in the cell. This information we now have, too. This is because it has been possible to isolate working organelles and analyse the reactions that go on in them and the enzymes they contain. In other words, investigations of the biochemical roles of organelles have been undertaken. Today we know about the structure and function of the cell organelles.

1 Cell surface membrane

The cell surface membrane is an extremely thin structure – less than 10 nm thick. It consists of a lipid bilayer in which proteins are embedded (Figure 4.3, page 76). At very high magnification it can be seen to have three layers – two dark lines (when stained) separated by a narrow gap (Figure 4.4, page 77).

The detailed structure and function of the cell surface membrane is the subject of a separate topic (Topic 4). In outline, the functions are as follows. Firstly, it retains the fluid cytosol. The cell surface membrane also forms the barrier across which all substances entering and leaving the cell must pass. In addition, it is where the cell is identified by surrounding cells.

2 Nucleus, nuclear envelope and nucleolus

The appearance of the nucleus in electron micrographs is shown in Figures 1.11 and 1.12 (page 13). The nucleus is the largest organelle in the eukaryotic cell, typically 10–20 μm in diameter. It is surrounded by two membranes, known as the **nuclear envelope**. The outer membrane is continuous with the endoplasmic reticulum. The nuclear envelope contains many pores. These are tiny, about 100 nm in diameter. However, they are so numerous that they make up about one third of the nuclear membrane's surface area. The function of the pores is to make possible speedy movement of molecules between nucleus and cytoplasm (such as messenger RNA), and between cytoplasm and the nucleus (such as proteins, ATP and some hormones).

The nucleus contains the **chromosomes**. These thread-like structures are made of DNA and protein, and are visible at the time the nucleus divides (page 107). At other times, the chromosomes appear as a diffuse network called **chromatin**.

Also present in the nucleus is a **nucleolus**. This is a tiny, rounded, darkly-staining body. It is the site where ribosomes (see below) are synthesised. Chromatin, chromosomes and the nucleolus are visible only if stained with certain dyes.

The everyday role of the nucleus in cell management, and its behaviour when the cell divides, are the subject of Topic 5. Here we can note that most cells contain one nucleus but there are interesting exceptions. For example, both the mature red blood cells of mammals and the sieve tube elements of the phloem of flowering plants are without a nucleus. Both lose their nucleus as they mature. The individual cylindrical fibres of voluntary muscle consist of a multinucleate sack (page 249). Fungal mycelia also contain multinucleate cytoplasm.

3 Endoplasmic reticulum

The endoplasmic reticulum consists of a network of folded membranes formed into sheets, tubes or sacs that are extensively interconnected. Endoplasmic reticulum 'buds off' from the outer membrane of the nuclear envelope, to which it may remain attached. The cytoplasm of metabolically active cells is commonly packed with endoplasmic reticulum. In Figure 1.15 we can see there are two distinct types of endoplasmic reticulum.

- **Rough endoplasmic reticulum (RER)** has ribosomes attached to the outer surface. At its margin, **vesicles** are formed from swellings. A vesicle is a small, spherical organelle bounded by a single membrane, which becomes pinched off as they separate. These tiny sacs are then used to store and transport substances around the cell. For example, RER is the site of synthesis of proteins that are 'packaged' in the vesicles. These vesicles then fuse with the Golgi apparatus and are then typically discharged from the cell. Digestive enzymes are discharged in this way.

- **Smooth endoplasmic reticulum (SER)** has no ribosomes. SER is the site of synthesis of substances needed by cells. For example, SER is important in the manufacture of lipids and steroids, and the reproductive hormones oestrogen and testosterone. In the cytoplasm of voluntary muscle fibres, a special form of SER is the site of storage of calcium ions, which have an important role in the contraction of muscle fibres.

SER and RER in cytoplasm, showing origin from outer membrane of nucleus

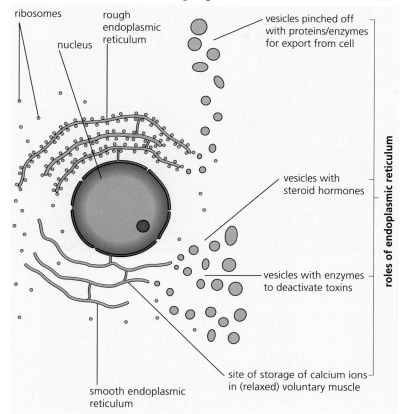

ribosomes

nucleus

rough endoplasmic reticulum

vesicles pinched off with proteins/enzymes for export from cell

vesicles with steroid hormones

vesicles with enzymes to deactivate toxins

roles of endoplasmic reticulum

smooth endoplasmic reticulum

site of storage of calcium ions in (relaxed) voluntary muscle

TEM of RER

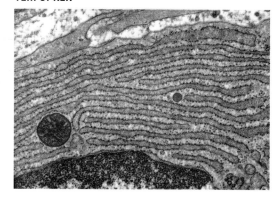

TEM of SER

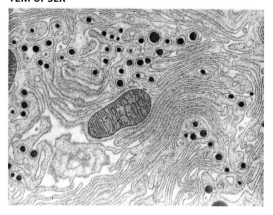

Figure 1.15 The structure of endoplasmic reticulum – rough (RER) and smooth (SER)

4 Golgi apparatus

The **Golgi apparatus** (Golgi body, Golgi complex) consists of a stack-like collection of flattened membranous sacs. One side of the stack of membranes is formed by the fusion of membranes of vesicles from RER or SER. At the opposite side of the stack, vesicles are formed from swellings at the margins that again become pinched off.

The Golgi apparatus occurs in all cells, but it is especially prominent in metabolically active cells – for example, secretory cells. More than one may be present in a cell. It is the site of synthesis of specific biochemicals, such as hormones, enzymes or others. Here specific proteins may be activated by addition of sugars (forming glycoprotein) or by the removal of the amino acid, methionine. These are then packaged into vesicles. In animal cells these vesicles may form **lysosomes**. Those in plant cells may contain polysaccharides for cell wall formation.

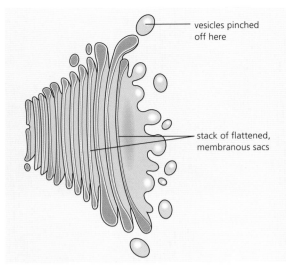

vesicles pinched off here

stack of flattened, membranous sacs

Figure 1.16 The structure of the Golgi apparatus

TEM of Golgi apparatus, in section

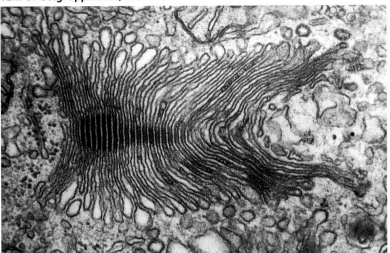

5 Mitochondria

a mitochondrion, cut open to show the inner membrane and cristae

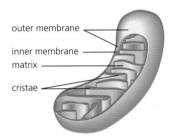

outer membrane

inner membrane

matrix

cristae

In the mitochondrion many of the enzymes of respiration are housed, and the 'energy currency' molecules adenosine triphosphate (ATP) are formed.

Mitochondria appear mostly as rod-shaped or cylindrical organelles in electron micrographs. Occasionally their shape is more variable. They are relatively large organelles, typically 0.5–1.5 μm wide, and 3.0–10.0 μm long. Mitochondria are found in all cells and are usually present in very large numbers. Metabolically very active cells will contain thousands of them in their cytoplasm – for example, in muscle fibres and hormone-secreting cells.

The mitochondrion also has a double membrane. The outer membrane is a smooth boundary, the inner is infolded to form **cristae**. The interior of the mitochondrion is called the **matrix**. It contains an aqueous solution of metabolites and enzymes, and small circular lengths of DNA, also. The mitochondrion is the site of the aerobic stages of respiration and the site of the synthesis of much ATP (see below).

Mitochondria (and chloroplasts) contain ribosomes, too. They appear as tiny dark dots in the matrix of the mitochondria, and are slightly smaller than the ribosomes found in the cytosol and attached to RER. The sizes of tiny objects like ribosomes are recorded in Svedberg units (S). This is a measure of their rate of sedimentation in centrifugation, rather than their actual size. The smaller ribosomes found in mitochondria are 70S, those in the rest of the cell 80S. We return to the significance of this discovery later in the topic (the endosymbiotic theory, page 25).

6 Ribosomes

Ribosomes are minute structures, approximately 25 nm in diameter. They are built of two sub-units, and do not have membranes as part of their structures. Chemically, they consist of protein and a nucleic acid known as RNA (ribonucleic acid). The ribosomes found free in the cytosol or bound to endoplasmic reticulum are the larger ones – classified as 80S. We have already noted that those of mitochondria and chloroplasts are slightly smaller – 70S.

Ribosomes are the sites where proteins are made in cells. The structure of a ribosome is shown in Figure 6.13, page 121, where their role in protein synthesis is illustrated. Many different types of cell contain vast numbers of ribosomes. Some of the cell proteins produced in the ribosomes have structural roles. Collagen is an example (page 48). A great many other cell proteins are enzymes. These are biological catalysts. They cause the reactions of metabolism to occur quickly under the conditions found within the cytoplasm.

TEM of a thin section of a mitochondrion

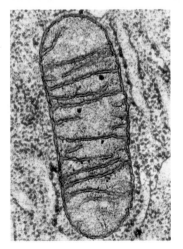

Figure 1.17 The structure of the mitochondrion

Question

10 Explain why the nucleus in a human cheek cell (see Figure 1.3, page 5) may be viewed by light microscopy in an appropriately stained cell but the ribosomes cannot.

7 Lysosomes

Lysosomes are tiny spherical vesicles bound by a single membrane. They contain a concentrated mixture of 'digestive' enzymes. These are correctly known as hydrolytic enzymes. They are produced in the Golgi apparatus or by the RER.

Lysosomes are involved in the breakdown of the contents of 'food' vacuoles. For example, harmful bacteria that invade the body are taken up into tiny vacuoles (they are engulfed) by special white cells called macrophages. Macrophages are part of the body's defence system (see Topic 11).

Any foreign matter or food particles taken up into these vacuoles are then broken down. This occurs when lysosomes fuse with the vacuole. The products of digestion then escape into the liquid of the cytoplasm. Lysosomes will also destroy damaged organelles in this way.

When an organism dies, the hydrolytic enzymes in the lysosomes of the cells escape into the cytoplasm and cause self-digestion, known as autolysis.

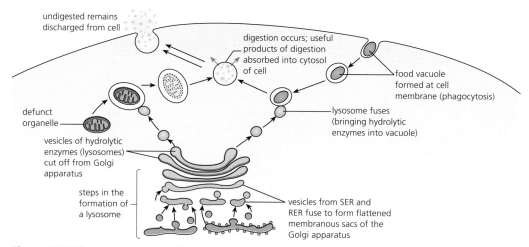

Figure 1.18 The structure and function of lysosomes

8 Centrioles and microtubules

A centriole is a tiny organelle consisting of nine paired **microtubules** arranged in a short, hollow cylinder. In animal cells, two centrioles occur at right angles, just outside the nucleus, forming the **centrosome**. Before an animal cell divides the centrioles replicate, and their role is to grow the spindle fibres – structures responsible for movement of chromosomes during nuclear division.

Microtubules themselves are straight, unbranched hollow cylinders, only 25 nm wide. They are made of a globular protein called tubulin. This is first built up and then later broken down in the cell as the microtubule framework is required in different places for different tasks.

The cells of all eukaryotes, whether plants or animals, have a well-organised system of these microtubules, which shape and support the cytoplasm. Microtubules are involved in movements of cell components within the cytoplasm, too, acting to guide and direct organelles.

9 Chloroplasts

Chloroplasts are large organelles, typically biconvex in shape, about 4–10 µm long and 2–3 µm wide. They occur in green plants, where most occur in the mesophyll cells of leaves. A mesophyll cell may be packed with 50 or more chloroplasts. Photosynthesis is the process that occurs in chloroplasts and is the subject of Topic 13.

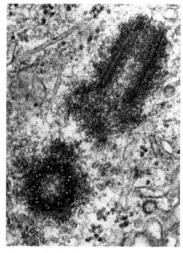

Figure 1.19 The structure of the centrosome

Look at the chloroplasts in the TEM in Figure 1.20. Each chloroplast has a double membrane. The outer layer of the membrane is a continuous boundary, but the inner layer becomes in-tucked to form a system of branching membranes called lamellae or **thylakoids**. In some regions of the chloroplast the thylakoids are arranged in flattened circular piles called **grana** (singular: **granum**). These look a little like a stack of coins. It is here that the **chlorophylls** and other pigments are located. There are a large number of grana present. Between them the branching thylakoid membranes are very loosely arranged. The fluid outside the thylakoid is the **stroma**, which contains the chloroplast DNA and ribosomes (70S), together with many enzymes.

Chloroplasts are the site of photosynthesis by which light is used as the energy source in carbohydrate and ATP synthesis (see below).

Chloroplasts are one of a larger group of organelles called **plastids**. Plastids are found in many plant cells but never in animals. The other members of the plastid family are leucoplasts (colourless plastids) in which starch is stored, and chromoplasts (coloured plastids), containing non-photosynthetic pigments such as carotene, and occurring in flower petals and the root tissue of carrots.

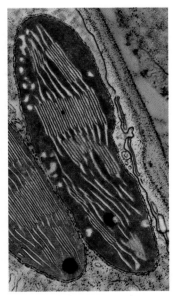

Figure 1.20 TEM of a thin section of chloroplasts

10 Cell wall and plasmodesmata

A cell wall is a structure external to a cell, and is therefore not an organelle, although it is the product of cell organelles. The presence of a rigid external cell wall is a characteristic of plant cells. Plant cell walls consist of cellulose together with other substances, mainly other polysaccharides, and are fully permeable. Cell walls are secreted by the cell they enclose, and their formation and composition are closely tied in with the growth, development and functions of the cell they enclose, support and protect.

Plasmodesmata (singular: **plasmodesma**) are the cytoplasmic connections between cells, running transversely through the walls, connecting the cytoplasm and adjacent cells. Plasmodesmata are formed when a cell divides and lays down new walls between the separating cell contents. Typically, the cytoplasm of plasmodesmata includes endoplasmic reticulum, and occupies the holes or pits between adjacent cells.

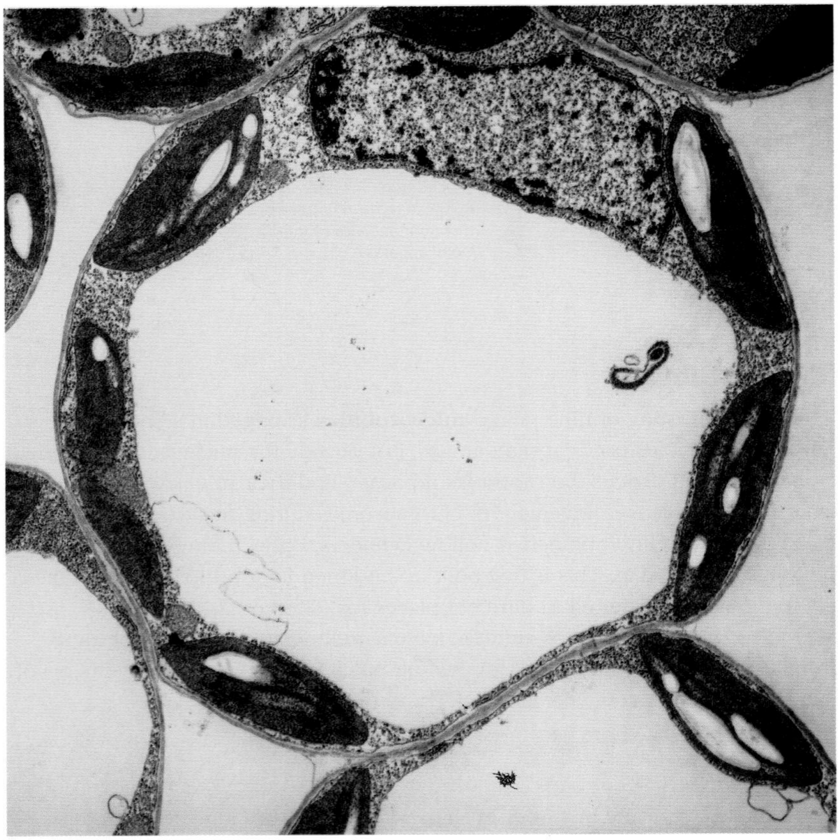

11 Permanent vacuole of plant cells and the tonoplast

We have seen that the plant cell is surrounded by a tough but flexible external cell wall. The cytoplasm and cell membrane are pressed firmly against the wall by a large, **fluid-filled permanent vacuole** that takes up the bulk of the cell. The vacuole is surrounded by a specialised membrane, the **tonoplast**. This is the barrier between the fluid contents of the vacuole (sometimes called 'cell sap') and the cytoplasm.

Figure 1.21 TEM of a mesophyll cell of *Zinnia*, showing a large central vacuole and chloroplasts within the cytoplasm (×3800)

Question

11 We have seen that analyses of TEM images of cells have played an important part in the discovery of cell ultrastructure. Carefully examine the image of a green plant cell shown in the TEM in Figure 1.21. Using the interpretive drawing in Figure 1.11 as a model and following the guidelines on biological recording on page 8, draw and label a representation of the green plant cell shown in Figure 1.21.

Extension

Cilia and flagella

Cilia and flagella are organelles that project from the surface of certain cells. For example, cilia occur in large numbers on the lining (epithelium) of the air tubes serving the lungs (bronchi). We shall discuss the role of these cilia in healthy lungs and how they respond to cigarette smoke in Topic 9. Flagella occur on certain small, motile cells, such as the sperm.

ATP, its production in mitochondria and chloroplasts, and its role in cells

ATP (**adenosine triphosphate**) is the universal energy currency molecule of cells. ATP is formed from **adenosine diphosphate** (**ADP**) and a phosphate ion (**Pi**) by the transfer of energy from other reactions. ATP is referred to as 'energy currency' because, like money, it can be used in different contexts, and it is constantly recycled. It occurs in cells at a concentration of $0.5–2.5$ mg cm^{-3}. ATP is a relatively small, water-soluble molecule, able to move easily around cells, that effectively transfers energy in relatively small amounts, sufficient to drive individual reactions.

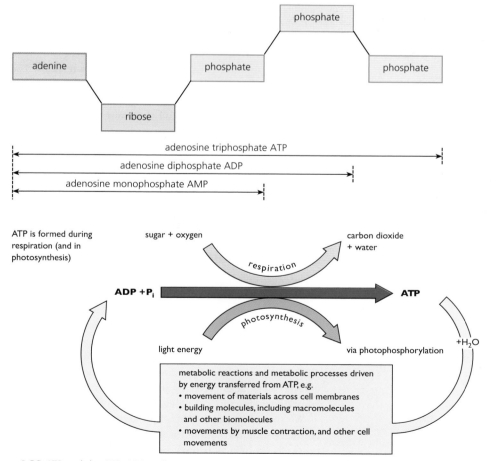

Figure 1.22 ATP and the ATP–ADP cycle

ATP is a **nucleotide** with an unusual feature. It carries three phosphate groups linked together in a linear sequence. ATP *may* lose both of the outer phosphate groups, but usually only one at a time is lost. ATP is a relatively small, soluble organic molecule.

ATP contains a good deal of chemical energy locked up in its structure. What makes ATP special as a reservoir of stored chemical energy is its role as a common intermediate between energy-yielding reactions and energy-requiring reaction and processes.

- Energy-yielding reactions include the photophosphorylation reactions of photosynthesis, and the reactions of cell respiration in which sugars are broken down and oxidised.
- Energy-requiring reactions include the synthesis of cellulose from glucose, the synthesis of proteins from amino acids, the contractions of muscle fibres, and the active transport of certain molecules across cell membranes, for example.
- The free energy available in the conversion of ATP to ADP is approximately 30–34 kJ mol^{-1}, made available in the presence of a specific enzyme. Some of this energy is lost as heat in a reaction, but much free energy is made available to do useful work, more than sufficient to drive a typical energy-requiring reaction of metabolism.
- Sometimes ATP reacts with water (a hydrolysis reaction) and is converted to ADP and Pi. Direct hydrolysis of the terminal phosphate groups like this happens in muscle contraction, for example.
- Mostly, ATP reacts with other metabolites and forms phosphorylated intermediates, making them more reactive in the process. The phosphate groups are released later, so both ADP and Pi become available for reuse as metabolism continues.

In summary, ATP is a molecule universal to all living things; it is the source of energy for chemical change in cells, tissues and organisms.

Question

12 Outline why ATP is an efficient energy currency molecule.

Prokaryotic and eukaryotic cells

Finally, we need to introduce a major division that exists in the structure of cells. The discovery of two fundamentally different types of cell followed on from the application of the electron microscope to the investigation of cell structure.

All plants, animals, fungi and protoctista (these are the single-celled organisms, such as *Amoeba* and the algae) have cells with a large, obvious nucleus. There are several individual chromosomes within the nucleus, which is a relatively large spherical sac bound by a nuclear envelope. The surrounding cytoplasm contains many different membranous organelles. These types of cells are called **eukaryotic cells** (literally meaning 'good nucleus') – the animal and plant cells in Figure 1.3 are examples.

On the other hand, bacteria contain no true nucleus but have a single, circular chromosome in the cytoplasm. Also, their cytoplasm does not have the organelles of eukaryotes. These are called **prokaryotic cells** (from *pro* meaning 'before' and *karyon* meaning 'nucleus').

Another key difference between the cells of the prokaryotes and eukaryotes is their **size**. Prokaryote cells are exceedingly small, about the size of organelles like the mitochondria and chloroplasts of eukaryotic cells.

Prokaryotic cell structure

Escherichia coli (Figure 1.23) is a bacterium of the human gut – it occurs in huge numbers in the lower intestine of humans and other endothermic (once known as 'warm blooded') vertebrates, such as the mammals, and it is a major component of their faeces. This tiny organism was named by a bacteriologist, Professor T. Escherich, in 1885. Notice the scale bar in Figure 1.23. This bacterium is typically about 1 µm × 3 µm in length – about the size of a mitochondrion in a eukaryotic cell. The cytoplasm lacks the range of organelles found in eukaryotic cells, and a nucleus surrounded by a double membrane is absent too. The DNA of the single, circular chromosome lacks proteins, and so is described as 'naked'. In Figure 1.23 the labels of the component structures are annotated with their function.

We should also note that all prokaryote cells are capable of extremely rapid growth when conditions are favourable for them. In such environments, prokaryote cells frequently divide into two cells (known as **binary fission**). New cells formed then grow to full size and divide again.

In summary, the key structural features of typical prokaryotic cells as seen in a typical bacterium are:

- they are unicellular
- typically 1–5 μm in diameter
- cell walls made of peptidoglycan, composed of polysaccharides and peptides combined together
- lack organelles surrounded by a double membrane in their cytoplasm
- have a single circular chromosome that is 'naked' (of DNA without associated proteins)
- ribosomes are present, but they are the smaller 70S variety.

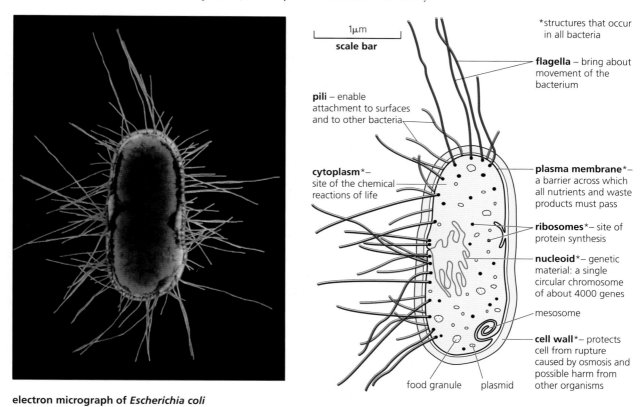

electron micrograph of *Escherichia coli*

Figure 1.23 The structure of *Escherichia coli*

Prokaryotic and eukaryotic cells compared

The fundamental differences in size and complexity of prokaryotic and eukaryotic cells are highlighted in Table 1.2.

Table 1.2 Prokaryotes and eukaryotes compared

Prokaryotes, e.g. bacteria, cyanobacteria	Eukaryotes, e.g. mammals, green plants, fungi
cells are extremely small, typically about 1–5 μm in diameter	cells are larger, typically 50–150 μm
nucleus absent: circular DNA helix in the cytoplasm, DNA not supported by histone protein	nucleus has distinct nuclear envelope (with pores), with chromosomes of linear DNA helix supported by histone protein
cell wall present (made of peptidoglycon – long molecules of amino acids and sugars)	cell wall present in plants (largely of cellulose) and fungi (largely of the polysaccharide chitin)
few organelles; membranous structures absent	many organelles bounded by double membrane (e.g. chloroplasts, mitochondria, nucleus) or single membrane (e.g. Golgi apparatus, lysosomes, vacuoles, endoplasmic reticulum)
proteins synthesised in small ribosomes (70S)	proteins synthesised in large ribosomes (80S)
some cells have simple flagella	some cells have cilia or flagella, 200 nm in diameter
some can fix atmospheric nitrogen gas for use in the production of amino acids for protein synthesis	none can metabolise atmospheric nitrogen gas but instead require nitrogen already combined in molecules in order to make proteins from amino acids (page 120)

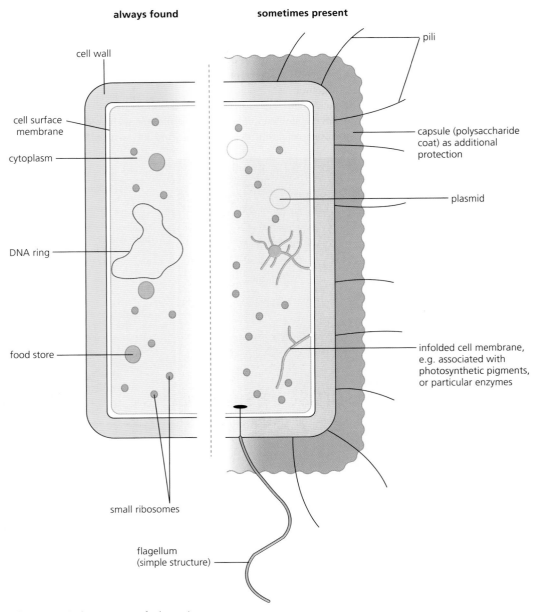

always found   **sometimes present**

cell wall

pili

cell surface
membrane

cytoplasm

capsule (polysaccharide
coat) as additional
protection

plasmid

DNA ring

food store

infolded cell membrane,
e.g. associated with
photosynthetic pigments,
or particular enzymes

small ribosomes

flagellum
(simple structure)

Figure 1.24 The structure of a bacterium

Question

13 Distinguish between the following terms:
    a cell wall and cell surface membrane
    b chromatin and chromosome
    c nucleus and nucleolus
    d prokaryote and eukaryote
    e centriole and chloroplast
    f plant cell and animal cell.

Extension

A possible origin for mitochondria and chloroplasts

Present-day prokaryotes are similar to many fossil prokaryotes. Some of these are 3500 million years old. By comparison, the earliest eukaryote cells date back only 1000 million years. Thus eukaryotes must have evolved surrounded by prokaryotes, many that were long-established organisms. How the eukaryotic cell arose is not known. However, there is evidence that the origin of the mitochondria and chloroplasts of eukaryotic cells was as previously independent-living prokaryotes.

It is highly possible that, in the evolution of the eukaryotic cell, prokaryotic cells (which at one stage were taken up into food vacuoles for digestion) came to survive as organelles inside the host cell, rather than becoming food items! If so, they have become integrated into the biochemistry of their 'host' cell, with time.

The evidence for this origin is that mitochondria and chloroplasts contain:
● a ring of DNA, like the circular chromosome a bacterial cell contains
● small (70S) ribosomes, like those of prokaryotes.

The present-day DNA and ribosomes of these organelles still function with roles in the synthesis of specific proteins, but the mitochondria and chloroplasts themselves are no longer capable of living independently.

It is these features that have led evolutionary biologists to propose this **endosymbiotic theory** of the origin of these organelles ('**endo**' = inside, '**symbiont**' = an organism living with another for mutual benefit).

Viruses as non-cellular structures

Viruses are disease-causing agents, rather than 'organisms'. The distinctive features of viruses are:
● they are not cellular structures, but rather consist of a core of nucleic acid (DNA or RNA) surrounded by a protein coat, called a **capsid**
● in some viruses there is an additional external envelope or membrane made of lipids and proteins (eg. HIV)
● they are extremely small when compared with bacteria. Most viruses are in a size range of 20–400 nm (0.02–0.4 mm). They become visible only by means of the electron microscope
● they can reproduce only inside specific living cells, so viruses function as **endoparasites** in their host organism
● they have to be transported in some way between host cells
● viruses are highly specific to particular host species, some to plant species, some to animal species and some to bacteria
● viruses are classified by the type of nucleic acid they contain.

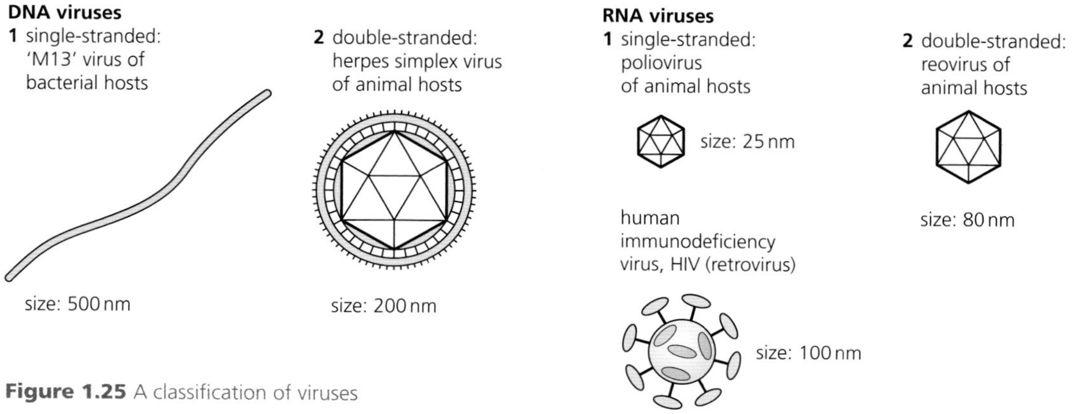

DNA viruses
1 single-stranded: 'M13' virus of bacterial hosts

2 double-stranded: herpes simplex virus of animal hosts

RNA viruses
1 single-stranded: poliovirus of animal hosts

size: 25 nm

human immunodeficiency virus, HIV (retrovirus)

2 double-stranded: reovirus of animal hosts

size: 80 nm

size: 500 nm

size: 200 nm

size: 100 nm

Figure 1.25 A classification of viruses

So are viruses 'living' at any stage?

Viruses are an assembly of complex molecules, rather than a form of life. Isolated from their host cell they are inactive, and are often described as 'crystalline'. However, within susceptible host cells they are highly active 'genetic programmes' that will take over the biochemical machinery of host cells. Their component chemicals are synthesised, and then assembled to form new viruses. On breakdown (lysis) of the host cell, viruses are released and may cause fresh infections. So, viruses are not living organisms, but may become active components of host cells.

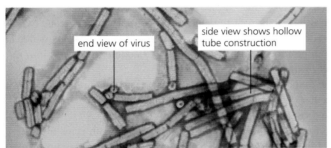

transmission electron micrograph of TMV (×40 000) negatively stained

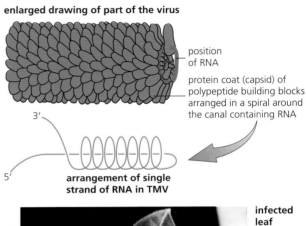

enlarged drawing of part of the virus

position of RNA

protein coat (capsid) of polypeptide building blocks arranged in a spiral around the canal containing RNA

arrangement of single strand of RNA in TMV

healthy leaves

infected leaf

Figure 1.26 Tobacco mosaic virus

Relative sizes of molecules, macromolecules, viruses and organisms

We now have a clear picture of prokaryotic and eukaryotic levels of cellular organisation and of the nature of viruses. Finally, we ought to reflect on the huge differences in size among organisms, viruses and the molecules they are built from.

In Figure 1.27 size relationships of biological and chemical levels of organisation on which we focus in this book are compared. Here the scale is logarithmic to accommodate the diversities in size in the space available. So each division is ten times larger than the division immediately below it. (In science, 'powers to ten' are used to avoid writing long strings of zeros). Of course we must remember that, although sizes are expressed by a single length or diameter, all cells and organisms are three-dimensional structures, with length, breadth and depth.

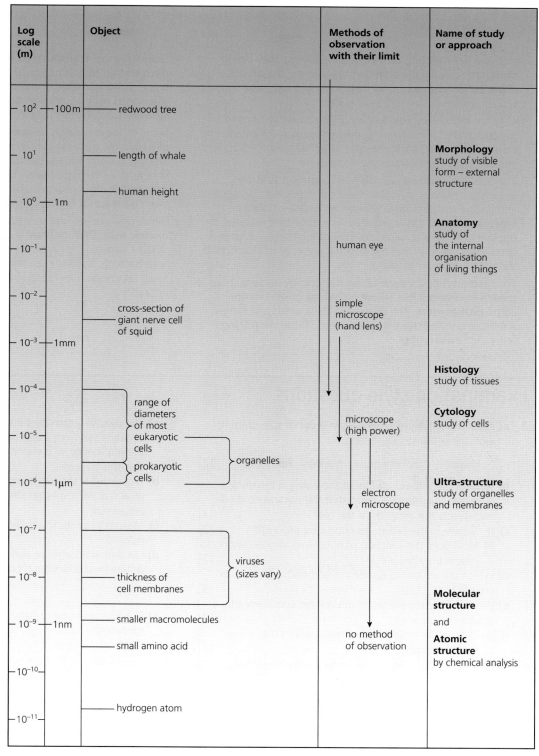

Figure 1.27 Size relationships on a logarithmic scale

Summary

- Cells are the **building blocks** of living things. They are derived from other cells by **division** and they are the site of all the **chemical reactions of life** (metabolism). A cell is the smallest unit of organisation we can say is alive.

- Cells are **extremely small**. They are measured in units of a thousandths of a millimetre (a micron – μm) and they must be viewed by microscopy. In the laboratory we view them by light microscopy using a compound microscope.

- The cells of plants and animals have **common features**, including a nucleus, cytoplasm and cell membrane. To observe and resolve the detailed structures within the cytoplasm, **electron microscopy** is required. The distinctive features of plant cells are a cellulose cell wall, the presence of large permanent vacuoles and the possible presence of chloroplasts, the site of photosynthesis.

- The simplest cellular organisation is shown by bacteria. Here there is no true nucleus. These unicellular organisms are called **prokaryotes**. The cells of plants, animals and fungi are larger and they have a true nucleus. These living things are called **eukaryotes**.

- Examination of **transmission electron micrographs** has revealed that the cytoplasm of the eukaryotic cell contains numerous **organelles**, some about the size of bacteria, suspended in an aquatic solution of metabolites (called the **cytosol**) surrounded by the cell surface membrane.

- Many of the organelles are **membrane-bound structures**, including the nucleus, mitochondria, chloroplasts, endoplasmic reticulum, Golgi apparatus and lysosomes. The organelles have specific roles in metabolism. The **biochemical roles of the organelles** are investigated by disrupting cells and isolating the organelles for further investigation.

- Viruses are non-cellular structures that consist of a core of **nucleic acid** (**DNA** or **RNA**) surrounded by a protein coat, called a **capsid**. They are extremely small when compared with bacteria, and they can reproduce only inside specific living cells, so viruses function as **endoparasites** in their host organism.

Examination style questions

1 a) Place the following organelles in order of size, starting with the smallest:

  lysosome   mitochondria   nucleus   ribosome   [2]

b) What is the most likely role of the pores in the nuclear envelope (membrane), given that the nucleus controls and directs the activities of the cell? [1]

c) i) Where in the cell are the ribosomes found? [2]
   ii) How does the function of ribosomes vary according to where they occur? [2]

d) Which cell structures synthesise and transport lipids in the cell? [3]

e) Which of the organelles of the cell are surrounded by a double membrane? [3]

f) i) What is the origin of the cristae within the mitochondria? [1]
   ii) What is the value of the surface area the cristae provide? [2]

g) Name a cell in which you would expect to find a large number of lysosomes and say why. [4]

[Total: 20]

2 By means of fully annotated diagrams only:
  a) Illustrate the general structure of a eukaryotic animal cell as seen by electron microscopy.
  b) Outline how this cell carries out:
   i) the packaging and export of polypeptides
   ii) the destruction of defective organelles.

3 a) The discovery that bacteria have a different level of cell organisation from the cells of animals and plants awaited the application of electron microscopy in biology. Why was this the case? [2]

b) Explain what the terms 'prokaryote' and 'eukaryote' tell us about the respective structures of cells of these organisms. [2]

c) Make a large, fully labelled diagram of a prokaryotic cell. [6]

d) List and annotate five ways in which a prokaryotic cell differs from animal and plant cells. [10]

[Total: 20]

4 a) The fine structure of cells is observed using the electron microscope.
   i) What features of cells are observed by electron microscopy that are not visible by light microscopy? [1]
   ii) State two problems that arise in electron microscopy because of the nature of an electron in relation to the living cell. [2]

b) i) What magnification occurs in a light microscope with a ×6 eyepiece lens and a ×10 objective lens? [1]
   ii) How many cells of 100 μm diameter will fit side by side along a millimetre? [1]
   iii) Explain what is meant by the resolution (resolving power) of a microscope. [2]

[Total: 7]

2 Biological molecules

This topic introduces carbohydrates, proteins and lipids: organic molecules that are important in cells. Nucleic acids are covered in a separate topic. Biological molecules are based on the versatile element carbon. This topic explains how macromolecules, which have a great diversity of function in organisms, are assembled from smaller organic molecules such as glucose, amino acids, glycerol and fatty acids.

Life as we know it would not be possible without water. Understanding the properties of this extraordinary molecule is an essential part of any study of biological molecules.

The emphasis in this topic is on the relationship between molecular structures and their functions. Some of these ideas are continued in other topics, for example, the functions of haemoglobin in gas transport in Transport in mammals, phospholipids in membranes in Cell membranes and transport and antibodies in Immunity.

2.1 Testing for biological molecules

Tests for biological molecules can be used in a variety of contexts, such as identifying the contents of mixtures of molecules and following the activity of digestive enzymes.

By the end of this section you should be able to:

a) carry out tests for reducing sugars and non-reducing sugars, the iodine in potassium iodide solution test for starch, the emulsion test for lipids and the biuret test for proteins to identify the contents of solutions

b) carry out a semi-quantitative Benedict's test on a reducing sugar using dilution, standardising the test and using the results (colour standards or time to first colour change) to estimate the concentration

Introducing the tests for biological molecules

The methods you will use to test for biological molecules are summarised in Figure 2.1. Each test exploits the chemical structure and properties of the groups of molecules it is designed to detect. Consequently, details of the tests are described in that context, later in Section 2.2 (for reducing sugars, non-reducing sugars, starch, and lipids) and in Section 2.3 (for proteins). Most tests are used to detect the presence or absence of particular molecules. The test for reducing sugar can be adapted to estimate concentration, and may be used in experiments that follow the activity of a digestive enzyme, for example – as explained on page 59.

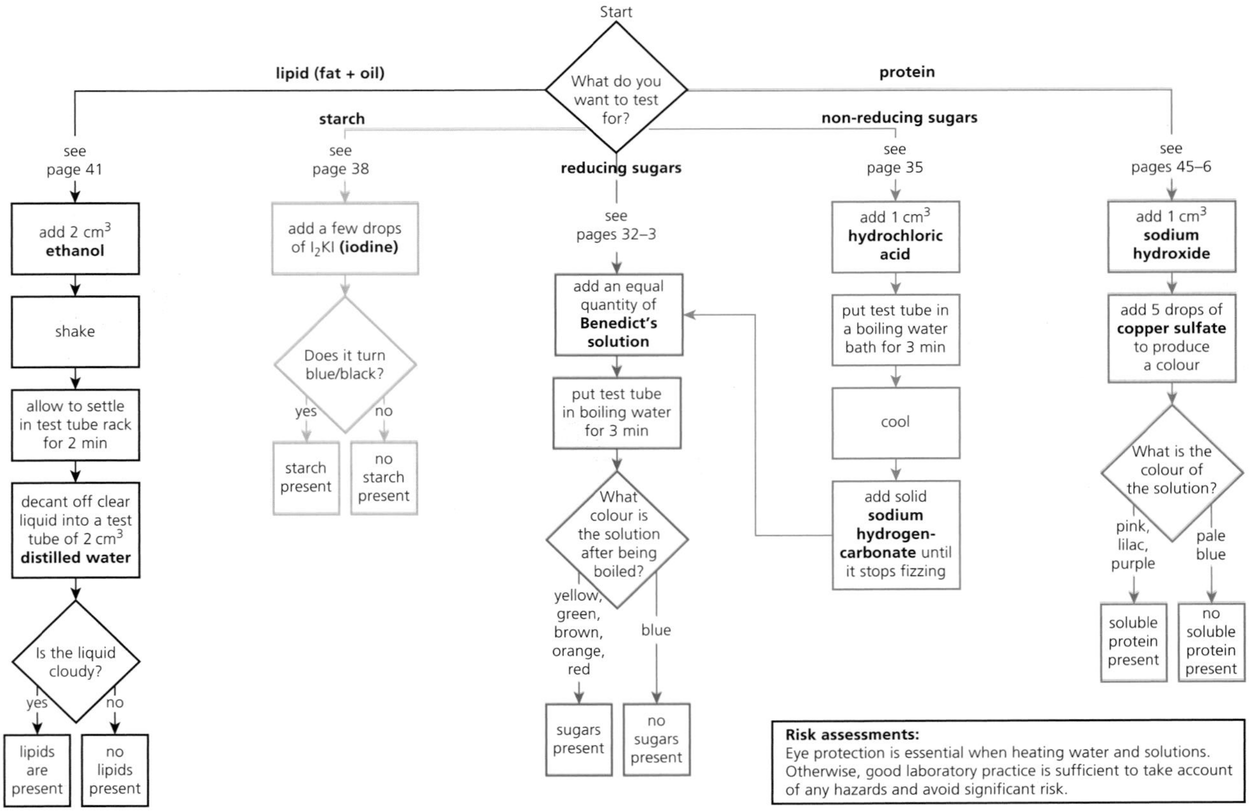

Start

What do you
want to test
for?

lipid (fat + oil)

protein

starch

non-reducing sugars

reducing sugars

see
page 41

see
page 38

see
pages 32–3

see
page 35

see
pages 45–6

add 2 cm³
ethanol

add a few drops
of I₂KI **(iodine)**

add an equal
quantity of
**Benedict's
solution**

add 1 cm³
**hydrochloric
acid**

add 1 cm³
**sodium
hydroxide**

shake

Does it turn
blue/black?

put test tube
in boiling water
for 3 min

put test tube in
a boiling water
bath for 3 min

add 5 drops of
copper sulfate
to produce
a colour

allow to settle
in test tube rack
for 2 min

yes   no

cool

What is the
colour of
the solution?

decant off clear
liquid into a test
tube of 2 cm³
distilled water

starch
present

no
starch
present

What
colour is
the solution
after being
boiled?

add solid
**sodium
hydrogen-
carbonate** until
it stops fizzing

pink,
lilac,
purple

pale
blue

Is the liquid
cloudy?

yellow,
green,
brown,
orange,
red

blue

soluble
protein
present

no
soluble
protein
present

yes   no

lipids
are
present

no
lipids
present

sugars
present

no
sugars
present

Risk assessments:
Eye protection is essential when heating water and solutions.
Otherwise, good laboratory practice is sufficient to take account
of any hazards and avoid significant risk.

Figure 2.1 Flow chart of the tests for biological molecules

2.2 Carbohydrates and lipids

Carbohydrates and lipids have important roles in the provision and storage of energy and for a variety of other functions such as providing barriers around cells: the phospholipid bilayer of all cell membranes and the cellulose cell walls of plant cells.

By the end of this section you should be able to:

a) describe the ring forms of α-glucose and β-glucose
b) define the terms monomer, polymer, macromolecule, monosaccharide, disaccharide and polysaccharide
c) describe the formation of a glycosidic bond by condensation, with reference both to polysaccharides and to disaccharides, including sucrose
d) describe the breakage of glycosidic bonds in polysaccharides and disaccharides by hydrolysis, with reference to the non-reducing sugar test
e) describe the molecular structure of polysaccharides including starch (amylose and amylopectin), glycogen and cellulose and relate these structures to their functions in living organisms
f) describe the molecular structure of a triglyceride with reference to the formation of ester bonds and relate the structure of triglycerides to their functions in living organisms
g) describe the structure of a phospholipid and relate the structure of phospholipids to their functions in living organisms

Introducing carbon compounds

Compounds built from carbon and hydrogen are called **organic compounds**. Examples include methane (CH_4) and glucose ($C_6H_{12}O_6$). Carbon is not a common element of the Earth's crust – it is quite rare compared to silicon and aluminium, for example. But in living things carbon is the third most abundant element by mass, after oxygen. In fact, organic compounds make up the largest number of molecules found in living things. This includes the carbohydrates, lipids and proteins.

Why is carbon so important to life?

The answer is that carbon has a unique collection of properties, so remarkable in fact, that we can say that they make life possible. If you are unfamiliar with these properties, then the special features of carbon are introduced in Appendix 1 (on the CD).

Question

1 Where do non-organic forms of carbon exist in the biosphere?

Carbohydrates

Carbohydrates are the largest group of organic compounds. They include sugars, starch, glycogen and cellulose. Carbohydrates are substances that contain only three elements carbon, hydrogen and oxygen. The hydrogen and oxygen atoms are present in the ratio 2:1 (as they are in water, H_2O). In fact, we represent carbohydrates by the **general formula $C_x(H_2O)_y$**.

We start by looking at the simplest carbohydrates.

Monosaccharides – the simple sugars

Monosaccharides are carbohydrates with relatively small molecules. They taste sweet and they are soluble in water. In biology, the monosaccharide **glucose** is especially important. All green leaves manufacture glucose using light energy, our bodies transport glucose in the blood and all cells use glucose in respiration. In cells and organisms glucose is the building block for many larger molecules, including cellulose.

Glucose exists in two ring forms. In solution, glucose molecules constantly change between the two ring structures.

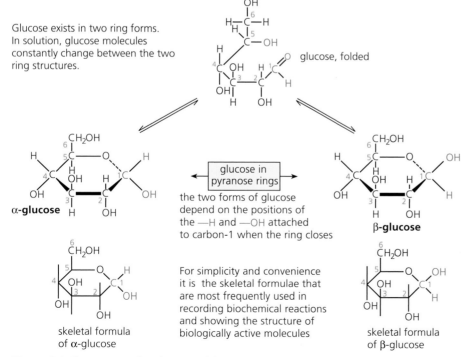

glucose, folded

glucose in pyranose rings

the two forms of glucose depend on the positions of the —H and —OH attached to carbon-1 when the ring closes

For simplicity and convenience it is the skeletal formulae that are most frequently used in recording biochemical reactions and showing the structure of biologically active molecules

α-glucose

skeletal formula of α-glucose

β-glucose

skeletal formula of β-glucose

Figure 2.2 The structure of α-glucose and β-glucose

The structure of glucose

Glucose has a chemical or **molecular formula** of **$C_6H_{12}O_6$**. This indicates the atoms present and their numbers in the molecule. So, glucose is a six-carbon sugar or **hexose**. But this molecular formula does not tell us the structure of the molecule.

Glucose can be written down on paper as a linear molecule however it cannot exist in this form. Rather, glucose is folded and takes a ring or cyclic form. This is its **structural formula**. The ring closes up when the oxygen on carbon-5 attaches itself to carbon-1. The glucose ring, containing five carbon atoms and an oxygen atom, is called a **pyranose ring**. Furthermore, the pyranose ring exists in two forms. These are the **α-form** and the **β-form**, depending on whether a –H atom was trapped 'up' (α-form) or 'down' (β-form) when the ring closed. So, there are two forms of glucose, known as α-glucose and β-glucose (Figure 2.2).

Incidentally, **fructose** is another hexose sugar found in cells. It has a ring structure different from that of glucose. Fructose forms into a ring with four carbon atoms and an oxygen atom (known as furanose ring). Fructose is by far the sweetest common sugar; fructose is used in the food industry in the manufacture of sweets and various confectioneries.

The functional groups of sugars

Fortunately, the huge number of organic compounds fit into a relatively small numbers of 'families' of compounds. These families are identified by a part of their molecule that is the **functional group**. It is the functional group of an organic compound that gives it characteristic **chemical properties**. The chemical structure of some important functional groups is shown in Figure 2.3.

The remainder of the organic molecule, referred to as the **R-group**, has little or no effect on the chemical properties of the functional group. However, the R-group does influence the **physical properties** of the molecule. These include whether a compound is a solid or liquid (its melting point) and at what temperature a liquid becomes vapour (its boiling point).

All the monosaccharides contain one of two functional groups. Some are **aldehydes**, like glucose, and are referred to as aldo-sugars or **aldoses**. Others are **ketones**, like fructose, and are referred to as keto-sugars or **ketoses**.

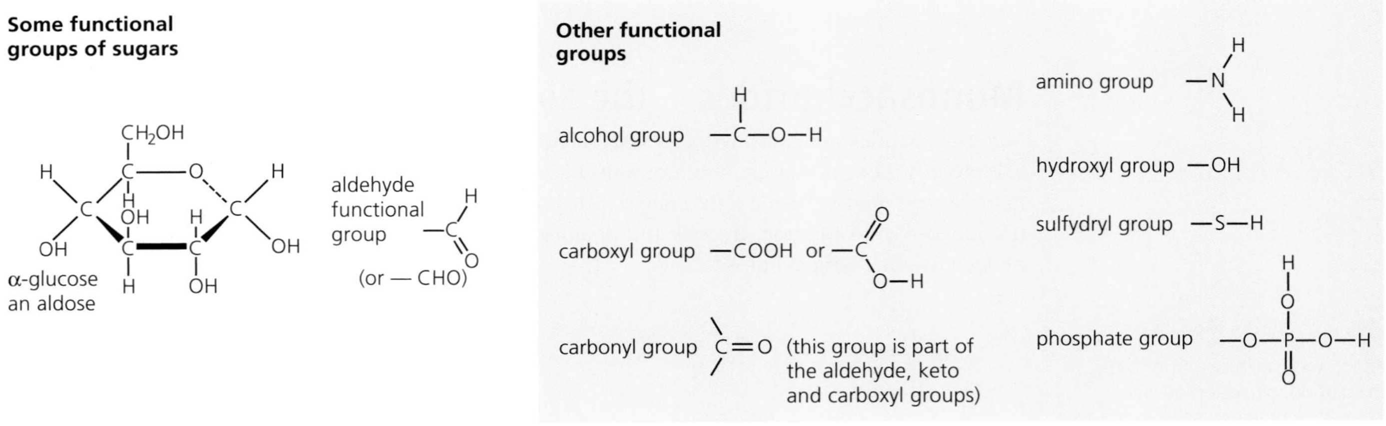

Some functional groups of sugars

Other functional groups

Figure 2.3 Some functional groups

Testing for reducing sugars

Aldoses and ketoses are **reducing sugars**. This means that, when heated with an alkaline solution of copper(II) sulfate (a blue solution, called Benedict's solution), the aldehyde or ketone group reduces Cu^{2+} ions to Cu^+ ions forming a brick-red precipitate of copper(I) oxide. In the process, the aldehyde or ketone group is oxidised to a carboxyl group (–COOH).

This reaction is used to test for reducing sugar and is known as **Benedict's test** (Figure 2.4). If no reducing sugar is present the solution remains blue.

Steps to the Benedict's test

1 Heat a water bath (large beaker) of water to 100 °C. Since this takes some time, set it up before you begin preparing for the testing.
2 From a 10% solution of glucose (18.0 g of glucose with 100 cm³ of distilled water):
   ● prepare 10 cm³ of a 0.1% solution of glucose (labelled Tube 2)
   ● prepare 10 cm³ of a 1% solution of glucose (Tube 3)
   ● and then place 10 cm³ of the 10% glucose solution in another test tube (Tube 4).

3 Finally, place 10 cm^3 of distilled water in a test tube (Tube 1) and 10 cm^3 of a 10% sucrose solution in a final test tube (Tube 5).

4 Add 5 cm^3 of Benedict's solution to each tube.

5 Place all five tubes in the water bath for 5 minutes, gently agitating them during the process. Finally, remove the tubes to a rack where any colour changes can be observed (Figure 2.4). Notice that:

● in the absence of reducing sugar the solution remains blue (Tube 1)

● sucrose is not a reducing sugar (Tube 5)

● in the presence of glucose, the colour change observed depends on the concentration of reducing sugar. The greater the concentration the more precipitate is formed, and the greater the colour changes:

$$\text{blue} \rightarrow \text{green} \rightarrow \text{yellow} \rightarrow \text{red} \rightarrow \text{brown}$$

5 cm^3 of Benedict's solution (blue) was added to 10 cm^3 of solution to be tested ⟶ test tubes were placed in a boiling water bath for 5 minutes ⟶ tubes were transferred to a rack and the colours compared

boiling water bath

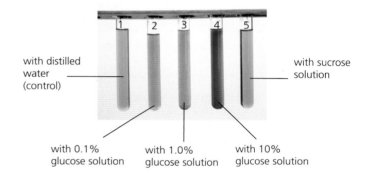

with distilled water (control)

with sucrose solution

with 0.1% glucose solution

with 1.0% glucose solution

with 10% glucose solution

Figure 2.4 The test for reducing sugar

Making the Benedict's test semi-quantitative

The colour sequence obtained with Benedict's solution with sugar solutions of increasing concentration can be used to make an approximate estimation of the concentration of reducing sugar in other solutions of unknown concentration. To discover the relationship between colour and concentration, accurate quantities need to be used. The steps to using Benedicts test semi-quantitatively are shown in Figure 2.5.

Look at the sequence of actions carefully. Note that the product of steps 1–6 is a serial dilution of glucose starting with 1% glucose. Each subsequent dilution is 50% less concentrated. What is the concentration of the glucose solution in Tube 6?

Steps 7 and 8 are the Benedict's tests, carried out on each tube. The tubes are held in the boiling water for a standard length of time, at least 3 minutes, before being removed and the colour recorded. The colour produced by an unknown glucose solution, tested with Benedict's in the same way, indicates the concentration of glucose present.

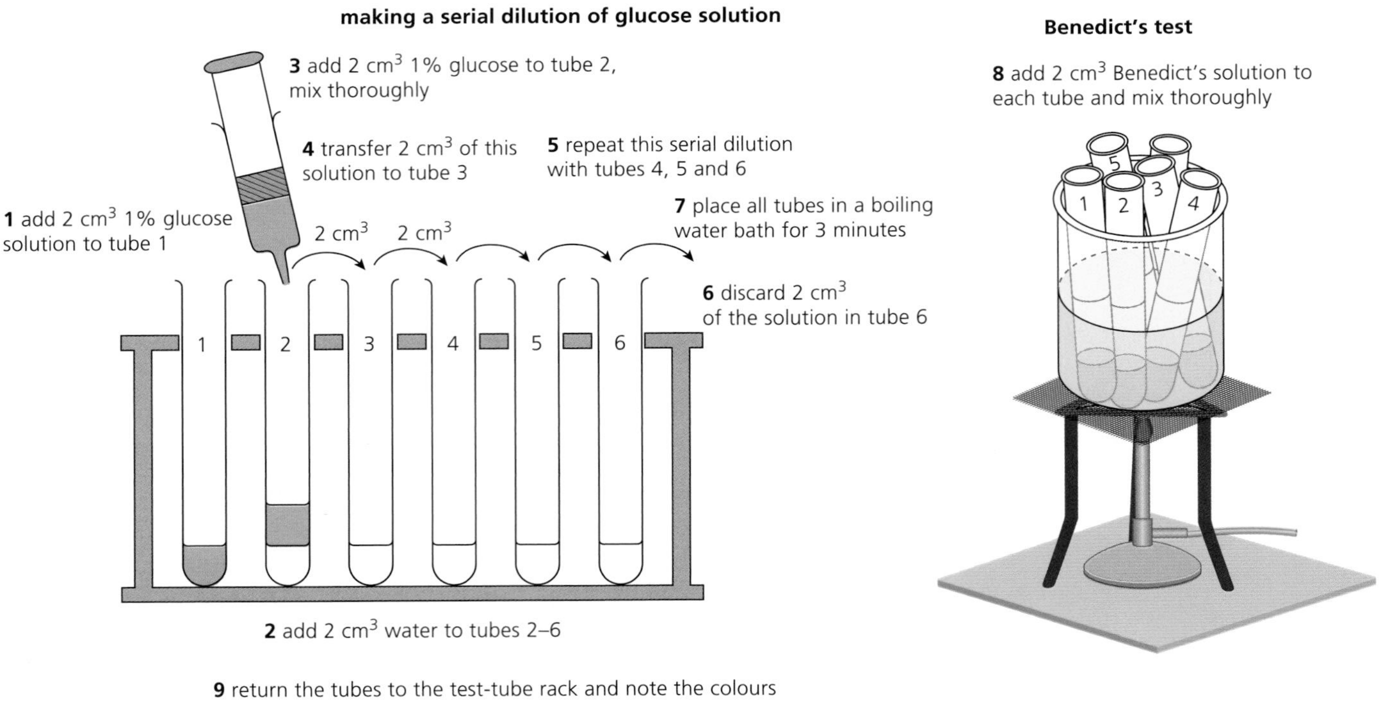
making a serial dilution of glucose solution
3 add 2 cm3 1% glucose to tube 2, mix thoroughly
4 transfer 2 cm3 of this solution to tube 3
5 repeat this serial dilution with tubes 4, 5 and 6
1 add 2 cm3 1% glucose solution to tube 1
2 cm3
2 cm3
7 place all tubes in a boiling water bath for 3 minutes
6 discard 2 cm3 of the solution in tube 6
1
2
3
4
5
6
2 add 2 cm3 water to tubes 2–6
Benedict's test
8 add 2 cm3 Benedict's solution to each tube and mix thoroughly
9 return the tubes to the test-tube rack and note the colours

Figure 2.5 Using the Benedict's test semi-quantitatively

Other monosaccharides of importance in living cells

There are several other sugars produced by cells and used in their chemical reactions. These include some three-carbon sugars (**trioses**) that are early products in photosynthesis in green plants (Topic 13). Five-carbon sugars (**pentoses**) are important components of nucleic acids (Topic 6).

Table 2.1 Other monosaccharides of importance in living cells

Length of carbon chain	Name of sugar	Molecular formula	Formula	Roles
3C = triose	glyceraldehyde	$C_3H_6O_3$	**a)** 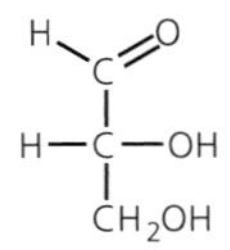	intermediate in respiration and photosynthesis
5C = pentoses	ribose	$C_5H_{10}O_5$	**b)** 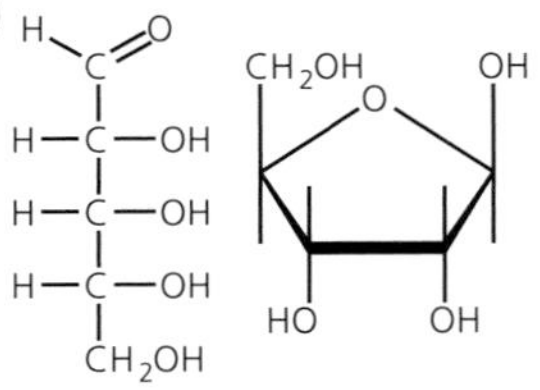	in RNA, ATP and hydrogen acceptors NAD and NADP
	deoxyribose	$C_5H_{10}O_4$	**c)** 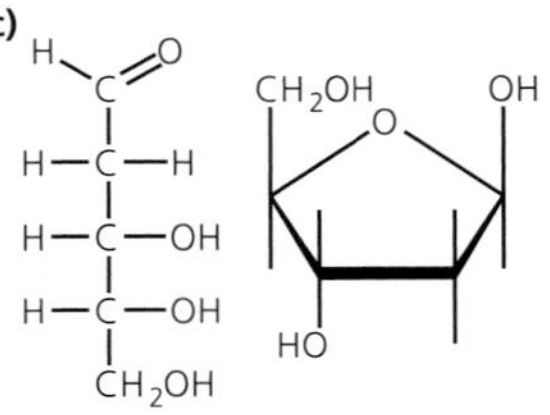	in DNA

Question

2 List the properties of the carbon atom responsible for the huge array of organic compounds.

Disaccharides

Disaccharides are carbohydrates made of two monosaccharides combined together. An example is sucrose.

Sucrose is formed from a molecule of α-glucose and a molecule of fructose. A molecule of water is removed when two monosaccharides combine, so the reaction is called a **condensation reaction**. The bond formed between the α-glucose and fructose molecules by the removal of water is called a **glycosidic bond** (Figure 2.6). This is another example of a strong, covalent bond.

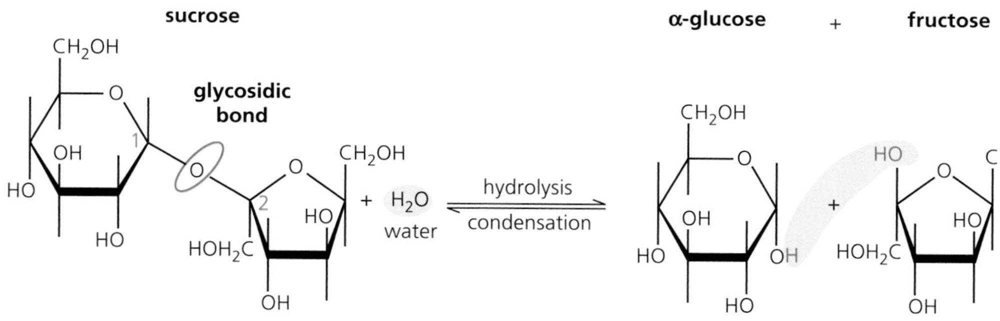

Figure 2.6 Sucrose (a disaccharide) and the monosaccharides that form it

In the reverse process, a disaccharide is 'digested' to the component monosaccharides in a **hydrolysis reaction**. This reaction involves adding a molecule of water ('*hydro-*') at the same time as the splitting ('-*lysis*') of the glycosidic bond occurs (Figure 2.6).

Sucrose is not a reducing sugar. This is because the aldehyde group of glucose and the ketone group of fructose (which are reducing groups) are used to form the glycosidic bond in sucrose. Neither is then available to reduce copper(II) to copper(I) in a Benedict's test. Benedict's solution will remain blue when heated with sucrose.

Testing for non-reducing sugars

To demonstrate that sucrose is a non-reducing sugar, two tests are required. To carry these out:

1 Dissolve a small spatula-load of sucrose in half a test tube of distilled water, and split the resulting solution equally between two test tubes, labelled A and B.
2 To A, add approximately $1\,cm^3$ of dilute hydrochloric acid, and to B add the same quantity of distilled water. Heat both tubes in a boiling water bath for 1 minute.
3 Remove both tubes and cool to room temperature. To tube A, cautiously add a small quantity of sodium hydrogencarbonate powder to neutralise any excess acid. When no more effervescence occurs on addition of the powder, all the acid will have been neutralised, and conditions will now be alkaline in the tube.
4 Add a small quantity of Benedict's solution to each tube and heat for a standard time in the water bath. Then remove the tubes and compare the final colours.
5 In tube A the sucrose will have been hydrolysed (by acid, in this case) to glucose and fructose, and these reducing sugars with give a positive result in the Benedict's test.
In tube B the sucrose present will have caused no colour change, but Tube B has established the initial absence of reducing sugar.

Question

3 What is meant by
  a a non-reducing sugar
  b a glycosidic bond?

Sucrose and other disaccharides

Sucrose is an important sugar. It is sucrose that is transported from leaves to the growing points of plant stems and roots or to the carbohydrate storage sites all over the plant. This occurs in the sieve tubes of phloem tissue. Sucrose is also the 'sugar' humans mostly prefer to use in foods and drinks. The sugar industry extracts and purifies sucrose from sugar beet and sugar cane.

There are other disaccharides formed in cells, depending on which monosaccharides are involved and whether they are in their α or β condition. Two other commonly occurring disaccharides are maltose and lactose.

Maltose (α-glucose + α-glucose) is a product of starch hydrolysis. The extraction of maltose from germinating barley (as malt extract) is an important industry. Malt is used in brewing and in food manufacture.

Lactose (β-galactose + glucose) is the sugar found in the milk of mammals.

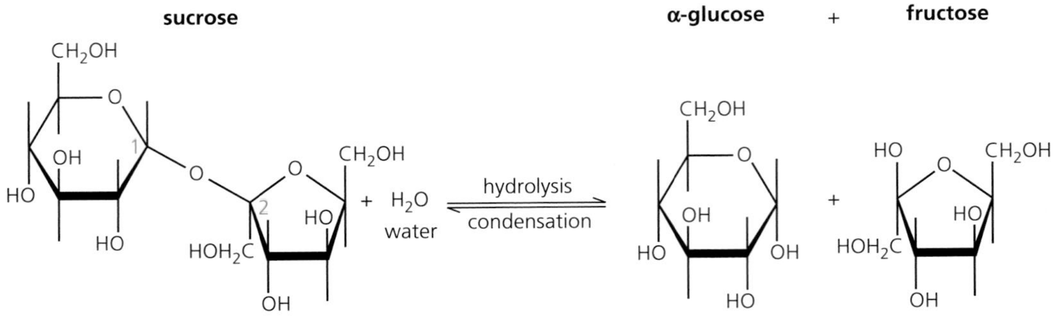

Figure 2.7 Sucrose (a disaccharide) and the monosaccharides that form it

Finally, a definition check:
- Monosaccharides are simple sugars; all are reducing sugars.
- Disaccharides are sugars that are condensation products of two monosaccharide molecules, with the elimination of water and the formation of a glycosidic bond. Some disaccharides are reducing sugars and some are non-reducing sugars.

Polysaccharides

Polysaccharides are built from very many monosaccharide molecules condensed together. Once they are part of a larger molecule, they are called residues. Each residue is linked by glycosidic bonds. 'Poly' means 'many' and, in fact, thousands of 'saccharide' residues make up a polysaccharide. So a polysaccharide is an example of a giant molecule, a **macromolecule**. Normally each polysaccharide contains only one type of **monomer**. Cellulose is a good example – built from the monomer glucose.

Bonds in polysaccharides

Atoms naturally combine together (they 'bond') to form molecules in ways that have a stable arrangement of electrons in the outer shells of each atom (see Appendix 1 on the CD). In **covalent bonding**, electrons are shared between atoms. Covalent bonds are the strongest bonds in biological molecules. In polysaccharides, the individual monomers are held by covalent bonds.

However, entirely different bonds, called **hydrogen bonds**, are also formed. We can illustrate how hydrogen bonds form by reference to the water molecule.

Water is composed of one atom of oxygen and two atoms of hydrogen, also combined by covalent bonding. The water molecule is *triangular* rather than linear, and the nucleus of the oxygen atom draws electrons (negatively charged) away from the hydrogen nuclei (positively charged) – with an interesting consequence. Although overall the water molecule is electrically neutral, there is a net negative charge on the oxygen atom (conventionally represented by Greek letter *delta* as δ^-) and a net positive charge on the hydrogen atoms (represented by δ^+). In other words, the water molecule carries an unequal distribution of electrical charge within it. This unequal distribution of charge is called a **dipole**. Molecules that contain groups with dipoles are known as **polar molecules**. Water is a polar molecule (see Figure 2.27, page 56).

Dipoles exist in many different molecules, especially where there are —OH, —C=O, or =N–H groups, so hydrogen bonds can form between these groups. Hydrogen bonds are weaker than covalent bonds, but never the less are very important in the structure and properties of carbohydrates (including polysaccharides) and proteins. For example, molecules containing these groups are attracted to water molecules because of their dipoles, and are therefore said to be 'water loving' or **hydrophilic**.

Cellulose

Cellulose is by far the most abundant carbohydrate – it makes up more than 50 per cent of all organic carbon. (Remember, the gas carbon dioxide, CO_2, and the mineral calcium carbonate, $CaCO_3$, are examples of inorganic carbon.)

Cellulose is a polymer of β-glucose molecules combined together by glycosidic bonds between carbon-4 of one β-glucose molecule and carbon-1 of the next. Successive glucose units are linked at 180° to each other (Figure 2.8). This structure is stabilised and strengthened by hydrogen bonds between adjacent glucose units in the same strand and, in fibrils of cellulose, by hydrogen bonds between parallel strands, too. In plant cell walls additional strength comes from the cellulose fibres being laid down in layers running in different directions.

The chemical test for cellulose is that it gives a purple colour when Schultz' solution is added. The cell walls of green plants and the debris of plants in and on the soil are where cellulose occurs. It is an extremely strong material – insoluble, tough and durable but slightly elastic. Cellulose fibres are straight and uncoiled. When it is extracted from plants, cellulose has many industrial uses. We use cellulose fibres as cotton, we manufacture them into paper, rayon fibres for clothes manufacture, nitrocellulose for explosives, cellulose acetate for fibres of multiple uses, and cellophane for packaging.

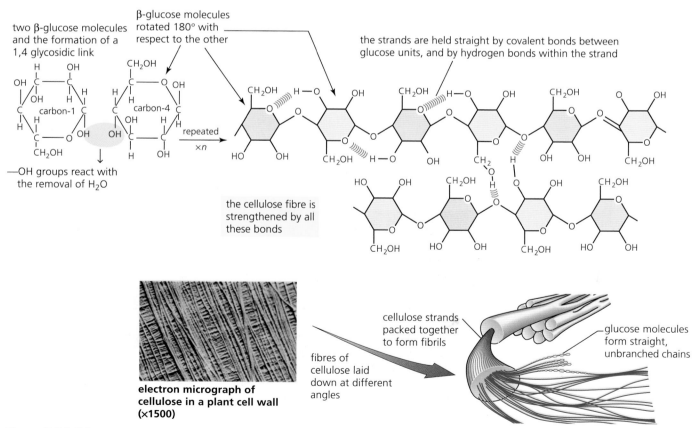

Figure 2.8 Cellulose

Starch

Starch is a mixture of two polysaccharides (Figure 2.9).

- **Amylose** is an unbranched chain of several thousand 1,4 linked α-glucose units.
- **Amylopectin** has shorter chains of 1,4 linked α-glucose units but, in addition, there are branch points of 1,6 links along its chains.

These covalent bonds between glucose residues in starch bring the molecules together as a **helix**. The whole starch molecule is stabilised by countless hydrogen bonds between parts of the component glucose molecules.

Starch is the major storage carbohydrate of most plants. It is laid down as compact grains in **plastids** called leucoplasts. Starch is an important energy source in the diet of many animals, too. Its usefulness lies in the compactness and insolubility of its molecule. However, it is readily hydrolysed to form sugar when required. We sometimes see 'soluble starch' as an ingredient of manufactured foods. Here the starch molecules have been broken down into short lengths, making them more easily dissolved.

Testing for starch

We test for starch by adding a solution of iodine in potassium iodide. Iodine molecules fit neatly into the centre of the starch helix, creating a blue–black colour (Figure 2.9).

> **Question**
>
> **4** Starch is a powdery material; cellulose is a strong, fibrous substance; yet both are made of glucose. What features of the cellulose molecule account for its strength?

Figure 2.9 Starch

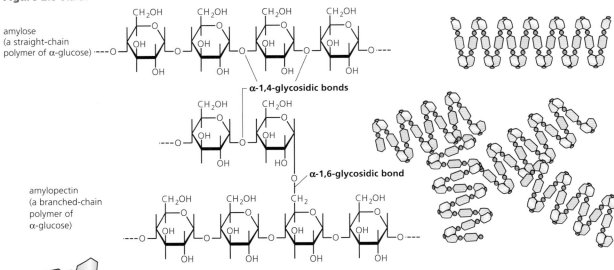

amylose (a straight-chain polymer of α-glucose)

α-1,4-glycosidic bonds

amylopectin (a branched-chain polymer of α-glucose)

α-1,6-glycosidic bond

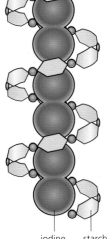

iodine molecules   starch chain

test for starch with iodine in potassium iodide solution; the blue–black colour comes from a starch/iodine complex

a) Test on a potato tuber cut surface

1% starch solution   0.1% starch solution   0.01% starch solution

b) Test on starch solutions of a range of concentrations

TEM of a liver cell (×7000)

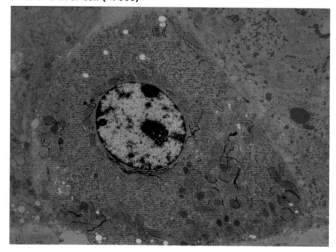

Figure 2.10 Glycogen granules (dark blue spots) in liver cells

Glycogen is a polymer of α-glucose, chemically very similar to amylopectin, although larger and more highly branched. Glycogen is one of our body's energy reserves and is drawn upon and respired as needed. Granules of glycogen are seen in liver cells (Figure 2.10), muscle fibres, and throughout the tissues of the human body. The one exception is in the cells of the brain. Here virtually no energy reserves are stored. Instead brain cells require a constant supply of glucose from the blood circulation.

Other sugar compounds

In addition to cellulose, plant cell walls contain small quantities of substances called **pectins** and **hemicelluloses**. These are mixtures of polysaccharides formed from organic acids, themselves made from sugars. These 'acids' occur as their calcium salts. For example, we have calcium pectate in plant cell walls. We can think of these pectins and hemicelluloses as the 'glues' that hold the cellulose of cell walls of plants together. At the junction between walls these substances appear as a layer called the **middle lamella**. Pectins and hemicelluloses also pass between the cellulose fibres, adding to the strength of the cell wall.

> **Question**
>
> **5** Define the terms 'monomer' and 'polymer', giving examples from the carbohydrates.

Lipids

Lipids contain the elements carbon, hydrogen and oxygen, as do carbohydrates. However, in lipids the proportion of oxygen is much less. Lipids are insoluble in water. In fact they generally behave as 'water-hating' molecules, a property described as **hydrophobic**. Lipids can be dissolved in organic solvents such as alcohol (for example, ethanol), propanone and ether. This property is made use of in the emulsion test for lipids (Figure 2.13).

Lipids occur in living things as animal **fats** and plant **oils**. They are also present as the **phospholipids** of cell membranes. In addition, there are other, more unusual forms of lipid. For example, the steroids from which many growth and sex hormones are produced are lipids. The waxes found on plants and animals are also lipids.

meat joint with fat layers around the blocks of muscle fibres   olive oil   nuts and seeds rich in oils

Figure 2.11 Fats and oils – these all leave a translucent stain on paper or fabric

Fats and oils

Fats and oils are compounds called **triglycerides** (strictly, triacylglycerols). They are formed by reactions in which water is removed. These condensation reactions occur between **fatty acids** (full name, monocarboxylic acids) and an alcohol called **glycerol**, as shown in Figure 2.12. Three fatty acid molecules combine with one glycerol molecule to form a triglyceride. Fats and oils are both formed in this way and they have the same chemical structure. The only difference between them is that at about 20°C (room temperature) **oils are liquids** and **fats are solid**.

Triglycerides are quite large molecules but they are small when compared to macromolecules like glycogen and starch. It is only because of their hydrophobic properties that triglyceride molecules clump together (aggregate) into huge globules that make them *appear* to be macromolecules.

The fatty acids present in fats and oils have long hydrocarbon 'tails'. These are typically of about 16 to 18 carbon atoms long but may be anything between 14 and 22 (Figure 2.12). The hydrophobic properties of triglycerides are due to these hydrocarbon tails.

We describe fatty acid molecules as 'acids' because their functional group (–COOH, carboxyl) tends to ionise (slightly) to produce hydrogen ions, which is the property of an acid.

$$-COOH \rightleftharpoons -COO^- + H^+$$

It is the carboxyl functional group of three organic acids that react with the three hydroxyl functional groups of glycerol to form a triglyceride (Figure 2.12). The bonds formed in this case are known **ester bonds**.

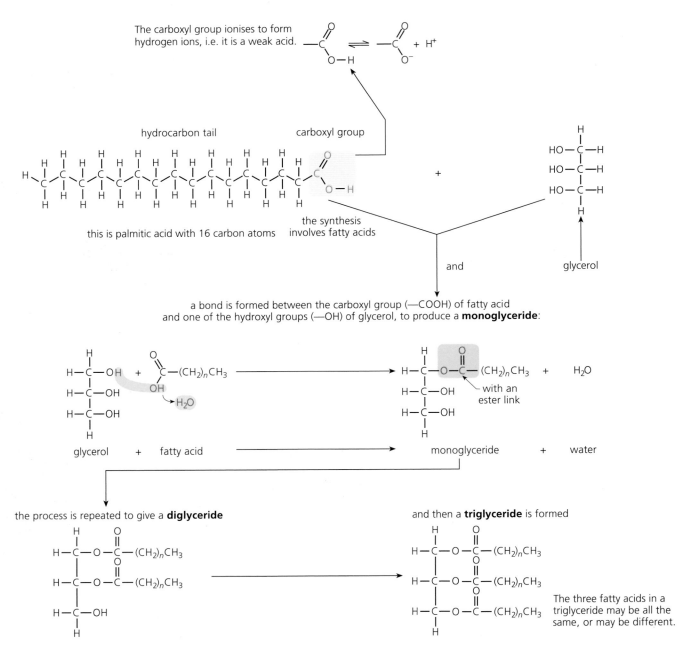

Figure 2.12 The formation of a triglyceride

The emulsion test for lipid

When organic matter is shaken up with a small quantity of absolute ethanol, if lipid is present it will become dissolved in this organic solvent (see Figure 2.13). Then, when an equal quantity of cold water is added, a cloudy white suspension will form, since lipid is totally immiscible with water. Alternatively, in the absence of lipid in the original material, alcohol and water will mix and the resulting solution will remain transparent.

Saturated and unsaturated triglycerides

Much attention is paid to the value of 'unsaturated fats' in the diet – especially among people who are well fed! Fats and oils which are **saturated** have no **double bonds** (–C=C–) between the carbon atoms in their hydrocarbon tails, whereas **unsaturated fats and oils** have one *or more* double bonds present in the hydrocarbon tails (Figure 2.14).

oil forms a layer on water, but when shaken together, froms an emulsion (milky in appearance) which may take a while to disperse

Figure 2.13 The emulsion test

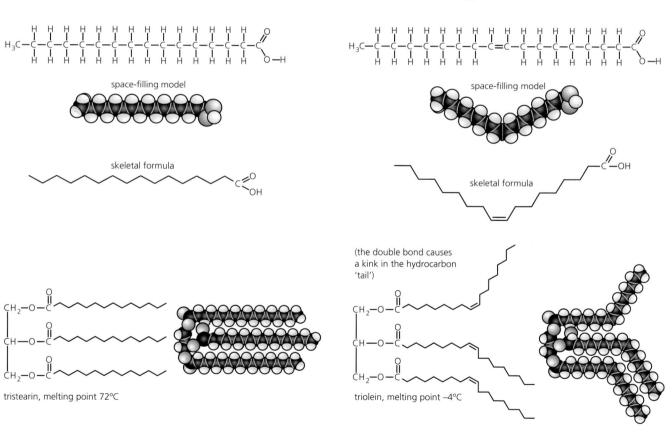

palmitic acid, $C_{15}H_{31}COOH$, a saturated fatty acid

space-filling model

skeletal formula

tristearin, melting point 72°C

oleic acid, $C_{17}H_{33}COOH$, an unsaturated fatty acid

space-filling model

skeletal formula

(the double bond causes a kink in the hydrocarbon 'tail')

triolein, melting point –4°C

Figure 2.14 Saturated and unsaturated fatty acids, and the triglycerides they form

Where several double bonds occur the resulting fat is called **polyunsaturated**. Fats with unsaturated fatty acids melt at a lower temperature than those with saturated fatty acids, because their unsaturated hydrocarbon tails do not pack so closely together. You can see why in Figure 2.14.

This difference between saturated and polyunsaturated fats is important in the manufacture of margarine and butter spreads, since the later perform better 'from the fridge'. There is disagreement about the value (or otherwise) of 'polyunsaturates' in the human diet. What seems clear is that it is better to eat less rather than more fat in our diets.

The roles of fats and oils in living things

Energy source and metabolic water source

When triglycerides are oxidised in respiration a lot of energy is transferred to make ATP (see page 239). Mass for mass, fats and oils release more than twice as much energy as carbohydrates do when they are respired. This is because fats are more reduced than carbohydrates (page 240). More of the oxygen in the respiration of fats comes from the atmosphere. In the oxidation of carbohydrate, more oxygen is present in the carbohydrate molecule itself. Fat therefore forms a concentrated, insoluble energy store.

Fat layers are typical of animals that endure long unfavourable seasons in which they survive by using the concentrated reserves of food stored in their bodies. Oils are often a major energy store in plants, their seeds and fruits, and it is common for fruits and seeds to be used commercially as a source of edible oils for humans, including maize, olives and sunflower.

Complete oxidation of fats and oils produces a large amount of water. This is far more than when the same mass of carbohydrate is respired. Desert animals like the camel and the desert rat retain much of this 'metabolic water' within the body, helping them survive when there is no liquid water for drinking. The development of the embryos of birds and reptiles whilst in their shells also benefits from metabolic water formed by the oxidation of the stored fat in their egg's yolk. Whales are mammals surrounded by salt water they cannot safely drink and they rely solely on metabolic water!

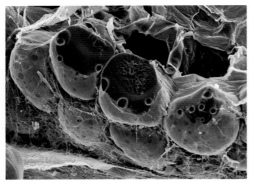

Figure 2.15 Adipose tissue

Fat as a buoyancy aid

Fat is stored in animals as **adipose tissue**, typically under the skin, where it is known as **subcutaneous fat**. Aquatic, diving mammals have so much it is identified as blubber. This gives buoyancy to the body, because fat is not as dense as muscle or bone. If fat reserves like these have a restricted blood supply and the heat of the body is not especially distributed to the fat under the skin (as is commonly the case), then the subcutaneous fat also functions as a heat insulation layer.

Waterproofing of hair and feathers

Oily secretions of the sebaceous glands, found in the skin of mammals, acts as a water repellent, preventing fur and hair from becoming waterlogged when wet. Birds have a preen gland that serves the same purpose for feathers.

Electrical insulation

Myelin lipid in the membranes of Schwann cells, forming the sheath around the axons of neurones (page 313), electrically isolates the cell surface membrane and facilitates the conduction of the nerve impulse there.

Phospholipids

A phospholipid has a very similar chemical structure to a triglyceride, except that one of the fatty acids is replaced by a phosphate group. The phosphate group is ionised and therefore water soluble. Consequently, phospholipids combine the hydrophobic properties of the hydrocarbon tails with the water-loving (**hydrophilic**) properties of the phosphate group.

A thin layer of phospholipid floats on water, forming a monolayer with the hydrophilic heads in water and the hydrophobic tails projecting. With more phospholipid present, the phospholipid molecules form a bilayer, with the hydrophobic tails aligned together. We return to this when looking into the structure of cell surface membranes (Topic 4).

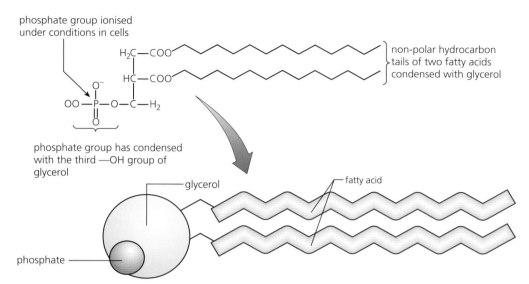

Question

6 Describe the properties given to a lipid when it combines with a phosphate group.

Figure 2.16 Phospholipids

Other lipids

If you compare the structure of a **steroid** such as cholesterol (Figure 2.17) with that of a triglyceride (Figure 2.14), you can see just how structurally different the lipids are. The 'skeleton' of a steroid is a set of complex rings of carbon atoms; the bulk of the molecule is hydrophobic but the polar hydroxyl group is hydrophilic. Steroids occur in both plants and animals, and one widespread form is the substance **cholesterol**.

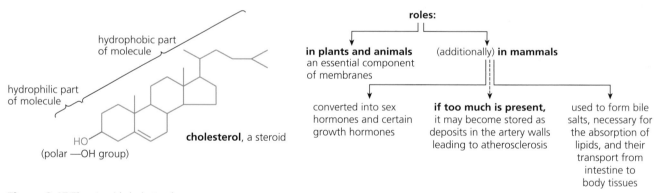

Figure 2.17 The steroid cholesterol

Cholesterol is an essential component of membranes of plants and animals. In mammals, the sex hormones progesterone, oestrogen and testosterone and also growth hormones are also produced from it. Bile salts (involved in lipid transport from the intestine to the tissues, in the blood) are also synthesised from cholesterol. Consequently, cholesterol is essential for normal, healthy metabolism. However, excess blood cholesterol may cause harmful deposits in artery walls, causing a disease called atherosclerosis.

Finally, **waxes** are esters formed from a fatty acid and a complex alcohol (in place of glycerol). They are produced by plants and used as the waxy cuticle that protects the plant stem and leaf epidermis from loss of water by evaporation. The same role is fulfilled by the wax that insects have on the outer surface of their cuticle. Bees use wax in the honeycomb cells they build for rearing their young (larvae) and storing honey.

2.3 Proteins and water

An understanding of protein structure and how it is related to function is central to many aspects of biology, such as enzymes, antibodies and muscle contraction.

Globular and fibrous proteins play important roles in biological processes such as the transport of gases and providing support for tissues.

Water is a special molecule with extraordinary properties that make life possible on this planet 150 million kilometres from the Sun.

By the end of this section you should be able to:

a) describe the structure of an amino acid and the formation and breakage of a peptide bond
b) explain the meaning of the terms primary structure, secondary structure, tertiary structure and quaternary structure of proteins and describe the types of bonding (hydrogen, ionic, disulfide and hydrophobic interactions) that hold these molecules in shape
c) describe the molecular structure of haemoglobin as an example of a globular protein, and of collagen as an example of a fibrous protein and relate these structures to their functions
d) explain how hydrogen bonding occurs between water molecules and relate the properties of water to its roles in living organisms

Proteins

Proteins make up about two-thirds of the total dry mass of a cell. They differ from carbohydrates and lipids in that they contain the element **nitrogen**, and usually the element sulfur, as well as carbon, hydrogen and oxygen.

Proteins are large molecules formed from many **amino acids** combined in a long chain. Typically several hundreds or even thousands of amino acid molecules are combined together to make a protein. (So a protein can also be described as a **macromolecule**.) Once the chain is constructed, a protein takes up a specific shape. The shape of a protein is closely related to the functions it performs. We will return to this aspect of protein structure, after we have examined the properties of individual amino acids.

Amino acids, as their name implies, contain two functional groups, a basic **amino group** ($-NH_2$) and an acidic **carboxyl group** ($-COOH$). In the naturally-occurring amino acids both functional groups are attached to the same carbon atom. Since the carboxyl group is acidic:

$$-COOH \rightleftharpoons -COO^- + H^+$$

and the amino group is basic:

$$-NH_2 + H^+ \rightleftharpoons -NH_3$$

an amino acid is both an acid and a base, a condition known as **amphoteric**. The remainder of the molecule, the side chain or **R-group**, is very variable.

The bringing of amino acids together in different combinations produces proteins with very different properties. This helps explain how the proteins in organisms are able to fulfil the very different biological functions they have.

Question

7 What does the symbol 'R' represent in a generalised formula of an organic compound?

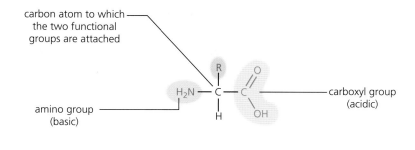

Questions

8 The possible number of different polypeptides (*P*) that can be assembled is given by:

$$P = A^n$$

where *A* = the number of different types of amino acids available and *n* = the number of amino acid monomers that make up the polypeptide molecule.

Given the naturally occurring pool of 20 different types of amino acids, calculate how many different polypeptides are possible if constructed from 5, 25 and 50 amino acid residues respectively?

9 Distinguish between condensation and hydrolysis reactions. Give an example of each.

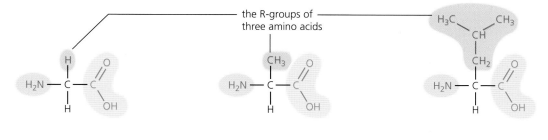

Figure 2.18 The structure of amino acids

The peptide bond

Two amino acids can react together with the loss of water to form a dipeptide. The amino group of one amino acid reacts with the carboxyl group of the other, forming a **peptide bond** (see Figure 2.19 on the next page). Long strings of amino acids linked by peptide bonds are called **polypeptides**. A peptide or protein chain is assembled, one amino acid at a time, as we shall see later.

The terms 'polypeptide' and 'protein' can be used interchangeably but, when a polypeptide is about 50 amino acid molecules long, it is generally agreed to be have become a protein.

Testing for proteins

The **biuret test** is used as an indicator of the presence of protein because it gives a purple colour in the presence of peptide bonds (–C–N–). To a protein solution, an equal quantity of sodium hydroxide solution is added and mixed. Then a few drops of 0.5 per cent copper(II) sulfate is introduced with gentle mixing. A distinctive purple colour develops without heating (see Figure 2.20 on the next page).

The structure of proteins

The **primary structure** of a protein is the long chain of amino acids in its molecule. Proteins differ in the variety, number and order of these amino acids. In the living cell, the sequence of amino acids in the polypeptide chain is controlled by the coded instructions stored in the DNA of the chromosomes in the nucleus (Topic 6). Just changing one amino acid in the sequence of a protein may alter its properties completely. This sort of 'mistake' or mutation does happen (see page 125).

amino acids combine together, the amino group of one with the carboxyl group of the other
carboxyl group
amino group
condensation reaction
hydrolysis reaction
peptide linkage
amino acid 1
amino acid 2
dipeptide
water
H_2O
for example, glycine and alanine can react like this:
peptide linkage
glycine
alanine
glycyl-alanine
but if the amino group of glycine reacts with the carboxyl group of alanine, a different polypeptide, alanyl-glycine, is formed
three amino acids combine together to form a tripeptide
glycine residue
alanine residue
cysteine residue

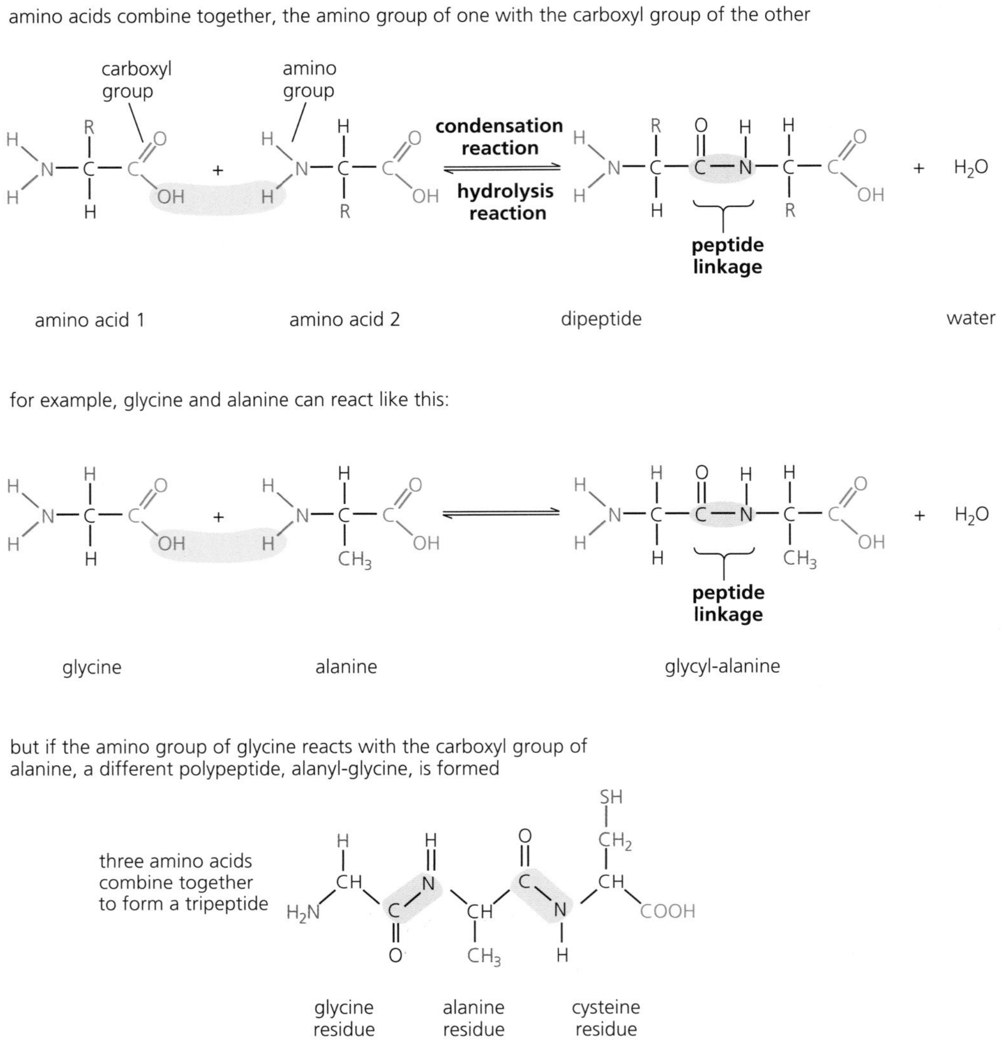

Figure 2.19 Peptide bond formation

to the solution to be tested add:
2M sodium hydroxide solution
then:
0.5% copper(II) sulfate solution (via dropping bottle)
egg white solution
wheat flour solution
1% starch solution (control)
purple colour develops
remains blue

Figure 2.20 The biuret test for protein

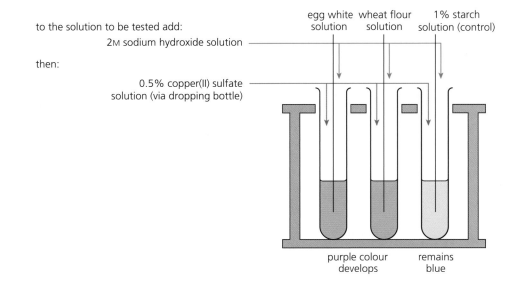

10 Describe three of the types of bonds that maintain the tertiary structure of proteins.

The **secondary structure** of a protein develops when parts of the polypeptide chain take up a particular shape, immediately after formation at the ribosome (Figure 2.21). Parts of the chain become folded or twisted, or both, in various ways. Two major structural forms are particularly stable and common. Either part or all of the peptide chain becomes coiled to produce an **α-helix** or it becomes folded into **β-sheets**. These shapes are permanent, held in place by hydrogen bonds.

The **tertiary structure** of a protein is the precise, compact structure, unique to that protein that arises when the molecule is further folded and held in a particular complex shape. This shape is made permanent by four different types of bonding, established between adjacent parts of the chain (Figure 2.22).

The primary, secondary and tertiary structure of the protein called lysozyme is shown in Figure 2.24.

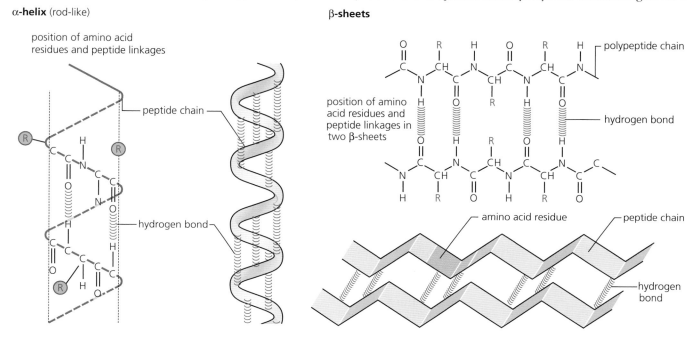

Figure 2.21 The secondary structure of proteins

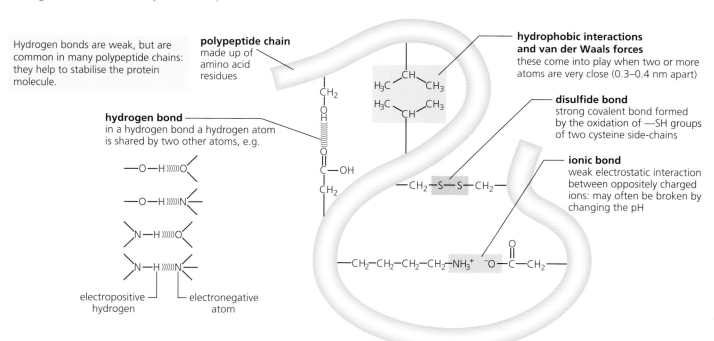

Figure 2.22 Cross-linking within a polypeptide

photomicrograph of collagen fibre (× 000) – many triple helices bound together

the chemical basis of the strength of collagen

three long polypeptide molecules, coiled together to form a triple helix

every third amino acid is glycine (the smallest amino acid) and the other two amino acids are mostly proline and hydroxyproline

covalent bonds form between the polypeptide chains – together with many hydrogen bonds

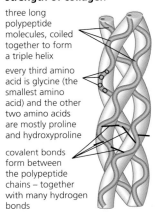

Figure 2.23 Collagen: an example of a fibrous protein

Fibrous and globular proteins

Some proteins take up a tertiary structure which is a long strand, and are called **fibrous proteins**. They have structural roles in the body. Examples of fibrous proteins are **fibrin**, a blood protein involved in the clotting mechanism, and **keratin**, found in hair, horn and nails.

Collagen

Collagen is the most abundant structural protein in animals in general. In our skin it forms a mat in the deepest layers. It occurs in tendons, cartilage, bone, teeth and in the walls of blood vessels. It also forms the cornea of the eye.

Chemically, a collagen **molecule** consists of three polypeptide chains, each in the shape of a helix. Each chain consists of about 1000 amino acids. They are wound together as a triple helix forming a stiff cable – a unique arrangement in proteins. These helices are stretched out, meaning this structure is not as tightly wound as an α-helix. This structure is only able to form if every third residue in each polypeptide chain is glycine – the smallest amino acid. In fact, along the helix a typical repeating sequence is glycine-proline-hydroxyproline.

The three helices are held together by numerous hydrogen bonds and have great mechanical strength. Many of these triple helices lie side-by-side, forming collagen **fibres**. Within the fibres, collagen molecules are held together by covalent cross-linkages between the carboxyl group of one amino acid and the amino group of another. Also, the ends of individual collagen molecules are staggered so there are no weak points in collagen fibres, giving the whole structure incredible tensile strength. For example, a load of at least 10 kg is needed to break a collagen fibre that is only 1 mm in diameter. This quality is essential in tendons, for example. Within the body, collagen also binds strongly to other substances, such as the calcium phosphate of bone. The structure of collagen is illustrated in Figure 2.23.

Other proteins take up a tertiary structure that is more spherical. They are called **globular proteins**. Enzymes are typically globular proteins, including lysozyme – an antibacterial enzyme present in animal excretions such as tears (see Figure 2.24).

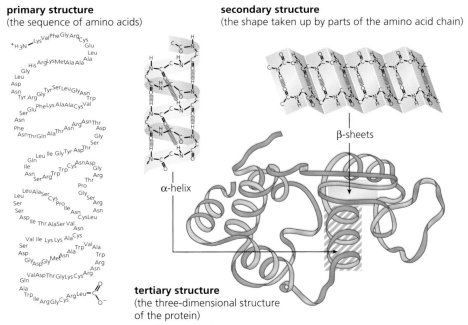

primary structure
(the sequence of amino acids)

secondary structure
(the shape taken up by parts of the amino acid chain)

β-sheets

α-helix

tertiary structure
(the three-dimensional structure of the protein)

Figure 2.24 Lysozyme: primary, secondary and tertiary structure

Finally, the **quaternary structure** of a protein arises when two or more proteins become held together, forming a complex, biologically active molecule. An example is **haemoglobin**, a globular protein (**globin**) consisting of four polypeptide chains (two **α-chains** and two **β-chains**). Each polypeptide chain is associated with a non-protein **haem group**, in which an **iron ion (Fe^{2+})** occurs (Figure 2.25). Since the pigment 'haem' is an integral part of the quaternary protein and yet is not itself made of amino acids, it is referred to as a **prosthetic group**. A molecule of **oxygen (O_2)** combines reversibly with each haem group in the 'pocket' where the iron ion occurs. This reaction is dependent upon the **partial pressure of oxygen** (high in the lungs, low in the respiring tissues of the body).

$$\text{haemoglobin (Hb)} + 4O_2 \rightleftharpoons \text{oxyhaemoglobin (HbO}_8)$$

Within the haemoglobin molecule the forces holding it together (and thereby maintaining the all important three-dimensional shape of haemoglobin) are inwardly-facing hydrophobic R-groups. On the exterior of the molecule it is the presence of hydrophilic R-groups that maintain the solubility of the protein. Vast numbers of the resulting more or less spherical haemoglobin molecules are normally located in the red blood cells.

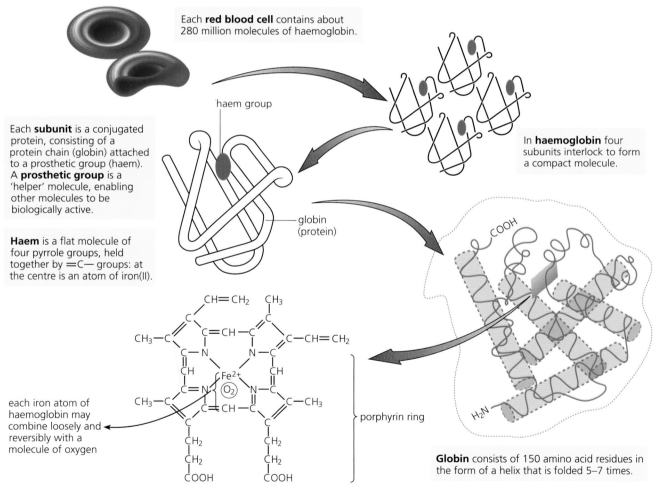

Each **red blood cell** contains about 280 million molecules of haemoglobin.

haem group

Each **subunit** is a conjugated protein, consisting of a protein chain (globin) attached to a prosthetic group (haem). A **prosthetic group** is a 'helper' molecule, enabling other molecules to be biologically active.

Haem is a flat molecule of four pyrrole groups, held together by =C— groups: at the centre is an atom of iron(II).

In **haemoglobin** four subunits interlock to form a compact molecule.

globin (protein)

each iron atom of haemoglobin may combine loosely and reversibly with a molecule of oxygen

porphyrin ring

Globin consists of 150 amino acid residues in the form of a helix that is folded 5–7 times.

Figure 2.25 Haemoglobin: the quaternary protein of red cells

An important exception to this arises in **sickle-cell anaemia**, a genetically controlled condition in which one of the amino acids that comprise the β-chain is replaced (non-polar valine is substituted for polar glutamic acid). The result is long fibrous haemoglobin molecules that distort and damage the red blood cells and which hinders their efficient delivery of oxygen to the tissues. However, the sickle cell condition may confer some resistance to malaria.

Denaturation of protein

Denaturation is the loss of the three-dimensional structure of a protein. It happens when the bonds that maintain the three-dimensional shape of the protein molecules are changed (Figure 2.22).

We have noted that many of the properties and uses of proteins within cells and organisms depend on their particular shape. When the shape of a protein changes the protein may cease to be useful. The biochemistry of cells and organisms is extremely sensitive to conditions that alter proteins in this way.

Exposure to heat (or radiation) causes atoms to vibrate violently, and this disrupts hydrogen and ionic bonds. Under these conditions, protein molecules become elongated, disorganised strands. We see this when a hen's egg is cooked. The translucent egg 'white' is a globular protein, **albumen**, which becomes irreversibly opaque and insoluble. Exposure to heavy metal ions, to organic solvents or to extremes of acidity or alkalinity will all trigger irreversible denaturations. These conditions sometime alter charges on the R-groups, too.

Small changes in the pH of the medium may alter the ionic charges of acidic and basic groups and temporarily cause change (see page 65). However, the tertiary structure may spontaneously reform. This suggests that it is the primary structure of a protein that determines cross-linking and the tertiary structure, given a favourable medium.

The making of macromolecules – a review

We began this topic by introducing four elements – carbon, hydrogen, oxygen and nitrogen. We saw that it is from atoms of these elements that the carbohydrates, lipids and proteins are made – the bulk of the molecules of life. These biological molecules exist in a huge range of sizes, all of them with important roles in working cells.

However, some of the molecules in cells are very large indeed. These are the giant molecules, known as **macromolecules**. Polysaccharides and proteins are the macromolecules that have been introduced in this topic. Chemically we have described them as **polymers**, made by linking together similar building blocks called **monomers**. Proteins are built from amino acids and polysaccharides from monosaccharide sugars.

Monomers are linked together by a reaction in which the elements of water are removed (a **condensation** reaction). So the monomer units that have made up a polymer are now called **residues**. The reverse reaction requires water and it results in the release of individual monomers again. It is called a **hydrolysis** reaction. The bonds between monomers are covalent bonds. Between monosaccharide residues of a polysaccharide they are **glycosidic bonds**. Between amino acid residues of a protein, **peptide bonds** are formed.

The other macromolecules found in cells are the **nucleic acids**. These are discussed in Topic 6. The most significant points to keep in mind about the properties of macromolecules are:
- the type of macromolecule depends upon the type of monomer from which it is built
- the order in which the monomers are combined decides the shape of the macromolecule
- the shape of the macromolecule decides its biological properties and determines its role in cells.

As the functions of each of the macromolecules of cells – polysaccharides, proteins and (later) nucleic acids are discussed, the significance of 'type', 'order' and 'shape' will become clear.

Question

11 Explain why polysaccharides and proteins are described as macromolecules but lipids are not.

Roles of proteins

We have seen that some proteins of organisms have a **structural role** in cells and organisms; others have **biochemical** or **physiological roles**. Whatever their roles are in metabolism, however, cell proteins are in a continuous state of flux. They are continuously built up, used and broken down again into their constituent amino acids, to be rebuilt or replaced by fresh proteins, according to the needs of the cells (Figure 2.26).

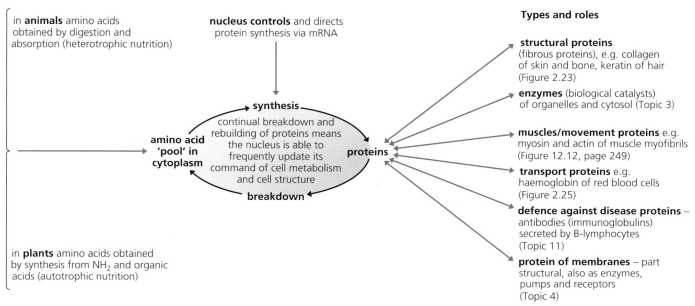

Figure 2.26 The dynamic state of cell proteins

Water

Living things are typically solid, substantial objects, yet water forms the bulk of their structures. Between 65 and 95 per cent by mass of most multicellular plants and animals (about 80 per cent of a human cell) consists of water. Despite this, water is a substance that is often taken for granted.

Water is composed of atoms of the elements hydrogen and oxygen. One atom of oxygen and two atoms of hydrogen combine by sharing of electrons (**covalent bonding**). However, the water molecule is *triangular* rather than linear and the nucleus of the oxygen atom draws electrons (which are negatively charged) away from the hydrogen nuclei (which are positively charged) – with an interesting consequence (Figure 2.27). Although overall the water molecule is electrically neutral, there is a net negative charge on the oxygen atom (conventionally represented by Greek letter *delta* as δ^-) and a net positive charge on the hydrogen atoms (represented by δ^+). In other words, the water molecule carries an unequal distribution of electrical charge within it. This unequal distribution of charge is called a **dipole**. Molecules which contain groups with dipoles are known as **polar molecules**. Water is a polar molecule.

Hydrogen bonds

With water molecules, the positively charged hydrogen atoms of one molecule are attracted to the negatively charged oxygen atoms of nearby water molecules by forces called **hydrogen bonds**. These are weak bonds compared to covalent bonds, yet they are strong enough to hold water molecules together. Hydrogen bonds largely account for the unique properties of water. We examine these properties next.

Meanwhile we can note that dipoles are found in many different molecules, especially where there occurs an –OH, –C=O or >N–H group. Hydrogen bonds can form between all these groups and also between these groups and water. Materials with an affinity for water are described as **hydophilic** (meaning 'water-loving', see below).

Question

12 Distinguish between ionic and covalent bonding.

Figure 2.27 The water molecule and the hydrogen bonds it forms

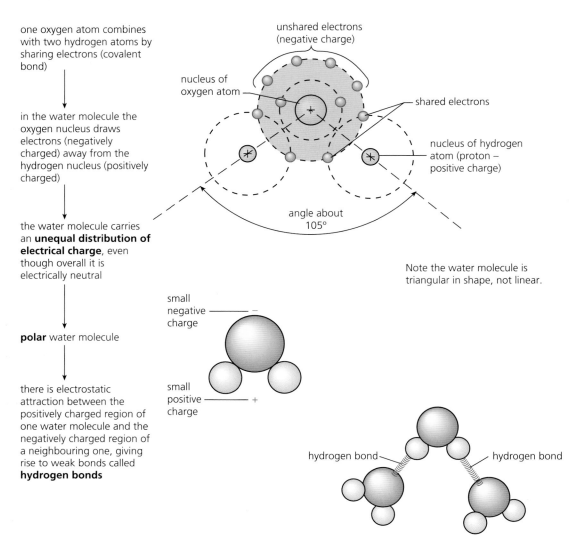

one oxygen atom combines with two hydrogen atoms by sharing electrons (covalent bond)

↓

in the water molecule the oxygen nucleus draws electrons (negatively charged) away from the hydrogen nucleus (positively charged)

↓

the water molecule carries an **unequal distribution of electrical charge**, even though overall it is electrically neutral

↓

polar water molecule

↓

there is electrostatic attraction between the positively charged region of one water molecule and the negatively charged region of a neighbouring one, giving rise to weak bonds called **hydrogen bonds**

unshared electrons (negative charge)

nucleus of oxygen atom

shared electrons

nucleus of hydrogen atom (proton – positive charge)

angle about 105°

Note the water molecule is triangular in shape, not linear.

small negative charge

small positive charge

hydrogen bond

hydrogen bond

The properties of water

Heat energy and the temperature of water

A lot of heat energy is required to raise the temperature of water. This is because much energy is needed to break the hydrogen bonds that restrict the movements of water molecules. This property of water is its **specific heat capacity**. The specific heat capacity of water is the highest of any known substance. Consequently, aquatic environments like streams, rivers, ponds, lakes and seas are very slow to change temperature when the surrounding air temperature changes. Aquatic environments have much more stable temperatures than do terrestrial (land) environments.

Another consequence is that cells and the bodies of organisms do not change temperature readily. Bulky organisms, particularly, tend to have a stable temperature in the face of rising or falling external temperatures.

Evaporation and heat loss

The hydrogen bonds between water molecules make it difficult for them to be separated and vapourised (i.e. evaporated). This means that much energy is needed to turn liquid water into water vapour (gas). This amount of energy is the **latent heat of vapourisation** and for water it is very high.

Consequently, the evaporation of water in sweat on the skin or in transpiration from green leaves causes marked cooling. The escaping molecules take a lot of energy with them. You experience this when you stand in a draught after a shower. And, since a great deal of heat is lost with the evaporation of a small amount of water, cooling by evaporation of water is economical on water too.

Ionic compounds like NaCl dissolve in water,

$$NaCl \rightleftharpoons Na^+ + Cl^-$$

with a group of orientated water molecules around each ion:

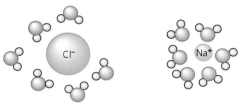

Sugars and alcohols dissolve due to hydrogen bonding between polar groups in their molecules (e.g. — OH) and the polar water molecules:

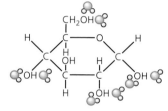

Figure 2.28 Water as universal solvent

The solvent properties of water

Water is a powerful solvent for polar substances. These include:

- ionic substances like sodium chloride (Na^+ and Cl^-). All **cations** (positively charged ions) and **anions** (negatively charged ions) become surrounded by a shell of orientated water molecules (Figure 2.28)
- carbon-containing molecules with ionised groups (such as the carboxyl group $-COO^-$ or the amino group $-NH^{3+}$). Soluble organic molecules like sugars dissolve in water due to the formation of hydrogen bonds with their slightly charged hydroxyl groups ($-OH$), for example.

Once they have dissolved, molecules (the **solute**) are free to move around in the water (the **solvent**) and, as a result, are more chemically reactive than when in the undissolved solid.

We have noted that molecules with an affinity for water are described as **hydrophilic**. On the other hand, non-polar substances are repelled by water, as in the case of oil on the surface of water. Non-polar substances do not contain dipoles and are **hydrophobic** (water-hating).

Table 2.2 Water and life – a summary

	Property	Benefit to life
1	A liquid at room temperature, water dissolves more substances than any other common liquid.	Liquid medium for living things and for the chemistry of life.
2	Much heat energy is needed to raise the temperature of water.	Aquatic environments are slow to change temperature. Bulky organisms have stable temperatures.
3	Evaporation requires a great deal of heat.	Evaporation causes marked cooling. Much heat is lost by the evaporation of a small quantity of water.
4	Water molecules adhere to surfaces. Water column does not break or pull apart under tension (see Topic 7).	Water adheres to the walls of xylem vessels as it is drawn up the stem to the leaves from the roots. Water can be lifted by forces applied at the top, and so can be drawn up the xylem vessels of a tree trunk by forces generated in the leaves.

Question

13 Water has a relative molecular mass of only 18 yet it is a liquid at room temperature. This contrasts with other small molecules which are gases. (For example, methane (CH_4) with a relative molecular mass of 16; ammonia (NH_3) with a relative molecular mass of 17 and carbon dioxide (CO_2) with a relative molecular mass of 44.) What property of water molecules may account for this?

Summary

- Tests can be carried out to identify **reducing sugars, non-reducing sugars, starch** and **proteins**. The Benedict's test can be modified to estimate the concentration of reducing sugar.

- **Organic compounds** contain the elements **carbon and hydrogen**, usually with oxygen and other elements. Carbon atoms form strong, **covalent bonds** (sharing electrons between atoms) with other carbon atoms and with atoms of other elements, forming a huge range of compounds. Many of the organic compounds of life fall into one of the four groups of compounds, **carbohydrates**, **lipids**, **proteins** or **nucleic acids**.

- **Carbohydrates** contain carbon, hydrogen and oxygen only, and have the general formula of $C_x(H_2O)_y$. They consist of the sugars (**monosaccharides** and **disaccharides**) and macromolecules built from sugars (**polysaccharides**).

- Monosaccharides include the six-carbon sugars (**hexoses**), such as glucose and fructose. They are important energy sources for cells. The hexoses have the same molecular formula ($C_6H_{12}O_6$) but different **structural formulae**. Two hexose monosaccharides combine together with the removal of water (a **condensation reaction**) to form a disaccharide sugar, held together by a **glycosidic bond**. For example, the disaccharide sucrose is formed from glucose and fructose.

- Monosaccharides (and some disaccharides) are **reducing sugars**. When heated with alkaline copper(II) sulfate (**Benedict's solution**), which is blue, they produce reduced copper(I) oxide, which is a **brick red precipitate**. The sugar is oxidised to a sugar acid. Sucrose is not a reducing sugar.

- Most **polysaccharides** are built from glucose units condensed together. They are examples of **polymers** consisting of a huge number of **monomers**, such as glucose, condensed together. They may be **food stores** (e.g. starch, glycogen) or **structural components** of organisms (e.g. cellulose, chitin).

- **Lipids** contain the elements carbon, hydrogen and oxygen but the **proportion of oxygen is low**. They are **hydrophobic**, non polar and insoluble in water but otherwise form a diverse group of compounds, including the **fats and oils**. These latter compounds leave a cloudy white suspension in the **emulsion test**.

- Fats are solids at room temperature; oils are liquids. They are formed by a **condensation reaction** between the hydroxyl groups of glycerol and three fatty acids, forming a **triglyceride**, with ester bonds. The fatty acids may be **saturated** or **unsaturated** (containing one or more **double bonds** between carbon atoms in the hydrocarbon tails). Fats and oils are effectively **energy stores**.

- In **phospholipids**, one of the hydroxyl groups of glycerol reacts with **phosphoric acid**. The product has a hydrocarbon 'tail' (**hydrophobic**) and an ionised phosphate group (**hydrophilic**) so they form **bilayers** on water and form the **membranes of cells** (along with proteins).

- Proteins contain **nitrogen** and (often) sulfur, in addition to carbon, hydrogen and oxygen. The building blocks of proteins are **amino acids**, and many hundreds or thousands of amino acid residues form a typical protein. Amino acids have a **basic amino group (–NH$_2$)** and an **acidic carboxyl group (–COOH)** attached to a carbon atom to which the **rest of the molecule (–R)** is also attached.

- Of the many amino acids that occur, only **20 are built up to make the proteins** of living things. Amino acids combine by **peptide bonds** between the carboxyl group of one molecule and the amino group of the other to form a **polypeptide chain**. Proteins, too, are polymers.

- The properties of polypeptides and proteins are determined by the amino acids they are built from and the **sequence in which the amino acids occur**. The sequence of amino acids is the **primary structure** of the protein and is indirectly controlled by the nucleus.

- The **shape the protein molecule** takes up also determines its properties. Part of a polypeptide chain may form into a helix and part into sheets (the **secondary structure** of a protein). The whole chain is held in position by bonds between parts folded together, forming the **tertiary structure** of a protein. Separate polypeptide chains may combine, sometimes with other substances, to form the **quaternary structure**.

- Some proteins have a **structural role** and are **fibrous** proteins, for example, collagen of skin, bone and tendon. Many proteins are enzymes or antibodies (and some are hormones) and these are **globular** proteins.

- Water is electrically neutral but, because of its shape, it carries an unequal distribution of charge – it is a **polar molecule** (oxygen with a small negative charge, hydrogen with a small positive charge). Consequently, water forms **hydrogen bonds** with other water molecules. These hydrogen bonds are responsible for the **properties of water important to life**.

Examination style questions

1 Polysaccharides, such as glycogen, amylopectin and amylose, are formed by polymerisation of glucose. Fig. 1.1 shows part of a glycogen molecule.

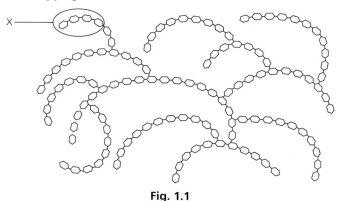

Fig. 1.1

a) With reference to Fig. 1.1,
  i) describe how the **structure** of glycogen differs from the structure of amylose. [2]
  ii) describe the advantages for organisms in storing polysaccharides, such as glycogen, rather than storing glucose. [3]

b) Glycogen may be broken down to form glucose.

Fig. 1.2 shows region X from the glycogen molecule in Fig. 1.1 in more detail.

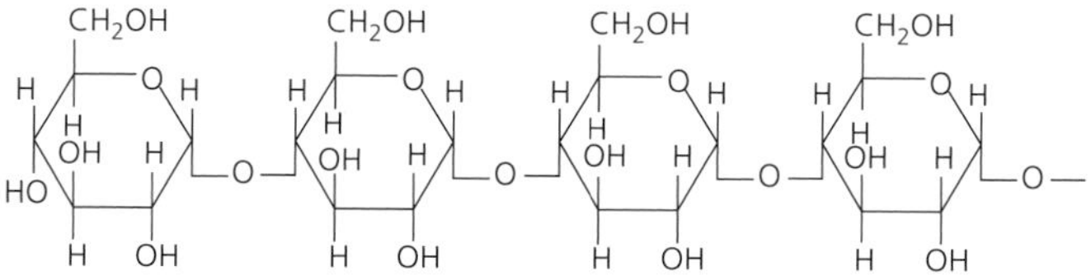

Fig. 1.2

Draw an annotated diagram to explain how a glucose molecule is formed from the free end of the glycogen molecule shown in Fig. 1.2. [3]

[Total: 8]

(Cambridge International AS and A Level Biology 9700, Paper 02 2nd variant Q2 November 2008)

2 a) Cellulose and collagen are macromolecules. What do you understand by the term 'macromolecule'? [2]
b) When cellulose and collagen are hydrolysed by appropriate enzymes, what are the products in each case? [2]

c) By means of fully annotated skeletal formulae of parts of adjacent cellulose molecules, explain how cellulose is formed from its component monosaccharide. Explain how this molecular structure is the basis of the distinctive property of cellulose. [6]
d) Describe the steps to the test you would need to show whether the monosaccharide formed by hydrolysis was a reducing sugar. What result would you expect? [3]
e) Where does collagen occur in the human body? [1]
f) By means of fully annotated skeletal formulae of parts of adjacent collagen molecules, explain how collagen is formed from its component monomers, and the basis of the distinctive property of collagen. [6]

[Total: 20]

3 a) The angle between the oxygen–hydrogen bonds in water is about 150°. By means of a fully annotated diagram only, explain why the existence of this angle causes the water molecule to be polar (although overall it is electrically neutral).
b) Outline **four** of the unusual properties of water that can be ascribed to the effect of hydrogen bonds and summarise the significance of each for living things.

4 a) Chemists often represent the structure of organic molecules as skeletal formulae in which the C and H atoms are left out. Write the skeletal formulae for two molecules of amino acid and show the reaction and products of a condensation reaction between them. On your diagram, identify and name the bond that is formed between the two molecules. [4]
b) To two test tubes, one with 2 cm³ of a dilute egg white solution and one with 2 cm³ of a dilute starch solution, was added 2 cm³ of a dilute sodium hydroxide solution and the contents gently shaken. Then a dilute copper(II) sulfate solution was added to each, drop by drop. The tubes were gently shaken between additions, and this procedure was continued until a definite colour could be seen. What colours would have appeared in the two tubes, and what do the results indicate? [4]
c) Explain concisely what is meant by the following structures of a protein, such as haemoglobin.
  i) Primary [2]
  ii) Secondary [2]
  iii) Tertiary [2]
d) In addition to hydrogen bonds, three other types of bond are important in the secondary and tertiary structure of a protein. Name and explain the chemical nature of these additional bonds. [6]

[Total: 20]

3 Enzymes

Enzymes are essential for life to exist. Their mode of action and the factors that affect their activity are explored in this topic. Prior knowledge for this topic is an understanding that an enzyme is a biological catalyst that increases the rate of a reaction and remains unchanged when the reaction is complete.

There are many opportunities in this topic for students to gain experience of carrying out practical investigations and analysing and interpreting their results.

 ## 3.1 Mode of action of enzymes

There are many different enzymes, each one specific to a particular reaction. This specificity is the key to understanding the efficient functioning of cells and living organisms.

By the end of this section you should be able to:

a) explain that enzymes are globular proteins that catalyse metabolic reactions
b) state that enzymes function inside cells (intracellular enzymes) and outside cells (extracellular enzymes)
c) explain the mode of action of enzymes in terms of an active site, enzyme/substrate complex, lowering of activation energy and enzyme specificity (the lock and key hypothesis and the induced fit hypothesis)
d) investigate the progress of an enzyme-catalysed reaction by measuring rates of formation of products or rates of disappearance of substrate

Introducing enzymes and their role in metabolism

Enzymes are globular proteins that catalyse the many thousands of metabolic reactions taking place within cells and organisms. **Metabolism** is the name we give to these chemical reactions of life. The molecules involved are collectively called **metabolites**. Many of these metabolites are made in organisms. Other metabolites have been imported from the environment, such as from food substances taken in, water and the gases carbon dioxide and oxygen.

Metabolism actually consists of chains (linear sequences) and cycles of enzyme-catalysed reactions, such as we see in respiration (page 242), photosynthesis (page 266), protein synthesis (page 120) and very many other pathways. These reactions may be classified as one of just two types, according to whether they involve the build-up or breakdown of organic molecules.

- In **anabolic reactions**, larger molecules are built up from smaller molecules. Examples of anabolism are the synthesis of proteins from amino acids and the synthesis of polysaccharides from simple sugars.
- In **catabolic reactions**, larger molecules are broken down. Examples of catabolism are the digestion of complex foods and the breakdown of sugar in respiration.

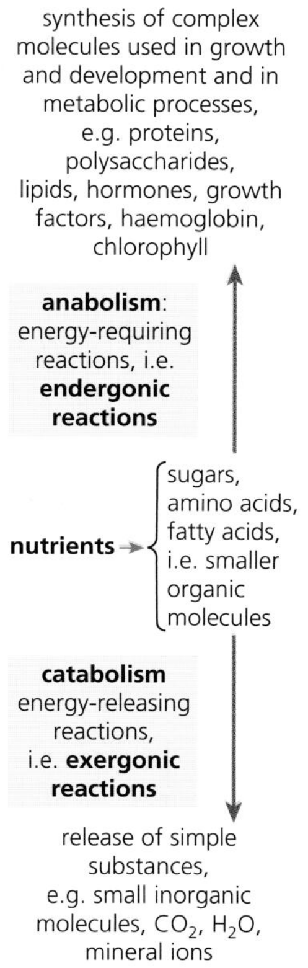

synthesis of complex molecules used in growth and development and in metabolic processes, e.g. proteins, polysaccharides, lipids, hormones, growth factors, haemoglobin, chlorophyll

anabolism: energy-requiring reactions, i.e. **endergonic reactions**

sugars, amino acids, fatty acids, i.e. smaller organic molecules

nutrients →

catabolism energy-releasing reactions, i.e. **exergonic reactions**

release of simple substances, e.g. small inorganic molecules, CO_2, H_2O, mineral ions

Figure 3.1 Metabolism: an overview

Enzymes as globular proteins

In Topic 2 we saw that the tertiary structure of **globular proteins** was typically spherical. In these molecules their linear chain of amino acids (primary structure) is precisely folded and held in a globular three-dimensional shape containing α-helices and β-sheets. Also, the R-groups of the amino acids present on the exterior of the molecule are hydrophilic groups, making the protein water soluble. Remember, this structure contrasted with that of the fibrous proteins, such as collagen.

Enzymes as biological catalysts

A catalyst is a molecule that speeds up a chemical reaction but remains unchanged at the end of the reaction. Most chemical reactions do not occur spontaneously. In a laboratory or in an industrial process, chemical reactions may be made to occur by applying high temperatures, high pressures, extremes of pH and by keeping a high concentration of the reacting molecules. If these drastic conditions were not applied, very little of the chemical product would be formed. In contrast, in cells and organisms, many of the chemical reactions of metabolism occur at exactly the same moment, at extremely low concentrations, at normal temperatures and under the very mild, almost neutral, aqueous conditions we find in cells.

How is this brought about?

For a reaction between two molecules to occur there must be a successful collision between them. The molecules must collide with each other in the right way and at the right speed. If the angle of collision is not correct, the molecules bounce apart. Alternatively, if the speed of the collision speed is wrong or the impact is too gentle, for example, then there will be insufficient energy for the rearrangement of electrons. Only if the molecules are lined up and collide with the correct energies does a reaction occur.

The 'right' conditions happen so rarely that the reaction doesn't happen to a significant extent normally. If we introduce extreme conditions, such as those listed above, we can cause the reaction to happen. On the other hand, if we introduce an enzyme for this particular reaction then the reaction occurs at great speed. Enzymes are amazing molecules in this respect.

Where do enzymes operate?

Some enzymes are exported from cells, such as the digestive enzymes. Enzymes, like these, that are parcelled up, secreted and work externally are called **extracellular enzymes**. However, very many enzymes remain within the cells and work there. These are **intracellular enzymes**. They are found inside organelles, in the membranes of organelles, in the fluid medium around the organelles (the cytosol) and in the cell surface membrane.

Enzymes control metabolism

There is a huge array of enzymes that facilitate the chemical reactions of the metabolism. Since these reactions can only take place in the presence of specific enzymes, we know that if an enzyme is not present then the reaction it catalyses cannot occur.

Many enzymes are always present in cells and organisms but some enzymes are produced only under particular conditions or at certain stages. By making some enzymes and not others, **cells can control what chemical reactions happen in the cytoplasm**. Sometimes it is the presence of the substrate molecule that triggers the synthesis of the enzyme. In Topic 4 we see how protein synthesis (and therefore enzyme production) is directly controlled by the cell nucleus.

How enzymes work

So, enzymes are biological catalysts made of protein. They speed up the rate of a chemical reaction. The general properties of catalysts are:
- they are **effective in small amounts**
- they remain **unchanged at the end of the reaction**.

The presence of enzymes enables reactions to occur at incredible speeds, in an orderly manner, yielding products that the organism requires, when they are needed. Sometimes reactions happen even though the reacting molecules are present in very low concentrations.

How is this brought about?

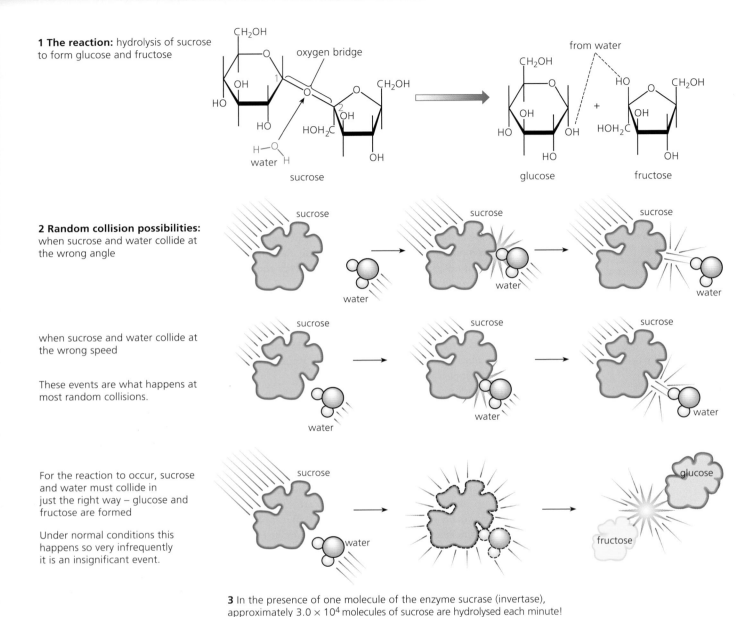

1 The reaction: hydrolysis of sucrose to form glucose and fructose

2 Random collision possibilities: when sucrose and water collide at the wrong angle

when sucrose and water collide at the wrong speed

These events are what happens at most random collisions.

For the reaction to occur, sucrose and water must collide in just the right way – glucose and fructose are formed

Under normal conditions this happens so very infrequently it is an insignificant event.

3 In the presence of one molecule of the enzyme sucrase (invertase), approximately 3.0×10^4 molecules of sucrose are hydrolysed each minute!

Figure 3.2 Can a reaction occur without an enzyme?

Enzymes lower the activation energy

As molecules react they become unstable, high energy intermediates, but they are in this state only momentarily. We say they are in a transition state because the products are formed immediately. The products have a lower energy level than the substrate molecules. Energy is needed to raise molecules to a transition state and the minimum amount of energy needed to do this is called the **activation energy**. It is an energy barrier that has to be overcome before the reaction can happen. Enzymes work by lowering the amount of energy required to activate the reacting molecules.

A model of what is going on is the boulder (the substrate) perched on a slope, prevented from rolling down by a small hump (the activation energy) in front of it. The boulder can be pushed over the hump. Alternatively, the hump can be dug away (the activation energy can be lowered), allowing the boulder to roll and shatter at a lower level (into the products).

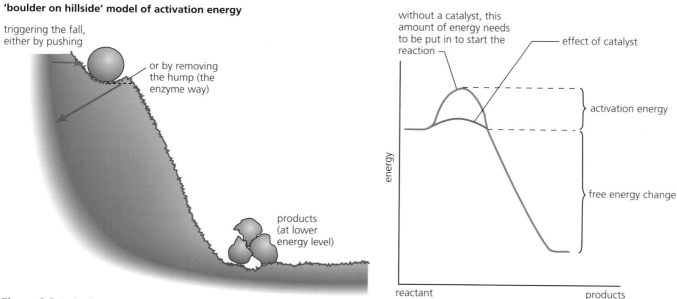

'boulder on hillside' model of activation energy

triggering the fall, either by pushing

or by removing the hump (the enzyme way)

products (at lower energy level)

without a catalyst, this amount of energy needs to be put in to start the reaction

effect of catalyst

activation energy

free energy change

energy

reactant

products

Figure 3.3 Activation energy

The enzyme has an active site

In a reaction catalysed by an enzyme, the starting substance is called the **substrate**. It is converted to the **product**. The way an enzyme works is for the substrate molecule to become attached (we say 'bind to') the enzyme at a specially formed pocket in the enzyme – very briefly. This binding point is called the **active site**. The active site takes up a relatively small part of the total volume of the enzyme.

So, an enzyme (**E**) works by binding to its substrate (**S**) molecule at a specially formed pocket in the enzyme. This concept is referred to the **'lock and key'** hypothesis of enzyme action. As the enzyme and substrate form a complex (**E–S**), the substrate is raised in energy to a transition state and then breaks down into products (**Pr**) plus unchanged enzyme.

$$E + S \rightarrow E\text{–}S \rightarrow Pr - E$$

Enzymes are typically large **globular protein** molecules. Most substrate molecules are quite small molecules by comparison. Even when the substrate molecules are very large, such as certain macromolecules like the polysaccharides, only one bond in the substrate is in contact with the active site of the enzyme.

Enzymes are highly specific

Enzymes are highly specific in their action. They catalyse only one type of reaction or only a very small group of very similar reactions.

The sequence of steps to an enzyme-catalysed reaction:

enzyme + substrate ⟶ **E–S complex** ⟶ **product + enzyme available for reuse**
(substrate raised to transition state)

$$E + S \longrightarrow E\text{–}S \longrightarrow Pr + E$$

substrate molecule – **the key**

active site (here the substrate molecule is held and reaction occurs) – **the lock**

enzyme (large protein molecule)

enzyme–substrate complex

product molecules

substrate molecule now at transition state

Figure 3.4 The lock and key hypothesis of enzyme action

This means that an enzyme 'recognises' a very small group of substrate molecules or even only a single type of molecule. This is because the active site where the substrate molecule binds has a **precise shape** and **distinctive chemical properties** (meaning the presence of particular chemical groups and bonds). Only particular substrate molecules can fit to a particular active site. All other substrate molecules are unable to fit and so cannot bind.

Catalysis by 'induced fit'

We have seen that enzymes are highly specific in their action. This makes them different from most inorganic catalysts. Enzymes are specific because of the way they bind with their substrate at a pocket or crevice in the protein. The 'lock and key' hypothesis, however, does not fully account for the combined events of 'binding' and simultaneous chemical change observed in most enzyme-catalysed reactions.

At the active site, the arrangement of certain amino acid molecules in the enzyme exactly matches certain groupings on the substrate molecule, enabling the enzyme–substrate complex to form. As the complex is formed, an essential, critical **change of shape** is caused in the enzyme molecule. It is this change of shape that is important in momentarily raising the substrate molecule to the transitional state. It is then able to react.

With a transitional state achieved, other amino acid molecules of the active site bring about the breaking of particular bonds in the substrate molecule, at the point where it is temporarily held by the enzyme. It is because different enzymes have different arrangements of amino acids in their active sites that each enzyme catalyses either a single chemical reaction or a group of closely related reactions.

Question

1 Explain why the shape of globular proteins that are enzymes is important in enzyme action?

Figure 3.5 The induced fit hypothesis of enzyme action

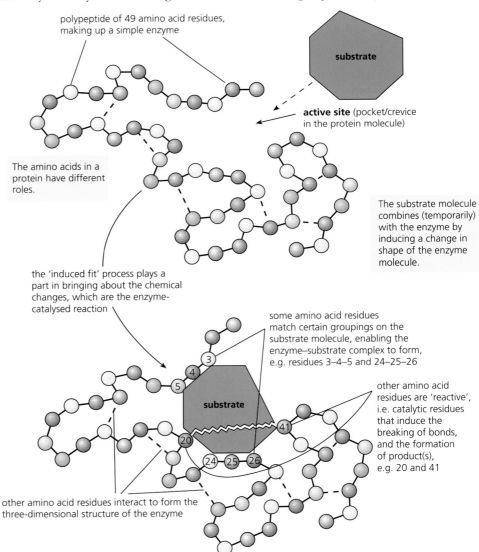

polypeptide of 49 amino acid residues, making up a simple enzyme

substrate

active site (pocket/crevice in the protein molecule)

The amino acids in a protein have different roles.

The substrate molecule combines (temporarily) with the enzyme by inducing a change in shape of the enzyme molecule.

the 'induced fit' process plays a part in bringing about the chemical changes, which are the enzyme-catalysed reaction

some amino acid residues match certain groupings on the substrate molecule, enabling the enzyme–substrate complex to form, e.g. residues 3–4–5 and 24–25–26

substrate

other amino acid residues are 'reactive', i.e. catalytic residues that induce the breaking of bonds, and the formation of product(s), e.g. 20 and 41

other amino acid residues interact to form the three-dimensional structure of the enzyme

Specificity:
- Some amino acid residues allow a particular substrate molecule to 'fit'
- Some amino acid residues bring about particular chemical changes.

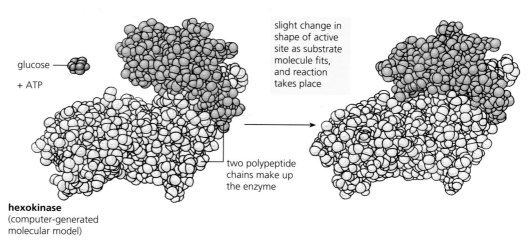

Figure 3.6 Computer-generated image of the induced fit hypothesis in action

Naming enzymes

Many enzymes have a name based on the name of their substrate, with the ending *-ase* added. For example, **lactase** hydrolyses lactose and **amylase** hydrolyses amylose.

Other enzymes have been given names that tell us little or nothing about what they do, such as many of the enzymes of digestion, for example, **pepsin**, **trypsin** and **rennin**.

Today, systematic naming of enzymes is based on an agreed classification of enzymes and on the name of the substrate catalysed. These types of names are long and detailed. They are outside the scope of this book. They are used in the communications of enzymologists but not in everyday usage. However, you are already familiar with certain enzymes. For example, the enzymes that catalyse the formation of two products from a larger substrate molecule by a hydrolysis reaction are classified as 'hydrolases'. Can you name a hydrolase?

Question

2 a Define the term 'catalyst'.
   b List two differences between inorganic catalysts and enzymes.

Studying enzyme-catalysed reactions

Enzyme-catalysed reactions are fast reactions. We know this from the very large numbers of substrate molecules converted to products by a mole of enzyme in one minute. The mole is a unit for the amount of a substance. It is explained in Appendix 1 on the CD.

The enzyme catalase catalyses the breakdown of hydrogen peroxide. This enzyme is a good example to use in studying the rate of enzyme-catalysed reactions.

$$2H_2O_2 \xrightarrow{\text{catalase}} 2H_2O + O_2$$

Catalase occurs widely in the cells of living things. It functions as a protective mechanism for the delicate biochemical machinery of cells. This is because hydrogen peroxide is a common by-product of some of the reactions of metabolism. Hydrogen peroxide is a very toxic substance – a very powerful oxidising agent, in fact (Appendix 1). Catalase inactivates hydrogen peroxide as soon as it forms, before damage can occur.

You can demonstrate the presence of catalase in fresh liver tissue by dropping a small piece into dilute hydrogen peroxide solution (Figure 3.7). Compare the result obtained with that from a similar piece of liver that has been boiled in water (high temperature denatures and destroys enzymes, including catalase). If you do not wish to use animal tissues, then you can use potato or soaked and crushed dried peas instead – the results will be equally dramatic!

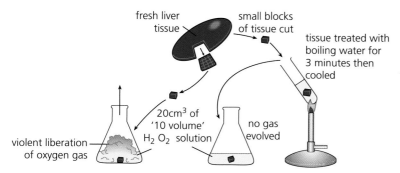

Figure 3.7 Liver tissue in dilute hydrogen peroxide solution

Measuring the rate of reaction

In practical terms, the rate of an enzyme-catalysed reaction is taken as the amount of substrate that has disappeared from a reaction mixture in a given time. Alternatively, the amount of product that has accumulated in a period of time can be measured. For example, in Figure 2.38, the enzyme amylase is used to digest starch to sugar, and here it is the rate that the substrate starch disappears from a reaction mixture that is measured.

Working with catalase, however, it is easiest to measure the rate at which the product (oxygen) accumulates. In the experiment illustrated in Figure 3.8, the volume of oxygen that has accumulated at half minute intervals is recorded on the graph.

In both of these examples we find that, over a period of time, the initial rate of reaction is not maintained but, rather, falls off quite sharply. This is typical of enzyme actions studied outside their location in the cell. Can you think of reasons why?

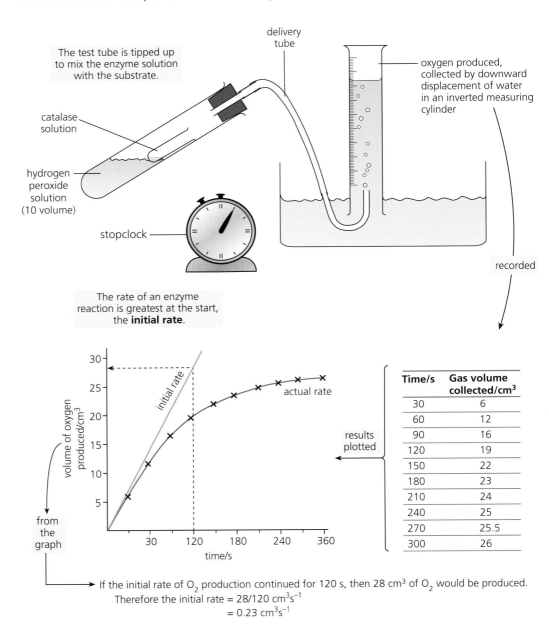

Time/s	Gas volume collected/cm³
30	6
60	12
90	16
120	19
150	22
180	23
210	24
240	25
270	25.5
300	26

If the initial rate of O_2 production continued for 120 s, then 28 cm³ of O_2 would be produced.
Therefore the initial rate = 28/120 cm³s⁻¹
= 0.23 cm³s⁻¹

Figure 3.8 Measuring the rate of reaction using catalase

The fall-off can be due to a number of reasons. Most commonly it is because the concentration of the substrate in the reaction mixture has fallen. Consequently, it is the initial rate of reaction that is measured. This is the slope of the tangent to the curve in the initial stage of reaction. How this is calculated is shown in Figure 3.8.

Question

3 In an investigation of oxygen production from hydrogen peroxide solution by the enzyme catalase the following results were obtained.

Time of readings (s)	0	20	40	60	80	100	120
Volume of O_2 produced (cm^3)	0	43	66	78	82	91	92

a Plot a graph to show oxygen production against time.

b From your graph, find the initial rate of the reaction.

3.2 Factors that affect enzyme action

Investigating the effects of factors on enzyme activity gives opportunities for planning and carrying out experiments under controlled conditions.

By the end of this section you should be able to:

a) investigate and explain the effects of the following factors on the rate of enzyme-catalysed reactions:
   - temperature
   - pH (using buffer solutions)
   - enzyme concentration
   - substrate concentration
   - inhibitor concentration

b) explain that the maximum rate of reaction (V_{max}) is used to derive the Michaelis-Menten constant (K_m) which is used to compare the affinity of different enzymes for their substrates

c) explain the effects of reversible inhibitors, both competitive and non-competitive, on the rate of enzyme activity

d) investigate and explain the effect of immobilising an enzyme in alginate on its activity as compared with its activity when free in solution

Investigating factors that affect the rate of reaction of enzymes

Enzymes are sensitive to environmental conditions – very sensitive, in fact. Many factors within cells affect enzymes and therefore alter the rate of the reaction being catalysed. Investigations of these factors, including temperature, pH and the effect of substrate concentration in particular, have helped our understanding of how enzymes work.

Temperature

Examine the investigation of the effect of temperature on the hydrolysis of starch by the enzyme amylase shown in Figure 3.9. When starch is hydrolysed by the enzyme amylase, the product is maltose, a disaccharide. Starch gives a blue–black colour when mixed with iodine solution (iodine in potassium iodide solution) but maltose gives a red colour. The first step in this experiment is to bring samples of the enzyme and the substrate (the starch solution) to the temperature of the water bath before being mixed – a step called 'pre-incubation'.

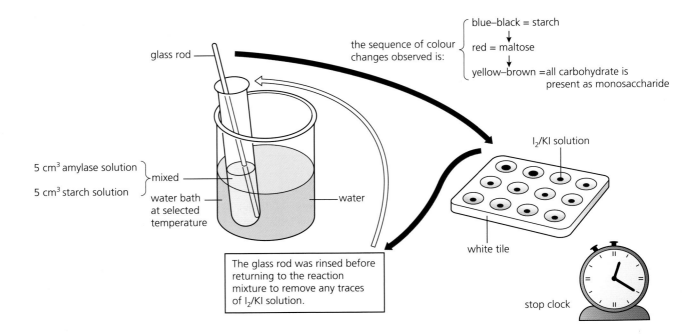

the sequence of colour changes observed is:
- blue–black = starch
- red = maltose
- yellow–brown = all carbohydrate is present as monosaccharide

glass rod

5 cm³ amylase solution
5 cm³ starch solution
} mixed

water bath at selected temperature

water

I₂/KI solution

white tile

The glass rod was rinsed before returning to the reaction mixture to remove any traces of I₂/KI solution.

stop clock

Figure 3.9 The effect of temperature on hydrolysis of starch by amylase

The experiment is **repeated at a range of temperatures,** such as at 10, 20, 30, 40, 50 and 60°C.

A **control tube** of 5 cm³ of starch solution + 5 cm³ of distilled water (in place of the enzyme) should be included and tested for the presence/absence of starch, at each temperature used.

Other variables – such as the concentrations of the enzyme and substrate solutions – were kept constant.

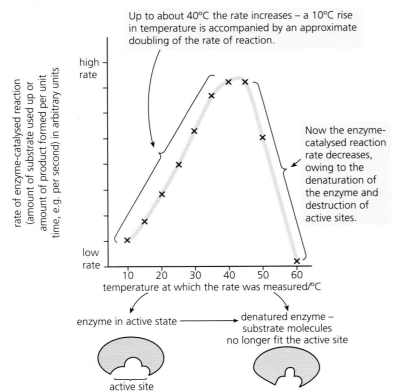

Up to about 40°C the rate increases – a 10°C rise in temperature is accompanied by an approximate doubling of the rate of reaction.

rate of enzyme-catalysed reaction (amount of substrate used up or amount of product formed per unit time, e.g. per second) in arbitrary units

high rate

low rate

Now the enzyme-catalysed reaction rate decreases, owing to the denaturation of the enzyme and destruction of active sites.

temperature at which the rate was measured/°C

enzyme in active state → denatured enzyme – substrate molecules no longer fit the active site

active site

Figure 3.10 The effect of temperature on the rate of an enzyme-catalysed reaction

The progress of the hydrolysis reaction is then followed by taking samples of a drop of the mixture on the end of the glass rod, at half minute intervals. These are tested with iodine solution on a white tile. Initially, a strong blue–black colour is seen confirming the presence of starch. Later, as maltose accumulates, a red colour forms. The end point of the reaction is when all the starch colour has disappeared from the test spot.

Using fresh reaction mixture each time, the investigation is repeated at a series of different temperatures, say at 10, 20, 30, 40, 50 and 60°C. The time taken for complete hydrolysis at each temperature is recorded and the rate of hydrolysis in unit time is plotted on a graph. A characteristic curve is the result – although the 'optimum' temperature varies from reaction to reaction and with different enzymes (Figure 3.10).

How is the graph interpreted? Look at Figure 3.10.

Question

4 In studies of the effect of temperature on enzyme-catalysed reactions, suggest why the enzyme and substrate solutions are pre-incubated to a particular temperature before they are mixed.

As the temperature is increased, molecules have more energy and reactions between them happen more quickly. The enzyme molecules are moving more rapidly and are more likely to collide and react. In chemical reactions, for every 10°C rise in temperature the rate of the reaction approximately doubles. This property is known as the **temperature coefficient (Q_{10})** of a chemical reaction.

However, in enzyme-catalysed reactions the effect of temperature is more complex because proteins are **denatured by heat**. The rate of denaturation increases at higher temperatures, too. So as the temperature rises above a certain point the amount of active enzyme progressively decreases and the rate is slowed. As a result of these two effects of heat on enzyme-catalysed reactions, there is an apparent optimum temperature for an enzyme.

Not all enzymes have the same optimum temperature. For example, the bacteria in hot thermal springs have enzymes with optimum temperatures between 80 and 100°C or higher, whereas seaweeds of northern seas and the plants of the tundra have optimum temperatures closer to 0°C. Humans have enzymes with optimum temperatures at or about normal body temperature. This feature of enzymes is often exploited in the commercial and industrial uses of enzymes today.

pH

Change in pH can have a dramatic effect on the rate of an enzyme-catalysed reaction. Each enzyme has a range of pH in which it functions efficiently – often at or close to neutrality. The pH affects the rate of reaction because the structure of a protein (and therefore the shape of the active site) is maintained by various bonds within the three-dimensional structure of the protein (Figure 2.22). A change in pH from the optimum value alters the bonding patterns. As a result, the shape of the enzyme molecule is progressively changed. The active site may quickly become inactive. However, the effects of pH on the active site are normally reversible (unlike temperature changes). That is, provided the change in surrounding acidity or alkalinity is not too extreme. As the pH is brought back to the optimum for that enzyme, the active site may reappear (Figure 3.11).

Some of the digestive enzymes of the gut have different optimum pH values from the majority of other enzymes. For example, those adapted to work in the stomach, where there is a high concentration of acid during digestion, have an optimum pH which is close to pH 2.0 (Figure 3.12).

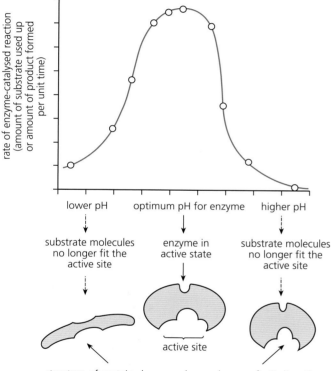

structure of protein changes when a change of pH alters the ionic charge on —COO⁻(acidic) and —NH₃⁺(basic) groups in the peptide chain, so the shape of the active site is lost

Figure 3.11 The effect of pH on enzyme shape and activity

Question

5 a Explain what a buffer solution is.
  b Why are they often used in enzyme experiments?

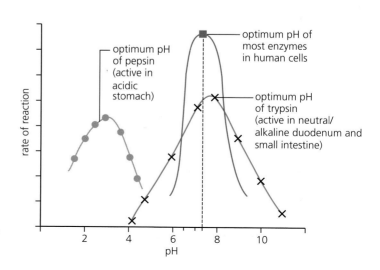

Figure 3.12 The optimum pH of different human enzymes

Substrate concentration

The effect of different concentrations of substrate on the rate of an enzyme-catalysed reaction can be shown using the enzyme catalase.

Look again at the investigation of the initial rate of reaction using catalyse (Figure 3.8). We saw that, when working with catalase, it is easy to measure the rate the product (oxygen) accumulated (recorded at half minute intervals).

To investigate the effect of substrate concentration on the rate of this enzyme-catalysed reaction, the experiment shown in Figure 3.8 is repeated **at different concentrations of substrate**. The initial rate of reaction plotted in each case. Other variables such as temperature and enzyme concentration are kept constant.

When the initial rates of reaction are plotted against the substrate concentration, the curve shows two phases. At lower concentrations the rate increases in direct proportion to the substrate concentration but at higher substrate concentrations, the rate of reaction becomes constant and shows no increase (Figure 3.13).

We can see that the enzyme catalase does work by forming a short-lived enzyme–substrate complex. At low concentrations of substrate, all molecules can find an active site without delay. Effectively, there is excess enzyme present. The rate of reaction is set by how much substrate is present – as more substrate is made available the rate of reaction increases.

At higher substrate concentrations there are more substrate molecules than enzyme molecules. Now, in effect, substrate molecules have to 'queue up' for access to an active site. Adding more substrate increases the number of molecules awaiting contact with an enzyme molecule. There is now no increase in the rate of reaction. This relationship between amount of substrate and the rate of reaction is shown in Figure 3.13.

Question

6   When there is an excess of substrate present in an enzyme-catalysed reaction, explain the effects on the rate of reaction of increasing the concentration of:
    a the substrate
    b the enzyme.

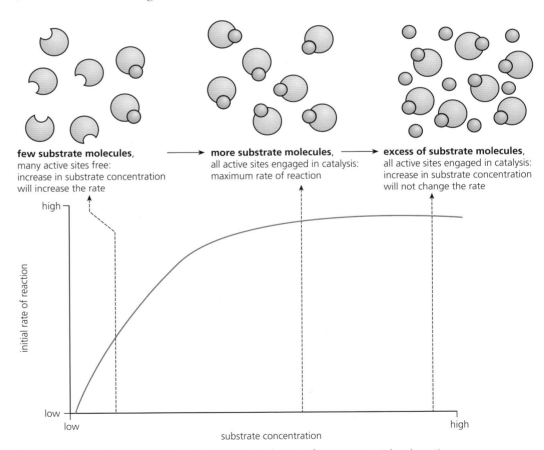

few substrate molecules, many active sites free: increase in substrate concentration will increase the rate

more substrate molecules, all active sites engaged in catalysis: maximum rate of reaction

excess of substrate molecules, all active sites engaged in catalysis: increase in substrate concentration will not change the rate

Figure 3.13 The effect of substrate concentration on the rate of an enzyme-catalysed reaction

Enzyme concentration

If there are plenty of substrate molecules in a reaction mixture, then the more enzyme that is added the faster the rate of reaction will be. This is the situation in a cell where an enzyme reaction occurs with a small amount of enzyme present. Any increase in enzyme production will lead to an increased rate of reaction, simply because more active sites are made available.

Introducing the Michaelis-Menton constant (K_m) and its significance

On page 62 we saw how the initial rate of an enzyme-catalysed reaction was measured and why (Figure 3.8).

Remind yourself 'how' and 'why' now.

When the initial rate of reaction of an enzyme is measured over a range of substrate concentrations (with a fixed amount of enzyme) and the results plotted on a graph, a typical example of the resulting curve is shown in Figure 3.14. You can see that, with increasing substrate concentration, the velocity increases – rapidly at lower substrate concentrations. However, the rate increase progressively slows, and above a certain substrate concentration, the curve has flattened out. No further increase in rate occurs. This tells us the enzyme is working at maximum velocity at this point. On the graph, this point of maximum velocity is shown as V_{max}.

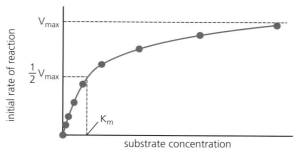

Figure 3.14 Graph of initial rate of an enzyme-catalysed reaction against substrate concentration

In 1913 the biochemists Michaelis and Menten studied this aspect of enzyme reactions and introduced a constant, now known as the Michaelis-Menten constant. The **Michaelis-Menten constant** ($\mathbf{K_m}$) is defined as the substrate concentration that sustains half maximum velocity ($\frac{1}{2}V_{max}$). It measures the degree of attraction or affinity of an enzyme for the substrate – the smaller the K_m the higher the affinity. K_m can be experimentally determined at a specified pH and temperature, and is expressed in units of molarity.

Actual values of K_m have been measured for a great many enzymes. These values fall between 10^{-3} and 10^{-5} mol dm^3 of substrate. This is a very wide range of concentrations. It means that some enzymes are able to work at maximum velocity at very low concentrations of substrate, whilst others only function effectively at much higher concentrations.

Take the case of two enzymes that catalyse the transformation of the same substrate molecule but in different reaction sequences. If the 'pool' or reserves of that substrate are low and its supply restricted, then the enzyme with the lowest K_m will claim more – and one particular metabolic pathway will benefit at the expense of the other. Knowing the K_m of enzymes is an important part of understanding metabolism, both quantitatively and qualitatively.

Inhibitors of enzymes

Certain substances present in cells (and some which enter from the environment) may react with an enzyme, altering the rate of reaction. These substances are known as **inhibitors**, since their effect is generally to lower the rate of reaction. Studies of the effects of inhibitors have helped our understanding of:

- the chemistry of the active site of enzymes
- the natural regulation of metabolism and which pathways operate
- the ways certain commercial pesticides and many drugs work (by inhibiting specific enzymes and preventing particular reactions).

For example, molecules that sufficiently resemble the substrate in shape may compete to occupy the active site. They are known as **competitive inhibitors**. For example, the enzyme that catalyses the reaction between carbon dioxide and the CO_2-acceptor molecule in photosynthesis, known as ribulose bisphosphate carboxylase (rubisco), is competitively inhibited by oxygen in the chloroplasts.

Because these inhibitors are not acted on by the enzyme and turned into 'products' as normal substrate molecular are, they tend to remain attached. However, if the concentration of the substrate molecule is raised, the inhibitor molecules are progressively displaced from the active sites (Figure 3.15).

Alternatively, an inhibitor may be unlike the substrate molecule, yet still combine with the enzyme. In these cases, the attachment occurs at some **other part** of the enzyme. This may be quite close to the active site. Here the inhibitor either partly blocks access to the active site by substrate molecules or it causes the active site to change shape so that it is then unable to accept the substrate.

These are called **non-competitive inhibitors**, since they do not compete for the active site. Adding excess substrate does not overcome their inhibiting effects (Figure 3.15). An example is the effect of the amino acid alanine on the enzyme pyruvate kinase in the final step of glycolysis (Topic 12).

When the initial rates of reaction of an enzyme are plotted against substrate concentration, the effects of competitive and non-competitive inhibitors are seen to be different.

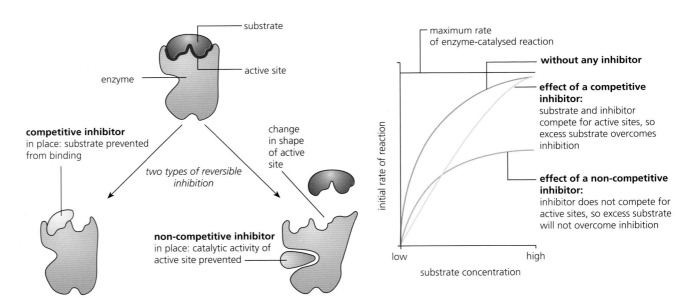

Figure 3.15 The principles of competitive and non-competitive inhibition

Table 3.1 A comparison of competitive and non-competitive inhibition of enzymes

Competitive inhibition	Non-competitive inhibition
Inhibitor chemically resembles the substrate molecule and occupies (blocks) the active site	Inhibitor chemically unlike the substrate molecule but reacts with the bulk of the enzyme, reducing access to the active site
With a low concentration of inhibitor, increasing concentration of substrate eventually overcomes inhibition as substrate molecules displace inhibitor	With a low concentration of inhibitor, increasing concentration of substrate cannot prevent binding of inhibitor – some inhibition remains at high substrate concentration
Example: O_2 competes with CO_2 for active site of rubisco	Example: alanine non-competitively inhibits pyruvate kinase

Certain **irreversible inhibitors** bind tightly and permanently to an enzyme and destroy its catalytic properties entirely. These drastic effects occur at low concentrations of inhibitor and we may describe these substances as **poisons**. Examples include:

- cyanide ions which block cytochrome oxidase in terminal oxidation in cell aerobic respiration
- the nerve gas sarin blocks a neurotransmitter (acetyl cholinesterase) in synapse transmission

Use of enzymes as industrial and laboratory catalysts

Enzymes as biological catalysts are important components of many industrial processes. Their use is widespread because they are:

- highly specific, catalysing changes in one particular compound or one type of bond
- efficient, in that a tiny quantity of enzyme catalyses the production of a large quantity of product
- effective at normal temperatures and pressures, and so a limited input of energy (as heat and high pressure) may be required.

The enzymes selected by industry are frequently produced from microorganisms – typically from species of fungi or bacteria. Table 3.2 lists some examples.

Table 3.2 Enzymes with industrial applications obtained from microorganisms

	Enzyme	Source	Application
Bacterial	Protease	*Bacillus*	'Biological' detergents
	Glucose isomerase	*Bacillus*	Fructose syrup manufacture
Fungal	Lactase	*Kluyveromyces*	Breakdown of lactose to glucose and galactose
	Amylase	*Aspergillus*	Removal of starch in woven cloth production

Using enzymes *in vitro*

Enzymes may be used as cell-free preparations added to a reaction mixture, or they may be immobilised enzymes, with the reactants passed over them.

Enzyme immobilisation involves the attachment of enzymes to insoluble materials, which then provide support for the enzyme. For example, the enzyme may be entrapped between inert fibres, or it may be covalently bonded to a matrix. In both cases the enzyme molecules are prevented from being leached away. The immediate advantages of using an immobilised enzyme are:

- it permits re-use of the enzyme preparation
- the product is obtained enzyme free
- the enzyme may be much more stable and long lasting, due to protection by the inert matrix.

Clearly, there are advantages in using industrial enzymes in the immobilised condition, where this is possible. We will return to this issue shortly. First, we can investigate the process of immobilisation of an enzyme and compare its activity when alternatively used free in a solution.

Immobilised enzyme – a laboratory demonstration

The steps to this demonstration are:

1 A solution of sodium alginate (a polysaccharide obtained from the walls of brown algae, capable of holding 200 times its own mass in water) is prepared by dispersing 2 g of the alginate in 100 cm^3 of distilled water at a temperature of 40 °C.

2 An alginate-enzyme solution is prepared by stirring 2 cm^3 of invertase concentrate into 40 cm^3 of the cooled alginate solution. (Remember, invertase catalyses the hydrolysis of sucrose [a disaccharide] to two monosaccharide molecules [glucose and fructose] from which it is formed in a condensation reaction.)

3 Immobilised enzyme pellets are produced by filling a syringe with the alginate-invertase solution and arranging for it to drip into a beaker of calcium chloride solution (100 cm^3 of CaCl$_2$, 0.1 mol dm^3). The insoluble pellets formed may be separated with a nylon/plastic sieve. They should be washed with distilled water at this stage, and allowed to harden.

4 These hardened pellets may be tested by placing them in a tube with a narrow nozzle at the base (the barrel of a large, plastic syringe is suitable). A quantity of sucrose solution (30 cm^3 of 5% sucrose w/v) may be slowly poured down the column of pellets and the effluent collected. Note the time taken for the solution to pass through.

5 Test the effluent solution for the presence of glucose, using Clinistix™ (see below).

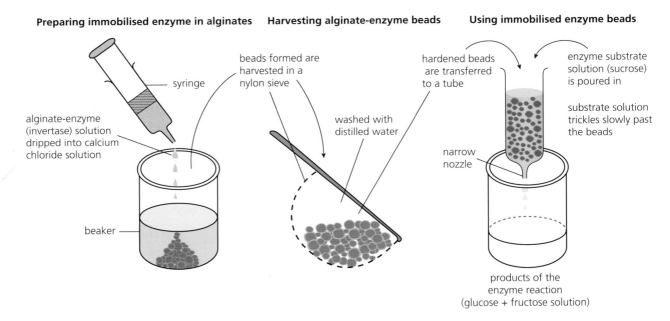

Figure 3.16 Preparing an immobilised enzyme

Investigating the efficiency of immobilised enzyme compared to its use free in solution

A comparison of the efficiency of immobilised invertase compared with free enzyme can be made by stirring into a second sample of the sucrose solution (again, 30 cm^3 of 5% sucrose w/v), 2 cm^3 of the invertase concentrate solution. The free enzyme should be allowed to catalyse hydrolysis of the sucrose for the same time period as it took for the earlier sucrose solution sample to pass through the alginate bead column. Once again, the solution should be tested for the presence of glucose, using Clinistix™ (see Figure 3.17).

Estimating the glucose concentration using Clinistix™

Clinistix™ exploit immobilised enzymes in dipsticks for the quantitative measurement of glucose. (An important medical application of the Clinistix™ has been the measurement of glucose in urine samples in patients with diabetes.)

The Clinistix™ strip contains two enzymes, glucose oxidase and peroxidase, together with a colourless hydrogen donor compound called chromogen. When the strip is dipped into a sample, if glucose is present it is oxidised to glucuronic acid and hydrogen peroxide. The second enzyme catalyses the reduction of hydrogen peroxide and the oxidation of chromogen. The product is water and the oxidised dye, which is coloured. The more glucose present in the sample the more coloured dye is formed. The colour of the test strip is then compared to the printed scale to indicate the amount of glucose present (Figure 3.17).

Note that this is another example of the industrial exploitation of biological catalysts.
How effective was the immobilised enzyme, compared to its activity, free in solution?

the principles

$$glucose + oxygen \xrightarrow{\text{glucose oxidase}} gluconic\ acid + H_2O_2$$

$$DH_2 + H_2O_2 \xrightarrow{\text{peroxidase}} 2H_2O + D$$

reduced chromagen (colourless)   chromagen (coloured)

the process

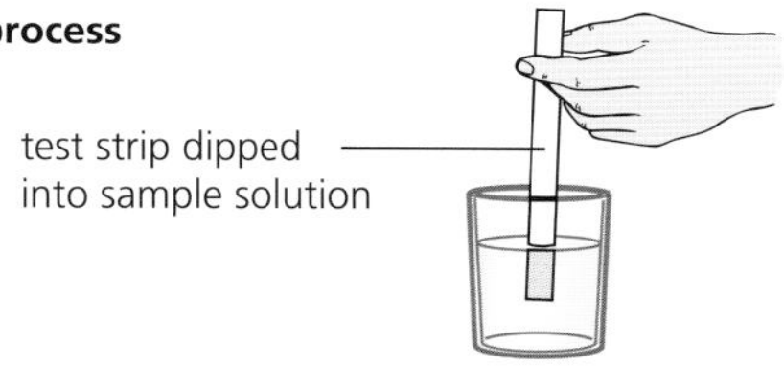

test strip dipped into sample solution

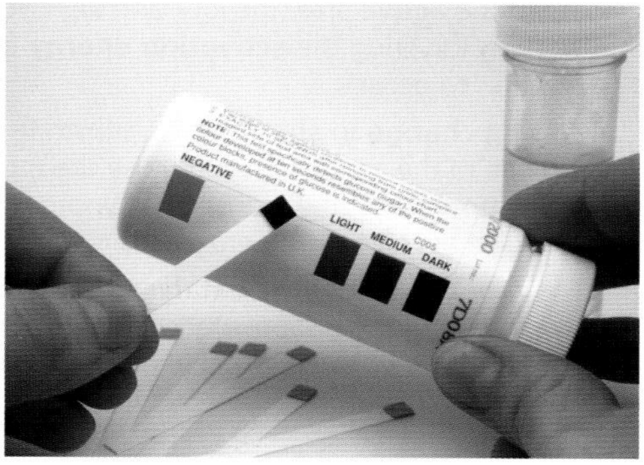

Figure 3.17 Measuring glucose in urine using a Clinistix™

What are the advantages and disadvantages of immobilisation?

The most obvious advantage is the recovery of the enzyme and its availability for re-use. Other issues are listed in Table 3.3.

Table 3.3 Advantages and disadvantages of immobilisation

Advantages	Disadvantages
The enzyme is held in a form that can be manipulated easily. It is absent from the product, so no purification steps are required.	An immobilisation mechanism that does not alter the shape or the catalytic ability of the enzyme must be selected.
The enzyme is available for multiple re-use, since it functions as an effective catalyst in pellet form.	The creation of stable, hardened pellets is an added expense that is inevitably reflected in the cost of the industrial product.
An immobilised enzyme is stable at the temperatures and pH at which it is held and used.	If the enzyme becomes detached it will appear in the product as a contaminant, possibly unnoticed.

Summary

- **Metabolism**, all the chemical reactions of life, consists of **anabolic reactions**, the build up of complex molecules from smaller ones, e.g. protein synthesis, and **catabolic reactions**, the breakdown of complex molecules, e.g. oxidation of sugar in respiration.

- All reactions of metabolism are made possible by **enzymes**. Enzymes are **biological catalysts** and most are made of globular protein. An enzyme is **highly specific** to the type(s) of substrate molecule and type of reaction that they catalyse.

- Enzymes work by forming a **temporary complex** with a **substrate** molecule at a special part of the enzyme surface, called the **active site** (the lock and key hypothesis). Enzymes work by lowering the **activation energy** needed for a reaction to occur.

- A slight **change in shape** of the substrate molecule when it binds to the active site helps raise the molecule to a **transition state** (the induced fit hypothesis), from which the products may form. The enzyme is released for reuse.

- The **rate of an enzyme-catalysed reaction** is found by measuring the disappearance of the substrate or the accumulation of the product in a given period of time. The

- **initial rate of reaction** is taken since the reaction rate falls with time under experimental conditions.

- The **factors that affect the rate of an enzyme-catalysed reaction** include **pH** and **temperature** – through their effects on protein structure. When molecules of substances recognised as **inhibitors** are in contact with enzyme molecules, the rate of reaction may be lowered in characteristic ways.

- **The Michaelis-Menten constant (K_m)** is the substrate concentration that sustains half maximum velocity ($\frac{1}{2} V_{max}$) of an enzyme-catalysed reaction. It is a measure of the degree of affinity of an enzyme for its substrate. K_m can be measured experimentally, and is expressed in units of molarity. Values of K_m have been measured for a great many enzymes, and it has been shown that some enzymes are able to work at maximum velocity at very low concentrations of substrate.

- Industries use enzymes as biological catalysts, often immobilised, with the reactants passed over them. In these cases, the advantages are the recovery of the enzyme and its availability for re-use. Being immobilised may affect the enzyme's efficiency compared to use of the same enzyme when free in the substrate solution.

Examination style questions

1 The enzyme sucrase catalyses the breakdown of the glycosidic bond in sucrose.

A student investigated the effect of increasing the concentration of sucrose on the rate of activity of sucrase. Ten test-tubes were set up with each containing $5\,cm^3$ of different concentrations of a sucrose solution. The test-tubes were placed in a water bath at 40°C for ten minutes. A flask containing a sucrase solution was also put into the water bath. After ten minutes, $1\,cm^3$ of the sucrase solution was added to each test-tube. The reaction mixtures were kept at 40°C for a further ten minutes.

After ten minutes, the temperature of the water bath was raised to boiling point. Benedict's solution was added to each test-tube. The time taken for a colour change was recorded and used to calculate rates of enzyme activity.
The results are shown in Fig. 1.1.

a) i) Name the type of reaction catalysed by sucrase. [1]
   ii) Explain why the temperature of the water was raised to boiling point. [2]
b) Describe **and** explain the results shown in Fig. 1.1. [5]

[Total: 8]

(Cambridge International AS and A Level Biology 9700, Paper 21 Q4 June 2011)

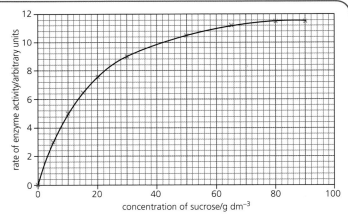

Fig. 1.1

2 a) Describe how enzymes take part in chemical reactions. [4]

Starch phosphorylase is an enzyme found in plant cells. In potato tuber cells, the enzyme takes part in the breakdown of starch when the tuber begins to grow.

starch phosphorylase

starch + phosphate ions → glucose-1-phosphate

A student investigated the effect of pH on this reaction using two buffer solutions.

The student prepared four test tubes, **A** to **D**, as shown in the table and described below.

The student made an extract of potato tissue that contained the enzyme. Some of this extract was boiled.

A solution of potassium dihydrogen phosphate was added to some tubes as a source of phosphate ions.

The test tubes were left for ten minutes in a water bath at 30 °C and then samples were tested with iodine solution.

b) i) State what the student would conclude from a positive result with iodine solution. [1]

ii) Explain why the student boiled some of the extract in this investigation. [2]

c) Explain the results shown in the table. [4]

[Total: 11]

(Cambridge International AS and A Level Biology 9700, Paper 02 Q2 June 2007)

Test tube	Contents					Results with iodine solution after 10 minutes
	Volume of starch solution / cm³	Volume of glucose-1-phosphate solution / cm³	Volume of potassium dihydrogen phosphate solution / cm³	pH of buffer solution	Enzyme extract	
A	2		0.5	6.5	unboiled	negative
B	2		0.5	2.0	unboiled	positive
C	2		0.5	6.5	boiled	positive
D		2		6.5	boiled	negative

3 The diagram shows an apparatus used in an investigation using immobilised enzymes. It is **not** expected that you will have done this investigation.

A solution of a substrate was poured into a burette containing an enzyme immobilised onto alginate beads. The liquid passing through the burette was collected into a beaker and the concentration of substrate and the concentration of the product measured. The table shows the results obtained by five students.

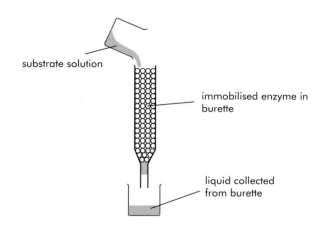

substrate solution

immobilised enzyme in burette

liquid collected from burette

	Enzyme concentration				Enzyme concentration			
	0.2 gdm⁻³		0.4 gdm⁻³		0.2 gdm⁻³		0.4 gdm⁻³	
	Substrate concentration / gdm⁻³				Product concentration / gdm⁻³			
	Repeat 1	Repeat 2	Repeat 1	Repeat 2	Repeat 1	Repeat 2	Repeat 1	Repeat 2
student A	24	26	14	13	32	33	60	64
student B	25	22	12	12	34	39	60	63
student C	22	23	10	13	35	32	59	61
student D	18	24	11	12	34	33	62	68
student E	25	28	13	18	30	32	65	64

a) Identify two variables and explain how each might be controlled. [2]

b) On a copy of the table above, indicate by placing a circle around the value, **two** results that are anomalous. [2]

c) A student drew the following conclusion from this investigation:
Doubling the enzyme concentration doubled the rate of reaction of the enzyme.

i) State **one** way in which the evidence in the table above supports the conclusion. [1]

ii) State **two** ways in which the reliability of the results might be improved. [2]

[Total: 7]

(Cambridge International AS and A Level Biology 9700, Paper 05 Q3 June 2007)

4 Cell membranes and transport

The fluid mosaic model introduced in 1972 describes the way in which biological molecules are arranged to form cell membranes. The model has stood the test of time as a way to visualise membrane structure and continues to be modified as understanding improves of the ways in which substances cross membranes, how cells interact and how cells respond to signals. The model also provides the basis for our understanding of passive and active movement between cells and their surroundings, cell to cell interactions and long distance cell signalling.

Investigating the effects of different factors on diffusion, osmosis and membrane permeability involves an understanding of the properties of phospholipids and proteins covered in the Topic 2 Biological molecules.

4.1 Fluid mosaic membranes

The structure of cell surface membranes allows movement of substances between cells and their surroundings and allows cells to communicate with each other by cell signalling.

By the end of this section you should be able to:

a) describe and explain the fluid mosaic model of membrane structure, including an outline of the roles of phospholipids, cholesterol, glycolipids, proteins and glycoproteins
b) outline the roles of cell surface membranes including references to carrier proteins, channel proteins, cell surface receptors and cell surface antigens
c) outline the process of cell signalling involving the release of chemicals that combine with cell surface receptors on target cells, leading to specific responses

The cell surface membrane

An organelle with many roles

The **cell surface membrane** is an organelle common to both eukaryotic and prokaryotic cells. It is an extremely thin structure, less than 10 nm thick, yet it has the strength to maintain the integrity of the cell. In addition to holding the cell's contents together, the cell surface membrane forms the barrier across which all substances entering or leaving the cell must pass. This membrane represents the 'identity' of the cell to surrounding cells. It is also the surface across which chemical messages (such as hormones and growth factors) communicate. In fact, movement of molecules across the cell surface membrane is continuous and very heavy. Figure 4.1 is a summary of all this membrane traffic.

The molecular components of membranes

The membranes of cells are made almost entirely of protein and lipid, together with a small and variable amount of carbohydrate. The lipid of membranes is **phospholipids**. The chemical structure of phospholipids is shown in Figure 2.16 on page 43.

Look at its structure again now.

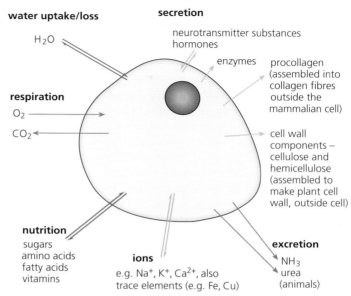

Figure 4.1 The movement of substances across the cell surface membrane

You can see that the phospholipid molecule has a 'head' composed of a glycerol to which is attached an ionised phosphate group. This latter part of the molecule has **hydrophilic properties** (water-loving). For example, **hydrogen bonds** readily form between the phosphate head and water molecules. The remainder of the phospholipid comprises two long, fatty acid residues consisting of hydrocarbon chains. These 'tails' have **hydrophobic properties** (water-hating).

The response of phospholipid to water

A small quantity of phospholipid in contact with a solid surface (a clean glass plate is suitable), remains as a discreet bubble; the phospholipid molecules do not spread. However, when a similar tiny drop of phospholipid is added to water it instantly spreads over the entire surface (as a monolayer of phospholipid molecules, in fact). The molecules float with their hydrophilic 'heads' in contact with the water molecules, and with their hydrocarbon tails exposed above and away from the water, forming a monolayer of phospholipid molecules (Figure 4.2).

When more phospholipid is added and shaken up with water, **micelles** may form. These are tiny spheres of lipids with hydrophilic heads outwards, in contact with the water, and with all the hydrophobic tails pointing inwards.

Alternatively, the phospholipid molecules arrange themselves as a **bilayer**, again with the hydrocarbon tails facing together (Figure 4.2). This is how the molecules of phospholipid are arranged in the cell surface membrane.

In the phospholipid bilayer, attractions between the hydrophobic hydrocarbon tails on the inside and between the hydrophilic glycerol–phosphate heads and the surrounding water on the outside make a stable, strong barrier.

Finally, the phospholipid bilayer has been found to contain molecules of **cholesterol**. Cholesterol has an effect on the flexibility of the cell surface membrane. We return to this feature shortly.

The permeability of a phospholipid membrane

We must note, however, that whilst lipids provide a cohesive structure, they also tend to act as a barrier to the passage of polar molecules and ions. This is particularly the case where close-packing of the hydrocarbon tails occurs. Potentially, water-soluble substances pass through the lipid bilayer with great difficulty. We will return to this issue shortly, too.

a phospholipid molecule has a **hydrophobic tail** – which repels water – and a **hydrophilic head** – which attracts water

Phospholipid molecules **in contact with water** form a **monolayer**, with heads dissolved in the water and the tails sticking outwards.

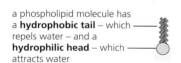

water

When **mixed with water**, phospholipid molecules arrange themselves into a **bilayer**, in which the hydrophobic tails are attracted to each other.

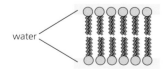

water

Figure 4.2 Phospholipid molecules arranged in a monolayer and a bilayer

The proteins and carbohydrates of membrane

The proteins of membranes are **globular proteins** which are buried in and across the lipid bilayer, with most projecting above the surfaces. Others are superficially attached on either surface of the lipid bilayer.

Proteins that occur partially or fully buried in the lipid bilayer are described as **integral proteins**. Those that are superficially attached on either surface of the lipid bilayers are known as **peripheral proteins**.

Membrane proteins have a range of roles. Some are enzymes, others are receptors, or antigens, and many are channels for transport of metabolites. Those that are involved in transport of molecules across membranes are in the spotlight in the next section. A summary of the various roles of membrane proteins is given in Figure 4.24 (page 94), after the movements of molecules across the membrane have been discussed.

The carbohydrate molecules of the cell surface membrane are relatively short chain **polysaccharides**, some attached to the proteins (**glycoproteins**) and some to the lipids (**glycolipids**) Glycoproteins and glycolipids are only found on the outer surface of the membrane. Together these form the **glycocalyx**. The roles of this glycocalyx are:

- cell–cell recognition
- as receptor sites for chemical signals, such as hormone messengers
- to assist in the binding together of cells to form tissues.

The fluid mosaic model of membrane structure

The molecular structure of the cell surface membrane, known as the **fluid mosaic model** by those who first suggested it, is shown in Figure 4.3. The membrane they described as *fluid* because the components (lipids and proteins) move around within their layer. In fact the movements of the lipid molecules are rapid, whereas mobile proteins move about more slowly. The word *mosaic* described the scattered pattern of the proteins, when viewed from above. This is the current view of the structure of the cell surface membrane. It is based on a range of evidence, in part from studies with the electron microscope (Figure 4.4).

Questions

1 In plants adapted to survive in very low winter temperatures, what seasonal change would you anticipate in the composition of the lipid bilayer of their membranes?
2 What is the difference between a lipid bilayer and the 'double membrane' of many organelles?

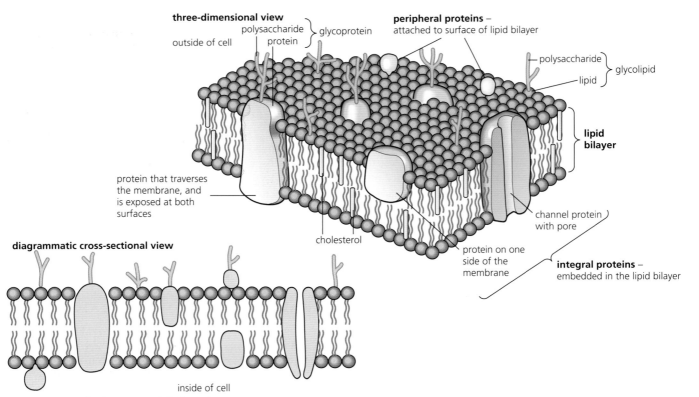

Figure 4.3 The fluid mosaic model of membrane structure

Composition of the membrane and control of how fluid it is

The phospholipids of the membrane are a mixture. Some have saturated fatty acid tails and some unsaturated fatty acid tails (Figure 2.14, page 41). An excess of unsaturated fatty acid tails makes the membrane more fluid. This is because the kinks in the tails prevent close-packing of the lipids. However, the presence of cholesterol among the phospholipid molecules has to be taken into account. The effect of cholesterol is to reduce fluidity by preventing or reducing the movements of the lipid molecules.

Membrane fluidity is an important factor. Membranes must be sufficiently fluid for many of the proteins present to move about and so to function correctly. If the temperature of a membrane falls it becomes less fluid. A point may be reached when the membrane will actually solidify. Some organisms have been found to vary the balance between saturated and unsaturated fatty acids and the amount of cholesterol in their membranes as ambient temperatures changes. In this way they maintain a properly functioning membrane, even at very low temperatures, for example.

The cell surface membrane and cell signalling

Cells respond to changes in their environment by receiving and integrating signals from other cells. They also send out messages. Most cell signals are chemical signals. For example, motile single-celled organisms detect nutrients in their environment and move towards the source. In multicellular organisms, chemical signals include growth factors and hormones that may have come from neighbouring cells or from more distant sources. Neurotransmitter signals cross a tiny gap from a neighbouring nerve cell (page 320).

Cells are able to detect these signals because of receptors in their cell surface membrane to which the 'chemical message' binds, and then trigger an internal response. Receptors are typically trans-membrane proteins, which on receipt of the signal, alert internal signalling pathways. However, some receptors occur within the cell, and possibly within the nucleus. These receptors are stimulated by signal chemicals that pass through the cell surface membrane, such as the gas nitrous oxide or a steroid hormone such as oestrogen. The activation of a receptor, whether internal or external, initiates a chain of reactions within the cell, possibly involving a second messenger. The outcome is some internal response.

TEM of the cell surface membrane of a red blood cell (×700 000)

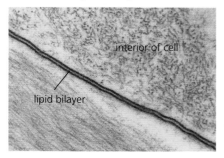

cell surface membrane in cross-section

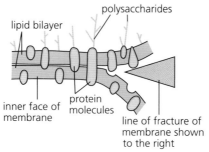

electron micrograph of the cell membrane (freeze-etched)

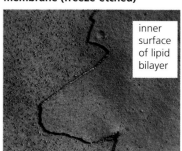

Figure 4.4 Membrane structure: evidence from the electron microscope

Summarising the rol of membrane proteins in cells

The many roles that cell membrane proteins play (as channels for transport of metabolites, as pumps for active transport, as electron carriers, carrier proteins, cells surface receptors, and as binding sites for specific hormone molecules and in antigen-antibody reactions) are outlined in Figure 4.24 on page 94.

Look at this diagram now.

The potential for complex communications between cells that may result as chemicals released from one cells combine with cell surface receptors on neighbouring cells is evident. It is the proteins of cell surface membranes that facilitate this 'cell signalling'.

4.2 Movement of substances into and out of cells

<table>
<tr><td colspan="2">

The fluid mosaic model allows an understanding of how substances enter and exit cells by a variety of different mechanisms.

Investigating the effect of increasing the size of model cells allows an understanding of the constraints of obtaining resources across the cell surface and moving substances out of cells.

</td><td>

By the end of this section you should be able to:

a) describe and explain the processes of diffusion, facilitated diffusion, osmosis, active transport, endocytosis and exocytosis

b) investigate simple diffusion using plant tissue and non-living materials, such as glucose solutions, Visking tubing and agar

c) calculate surface areas and volumes of simple shapes (e.g. cubes) to illustrate the principle that surface area to volume ratios decrease with increasing size

d) investigate the effect of changing surface area to volume ratio on diffusion using agar blocks of different sizes

e) investigate the effects of immersing plant tissues in solutions of different water potential, using the results to estimate the water potential of the tissues

f) explain the movement of water between cells and solutions with different water potentials and explain the different effects on plant and animal cells

</td></tr>
</table>

Into and out of cells passes **water**, **respiratory gases** (O_2 and CO_2), **nutrients**, **essential mineral ions** and **excretory products**. Cells may secrete substances such as **hormones** and **enzymes**. They may receive **growth substances** and **hormones**, too. Plants secrete the chemicals that make up their walls through their membranes, and assemble and maintain the wall outside the membrane. Certain mammalian cells secrete **structural proteins** such as collagen in a form that can be assembled outside the cells.

In addition, the membrane at the cell surface is where the cell is identified by surrounding cells and organisms. For example, protein **receptor sites** are recognised by hormones, neurotransmitter substances (from nerve cells), as well as other chemicals, sent from other cells.

We will now look into the **mechanisms of membrane transport**. Figure 4.5 is a summary of the mechanisms of transport across membranes.

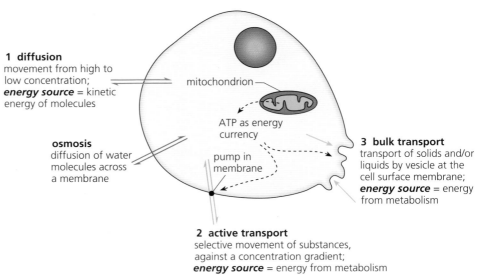

1 diffusion
movement from high to low concentration;
energy source = kinetic energy of molecules

mitochondrion

ATP as energy currency

osmosis
diffusion of water molecules across a membrane

pump in membrane

3 bulk transport
transport of solids and/or liquids by vesicle at the cell surface membrane;
energy source = energy from metabolism

2 active transport
selective movement of substances, against a concentration gradient;
energy source = energy from metabolism

Figure 4.5 Mechanisms of movement across membranes

Movement by diffusion

> **Diffusion**: the net movement of particles such as molecules from a region where they are at a higher concentration to a region with a lower concentration, using energy from the random movements of particles. This includes diffusion of small non-polar molecules (such as oxygen and carbon dioxide) through the cell surface membrane, as well as diffusion of fat-soluble molecules (such as vitamin A) through the cell surface membrane.

The atoms, molecules and ions of liquids and gases undergo continuous random movements. These movements result in the even distribution of the components of a gas mixture and of the atoms, molecules and ions in a solution. This is why we are able to take a tiny random sample from a solution and analyse it to find the concentration of dissolved substances in the whole solution. Every sample has the same composition as the whole. Similarly, every breath we take has the same amount of oxygen, nitrogen and carbon dioxide as the atmosphere as a whole.

Continuous random movements of all molecules ensures complete mixing and even distribution of molecules, given time, in solutions and gases. The energy for diffusion comes from the **kinetic energy** of molecules. 'Kinetic' means that a particle has this energy because it is in continuous motion.

Diffusion is the free passage of molecules (and atoms and ions) from a region of their high concentration to a region of low concentration. Where a difference in concentration has arisen in a gas or liquid, random movements carry molecules from a region of high concentration to a region of low concentration. As a result, the particles become evenly dispersed. The factors affecting the rate of diffusion are listed in Table 4.1.

Table 4.1 Factors affecting the rate of diffusion

Factors affecting the rate of diffusion	The conditions needed to achieve rapid diffusion across a surface
Concentration gradient – the greater the difference in concentration between two regions, the greater the amount that diffuses in a given time.	A fresh supply of substance needs to reach the surface and the substance that has crossed needs to be transported away.
Distance over which diffusion occurs – the shorter the distance, the greater the rate of diffusion.	The structure needs to be thin.
Area across which diffusion occurs – the larger the area, the greater the diffusion.	The surface area needs to be large.
Structure through which diffusion occurs – pores or gaps in structures may enhance diffusion.	A greater number of pores and a larger size of pores may enhance diffusion.
Size and type of diffusing molecules – smaller molecules and molecules soluble in the substance of a barrier will both diffuse more rapidly.	Oxygen may diffuse more rapidly than carbon dioxide. Fat-soluble substances diffuse more rapidly through the lipid bilayer than water-soluble substances.

Demonstrating diffusion

Diffusion in a liquid can be illustrated by adding a crystal of a coloured mineral to distilled water. Even without stirring, the ions become evenly distributed throughout the water. The process takes time, especially as the solid has first to dissolve (see Figure 4.6 on the next page).

Cell surface : volume ratio, cell size and diffusion

As a cell grows and increases in size an important difference develops between the surface area available for exchange by diffusion and the volume of the cytoplasm in which the chemical reactions of life occur. The volume increases faster than surface area; the surface area : volume ratio falls (see Figure 4.7 on the next page). So, with increasing size of a cell, less and less of the cytoplasm has access to the cell surface for exchange of gases, supply of nutrients and loss of waste products.

Put another way, we can say that the smaller the cell is, the more quickly and easily can materials be exchanged between its cytoplasm and environment by diffusion. One consequence of this is that cells cannot continue growing larger, indefinitely. When a maximum size is reached cell growth stops. The cell may then divide.

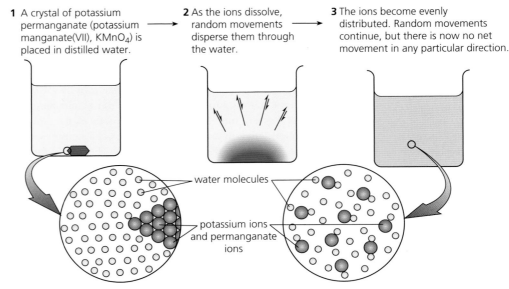

1 A crystal of potassium permanganate (potassium manganate(VII), KMnO₄) is placed in distilled water.

2 As the ions dissolve, random movements disperse them through the water.

3 The ions become evenly distributed. Random movements continue, but there is now no net movement in any particular direction.

water molecules

potassium ions and permanganate ions

Figure 4.6 Diffusion in a liquid

3 For imaginary cubic 'cells' with sides 1, 2, 4 and 6 mm:

a calculate the volume, surface area and ratio of surface area to volume for each

b plot a graph of the surface area of these cells against their volume

c state the effect on the SA:V ratio of a cell as it increases in size

d explain the effect of increasing cell size on the efficiency of diffusion in the removal of waste products from cell cytoplasm.

cubic cell of increasing size

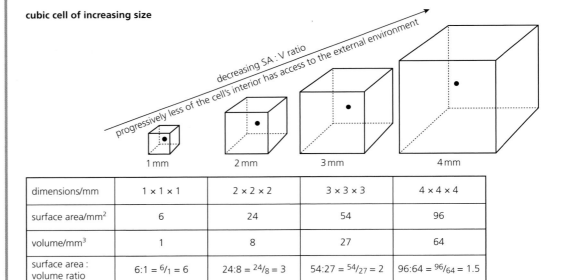

decreasing SA : V ratio

progressively less of the cell's interior has access to the external environment

dimensions/mm	1 × 1 × 1	2 × 2 × 2	3 × 3 × 3	4 × 4 × 4
surface area/mm²	6	24	54	96
volume/mm³	1	8	27	64
surface area : volume ratio	6:1 = ⁶/₁ = 6	24:8 = ²⁴/₈ = 3	54:27 = ⁵⁴/₂₇ = 2	96:64 = ⁹⁶/₆₄ = 1.5

Figure 4.7 The effect of increasing size on the surface area : volume ratio

Demonstrating the connection between surface area : volume ratio and diffusion

Working with a block of gelatine containing the acid-base indicator cresol red (coloured red in alkali, yellow in acid), small, regular-shaped blocks of known dimension can be cut, and then immersed in acid solution. The time taken for the red colour to completely disappear can be recorded in a table, against the dimensions of the blocks, listed in increasing size. The results of one such investigation are recorded in Question 4. Examine the data there, and answer the questions that follow.

4 Cubes of slightly alkaline gelatin of different dimensions, containing an acid–alkali indicator (red in alkalis but yellow in acids) were prepared. The cubes were then placed in dilute acid solution and the time taken for the colour in the gelatin to change from red to yellow was measured.

Dimensions/mm	Surface area/mm^2	Volume/mm^3	Time/minutes
10 × 10 × 10	600	1000	12
5 × 5 × 5	150	125	4.5
2.5 × 2.5 × 2.5	37.5	15.6	4.0

a For each block, calculate the ratio of surface area to volume (SA/V).

b Explain why the colour changes more quickly in some blocks than others.

Diffusion in cells

Diffusion across the cell surface membrane occurs where:

- the membrane is fully permeable to the substance concerned. The lipid bilayer of the cell surface membrane is permeable to non-polar substances, including steroids and glycerol, and also oxygen and carbon dioxide in solution, all of which diffuse quickly via this route. Note that the net diffusion of different types of particle can take place in opposite directions without hindrance, too. So, at the lung surface, oxygen diffuses into the blood whilst carbon dioxide diffuses out.

- the pores in the membrane are large enough for a substance to pass through. Water diffuses across the cell surface membrane by means of the protein-lined pores of the membrane. The tiny spaces between the phospholipid molecules are another route for water molecules. This latter occurs more easily where the fluid-mosaic membrane contains phospholipids with unsaturated hydrocarbon tails, for here these hydrocarbon tails are spaced more widely. In this state, the membrane is consequently especially 'leaky' to water, for example.

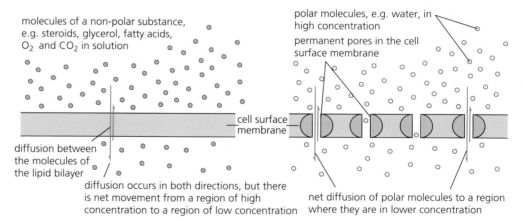

Figure 4.8 Diffusion across the cell surface membrane

Facilitated diffusion

In facilitated diffusion, a substance that otherwise is unable to diffuse across the cell surface membrane does so. This is as a result of its effect on a particular protein in the membrane. These are globular proteins that can form into pores large enough for diffusion. Those pores close up again when that substance is no longer present (Figure 4.9).

Facilitated diffusion: the diffusion of ions and polar (water-soluble) molecules through cell membranes using specific protein channels or carriers, down a concentration gradient (from regions where they are at higher concentration to regions where they are at lower concentration).

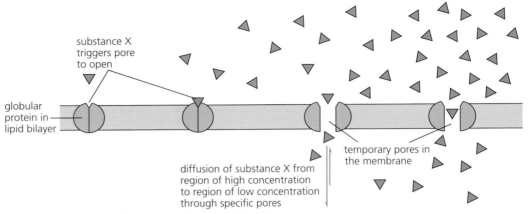

substance X triggers pore to open

globular protein in lipid bilayer

diffusion of substance X from region of high concentration to region of low concentration through specific pores

temporary pores in the membrane

Figure 4.9 Facilitated diffusion

Question

5 Distinguish between 'diffusion' and 'facilitated diffusion'.

Osmosis: the diffusion of water molecules from a region where water is at a higher water potential through a partially permeable membrane to a region with a lower water potential.

Alternatively a channel protein may be selective. For example, some channel proteins are 'gated' and open to allow the passage of ions only under particular conditions. There are different channels for potassium ions and sodium ions in the membrane of nerve fibres, for example (Figure 15.6, page 317).

Such proteins undergo rapid shape changes when in contact with a particular solute molecule. Facilitated diffusion follows. The movement of glucose into red blood cells occurs in this way. So does the movement of ADP into mitochondria and the movement of ATP from mitochondria into the cytosol.

Finally, we must remember that in facilitated diffusion the energy comes from the kinetic energy of the molecules involved, as is the case in all forms of diffusion. Energy from metabolism is not required.

Osmosis – a special case of diffusion

First, look at the experiment shown in Figure 4.10. The bag shown here is made from a short length of dialysis tubing. Some sucrose solution is added, perhaps by using a small plastic syringe. This is an experiment that is easy to set up. When this partially filled bag is lowered into a beaker of water it quite quickly becomes stretched and turgid. Obviously there is a huge inflow of water. This is a simple but dramatic demonstration of osmosis.

Why does osmosis happen? Remember, in a sucrose solution the sucrose molecules are the solute and the water molecules are the solvent.

$$\text{a solution} = \text{solute} + \text{solvent}$$

We saw in Topic 2 that in a solution, a dissolved substance attracts polar water molecules around it (Figure 2.28, page 53). The forces holding those water molecules in this way are **hydrogen bonds**. The effect of the presence of a 'cloud' of water molecules around each solute molecule is to hamper and restrict their movements. Organic substances like sugars, amino acids, polypeptides and proteins, and inorganic ions like Na^+, K^+, Cl^- and NO_3^-, all have this effect on the water molecules around them. On the other hand, in pure water, all of the water molecules are free to move about randomly, and do so – all the time. We say that here they diffuse freely.

In the experiment in Figure 4.10, the solution of sugar was separated from pure water by a membrane – the walls of a dialysis tube. This is described as **partially permeable** because it only allows certain molecules to pass, not all of them. Water molecules move across the membrane in both directions, by diffusion. At the same time, the solute molecules with their cloud of water molecules are too large to pass through. In the mean time, there is a continuing net movement of water molecules from the region of high concentration of free water molecules (the molecules of pure water in the beaker) to the region of low concentration of free water molecules (the water molecules of the sucrose solution). This causes the bag to fill up and become stretched and turgid.

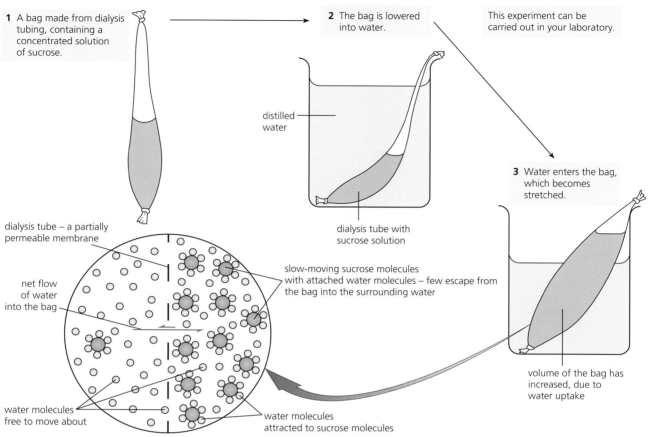

1 A bag made from dialysis tubing, containing a concentrated solution of sucrose.

2 The bag is lowered into water.

This experiment can be carried out in your laboratory.

distilled water

3 Water enters the bag, which becomes stretched.

dialysis tube with sucrose solution

dialysis tube – a partially permeable membrane

net flow of water into the bag

slow-moving sucrose molecules with attached water molecules – few escape from the bag into the surrounding water

water molecules free to move about

water molecules attracted to sucrose molecules

volume of the bag has increased, due to water uptake

Figure 4.10 Osmosis

Water potential

The name given to the tendency of water molecules to move about is **water potential**. 'Water potential' is really a measure of the free kinetic energy of the water molecules. The Greek letter *psi* (symbol ψ) is used to represent water potential.

Water moves from a region of higher water potential to a region of lower water potential. We say water moves down a **water potential gradient**. Equilibrium is reached only if or when the water potential is the same in both regions. At this point, there would be no net movement of water from one region to another, but random movements of water molecules continue, of course.

Next we will examine the effects of dissolved solutes, and then of mechanical pressure on water potential, before returning to a re-statement of water potential.

Questions

6 What is meant when we say a membrane is partially permeable?

7 When a concentrated solution of glucose is separated from a dilute solution of glucose by a partially permeable membrane, which solution will show a net gain of water molecules? Why do we speak of a 'net' gain?

8 Jam is made of equal weights of fruit and sucrose. What happens to a fungal spore that germinates after landing on jam?

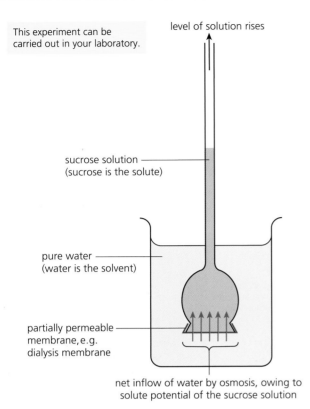

level of solution rises

This experiment can be carried out in your laboratory.

sucrose solution (sucrose is the solute)

pure water (water is the solvent)

partially permeable membrane, e.g. dialysis membrane

net inflow of water by osmosis, owing to solute potential of the sucrose solution

Figure 4.11 An osmometer

The concentration of solute molecules and water potential

Pure water obviously has the highest water potential. By convention, this is set at zero. Once a solute is dissolved in water, the water molecules are immediately less likely to diffuse (they are less mobile). So the effect of dissolving solute in water is to lower its water potential. Consequently solutions at atmospheric pressure must have a negative value of water potential (since pure water is set at zero). Also, the stronger the solution (i.e. the more solute dissolved per volume of water) the larger the number of water molecules that are slowed up and held almost stationery. So, in a very concentrated solution, very many more of the water molecules have restricted movements than do in a dilute solution.

The amount of dissolved solute present in a solution is known as the **solute potential** of the solution. It is given the symbol ψ_s. A simple osmometer is shown in Figure 4.11. We use an osmometer to demonstrate the solute potential of a solution. Once the osmometer is lowered into the beaker of water, very many more water molecules diffuse across the membrane into the solution than move in the opposite direction. The solution is diluted and it rises up the attached tube. An osmometer like this could be used to compare the solute potentials of solutions with different concentrations.

Pressure potential

The other factor that may influence osmosis is mechanical pressure acting on the solution. This factor is given the name **pressure potential**. It is represented by the symbol ψ_p.

If a pressure greater than atmospheric pressure is applied to a solution, then this is an example of a pressure potential being created in a solution. In the demonstration in Figure 4.12 a short length of dialysis tubing has been set up to show how the pressure potential of a solution can become large enough to stop the osmotic uptake of water altogether.

> ### Question
>
> **9** When a concentrated solution of sucrose is separated from a dilute solution of sucrose by a partially permeable membrane, which solution:
>   **a** has a higher concentration of water molecules
>   **b** has lower water potential
>   **c** will experience a net gain of water molecules?

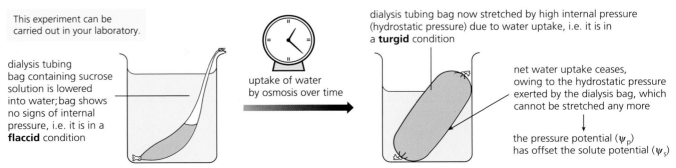

This experiment can be carried out in your laboratory.

dialysis tubing bag containing sucrose solution is lowered into water; bag shows no signs of internal pressure, i.e. it is in a **flaccid** condition

uptake of water by osmosis over time

dialysis tubing bag now stretched by high internal pressure (hydrostatic pressure) due to water uptake, i.e. it is in a **turgid** condition

net water uptake ceases, owing to the hydrostatic pressure exerted by the dialysis bag, which cannot be stretched any more

the pressure potential (ψ_p) has offset the solute potential (ψ_s)

Figure 4.12 Pressure potential at work

Water potential = solute potential + pressure potential

The concept of 'water potential' allows the effects of the two factors acting on water in a system (such as a dialysis tubing 'bag' or a living cell) to be brought together in a single equation:

$$\text{water potential} = \text{solute potential} + \text{pressure potential}$$
$$\psi = \psi_s + \psi_p$$

We have seen that the water potential of a solution is the name we give to this tendency of water molecules to enter or leave solutions by osmosis, and that the Greek letter psi (symbol ψ) is used to represent the water potential.

Since 'water potential' is really a measure of the free kinetic energy of the water molecules, pure water obviously has the highest water potential. By definition, this is set at zero. We have also seen that once a solute is dissolved in water, the water molecules are immediately less likely to diffuse (they are less mobile). So, the effect of dissolving a solute in water is to lower its water potential. Consequently solutions at atmospheric pressure have a negative value of water potential. For example, a solution containing:

3.42 g of sucrose in a 100 cm^3 has a water potential of −270 kPa, and
34.2 g of sucrose in a 100 cm^3 has a water potential of −3510 kPa.

Do negative values like these give you problems?

Not really, perhaps. You use them in daily life. Imagine an international weather forecast that reports the temperature in Siberia has changed from −10 °C to −25 °C. You immediately know this means it's much colder there. In other words, −25 °C is a much lower temperature than −10 °C, even though 25 is a larger number than 10.

In the same way −3510 kPa is a smaller water potential than −270 kPa, although 3510 is larger number than 270.

Questions

10 When a concentrated solution of glucose is separated from a dilute solution of glucose by a partially permeable membrane, which solution:

  a has a higher water potential

  b has a higher concentration of water molecules

  c will show a net gain of water molecules?

Osmosis in cells and organisms

Since water makes up 70–90 per cent of living cells and cell surface membranes are partially permeable membranes, osmosis is very important in living things. We can illustrate this first in plant cells.

Osmosis in plants

We need to think about the effects of osmosis in an individual plant cell first (Figure 4.13). There are two factors to consider.

● In a plant cell there is a cellulose cell wall. This is not present in an animal cell.
● In any cell the net direction of water movement depends upon whether the water potential of the cell solution is more or less negative than the water potential of the external solution.

When the external solution is **less negative** (that is, there is little or no dissolved solutes), there is a net flow of water **into** the cell. The cell solution becomes diluted. The volume of the cell is expanded by water uptake. As a result, the cytoplasm may come to press hard against the cell wall. If this happens the cell is described as **turgid**. The pressure potential that has developed (due to the stretching of the wall) offsets the solute potential of the cell solution completely and further net uptake of water stops. Incidentally, the cell wall has actually protected the delicate cell contents from damage due to osmosis.

When the external solution is **more negative** (that is, much dissolved solutes), there is a net flow of water **out** of the cell. The cell solution becomes more concentrated. As the volume of cell solution decreases, the cytoplasm starts to pull away from parts of the cell wall (contact with the cell wall is maintained at points where there are cytoplasmic connections between cells). The cells are said to be **plasmolysed** (*lysis* means splitting) and are **flaccid**.

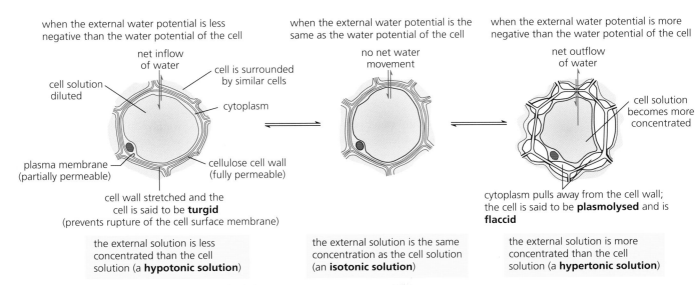

when the external water potential is less negative than the water potential of the cell

net inflow of water

cell solution diluted

cell is surrounded by similar cells

cytoplasm

plasma membrane (partially permeable)

cellulose cell wall (fully permeable)

cell wall stretched and the cell is said to be **turgid** (prevents rupture of the cell surface membrane)

the external solution is less concentrated than the cell solution (a **hypotonic solution**)

when the external water potential is the same as the water potential of the cell

no net water movement

the external solution is the same concentration as the cell solution (an **isotonic solution**)

when the external water potential is more negative than the water potential of the cell

net outflow of water

cell solution becomes more concentrated

cytoplasm pulls away from the cell wall; the cell is said to be **plasmolysed** and is **flaccid**

the external solution is more concentrated than the cell solution (a **hypertonic solution**)

Figure 4.13 A plant cell in changing external solutions

Estimation of the water potential of plant tissue

The net direction of water movement in a cell depends on whether the water potential (ψ) of the cell solution is more negative or less negative than the water potential of the external solution. We exploit this in our measurement of the water potential of plant tissues. Representative samples of tissue are bathed in solutions of a range of water potentials so that the solution which causes no net movement of water can be found. This technique may be applied to plant tissue from which you can cut reproducible-sized cylinders that will fit into a test tube or boiling tube. The tissue should be fully turgid at the outset; soak in water to ensure this. Examples of suitable tissues would include beetroot, potato tuber and carrot root. Cut the cylinders with a cork borer, wash them in tap water, and finally measure their length (or cut them all to a standard length) (Figure 4.14).

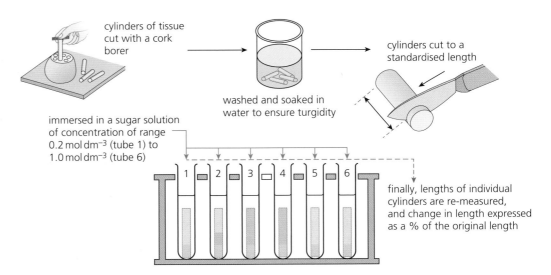

cylinders of tissue cut with a cork borer

washed and soaked in water to ensure turgidity

cylinders cut to a standardised length

immersed in a sugar solution of concentration of range $0.2\,mol\,dm^{-3}$ (tube 1) to $1.0\,mol\,dm^{-3}$ (tube 6)

1 2 3 4 5 6

finally, lengths of individual cylinders are re-measured, and change in length expressed as a % of the original length

Figure 4.14 Measuring water potential of plant tissue

Table 4.2 The water potential of sucrose solutions

Sucrose (mol dm⁻³)	Water potential (kPa)
0.2	−540
0.4	−1120
0.6	−1800
0.8	−2580
1.0	−3510

Steps to the experiment:

● Immerse (at least) one cylinder in each of the range of five sucrose solutions, 0.2 mol dm⁻³ to 1.0 mol dm⁻³ in a tube. Leave them immersed for one day. Set up a table to record the lengths of the tissue cylinders from each of the five tubes.
● Re-measure the length of each cylinder and record this in your table. Calculate the change in length as a percentage of the original length.
● Plot a graph of the percentage change in length against the molarity of the sucrose solution. Read off the molarity of sucrose that causes no change in length of the tissue.
● Using the conversion table below, quote the water potential of the tissue you have investigated.

In evaluating the experiment, consider the significant potential causes of error or inaccuracy in the technique. *What improvements could you make?*

Movement of water between plant cells

We have seen that water flows from a less negative to a more negative water potential. For example, consider two adjacent cells that have initial solute potentials and pressure potentials (values in kilopascals) as shown:

cell A	cell B
$\Psi s = -1400$	$\Psi s = -2100$
$\Psi p = 600$	$\Psi p = 900$

The water potential of each cell can be calculated ($\psi = \psi_s + \psi_p$):

$$\psi \text{ of cell A} = -800\,\text{kPa}$$

$$\psi \text{ of cell B} = -1200\,\text{kPa}$$

So net water flow is from cell A to cell B until an equilibrium in water potentials is established.

When we discuss water movement through plants (Topic 7, page 135) we will see the flow of water from cell to cell in roots. From root hair cells, where water uptake occurs from a soil solution (of high water potential, i.e. less negative ψ), to the cells at the centre of the root (of lower water potential, i.e. more negative ψ).

Plasmolysis in plant cells

Plasmolysis is most easily observed in cells with a coloured vacuolar solution. Examples include beetroot tissue or the epidermis of rhubarb leaf stalks. There may be others local to you (Figure 4.15).

Plasmolysed plant cells
In cells in which the solution in the vacuoles is coloured, they can be seen under the microscope whithout staining. → When they are placed in a solution of water potential *greater than that of the cell solution*, plasmolysis of the cells can be observed by microscopy.

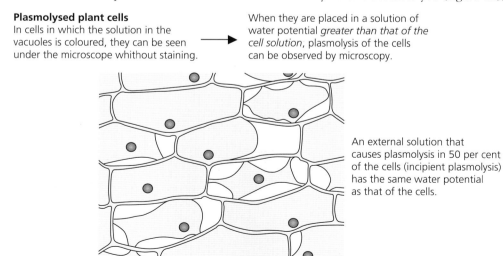

An external solution that causes plasmolysis in 50 per cent of the cells (incipient plasmolysis) has the same water potential as that of the cells.

Figure 4.15 Plasmolysed plant cells

In the plasmolysed cell, the wall exerts no pressure at all. As a plasmolysed cell starts to take up water the contents enlarge. At the point where the cytoplasm starts to push against the wall, a slight pressure potential is produced (Figure 4.16). As water uptake continues, the pressure potential eventually becomes large enough to reduce the cell's water potential to zero. At that point, water uptake stops.

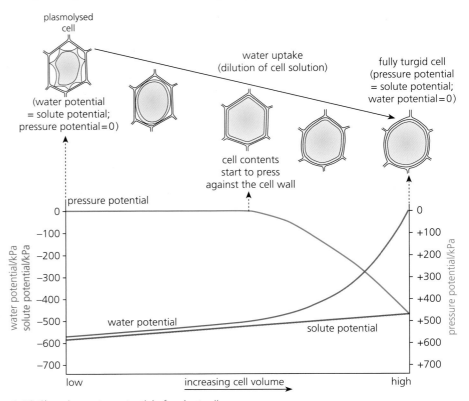

Figure 4.16 Changing water potential of a plant cell

Questions

11 As an herbaceous plant wilts, its cells change from being turgid to flaccid. In this situation, what happens to:

 a the solute potential of the vacuolar solution in the cells

 b the pressure potential of these cells?

Plasmolysis and wilting

Typically, the cytoplasm of plant cells is undamaged by a period of plasmolysis. Cells promptly recover when water becomes available. In non-woody plants (they are called **herbaceous plants**, to tell them apart from trees and shrubs) a state of turgor is the normal condition. The turgidity of all the cells plays a key part in the support of the plant body. If herbaceous plants become short of water for a prolonged period, the plant wilts. Ultimately, a wilted plant will die if it does not receive water.

well-watered potted plant with turgid cells

water loss continues; many cells of the leaves and stem become plasmolysed; flaccid cells provide no support

Figure 4.17 Wilting

Osmosis in animals

An animal cell lacks the protection of a cellulose cell wall. Consequently, compared with a plant cell, it is very easily damaged by changes in the external solution. If a typical animal cell is placed in pure water, the cell surface membrane will break apart quite quickly. The cell will have been destroyed by the pressure potential generated.

Figure 4.18 shows what happens to an animal cell (a red blood cell) when the external solution has a water potential that is both less negative and more negative than the cell solution.

In most animals (certainly in mammals) these problems are avoided because the osmotic concentration of body fluids – the blood plasma and tissue fluid – is very precisely regulated. The water potential is regulated both inside and outside body cells (to maintain isotonic conditions). This process is known as **osmoregulation** (Topic 14).

Figure 4.18 A red blood cell in changing external solutions

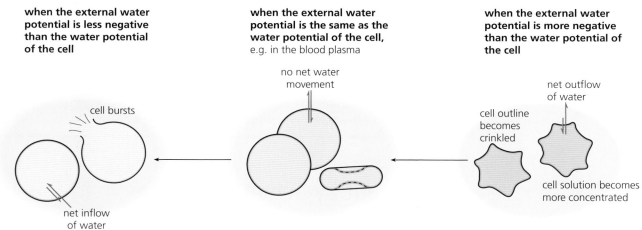

when the external water potential is less negative than the water potential of the cell

cell bursts

net inflow of water

when the external water potential is the same as the water potential of the cell, e.g. in the blood plasma

no net water movement

when the external water potential is more negative than the water potential of the cell

net outflow of water

cell outline becomes crinkled

cell solution becomes more concentrated

Questions

12 What is the outcome of immersing an animal cell in a solution of substantially lower water potential than that of the cell? What would be the effect on a plant cell?

13 Two adjacent animal cells, M and N, are part of a compact tissue. They have:

cell M	
ψ_s	−580 kPa
ψ_p	410 kPa
cell N	
ψ_s	−640 kPa
ψ_p	420 kPa

Will net water flow be from cell M to cell N? Explain your answer.

Aquatic, unicellular animals

Many unicellular animals live in aquatic environments and survive even though their surroundings have a water potential substantially less negative than the cell solution. All the time, water is flowing into these cells by osmosis. They are in constant danger that their membrane will burst because of high internal pressure. In fact the problem is overcome – or rather, prevented.

The protozoan *Amoeba* is one example. The cytoplasm of an amoeba contains a tiny water pump, known as a **contractile vacuole**. This works continuously to pump out the excess water. How important the contractile vacuole is to the organism is fatally demonstrated if the cytoplasm is temporarily anaesthetised – the *Amoeba* quickly bursts.

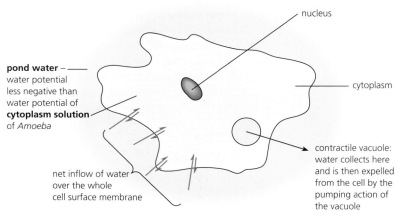

nucleus

cytoplasm

pond water – water potential less negative than water potential of **cytoplasm solution** of *Amoeba*

contractile vacuole: water collects here and is then expelled from the cell by the pumping action of the vacuole

net inflow of water over the whole cell surface membrane

Figure 4.19 The role of the contractile vacuole in *Amoeba*

Extension

Polar water molecules and the lipid bilayer – a recap

We have seen that water enters and leaves cells by osmosis, yet the lipid bilayer of the cell surface membrane is at least theoretically impermeable to polar water molecules. However, water molecules are very small and it is likely that there are many tiny spaces open between the giant molecules of the lipid bilayer. Remember, the lipid molecules are also constantly moving about. In addition, quite small, protein-lined pores exist in the membrane and are permanently open. These proteins allow unrestricted water movement because the pore is lined by hydrophilic amino acid residues.

small, protein-lined pores provide permanent channels for diffusion of water molecules

molecules of the lipid bilayer are always on the move, producing tiny (short-lived) pores that are large enough for water molecules to pass through

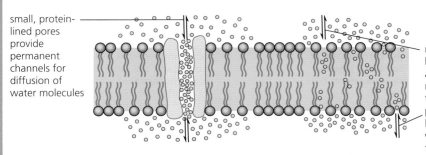

Figure 4.20 How polar water molecules cross the lipid bilayer

Movement by active transport

Active transport is the selective movement of molecules across a membrane. In active transport, metabolic energy produced by the cell and held as ATP is used to drive the transport of molecules and ions across membranes. Active transport has characteristic features distinctly different from those of movement by diffusion. These are as follows.

1 Active transport may occur against a concentration gradient

Molecules can be moved by active transport from a region of low concentration to a region of higher concentration. The cytosol of a cell normally holds some reserves of molecules valuable in metabolism, like nitrate ions in plant cells or calcium ions in muscle fibres. The reserves of useful molecules and ions do not escape; the cell surface membrane retains them inside the cell. However, when more useful molecules or ions become available for uptake, they are actively absorbed into the cells. This uptake occurs even though the concentration outside is lower than the concentration inside.

2 Active uptake is a highly selective process

For example, in a situation where potassium chloride, consisting of potassium (K^+) and chloride (Cl^-) ions, is available to an animal cell, it is the potassium ions that are more likely to be absorbed, since they are needed by the cell. Where sodium nitrate, consisting of sodium (Na^+) and nitrate (NO_3^-) ions, is available to a plant cell, it is likely that more of the nitrate ions are absorbed than the sodium ions, since this too reflects the needs of these cells.

3 Active transport involves special molecules of the membrane called 'pumps'

The pump molecule picks up particular molecules and transports them to the other side of the membrane where they are then released. The pump molecules are globular proteins that traverse the lipid bilayer. Movements by these pump molecules require reaction with ATP. By means of this reaction, metabolic energy is supplied to the process. Most membrane pumps are specific to particular molecules. This is the way *selective* transport is brought about. If the pump molecule for a particular substance is not present, the substance will not be transported.

Active transport is a feature of most living cells. We meet examples of active transport in the gut where absorption occurs, in the active uptake of ions by plant roots, in the kidney tubules where urine is formed, and in nerves fibres where an impulse is propagated. We discuss some of these examples later.

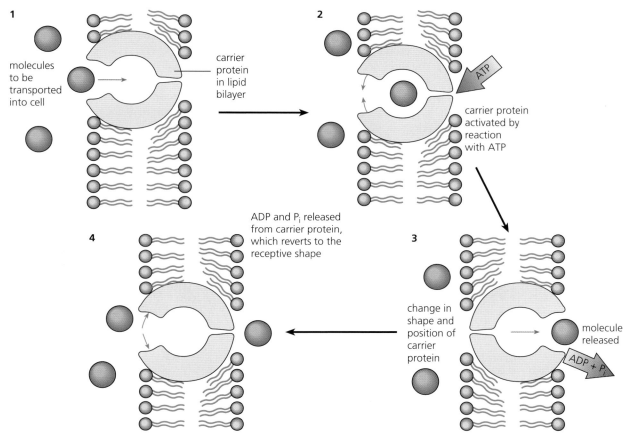

Figure 4.21 Active transport of a single substance

The protein pumps of cell surface membranes are of different types. Some transport a particular molecule or ion in one direction (Figure 4.21). Occasionally, two substances are transported in the same direction; for example, Na^+ and glucose. Others transport two substances (like Na^+ and K^+) in opposite directions (Figure 4.22).

A case study: the structure and function of sodium–potassium pumps in axons

A nerve impulse is transmitted along the axon of a nerve cell by a momentary reversal in electrical potential difference in the axon membrane, brought about by rapid movements of sodium and potassium ions. You can see the structure of a nerve cell and its axon in Figure 15.3, page 313.

Sodium–potassium pumps are globular proteins that span the axon membrane. In the preparation of the axon for the passage of the next nerve impulse, there is active transport of potassium (K^+) ions in across the membrane and sodium (Na^+) ions out across the membrane. This activity of the $Na^+–K^+$ pump involves transfer of energy from ATP. The outcome is that potassium and sodium ions gradually concentrate on opposite sides of the membrane. The steps to the cyclic action of these pumps are:

1 With the interior surface of the pump open to the interior of the axon, three sodium ions are loaded by attaching to specific binding sites.
2 The reaction of the globular protein with ATP now occurs, resulting in the attachment of a phosphate group to the pump protein. This triggers the pump protein to close to the interior of the axon and open to the exterior.
3 The three sodium ions are now released, and simultaneously, two potassium ions are loaded by attaching to specific binding sites.
4 With the potassium ions loaded, the phosphate group detaches. This triggers a reversal of the shape of the pump protein; it now opens to the interior, again, and the potassium ions are released.
5 The cycle is now repeated again.

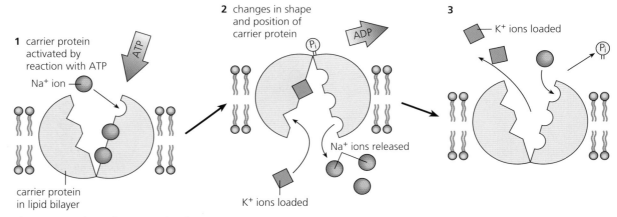

Figure 4.22 The sodium–potassium ion pump

Question

14 Samples of five plant tissue disks were incubated in dilute sodium chloride solution at different temperatures. After 24 hours it was found that the uptake of ions from the solutions were (in arbitrary units):

	Sodium ions	Chloride ions
Tissue at 5°C	80	40
Tissue at 25°C	160	80

Comment on how absorption of sodium chloride occurs, giving your reasons.

Movement by bulk transport

Bulk transport occurs by movements of **vesicles** of matter (solids or liquids) across the cell surface membrane. The flexibility of the fluid mosaic membrane makes this activity possible, and energy from metabolism (ATP) is required. Figure 4.23 shows this form of transport across a cell surface membrane, in action.

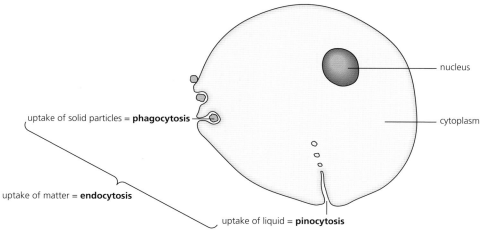

Figure 4.23 Movement by bulk transport

Questions

15 Phagocytic cells are also found in the airways of the lungs. What is their role and how do they carry it out?

16 Distinguish between the following pairs.

a proteins and lipids in cell membranes

b active transport and bulk transport

c hypotonic and hypertonic solutions

d endocytosis and exocytosis

Exocytosis is the process by which cells export products such as enzymes by means of vesicles. We have seen that vesicles may be budded off from the **Golgi apparatus** (Figure 1.16, page 18). The vesicles are then guided to the cell surface membrane by the network of microtubules in the cytosol. Here they merge with the membrane and their contents are discharged to the outside of the cell.

Alternatively, waste matter may be disposed of at the cell surface, in a similar way.

By the reverse process, **endocytosis**, substances may be imported into the cell. Here, fresh vesicles are formed at the cell surface. This occurs when part of the cell surface membrane forms a 'cup' around a particle or a drop of fluid, and encloses it to form a vesicle. This vesicle is then brought into the cytosol.

The wholesale import of solid matter in this way is termed **phagocytosis**. In the human body, there is a huge force of phagocytic cells, called the **macrophages**. Some of these mop up debris of damaged or dying cells and break it down. For example, we dispose of about 3×10^{11} red blood cells each day! All of these are ingested and disposed of by macrophages.

Pinocytosis is bulk import of fluids and occurs in the same way. Bulk uptake of lipids by cells lining the gut occurs as part of the digestion process, for example.

Exocytosis: secretion of materials out of cells by cytoplasmic vesicles fusing with the cell surface membrane and releasing the contents of the vesicle into the fluid around the cell, using ATP to move the cytoplasm.

Endocytosis: uptake of materials into cells by inward foldings of the cell surface membrane to form sacs of membrane that separate from the membrane to form vesicles within the cytoplasm, using energy from ATP to move the cytoplasm around. The process may involve liquid solutions or suspensions (pinocytosis) or solid macromolecules or cells (phagocytosis).

A summary of the functions of membrane proteins

We can now see that the proteins of the cell surface membrane, scattered about and across the lipid bilayer as they are, show diversity in function – all highly significant in the life of a cell. These functions are summarised in Figure 4.24.

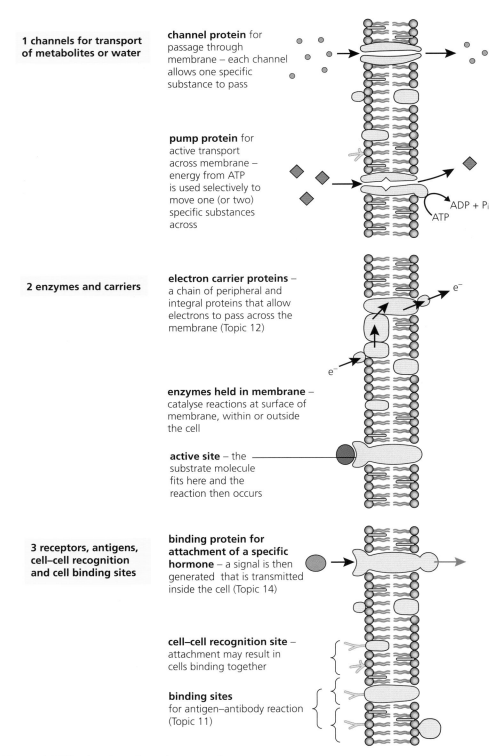

1 channels for transport of metabolites or water

channel protein for passage through membrane – each channel allows one specific substance to pass

pump protein for active transport across membrane – energy from ATP is used selectively to move one (or two) specific substances across

ADP + P_i
ATP

2 enzymes and carriers

electron carrier proteins – a chain of peripheral and integral proteins that allow electrons to pass across the membrane (Topic 12)

e^-

e^-

enzymes held in membrane – catalyse reactions at surface of membrane, within or outside the cell

active site – the substrate molecule fits here and the reaction then occurs

3 receptors, antigens, cell–cell recognition and cell binding sites

binding protein for attachment of a specific hormone – a signal is then generated that is transmitted inside the cell (Topic 14)

cell–cell recognition site – attachment may result in cells binding together

binding sites for antigen–antibody reaction (Topic 11)

Figure 4.24 The functions of membrane proteins – a summary

Summary

- The **cell surface membrane** is an **organelle** that surrounds and **contains the cell contents**. Across it there is a continuous **movement of the molecules** needed by cells or being disposed of by them.

- The membrane consists of **lipids and proteins** with a small amount of **carbohydrate**. The lipids are **phospholipids**, molecules with a hydrophilic head and a hydrophobic tail, organised into a bilayer with the heads on the outsides. The proteins are **globular proteins** arranged in, across and on the surface of the lipid bilayer. Short carbohydrate chains are attached to some of the proteins and lipids on the outside. In addition, a variable quantity of molecules of **cholesterol** occurs, inserted between some of the lipid molecules.

- The membrane is described as **a fluid mosaic**, as the components (particularly the lipids) move about laterally, relative to each other, and the proteins appear to be randomly scattered about. The cholesterol regulates membrane fluidity.

- There are three **mechanisms of transport** across membranes. In **diffusion** (and **osmosis**, a special case of diffusion) the energy comes from the kinetic energy of matter. In **active transport** (i.e. pumping) and in **bulk transport**, the energy is provided indirectly from cell metabolism, via ATP.

- **Diffusion** occurs from a region of high concentration to a region of low concentration. The lipid bilayer of the cell surface membrane acts as a barrier, at least to polar substances. Polar substances such as water diffuse through tiny pores or small spaces in the bilayer. Some molecules react with a membrane protein to open a specific channel (**facilitated diffusion**).

- The name given to the tendency of water molecules to move about is **water potential**. The Greek letter *psi* (symbol ψ) is used to represent water potential. Water moves from a region of higher water potential to a region of lower water potential – down a **water potential gradient**. Equilibrium is reached only if or when the water potential is the same in both regions. At this point, there would be no net movement of water from one region to another, but random movements of water molecules continue.

- **Solute molecules** in solution are surrounded by water molecules held by **hydrogen bonds**. In a concentrated solution, few water molecules are free to move. Therefore there is net diffusion of **free water molecules** in pure water or a dilute solution (free water molecules in high concentration), through a **partially permeable membrane** (such as the cell surface membrane), into a more concentrated solution (where free water molecules are in low concentration) – a process known as **osmosis**.

- **Active uptake** is a **selective process** and involves the accumulation of ions and organic substances **against a concentration gradient**. The process is powered by metabolic energy and involves highly specific **protein molecules in the membrane** that act as **pumps**.

- Bulk transport is the movement of the contents of **vesicles** (tiny, membrane-bound sacs) of matter, solid or liquid, into (**endocytosis**) or out of (**exocytosis**) the cell.

Examination style questions

1 Fig. 1.1 is a diagram of a cell surface membrane.

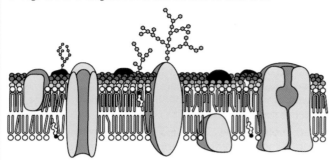

Fig. 1.1

a) Use a label line and the appropriate letter to label each of the following on a copy of Fig. 1.1.

P protein for active uptake of potassium ions

Q protein for facilitated diffusion of polar molecules

R receptor site for a hormone

S hydrophilic heads of phospholipids on the internal surface of the membrane

T molecule that modifies the fluidity of the membrane [5]

b) Some cells take in bacteria by endocytosis.

Explain how endocytosis occurs at a cell surface membrane. [3]

[Total: 8]

(Cambridge International AS and A Level Biology 9700, Paper 21 Q1 November 2011)

2 Fig. 2.1 shows a flatworm which lives in ponds, streams and rivers. The dimensions of the flatworm are 12.5 mm long by 3.0 mm wide. Its volume was estimated as 12.6 mm³. Flatworms do not have a transport system for the respiratory gases, oxygen and carbon dioxide.

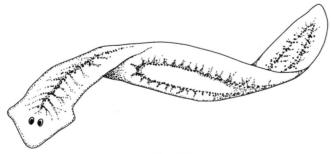

Fig. 2.1

a) With reference to Fig. 2.1 and the information above, explain how flatworms survive without a transport system for respiratory gases [4]

b) This flatworm lives in freshwater which has a low concentration of sodium ions. The flatworm's body fluids have a higher concentration of sodium ions than the surrounding water.

   i) Suggest how the flatworm retains sodium ions in its body fluids. [2]

   ii) State **one** role of sodium ions in organisms. [1]

[Total: 7]

(Cambridge International AS and A Level Biology 9700, Paper 22 Q3 June 2009)

3 Thirty strips of plant tissue of measured, standard length were cut, quickly washed, and then blotted dry. To each of three small beakers were added 10 strips and the tissue samples were immersed in sucrose solution of different concentrations. After a period of 30 minutes, the tissue strips were removed and their length remeasured. The results were as follows:

Beaker	1	2	3
Sucrose solution (mol dm³)	0.1	0.4	0.8
Change in length of tissue strips	increase	no change	decrease

a) Why were the tissue strips briefly washed and blotted dry before measurement? [1]

b) Why did the experimenter use a sample of 10 strips in each treatment? [2]

c) Explain what events in the cells of the tissue strips in beaker 1 during the experiment bring about their change in length. [6]

d) Explain what events in the cells of the tissue strips in beaker 3 during the experiment bring about their change in length. [6]

e) What can you conclude from the results from the tissue sample in beaker 2? [3]

f) For what reason is osmosis described as a special case of diffusion? [2]

[Total: 20]

4 Design and outline a laboratory experiment to investigate the effect of temperature on the cell surface membrane. Use samples of plant tissue that contains an intensely red, water-soluble pigment in the cell vacuoles (such as beetroot).

a) What steps would you take to ensure the significance of your results?

b) What outcomes to your experiment do you expect?

5 The mitotic cell cycle

When body cells reach a certain size they divide into two. Nuclear division occurs first, followed by division of the cytoplasm. The mitotic cell cycle of eukaryotes involves DNA replication followed by nuclear division. This ensures the genetic uniformity of all daughter cells.

5.1 Replication and division of nuclei and cells

During the mitotic cell cycle, DNA is replicated and passed to daughter cells.
   Stem cells in bone marrow and the skin continually divide by mitosis to provide a continuous supply of cells that differentiate into blood and skin cells.

By the end of this section you should be able to:

a) describe the structure of a chromosome, limited to DNA, histone proteins, chromatids, centromere and telomeres
b) explain the importance of mitosis in the production of genetically identical cells, growth, cell replacement, repair of tissues and asexual reproduction
c) outline the cell cycle, including interphase (growth and DNA replication), mitosis and cytokinesis
d) outline the significance of telomeres in permitting continued replication and preventing the loss of genes
e) outline the significance of mitosis in cell replacement and tissue repair by stem cells and state that uncontrolled cell division can result in the formation of a tumour

The division of cells

New cells arise by division of existing cells. In this process, the first step is for the nucleus to divide. The cytoplasm then divides around the daughter nuclei.

Unicellular organisms grow quickly under favourable conditions. They then divide in two. This cycle of growth and division is repeated rapidly, at least whilst conditions remain supportive.

In **multicellular organisms** the life cycle of individual cells is more complex. Here, life begins as a single cell which grows and divides, forming many cells. These eventually make up the adult organism. Certain of these cells retain the ability to grow and divide throughout life. They are able to replace old or damaged cells. However, the majority of the cells of multicellular organisms become specialised. Most are then unable to divide further.

The importance and role of the nucleus

The structure of the nucleus was introduced in Topic 1 (page 16). The nucleus is the organising centre of a cell, and has a double function. Firstly, it controls all the activities of the cell throughout life. Secondly, it is the location within the cell of the hereditary material, which is passed from generation to generation during reproduction. Both of the functions depend on 'information'. The information in the nucleus is contained within structures called **chromosomes**. These uniquely:

● control cell activities
● are copied from cell to cell when cells divide
● are passed into new individuals when sex cells fuse together in sexual reproduction.

So, the nucleus contains the chromosomes of the cell, and the chromosomes contain the coded instructions for the organisation and activities of cells and for the whole organism. It is on the structure of the chromosomes that we focus first.

Introducing the chromosomes

At the time a nucleus divides the chromosomes become compact, much-coiled structures. Only in this **condensed** state do the chromosomes become clearly visible in cells, provided they have been treated with certain dyes, At all other times, the chromosomes are very long, thin, uncoiled threads. In this condition they give the stained nucleus a granular appearance. The granules are called **chromatin**.

The information the nucleus holds on its chromosomes exists as a nucleic acid called **deoxyribonucleic acid** (**DNA**). DNA is a huge molecule made up of two paired strands in the form of a double helix (Figure 6.3, page 113). A single, enormously long DNA molecule runs the full length of each chromosome.

Look up the structure of DNA now.

The chromosomes are effectively a linear series of genes. A particular gene always occurs on the same chromosome in the same position. We can define 'gene' in different ways. For example, a gene is:
● a specific region of a chromosome that is capable of determining the development of a specific characteristic of an organism
● a specific length of the DNA double helix, hundreds or (more typically) thousands of base pairs long, which codes for a protein
● a unit of inheritance.

Meanwhile, there are four features of the chromosomes that are helpful to note at the outset:
● **The shape of a chromosome is characteristic**. Chromosomes are long, thin structures of a particular, fixed length. Somewhere along the length of the chromosome occurs a short, constricted region called the **centromere**. A centromere may occur anywhere along the chromosome, but it is always in the same position on any given chromosome. The position of the centromere, as well as the length of a chromosome is how they are identified in photomicrographs.
● **The number of chromosomes per species is fixed**. The number of chromosomes in the cells of different species varies, but in any one species the number of chromosomes per cell is normally constant. For example, the mouse has 40 chromosomes per cell, the onion has 16, humans have 46, and the sunflower 34. These are the chromosome numbers for the species. Please note that these are all even numbers.
● **Chromosomes occur in pairs**. The chromosomes of a cell occur in pairs, called homologous pairs (meaning 'similar in structure'). One of each pair came originally from each parent. So, for example, the human has 46 chromosomes, 23 coming originally from each parent in the process of sexual reproduction. This is why chromosomes occur in homologous pairs.
● **Chromosomes are copied**. Between nuclear divisions, whilst the chromosomes are uncoiled and cannot be seen, each chromosome is copied. This copying process occurs in the cell before nuclear division occurs. The two identical structures formed are called **chromatids**. The chromatids remain attached by their centromeres until they are divided during nuclear division. Then the **centromeres** and the chromatids separate. After this separation, the chromatids are recognised as chromosomes again. Of course, when chromosomes copy themselves the critical event is the copying of the DNA double helix that runs the length of the chromosome. This is known as **replication** of the DNA. You will discover how this replication process is brought about in Topic 6.

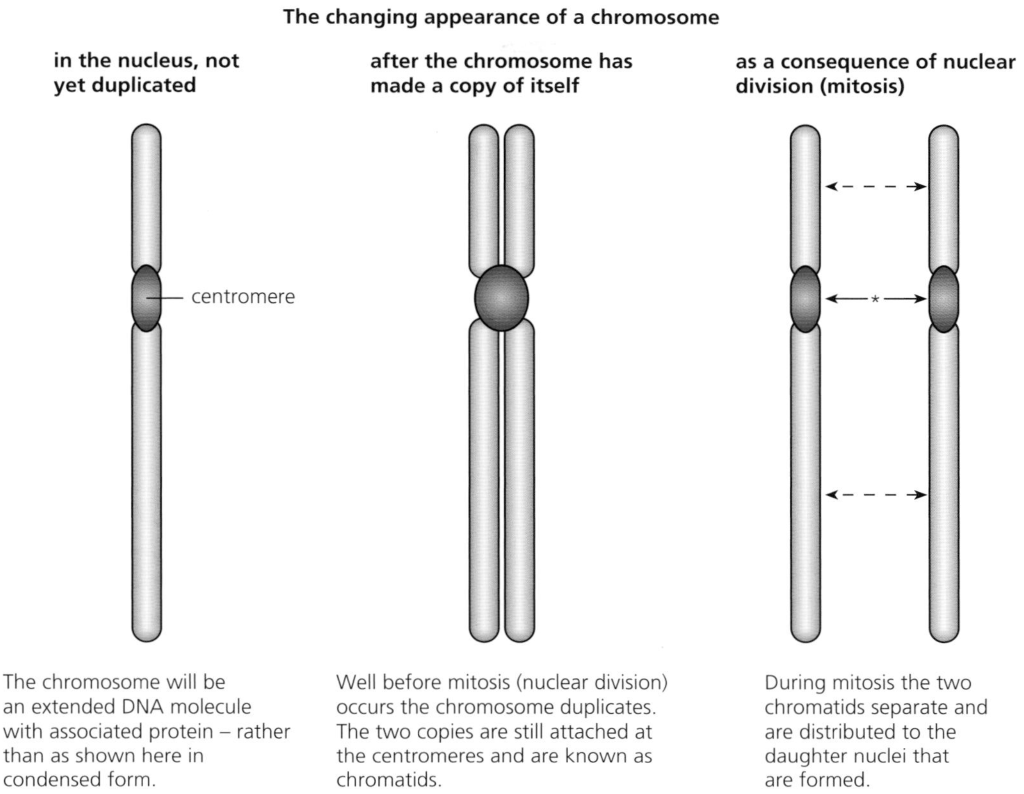

The changing appearance of a chromosome

in the nucleus, not yet duplicated	after the chromosome has made a copy of itself	as a consequence of nuclear division (mitosis)

centromere

The chromosome will be an extended DNA molecule with associated protein – rather than as shown here in condensed form.

Well before mitosis (nuclear division) occurs the chromosome duplicates. The two copies are still attached at the centromeres and are known as chromatids.

During mitosis the two chromatids separate and are distributed to the daughter nuclei that are formed.

Figure 5.1 One chromosome as two chromatids

Question

1 Explain why chromosomes occur in homologous pairs in cells.

The structure of chromosomes

We have already noted that each chromosome consists of a macromolecule of DNA in the form of a double helix. This runs the full length of the chromosome. However, it is supported by protein. About 50 per cent of a chromosome is built of protein, in fact. Some of these proteins are enzymes that are involved in copying and repair reactions of DNA. However, the bulk of chromosome protein has a support and packaging role for DNA.

Why is packaging necessary?

Well, take the case of human DNA. In each nucleus, the total length of the DNA of the chromosomes is over 2 metres. We know this is shared out between 46 chromosomes.

Chromosomes are of different lengths, but we can estimate that within a typical chromosome of 5 μm length, there is a DNA molecule approximately 5 cm long. This means that about 50 000 μm of DNA is packed into 5 μm of chromosome.

This phenomenal packaging problem is achieved by the much-coiled DNA double helix being looped around protein beads consisting of a packaging protein called **histone**. This is a basic (positively charged) protein containing a high concentration of amino acid residues with additional base groups ($-NH_2$), such as lysine and arginine. Eight histone molecules combine to make a single bead. Around each bead, the acidic (negatively charged) DNA double helix is wrapped in a double loop.

At times of nuclear division the whole beaded thread is itself coiled up, forming the chromatin fibre. The chromatin fibre is again coiled, and the coils are looped around a 'scaffold' protein fibre, made of a **non-histone protein**. This whole structure is folded (supercoiled) into the much-condensed chromosome, shown in Figure 5.2. This arrangement enables the safe storage of these phenomenal lengths of DNA that are packed in the nuclei. However, it also allows access to selected lengths of the DNA (particular genes) during transcription – a process you will encounter in Topic 6.

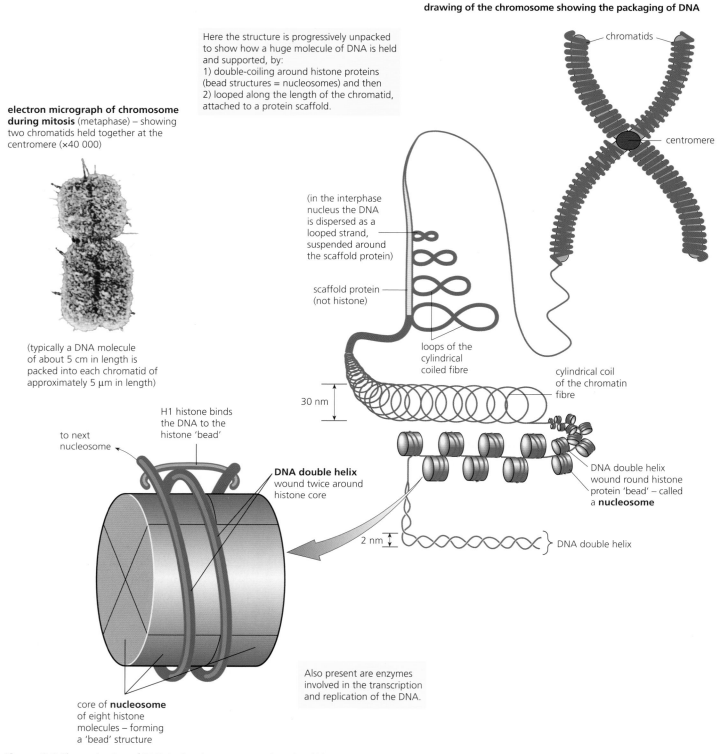

drawing of the chromosome showing the packaging of DNA

Here the structure is progressively unpacked to show how a huge molecule of DNA is held and supported, by:
1) double-coiling around histone proteins (bead structures = nucleosomes) and then
2) looped along the length of the chromatid, attached to a protein scaffold.

chromatids

centromere

electron micrograph of chromosome during mitosis (metaphase) – showing two chromatids held together at the centromere (×40 000)

(in the interphase nucleus the DNA is dispersed as a looped strand, suspended around the scaffold protein)

scaffold protein (not histone)

loops of the cylindrical coiled fibre

cylindrical coil of the chromatin fibre

(typically a DNA molecule of about 5 cm in length is packed into each chromatid of approximately 5 µm in length)

30 nm

H1 histone binds the DNA to the histone 'bead'

to next nucleosome

DNA double helix wound twice around histone core

DNA double helix wound round histone protein 'bead' – called a **nucleosome**

2 nm

DNA double helix

Also present are enzymes involved in the transcription and replication of the DNA.

core of **nucleosome** of eight histone molecules – forming a 'bead' structure

Figure 5.2 The packaging of DNA in the chromosome – the role of histone protein

Question

2 Deduce the significance of the positively charged histone protein and the negatively charged DNA.

This is the arrangement in the nuclei of eukaryotes. We can conclude that the single circular chromosome of prokaryotes does not present the same packaging problem, for in these organisms the DNA is described as 'naked'. It occurs without the support of any protein.

The nature and significance of telomeres

Each time the DNA of the chromosomes is copied (replicated), prior to mitosis and cell division, the ends of the DNA molecules cannot be copied. As a result, a few terminal nucleotide sequences are lost. With repeated replications comes progressive shortening on the DNA of each chromosome. If the ends were not protected in some way, chromosomes would lose genes or parts of genes. In fact the ends are protected by special, non-coding nucleotide sequences called telomeres. Typically, a telomere consists of a six-nucleotide sequence such as 'TTAGGG', repeated up to a thousand times. It is parts of this non-coding sequence that are lost at each replication of the DNA, without any harm to the genes the chromosome carries.

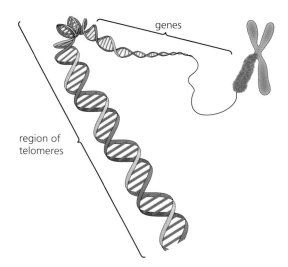

Figure 5.3 The position of telomeres

The chromosomes in nuclear division

Divisions of the nucleus are very precise processes, ensuring the correct distribution of chromosomes between the daughter cells. There are two types of nuclear division, known as mitosis and meiosis.

In **mitosis**, the daughter cells produced have the same number of chromosomes as the parent cell, typically two of each type and this is known as the **diploid (2*n*)** state. Mitosis is the nuclear division that occurs when cells grow or when cells need to be replaced and when an organism reproduces asexually.

In **meiosis**, the daughter cells contain half the number of chromosomes of the parent cell. That is, one chromosome of each type is present in the nuclei formed and this is known as the **haploid (*n*)** state. Meiosis is the nuclear division that occurs when sexual reproduction occurs, typically during the formation of the sex cells (called **gametes**) (see pages 345–8).

Whichever division takes place, it is normally followed by division of the cytoplasm to form separate cells. This process is called **cytokinesis**.

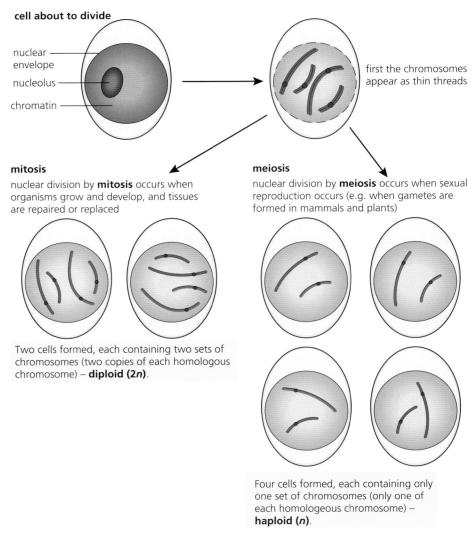

cell about to divide

nuclear envelope

nucleolus

chromatin

first the chromosomes appear as thin threads

mitosis

nuclear division by **mitosis** occurs when organisms grow and develop, and tissues are repaired or replaced

meiosis

nuclear division by **meiosis** occurs when sexual reproduction occurs (e.g. when gametes are formed in mammals and plants)

Two cells formed, each containing two sets of chromosomes (two copies of each homologous chromosome) – **diploid (2*n*)**.

Four cells formed, each containing only one set of chromosomes (only one of each homologeous chromosome) – **haploid (*n*)**.

Figure 5.4 Mitosis and meiosis: the significant differences

Question

3 a What is a haploid cell?
  b Where in the human body would you find cells undergoing:
   i meiosis
   ii mitosis?

Mitosis and the cell cycle

The cell cycle is the sequence of events that occur between one division and the next (Figure 5.5). It consists of three main stages: **interphase**, **nuclear division** and finally **cell division** (also called **cytokinesis**). The length of the cycle depends partly upon conditions external to the cell, such as temperature, supply of nutrients and oxygen. Its length also depends upon the type of cell. In cells at the growing point of a young stem or of a developing human embryo, the cycle is completed in less than 24 hours. The epithelium cells that line the gut typically divide every 10 hours. Liver cells divide every year or so. Nerve cells never divide again, after they have differentiated. Here the nucleus remains at interphase. In specialised cells the genes, other than those needed for the specific function, are 'switched off', so they cannot divide.

Interphase

Interphase is always the longest part of the cell cycle but it is of extremely variable length. At first glance the nucleus appears to be 'resting' but this is not the case at all. The chromosomes, previously visible as thread-like structures, have become dispersed. Now they are actively involved

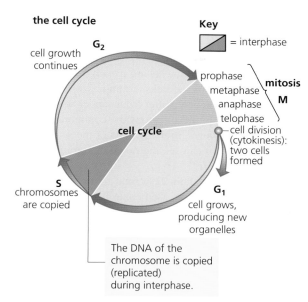

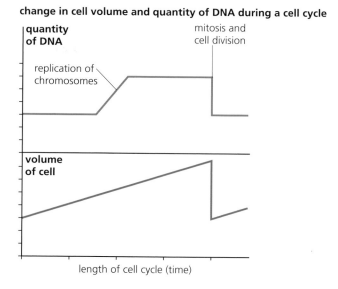

Figure 5.5 The stages of the cell cycle

in protein synthesis – at least for most of interphase. From the chromosomes, copies of the information of particular genes or groups of genes are taken in the form of ribonucleic acid (RNA), ('messenger RNA', page 123), for use in the cytoplasm. In the cytoplasm, ribosomes assemble proteins from amino acids, combining them in sequences dictated by the information from the gene. We return to this process later in this topic.

Look at Figure 5.5. The sequence of the events of interphase is illustrated there. The changes that occur are summarised in Table 5.1.

Table 5.1 The events of interphase

G_1: the 'first gap' phase	The cell grows by producing proteins and cell organelles, including mitochondria and endoplasmic reticulum.
S: synthesis	Growth of the cell continues as replication of DNA occurs (see Topic 6). Protein molecules called histones are synthesised and attach to the DNA. Each chromosome becomes two chromatids.
G_2: the 'second gap' phase	Cell growth continues by protein and cell organelle synthesis. Mitochondria and chloroplasts divide. The spindle begins to form.
Nuclear division, mitosis (M), follows.	

The significance of mitosis

Each daughter cell produced by mitosis has a full set of chromosomes, identical to those of the parent cell. No variation in genetic information arises by mitosis.

In the **growth and development of an embryo** it is essential that all cells carry the same genetic information (the same chromosomes) as the existing cells of the organism. Similarly, when **repair of damaged or worn out cells** occurs, the new cells must be exact copies of the cells being replaced. Mitosis ensures this. Otherwise, in growth, development and repair activities, different parts of the body of an organism might start working to conflicting blueprints. The results would be chaos!

In **asexual reproduction**, of which various forms exist, the offspring produced are identical to the parent since they are produced as a result of mitotic cell division. As a consequence, the offspring have all the advantages of the parents in mastering the same habitat – and any disadvantages, too. The offspring produced by asexual reproduction are often described as clones.

Question

4 What are the essential differences between mitosis and meiosis?

Cancer – diseases of uncontrolled cell division

Cancer is the result of the ill-regulated proliferation of cells, typically resulting in the formation of a solid tumour. There are many different forms of cancer, affecting different tissues of the body – cancer is not thought of as a single disease. Cancers are usually fatal if untreated. Today, many cancers are treatable, and in these cases, survival rates are improving markedly.

Today, in developed countries, one in three people suffer from cancer at some point in their life and approximately one in four will die from it. In these regions the commonest cancers are of the lung in males and of the breast in females. However, in many parts of the world, cancer rates are different – often they are significantly lower. You can see the range of common cancers and their incidences worldwide at: **http://globocan.iarc.fr/**

So, cancer arises when the cell cycle operates without its normal controls. In a healthy cell the cell cycle is regulated by a molecular control system, itself controlled by specific genes. However, cells may start to divide by mitosis, repeatedly, without control or regulation. Now the rate of cell multiplication is very much faster than the rate of cell death, and an irregular mass of cells, a tumour, is formed (Figure 5.6).

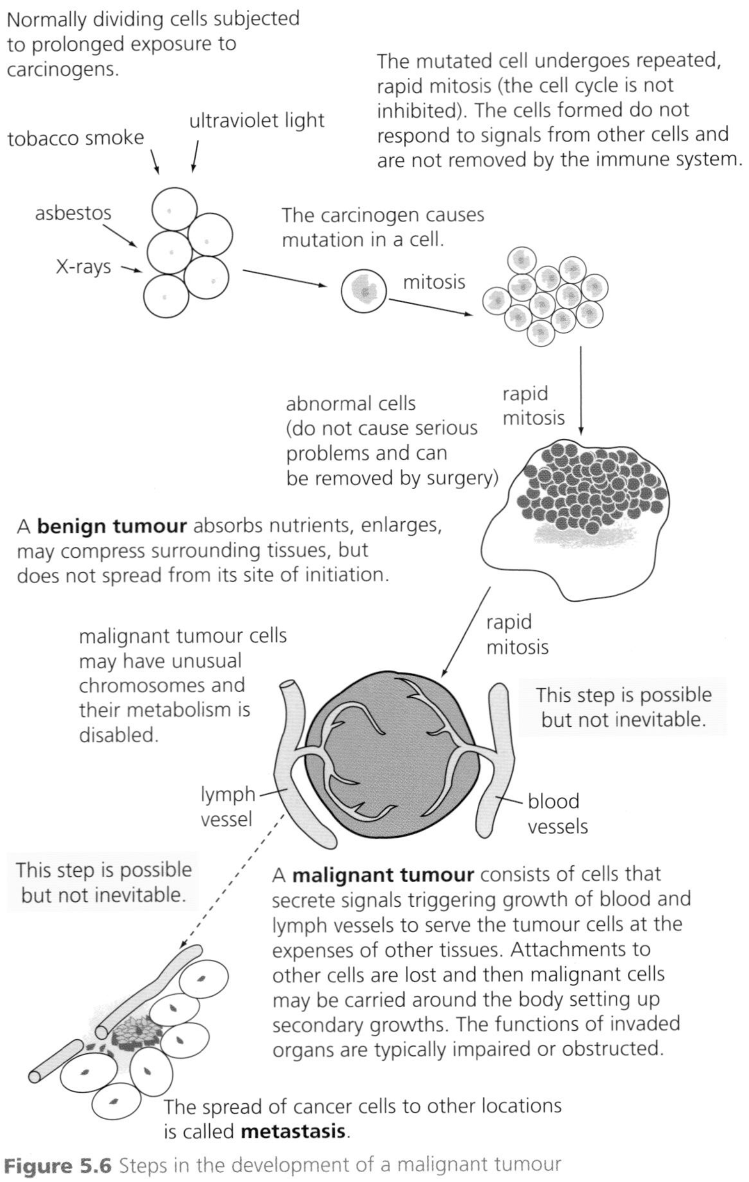

Cancers start when changes occur in the genes that regulate the cell cycle. This is where carcinogens may enter the story. A carcinogen is any agent that may cause cancer. Some highly significant carcinogens are identified in Table 5.2.

So, carcinogens are highly likely to cause damage to the DNA molecules of chromosomes. The result is a mutation – a change in the amount or chemical structure of DNA of a chromosome. Mutations of different types build up in the DNA of body cells over time. A single mutation is unlikely to be responsible for triggering cancer. With time, mutations may accumulate in the body and then trigger a disease. This explains why the majority of cancers arise in older people. Also, mutation may occur in the cells of different body tissues, and hence cancer is not one single disease.

Whatever the cause, two types of genes play a part in initiating a cancer, if they mutate:

- **Proto-oncogenes**. These are genes that code for the proteins that play a part in the control of the cell cycle. When a proto-oncogene mutates, an oncogene results. In this state, the gene permanently switches on cell division, causing cells in the tissue concerned to become 'immortal', if their nutrient supply is maintained.

- **Tumour-suppressing genes**. These are normal genes that either slow down cell division, repair mistakes in DNA when they arise, or instruct the cell to die (by programmed cell death). These genes, too, are vulnerable to carcinogens. For example, a mutation may inactivate the gene that codes for a protein known as p53. When protein p53 is present, copying of faulty DNA is stopped. Then, other enzymes are able to repair the DNA and correct the fault, or programmed cell death is switched on in the faulty cell, so cancer is avoided. Abnormalities in the p53 gene have been found in more than 50 per cent of human cancers.

Normally dividing cells subjected to prolonged exposure to carcinogens.

tobacco smoke · ultraviolet light · asbestos · X-rays

The carcinogen causes mutation in a cell.

mitosis

The mutated cell undergoes repeated, rapid mitosis (the cell cycle is not inhibited). The cells formed do not respond to signals from other cells and are not removed by the immune system.

abnormal cells (do not cause serious problems and can be removed by surgery)

rapid mitosis

A **benign tumour** absorbs nutrients, enlarges, may compress surrounding tissues, but does not spread from its site of initiation.

malignant tumour cells may have unusual chromosomes and their metabolism is disabled.

rapid mitosis

This step is possible but not inevitable.

lymph vessel · blood vessels

This step is possible but not inevitable.

A **malignant tumour** consists of cells that secrete signals triggering growth of blood and lymph vessels to serve the tumour cells at the expenses of other tissues. Attachments to other cells are lost and then malignant cells may be carried around the body setting up secondary growths. The functions of invaded organs are typically impaired or obstructed.

The spread of cancer cells to other locations is called **metastasis**.

Figure 5.6 Steps in the development of a malignant tumour

Table 5.2 Common carcinogens known to increase mutation rates and the likelihood of cancer

Ionising radiation	Ionising radiation includes X-rays and radiation (gamma rays, α particles, β particles) from various radioactive sources. These may trigger the formation of damaging ions inside the nucleus – leading to the break-up of the DNA.
Non-ionising radiation	Non-ionising radiations such a UV light. This is less penetrating than ionising radiation, but if it is absorbed by the nitrogenous bases of DNA, may modify it – causing adjacent bases on the DNA strand to bind to each other, instead of binding to their partner on the opposite strand (page 108).
Chemicals	Several chemicals that are carcinogens are present in tobacco smoke (page 171). Also, prolonged exposure to asbestos fibres may trigger cancer in the linings of the thorax cavity (pleural membranes, Figure 9.XX, page 000) – the harm usually becomes apparent only many years later.

Question

5 Describe three environmental conditions that may cause normal cells to become cancerous cells.

5.2 Chromosome behaviour in mitosis

The events that occur during mitosis can be followed by using a light microscope.

By the end of this section you should be able to:

a) describe, with the aid of photomicrographs and diagrams, the behaviour of chromosomes in plant and animal cells during the mitotic cell cycle and the associated behaviour of the nuclear envelope, cell surface membrane and the spindle

b) observe and draw the mitotic stages visible in temporary root tip squash preparations and in prepared slides of root tips of species such as those of *Vicia faba* and *Allium cepa*

Mitosis, the steps

In mitosis, the chromosomes, present as the chromatids formed during interphase, are separated and accurately and precisely distributed to two daughter nuclei. Here, mitosis is presented and explained as a process in four phases but this is for convenience of description only. Mitosis is one continuous process with no breaks between the phases.

Follow the phases of mitosis in Figure 5.8.

In **prophase** the chromosomes become visible as long thin threads. During interphase the DNA had been extended to allow for activity of the genes and for replication. Now, in nuclear division, the DNA must become tightly packed for the chromosomes to move about and to separate. They increasingly shorten and thicken by a process of supercoiling. Only at the end of prophase is it possible to see that the chromosomes consist of two chromatids held together at the centromere. At the same time, the nucleolus gradually disappears and the nuclear envelope breaks down. In animals, the **centrosome** divides and the two centrioles replicate (make copies) to form two centrosomes.

In **metaphase** the two centrosomes move to opposite ends of the cell. Microtubules of the cytoplasm start to form into a spindle, radiating out from the centrioles (page 19). Each pair of chromatids is attached to a microtubule of the spindle and is arranged at the equator of the spindle. (Note: in plant cells, a spindle of exactly the same structure is formed but without the presence of the centrosomes).

In **anaphase** the centromeres separate, the spindle fibres shorten and the chromatids are pulled by their centromeres to opposite poles. Once separated, the chromatids are referred to as chromosomes.

In **telophase** a nuclear envelope reforms around both groups of chromosomes at opposite ends of the cell. The chromosomes 'decondense' by uncoiling, becoming chromatin again. The nucleolus reforms in each nucleus. Interphase follows division of the cytoplasm.

Question

6 As the contents of a nucleus in the human body prepares to undergo mitosis, how many chromatids does it contain?

Cytokinesis

Cytokinesis is the division of the cytoplasm that follows telophase. During cytokinesis, cell organelles such as mitochondria and chloroplasts become evenly distributed between the cells. In animal cells, the cytoplasm separates by in-tucking of the cell surface membrane at the equator of the spindle, 'pinching' the cytoplasm in half (Figure 5.8).

In plant cells, the Golgi apparatus forms vesicles of new cell wall materials which collect along the line of the equator of the spindle, known as the cell plate. Here the vesicles merge, forming the new cell surface membranes and the cell walls between the two cells (Figure 5.7).

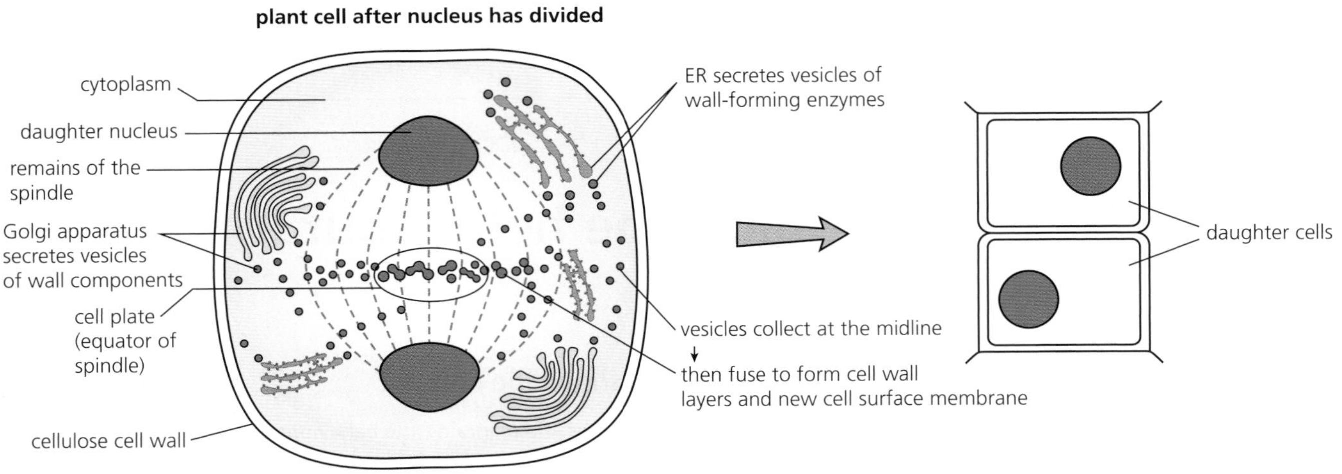

plant cell after nucleus has divided

Figure 5.7 Cytokinesis in a plant cell

Mitosis – a summary

Daughter cells produced by mitosis have a set of chromosomes identical to each other and to the parent cell from which they were formed because:

- an **exact copy** of each chromosome is made by accurate copying during interphase, when two chromatids are formed
- **chromatids remain attached** by their centromeres during metaphase of mitosis, when each becomes attached to a spindle fibre at the equator of the spindle
- centromeres then separate during anaphase and the chromatids of each pair are pulled apart to **opposite poles** of the spindle. Thus, one copy of each chromosome moves to the poles of the spindle
- the chromosomes at the poles **form the new nuclei** – two to a cell at this point
- two cells are then formed by separation of the cytoplasm at the mid-point of the cell, **each with an exact copy of the original nucleus**.

Question

7 Using slides they had prepared to observe chromosomes during mitosis in a plant root tip (Figure 5.9), five students observed and recorded the number of nuclei at each stage in mitosis in 100 cells.

Stage of mitosis	Number of nuclei counted				
	Student 1	Student 2	Student 3	Student 4	Student 5
Prophase	64	70	75	68	73
Metaphase	13	10	7	11	9
Anaphase	5	5	2	8	5
Telophase	18	15	16	13	13

a Calculate the mean percentage of dividing cells at each stage of mitosis. Present your results as a pie chart.

b Assuming that mitosis takes about 60 minutes to complete in this species of plant, deduce what these results imply about the lengths of the four steps?

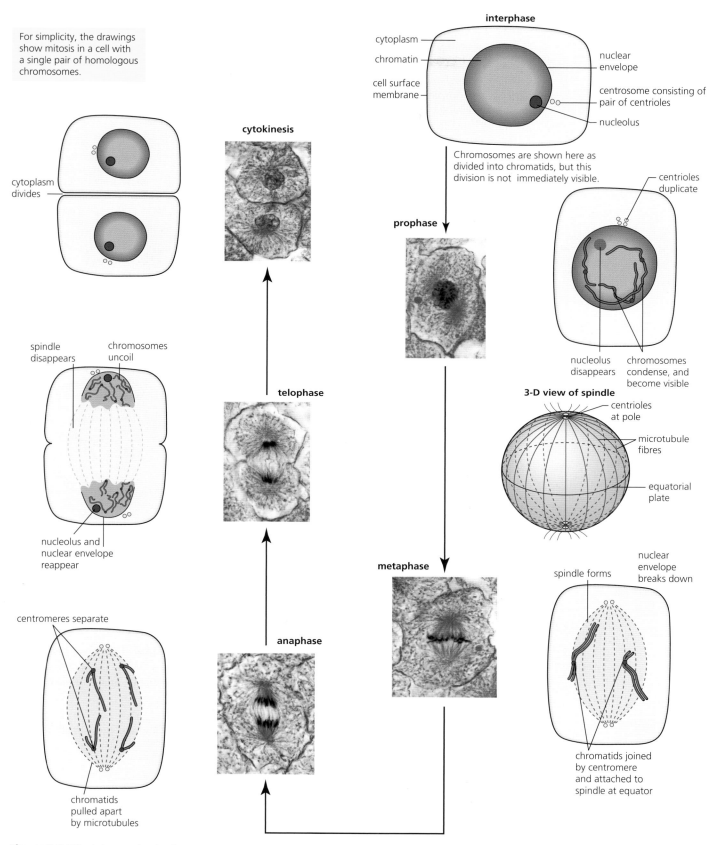

For simplicity, the drawings show mitosis in a cell with a single pair of homologous chromosomes.

interphase

cytoplasm

chromatin

cell surface membrane

nuclear envelope

centrosome consisting of pair of centrioles

nucleolus

Chromosomes are shown here as divided into chromatids, but this division is not immediately visible.

prophase

centrioles duplicate

nucleolus disappears

chromosomes condense, and become visible

3-D view of spindle

centrioles at pole

microtubule fibres

equatorial plate

cytokinesis

cytoplasm divides

telophase

spindle disappears

chromosomes uncoil

nucleolus and nuclear envelope reappear

anaphase

centromeres separate

chromatids pulled apart by microtubules

metaphase

spindle forms

nuclear envelope breaks down

chromatids joined by centromere and attached to spindle at equator

Figure 5.8 Mitosis in an animal cell

Observing chromosomes during mitosis

Actively dividing cells, such as those at the growing points of the root tips of plants, include many cells undergoing mitosis. This tissue can be isolated, stained with an ethano-orcein (aceto-orcein) stain, squashed and then examined under the high power of the microscope. Nuclei at interphase appear red–purple with almost colourless cytoplasm. The chromosomes in cells undergoing mitosis will be visible, too, rather as they appear in the electron micrographs in Figure 5.8. The procedure is summarised in the flow diagram in Figure 5.9.

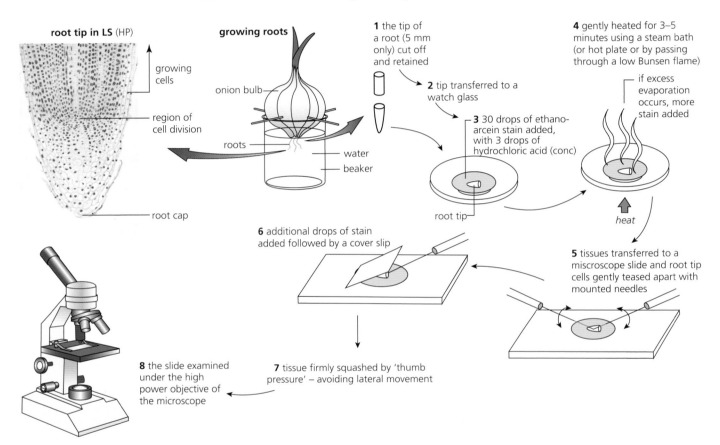

Figure 5.9 Preparing an onion root tip squash with ethano-orcein stain

Drawing the stages of mitosis, using temporary mounts

The temporary mounts you have produced from root tip squashes are likely to contain one or more cells at each stage of mitosis, when carefully examined.

Follow the instructions on biological drawing, on page 8, in producing an image of each stage. These should show, drawn to scale:

- the cell surface membrane
- the presence or absence of the nuclear envelope, the spindle fibres and the equator of the spindle, according to the stage of mitosis observed
- the position, number and shapes of the chromatids or chromosomes
- evidence of early stages in cytokinesis.

Note that, as you are observing plant cells, the centrioles seen in the photomicrographs in Figure 5.8 will be absent.

Summary

- The **nucleus** controls the activities of the cell throughout life. It is the location of the hereditary material which is passed from generation to generation in reproduction. The division of the nucleus is the first step in **cell division**.

- **Chromosomes** occur in the nucleus but are only visible, when stained, at times of nuclear division. The number of chromosomes per nucleus in each species is fixed. Chromosomes occur in pairs and have a characteristic shape.

- **Mitosis** is the replicative nuclear division in which, after cell division, the nucleus present in each of two daughter cells formed has exactly the same number of chromosomes as the parent cell, typically two of each type. It is in the **diploid (2n)** state. Mitosis is associated with cell divisions of growth and asexual reproduction.

- **Nucleic acids** are long, thread-like **macromolecules** with a uniform and unvarying 'backbone' of alternating sugar and phosphate molecules. Attached to each sugar molecule is a nitrogenous base which projects sideways. Since the bases vary, they represent a unique sequence that carries the **coded information** held by the nucleic acid.

- **DNA** exists as two strands arranged as a **double helix**. It occurs in the chromosomes in a highly coiled state, supported by a **protein framework**. Within a chromosome, specific lengths of the DNA code for particular polypeptides and are called **genes**.

- **Replication of the DNA** of the chromosomes occurs in the **interphase** of the **cell cycle** when the chromosomes are less coiled, well before nuclear division occurs.

- Each time the DNA of the chromosomes is copied (replicated) prior to mitosis and cell division, a part of the end of the DNA molecules of chromosomes is lost. Ultimately, chromosomes would lose genes or parts of genes by this process, if they were not protected by the presence of **telomeres**. Telomeres are long, seemingly disposable non-coding sequences of nucleotide bases at the ends of the DNA molecules of chromosomes that fulfil this important function.

- **Mutations** are unpredictable changes in the genetic make-up of a cell involving the sequence of particular bases at one location (gene mutation).

- **Cancers** are initiated when cells start to divide repeatedly, by mitosis, without control or regulation. An irregular mass of cells is formed, called a **tumour**. There are many forms of the disease, affecting different tissues of the body. Often cancers are the result of numerous chance mutations triggered by agents known as **carcinogens**. These include ionising radiation, tobacco smoke.

Examination style questions

1 a) When does mitosis occur in the life cycle of an animal from its development from a fertilised egg cell until the point when it forms gametes? [2]

b) What do you understand by the 'cell cycle'? [2]

c) By means of fully annotated diagrams explain the difference between the phases of the cycle and the events that characterise each phase. [12]

d) What changes occur during the phases of the cell cycle to

   i) the volume of a cell

   ii) the quantity of DNA during the phases of the cell cycle? [4]

   [Total: 20]

2 a) Explain how uncontrolled cell division can result in the formation of a tumour.

b) What do you understand by non-malignant and malignant tumours?

c) Outline the range of factors that can increase the chances of cancerous growths.

3 Muntjac are small deer found throughout Asia. Cells at the base of the epidermis in the skin continually divide by mitosis. Fig. 3.1 shows the chromosomes from a skin cell of a female Indian muntjac deer at metaphase of mitosis.

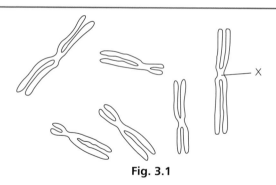

Fig. 3.1

a) i) State the diploid chromosome number of the female Indian muntjac deer. [1]

   ii) Name **X** and state its role in mitosis. [2]

   iii) On a copy of Fig. 3.1, **shade in** a pair of homologous chromosomes. [1]

   iv) Draw one of the chromosomes shown in Fig. 3.1 as it would appear during **anaphase** of mitosis. [2]

b) Outline what happens to **a chromosome** between the end of anaphase and the start of the next mitosis. [3]

c) During the formation of eggs in the ovary of the female Indian muntjac deer, the chromosome number changes.

   State what happens to the chromosome number and explain why this change is necessary. [2]

   [Total: 11]

(Cambridge International AS and A Level Biology 9700, Paper 02 Q3 June 2007)

6 Nucleic acids and protein synthesis

Nucleic acids have roles in the storage and retrieval of genetic information and in the use of this information to synthesise polypeptides. DNA is an extremely stable molecule that cells replicate with extreme accuracy. The genetic code is used by cells for assembling amino acids in correct sequences to make polypeptides. In eukaryotes this involves the processes of transcription in the nucleus to produce short-lived molecules of messenger RNA followed by translation in the cytoplasm.

↻ 6.1 Structure and replication of DNA

Understanding the structure of nucleic acids allows an understanding of their role in the storage of genetic information and how that information is used in the synthesis of proteins.

By the end of this section you should be able to:

a) describe the structure of nucleotides, including the phosphorylated nucleotide ATP
b) describe the structure of RNA and DNA and explain the importance of base pairing and the different hydrogen bonding between bases
c) describe the semi-conservative replication of DNA during interphase

Nucleic acids – the information molecules

Nucleic acids are the **information molecules** of cells; they are the genetic material of all living organisms and also of viruses. Within the structure of nucleic acid are coded the 'instructions' that govern all cellular activities. This code (known as the **genetic code**) is a universal one – it makes sense in all organisms.

There are two types of nucleic acids found in living cells, **deoxyribonucleic acid (DNA)**, and **ribonucleic acid (RNA)**. DNA is the genetic material and occurs in the chromosomes of the nucleus. But, whilst some RNA also occurs in the nucleus, most is found in the cytoplasm – particularly in the ribosomes. Both DNA and RNA have roles in the day-to-day control of cells and organisms, as we shall see shortly.

Nucleic acids are giant molecules known as **macromolecules**. Chemically they are described as **polymers** because they are made by linking together smaller building blocks called **monomers**. These smaller molecules that make up DNA and RNA are the **nucleotides**, and it is these molecules that are condensed together to form long, thread-like nucleic acids. First you will look at the structure of the nucleotides, themselves.

Nucleotides–the building blocks

A nucleotide consists of three components combined together:
- a **nitrogenous base**, either cytosine (C), guanine (G), adenine (A), thymine (T) or uracil (U)
- a **pentose sugar**, either deoxyribose (in DNA) or ribose (in RNA)
- **phosphoric acid**.

The nitrogenous bases are derived from one of two parent compounds, purine or pyrimidine. The purine bases are **cytosine** or **guanine**; the pyrimidine bases are **adenine**, **thymine** or **uracil**. These molecules are illustrated in Figure 6.1. You do not need to know their structural formulae. However, you should note that purines have a double ring structure and pyrimidines have a single ring structure. We will note the special significance of this size difference, shortly. It has become traditional in biology to abbreviate the names of these five bases to their first letters, **C**, **G**, **A**, **T** and **U**.

The three components, nitrogenous base, pentose sugar and phosphoric acid, are combined by **condensation reactions** to form a nucleotide with the formation of two molecules of water. The structure of these components and how they are combined together are illustrated in Figure 6.1. Here a diagrammatic way of representation has been used to emphasise their spatial arrangement.

the components:

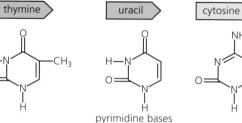

condensation to form a nucleotide:

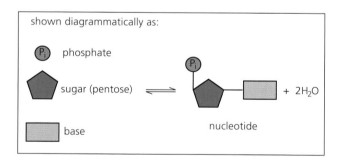

Question

1 Distinguish between a 'nitrogenous base' and a 'base' found in inorganic chemistry.

Figure 6.1 The components of nucleotides

ATP – a nucleotide with unusual features

Energy made available within the cytoplasm may be transferred to a molecule called **adenosine triphosphate** (**ATP**). This substance occurs in all cells at a concentration of 0.5–2.5 mg cm^{-3}. It is a relatively small, soluble organic molecule – a **nucleotide**, which carries three phosphate groups linked together in a linear sequence (Figure 1.22, page 21).

Look at Figure 1.22 now.

ATP is formed from **adenosine diphosphate** (**ADP**) and a **phosphate ion** (**Pi**) by transfer of energy from other reactions. ATP is referred to as '**the universal energy currency**' because, like money, it can be used in different contexts, and it is constantly recycled. ATP contains a good deal of chemical energy locked up in its structure. What makes ATP special as a reservoir of stored chemical energy is its role as a common intermediate between energy-yielding reactions and energy-requiring reactions and processes.

Energy-yielding reactions include the photophosphorylation reactions of photosynthesis (page 267), and many of the reactions of cell respiration (page 243).

Energy-requiring reactions include the synthesis of cellulose from glucose, the synthesis of proteins from amino acids and the contraction of muscle fibres.

The free energy available in ATP is approximately 30–$34\,kJ\,mol^{-1}$, made available in the presence of a specific enzyme, referred to as ATPase. Some of this energy is lost as heat in a reaction, but much free energy is made available to do useful work, more than sufficient to drive a typical energy-requiring reaction of metabolism.

- Sometimes ATP reacts with water (a hydrolysis reaction) and is converted to ADP and Pi. Direct hydrolysis of the terminal phosphate groups like this happens in muscle contraction, for example.
- Mostly, ATP reacts with other metabolites and forms phosphorylated intermediates, making them more reactive in the process. The phosphate groups are released later, so both ADP and Pi become available for reuse as metabolism continues.

Nucleotides become nucleic acid

Nucleotides themselves then combine together, one nucleotide at a time, to form huge molecules called nucleic acids or **polynucleotides** (Figure 6.2). This occurs by a condensation reaction between the phosphate group of one nucleotide and the sugar group of another nucleotide. Up to 5 million nucleotides, condensed together, form a polynucleotide. So, nucleic acids are very long, thread-like macromolecules with alternating sugar and phosphate molecules forming the 'backbone'. This part of the nucleic acid molecule is uniform and unvarying. However, also attached to each of the sugar molecules along the strand is one of the bases, and these project sideways. Since the bases vary, they represent a unique sequence that carries the coded information held by the nucleic acid.

Introducing DNA

The **DNA molecule** consists of two polynucleotide strands, paired together. These strands are held by hydrogen bonds between their paired bases. The two strands are of the order of several million nucleotides in length and they take the shape of a **double helix** (Figure 6.3). In all DNA molecules the nucleotides contain deoxyribose.

The other chemical feature that distinguishes DNA from RNA is the organic bases present and the way they pair up. In DNA the nitrogenous bases are **cytosine (C)**, **guanine (G)**, **adenine (A)** and **thymine (T)**, but never uracil (U). The pairing of bases is between adenine and thymine (**A** and **T**) and between cytosine and guanine (**C** and **G**). This is because there is just enough space between the two sugar–phosphate backbones for one purine and one pyrimidine molecule. So a purine in one strand must be opposite a pyrimidine in the other, and vice versa. These are the only combinations that fit together along the helix. Pairing of bases is accompanied by the formation of hydrogen bonds; two are formed between adenine and thymine and three are formed between cytosine and guanine. This pairing, known as **complementary base pairing**, also makes possible the very precise way that DNA is copied in a process called replication.

We have noted there is direction in a nucleic acid molecule. *Look again at Figure 6.2 to make certain you understand what 'direction' means.*

So, in the DNA molecule, the phosphate groups along each strand are bridges between carbon-3 of one sugar molecule and carbon-5 of the next. Furthermore, within the DNA double helix the two chains are arranged **antiparallel**. In Figure 6.3 you can see that one chain runs from 5′ to 3′ whilst the other runs from 3′ to 5′ – which is what 'antiparallel' means.

The existence of direction in DNA strands becomes important in replication and also when the genetic code is transcribed into messenger RNA.

Direction in the nucleic acid molecule:

in a nucleic acid molecule, the phosphate groups are bridges between **carbon-3** of one sugar molecule and **carbon-5** of the next. Consequently, we can identify '*direction*' in the nucleic acid molecule, indicating one end as 5′ and the other end as 3′.

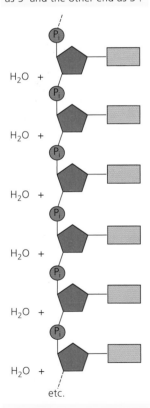

When the polynucleotide chain grows by addition of another nucleotide, the nucleotides are always added to the 3′ end.

Figure 6.2 'Direction' in the nucleic acid molecule

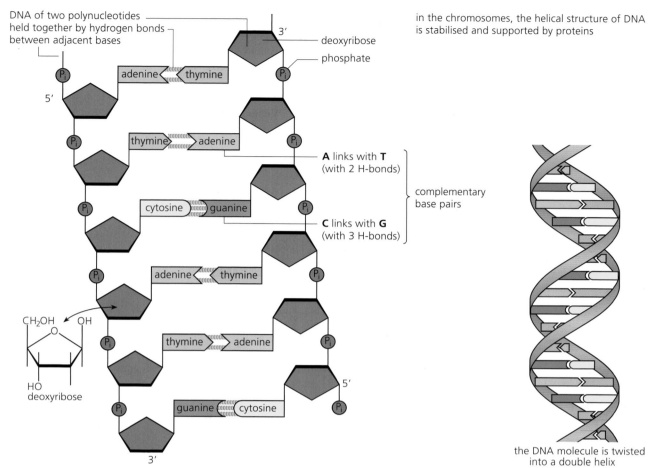

Figure 6.3 The structure of DNA

in the chromosomes, the helical structure of DNA is stabilised and supported by proteins

the DNA molecule is twisted into a double helix

Figure 6.5 The structure of RNA

Introducing RNA

The **RNA molecule** is relatively short in length, compared with DNA. In fact, RNAs tend to be between one hundred and one thousand nucleotides long, depending on the particular role they have. In all RNA molecules the nucleotides contain ribose.

The other chemical feature that distinguishes RNA is that the bases are cytosine (C), guanine (G), adenine (A) and **uracil (U)**, but never thymine (T). RNA occurs as a single strand.

In the 'information business' of cells there are two functional types of RNA. These are **messenger RNA (mRNA)** and **transfer RNA (tRNA)**. Their roles are in the transfer of information from the nucleus to the cytoplasm and in the construction of proteins at the ribosomes, located in the cytoplasm. We will return to this process, shortly.

Table 6.1 The differences between DNA and RNA

	DNA	**RNA**
Length	Very long strands, several million nucleotides long	Relatively short strands
Sugar	Deoxyribose	Ribose
Bases	C, G, A and T (not U)	C, G, A and U (not T)
Forms	Consists of two polynucleotide strands in the form of a double helix with complementary base pairs: • C with G and A with T • held by hydrogen bonds.	Consists of single strands and exists in two functional forms: • messenger RNA (mRNA) • transfer RNA (tRNA).

Extension

The existence of the DNA double helix was discovered, and the way DNA holds information was suggested by Francis Crick and James Watson in 1953, for which a Nobel Prize was awarded (Figure 6.4).

Francis Crick (1916–2004) and **James Watson** (1928–) laid the foundations of a new branch of biology – cell biology – and achieved this while still young men. Within two years of their meeting in the Cavendish Laboratory, Cambridge (1951), Crick and Watson had achieved their understanding of the nature of the gene in chemical terms.

Crick and Watson brought together the experimental results of many other workers, and from this evidence they deduced the likely structure of the DNA molecule.

- **Erwin Chergaff** measured the exact amount of the four organic bases in samples of DNA, and found the ratio of A:T and of C:G was always close to 1. Chergaff's results suggest consistent base pairing in DNA from different organisms.

Organism	Ratio of bases in DNA samples	
	Adendine : Thymine	**Guanine : Cytosine**
Cow	1.04	1.00
Human	1.00	1.00
Salmon	1.02	1.02
Escherichia coli	1.09	0.99

- **Rosalind Franklin** and **Maurice Wilkins** produced X-ray diffraction patterns by bombarding crystalline DNA with X-rays.

Crick and Watson concluded that DNA is a double helix consisting of:
- two polynucleotide strands with nitrogenous bases stacked on the inside of the helix (like rungs on a twisted ladder)
- parallel strands held together by hydrogen bonds between the paired bases (A–T, C–G)
- ten per turn of the helix
- two antiparallel strands of a double helix.

They built a model. See Figure 6.3 for a simplified model of the DNA double helix.

a) Watson and Crick with their demonstration model of DNA

b) Rosalind Franklin produced the key X-ray diffraction pattern of DNA at Kings College, London

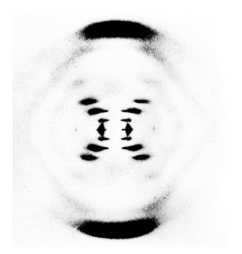

c) X-ray diffraction pattern of DNA

Figure 6.4 The discovery of the role of DNA

Replication – how DNA copies itself

Some features of the DNA of the chromosome and how it is copied or **replicated** have already been established.

- Replication must be an extremely accurate process since DNA carries the genetic message.
- Replication is quite separate from cell division, for replication of DNA takes place in the nucleus during **interphase**, well before the events of nuclear division.
- Strands of the DNA double helix are built up individually from free **nucleotides**. The structure of nucleotides is shown in Figure 6.1.

Before nucleotides can be replicated, the DNA double helix has to **unwind** and the hydrogen bonds holding the strands together must be broken, allowing the two strands of the helix to separate. The enzyme **helicase** brings about the unwinding process and holds the strands apart for replication to occur. Each point along the DNA molecule where this occurs is called a **replication fork** (Figure 6.6). At a replication fork, both strands act as **templates**. A molecule of the enzyme DNA polymerase binds to each of the DNA strands and begins to move along them. As each base along a strand is reached, a free nucleotide that is complementary is paired with it (A with T, C with G). These free nucleotides are already 'activated' by the presence of extra phosphates. Each free nucleotide is held in place by the hydrogen bonds that form between complementary bases. Then, from each free nucleotide, the extra phosphates are broken off. This provides the energy by which the remaining phosphate is joined to the sugar molecule of the neighbouring nucleotide. This condensation reaction is brought about by DNA polymerase. The new strands of DNA 'grow' in this way.

DNA polymerase moves along the single strands of DNA in the 5′ → 3′ direction. If you look at the illustration of replication in Figure 6.6, you will see that only one strand can be copied continuously (the strand where the DNA polymerase enzyme is travelling in the same direction as

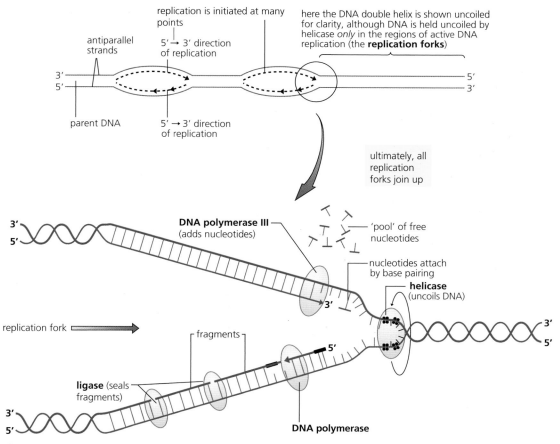

Figure 6.6 DNA replication

Question

2 Why are many replication forks required in the replication of a single strand of DNA?

the helicase enzyme uncoils the DNA). The copying of the other strand is moving away from the unwinding site. Here, copying has to be frequently restarted. The resulting fragments in this new DNA strand have to be joined up. This is done by another enzyme, known as **ligase**.

DNA polymerase also has a role in **'proof reading'** the new strands. Any 'mistakes' that start to happen (such as the wrong bases attempting to pair up) are immediately corrected so that each new DNA double helix is exactly like the original. Then each pair of double strands winds up into a double helix. The outcome is that one strand of each new double helix came from the parent chromosome and one is a newly synthesised strand. This arrangement known as **semi-conservative replication** because half the original molecule is conserved. We return to this issue shortly.

The evidence for semi-conservative replication of DNA

By semi-conservative replication we mean that in each DNA molecule formed, one strand is new and one strand is from the original DNA molecule – so is 'old'. On the other hand, if replication was 'conservative', then one DNA molecule formed would be of two new strands and the other molecule would consist of the two originals strands.

Crick and Watson suggested replication of DNA would be 'semi-conservative', and this has since been shown experimentally, using DNA of bacteria 'labelled' with a 'heavy' nitrogen isotope.

In **semi-conservative replication** one strand of each new double helix comes from the parent chromosome and one is a newly synthesised strand (i.e. half the original molecule is conserved).

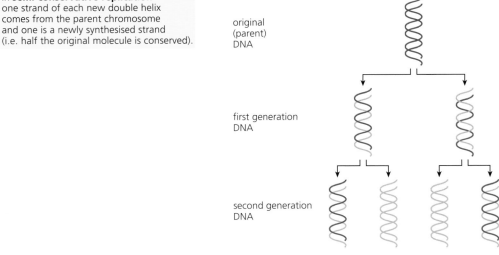

original (parent) DNA

first generation DNA

second generation DNA

If an entirely new double helix were formed alongside the original, then one DNA double helix molecule would be conserved without unzipping in the next generation (i.e. **conservative replication**).

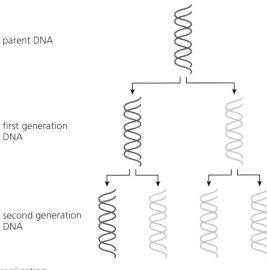

parent DNA

first generation DNA

second generation DNA

Figure 6.7 Semi-conservative versus conservative replication

The confirmation that DNA is replicated semi-conservatively came from an investigation by Meselson and Stahl. In this experiment, a culture of a bacterium (*Escherichia coli*, page 2) was grown in a medium (food source) where the available nitrogen contained *only* the heavy nitrogen isotope (nitrogen-15, ^{15}N) for many generations. The DNA of the bacterium became entirely 'heavy' as a result. (Isotopes are explained in Appendix I, on the CD).

These bacteria were then transferred to a medium of the normal (light) isotope (nitrogen-14, ^{14}N). The result was that the new DNA manufactured by the cells was now made of nitrogen-14. The experimenters then observed and measured the change in concentration of nitrogen-15 and nitrogen-14 in the DNA of succeeding generations. Note that these observations were possible because the bacterial cell divisions in a culture of *E. coli* are naturally synchronised; every 60 minutes they all divide. The DNA was extracted from samples of the bacteria from each succeeding generation and the DNA in each sample was separated.

Separation of DNA was carried out by placing each sample on top of a salt solution of increasing density in a centrifuge tube. On being centrifuged, the different DNA molecules were carried down to the level where the salt solution was of the same density. Thus, DNA with 'heavy' nitrogen ended up nearer the base of the tubes, whereas DNA with 'light' nitrogen stayed near the top of the tubes. Figure 6.8 shows the results that were obtained.

1 Meselson and Stahl 'labelled' nucleic acid (i.e. DNA) of the bacterium *Escherichia coli* with 'heavy' nitrogen (^{15}N), by culturing in a medium where the only nitrogen available was as $^{15}NH_4^+$ ions, for several generations of bacteria.

2 When DNA from labelled cells was extracted and centrifuged in a density gradient (of different salt solutions) all the DNA was found to be 'heavy'.

3 In contrast, the DNA extracted from cells of the original culture (before treatment with ^{15}N) was 'light'.

4 Then a labelled culture of *E.coli* was switched back to a medium providing unlabelled nitrogen only, i.e. $^{14}NH_4^+$. Division in the cells was synchronised, and:
- after **one generation** all the DNA was of intermediate density (each of the daughter cells contained (i.e. *conserved*) one of the parental DNA strands containing ^{15}N alongside a newly synthesised strand containing DNA made from ^{14}N)
- after **two generations** 50% of the DNA was intermediate and 50% was 'light'. This too agreed with semi-conservative DNA replication, given that labelled DNA was present in only half the cells (one strand per cell).

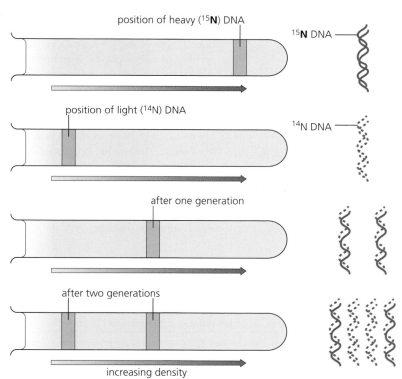

position of heavy (^{15}N) DNA

^{15}N DNA

position of light (^{14}N) DNA

^{14}N DNA

after one generation

after two generations

increasing density

Figure 6.8 Experiment to show that DNA replication is semi-conservative

Question

3 a Predict the experimental results you would expect to see if the Meselson–Stahl experiment (Figure 6.8) was carried on for three generations.
  b If replication was by a 'conservative' process, what would be the result of this experiment after one generation in nitrogen-14 (^{14}N)?

6.2 Protein synthesis

The genetic code specifies the amino acids that are assembled to make polypeptides. The way that DNA codes for polypeptides is central to our understanding of how cells and organisms function.

By the end of this section you should be able to:

a) state that a polypeptide is coded for by a gene and that a gene is a sequence of nucleotides that forms part of a DNA molecule

b) state that a gene mutation is a change in the sequence of nucleotides that may result in an altered polypeptide

c) describe the way in which the nucleotide sequence codes for the amino acid sequence in a polypeptide with reference to the nucleotide sequence for Hb^A (normal) and Hb^S (sickle cell) alleles of the gene for the β-globin polypeptide

d) describe how the information in DNA is used during transcription and translation to construct polypeptides, including the role of messenger RNA (mRNA), transfer RNA (tRNA) and the ribosomes

DNA and the genetic code

Genetic dictionary: a list of the particular base sequences that correspond with particular amino acids. This will vary depending on whether messenger RNA, transfer RNA or either of the two DNA base sequences is given.

In effect, the role of DNA is to instruct the cell to make specific polypeptides and proteins. The huge length of the DNA molecule in a single chromosome carries the codes for a very large number of polypeptides. Within this extremely long molecule, the relatively short length of DNA that codes for a single polypeptide is called a **gene**.

Polypeptides are very variable in size and, therefore, so are genes. A very few genes are as short as 75–100 nucleotides long. Most are at least one thousand nucleotides in length, and some are more.

Polypeptides are built up by condensation of individual amino acids (Figure 2.19, page 46). There are only 20 or so amino acids which are used in protein synthesis; all cell proteins are built from them. The unique properties of each protein lie in:

● which amino acids are involved in its construction
● the sequence in which these amino acids are joined.

The DNA code is in the form of a sequence of the four bases, cytosine (C), guanine (G), adenine (A) and thymine (T). This sequence dictates the order in which specific amino acids are assembled and combined together.

The code lies in the sequence in one of the two strands of the DNA double helix. This is called the **coding strand**. The other strand is complementary to the coding strand. The coding strand is always read in the same direction (from $5' \rightarrow 3'$).

Despite the fact that there are only four 'letters' to the base code 'alphabet' with which to code 20 amino acids, it was immediately clear that combinations of three bases would be the most likely to code for the amino acids (Figure 6.9). The reason for this is shown in Table 6.2.

So the fact that the code was a three-base one was first deduced. Since then experiments have established that the code is a three-base or **triplet code**. Each sequence of three of the four bases codes for one of the 20 amino acids.

Table 6.2 The case for a triplet code

DNA contains four bases: Adenine, **T**hymine, **G**uanine, **C**ytosine. These are the alphabet of the code.
If one base = one amino acid (singlet code), 4 amino acids can be coded.
If two bases = one amino acid (doublet code), 4^2 (= 16) amino acids can be coded.
If three bases = one amino acid (triplet code), 4^3 (= 64) amino acids can be coded.
If four bases = one amino acid (quadruplet code), 4^4 (= 256) amino acids can be coded.
Conclusion Whilst the doublet code has too few combinations to code for 20 amino acids, the triplet code has too many (44 'spares'!). Each sequence of three bases   ● in DNA is called a triplet code   ● in messenger RNA is called a **codon**   ● in transfer RNA is called an **anticodon**.   It has been found that some triplet codes are punctuations (signifying a stop or the end of a gene) and some amino acids are coded by more than one triplet code.

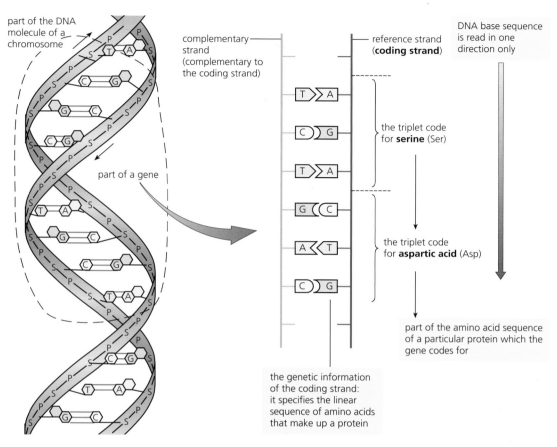

Figure 6.9 Part of a gene and how its DNA codes for amino acids

Amino acid	Abbreviation
alanine	Ala
arginine	Arg
asparagine	Asn
aspartic acid	Asp
cysteine	Cyc
glutamine	Gln
glutamic acid	Glu
glycine	Gly
histidine	His
isoleucine	Ile
leucine	Leu
lysine	Lys
methionine	Met
phenylalanine	Phe
proline	Pro
serine	Ser
threonine	Thr
tryptophan	Trp
tyrosine	Tyr
valine	Val

Most amino acids have two or three similar codons that code for them (Figure 6.10). The DNA triplet code 'TAC' codes for the amino acid methionine. It is always the first triplet code of a gene, so a polypeptide starts with this amino acid. Three of the DNA triplet codes are 'stop' codes, as we shall see.

		Second base			
		A	**G**	**T**	**C**
First base	**A**	AAA Phe AAG Phe AAT Leu AAC Leu	AGA Ser AGG Ser AGT Ser AGC Ser	ATA Tyr ATG Tyr ATT *Stop* ATC *Stop*	ACA Cys ACG Cys ACT *Stop* ACC Trp
	G	GAA Leu GAG Leu GAT Leu GAC Leu	GGA Pro GGG Pro GGT Pro GGC Pro	GTA His GTG His GTT Gln GTC Gln	GCA Arg GCG Arg GCT Arg GCC Arg
	T	TAA Ile TAG Ile TAT Ile TAC Met	TGA Thr TGG Thr TGT Thr TGC Thr	TTA Asn TTG Asn TTT Lys TTC Lys	TCA Ser TCG Ser TCT Arg TCC Arg
	C	CAA Val CAG Val CAT Val CAC Val	CGA Ala CGG Ala CGT Ala CGC Ala	CTA Asp CTG Asp CTT Glu CTC Glu	CCA Gly CCG Gly CCT Gly CCC Gly

Figure 6.10 The amino acids used in proteins and their DNA triplet codes

Protein synthesis–the stages

So, in the working cell, the information in DNA dictates the structure of polypeptides and proteins, many of which form the enzymes in metabolic process. There are three stages to the process.

Stage one: transcription

This occurs in the nucleus. A complementary copy of the code is made by the building of a molecule of **messenger RNA (mRNA)**. In the process, the DNA **triplet codes** are transcribed into **codons** in the messenger RNA. This process is called **transcription**. It is catalysed by the enzyme **RNA polymerase** (Figure 6.11).

In transcription, the DNA double helix first unwinds and the hydrogen bonds are broken at the site of the gene being transcribed. There is a pool of free nucleotides nearby. Then, one strand of the DNA molecule, the coding strand, is the template for transcription. This occurs by complementary base pairing. First, the enzyme RNA polymerase recognises and binds to a 'start' signal. The polymerase enzyme then matches free nucleotides by base pairing (A with U, C with G), working in the 5′ → 3′ direction. Note that in RNA synthesis it is uracil which pairs with adenine. Hydrogen bonds then form between complementary bases, holding the nucleotides in place. Each free nucleotide is then joined by a condensation reaction between the sugar and the phosphate groups of adjacent nucleotides of the RNA strand. This is brought about by the action of the enzyme RNA polymerase.

Once the messenger RNA strand is formed it leaves the nucleus through pores in the nuclear envelope and passes to tiny structures in the cytoplasm called **ribosomes** where the information can be 'read' and is used. Then the DNA double stand once more re-forms into a compact helix at the site of transcription.

> **Question**
>
> **4** State the sequence of changes catalysed by RNA polymerase.

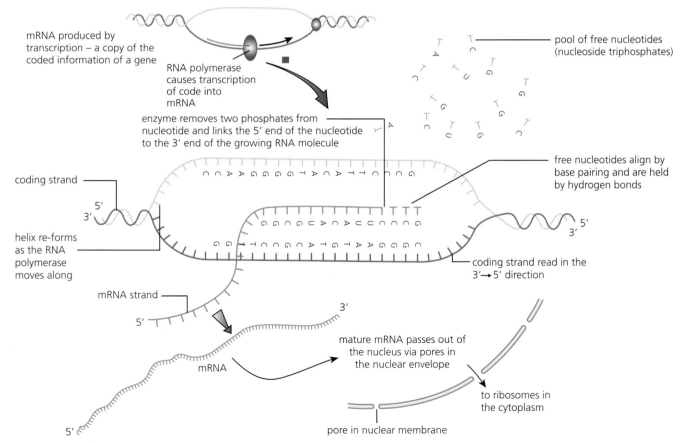

Figure 6.11 Transcription

Stage two: amino acid activation

This occurs in the cytoplasm. In stage two, amino acids are activated by combining with short lengths of a different sort of RNA, called **transfer RNA (tRNA)**. Whilst transfer RNAs are relatively short, they are long enough to be folded into the shape of a clover leaf. Their most important feature is that there is a different transfer RNA for each of the 20 amino acids involved in protein synthesis.

So each amino acid has its 'own' type of transfer RNA. At one end of each transfer RNA molecule is a site where the particular amino acid can be joined. The other end of the transfer RNA molecule is folded into a loop. Here, a sequence of three bases forms an **anticodon**. This anticodon is complementary to the codon of messenger RNA that codes for the specific amino acid.

The amino acid is attached to its transfer RNA by an enzyme. These enzymes, too, are specific to the particular amino acids (and types of transfer RNA) to be used in protein synthesis. The specificity of the enzymes is a further way of ensuring the code is implemented correctly and the amino acids are used in the correct sequence.

ATP provides the energy for the attachment of amino acid to its specific tRNA.

Each amino acid is linked to a specific transfer RNA (tRNA) before it can be used in protein synthesis. This is the process of amino acid activation. It takes place in the cytoplasm.

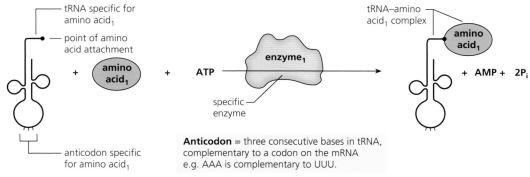

Anticodon = three consecutive bases in tRNA, complementary to a codon on the mRNA e.g. AAA is complementary to UUU.

Figure 6.12 Amino acid activation

Stage three: translation

In stage three a protein chain is assembled by a process called **translation**. This occurs in the **ribosome** (Figure 6.13).

You can see that the ribosome consists of a large and a small subunit. Both parts of the ribosome are intimately involved in the process of translation.

- On arrival at the ribosome, the messenger RNA molecule binds to the small subunit at an attachment site. In this position there are six bases (two codons) of the messenger RNA exposed to the large subunit at any time.
- The first three exposed bases (codon) of messenger RNA are always AUG. A molecule of transfer RNA with the complementary anticodon UAC forms hydrogen bonds with this codon. The amino acid methionine is attached to this transfer RNA molecule.
- A second transfer RNA molecule bonds with the next three bases of the messenger RNA molecule, bringing another amino acid alongside the methionine molecule. Which amino acid this is depends on the second codon, of course.
- Whilst the two amino acids are held close together within the ribosome, a peptide bond is formed between them by condensation reaction. This reaction is catalysed by an enzyme found in the large subunit. A dipeptide has been formed.
- Now the ribosome moves along the messenger RNA molecule in the 5' → 3' direction and the next codon is read. At the same time, the first transfer RNA molecule (with anticodon UAC) leaves the ribosome, minus it amino acid.

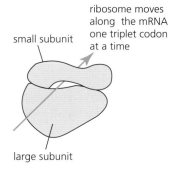

small subunit

ribosome moves along the mRNA one triplet codon at a time

large subunit

Figure 6.13 Ribosome structure

Question

5 Draw and label the structure of a peptide bond between two amino acids.

- Then a third transfer RNA molecule bonds with the next codon of the messenger RNA molecule, bringing another amino acid alongside the second amino acid residue of the dipeptide. Immediately, a peptide bond is formed between them by a condensation reaction. A tripeptide has been formed and starts to emerge from a hole within the large subunit.
- Again, the ribosome moves along the messenger RNA molecule and the next codon is read. At the same time, the second transfer RNA molecule leaves the ribosome, minus its amino acid.
- A fourth transfer RNA brings another amino acid to lie alongside the third amino acid residue. Whilst these amino acids are held close together another peptide bond is formed.
- By these steps, constantly repeated, a polypeptide is formed and emerges from the large subunit. Eventually a 'stop' codon is reached. This takes the form of one of three codons – UAA, UAG or UGA. At this point the completed polypeptide is released from the ribosome into the cytoplasm. The steps of translation are shown in Figure 6.15.

Where ribosomes occur in cells

Question

6 What different forms of RNA are involved in 'transcription' and 'translation' and what are the roles of each?

Many ribosomes occur freely in the cytoplasm. These are the sites of synthesis of proteins that are to remain in the cell and fulfil particular roles there. It is common for several ribosomes to move along the messenger RNA at one time; the structure (messenger RNA, ribosomes and their growing protein chains) is called a **polysome** (Figure 6.14).

Other ribosomes are bound to the membranes of the endoplasmic reticulum (known as rough endoplasmic reticulum, RER) and these are the site of synthesis of proteins that are subsequently secreted from cells or packaged in lysosomes there.

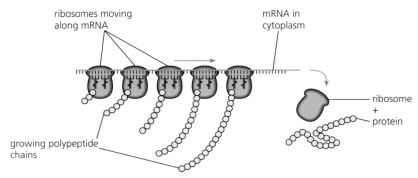

Figure 6.14 A polysome

Triplet codes, codons and anticodons

The sequence of bases along DNA molecules is the **genetic code**. This sequence ultimately determines the sequence of amino acids from which polypeptides are constructed.

- The sequence of bases in the DNA molecule is transcribed to a complementary sequence of bases in messenger RNA. DNA 'codes' are referred to as **triplet codes** and those in messenger RNA are called **codons**.
- Free amino acids in the cytoplasm attach to specific transfer RNA molecules. There is one transfer RNA molecule for each of the twenty different amino acids. The triplet of bases of a transfer RNA molecule is known as an **anticodon**.
- In ribosomes, transfer RNA anticodons align with RNA codons by **complementary base pairing**. This brings amino acids side by side and peptide bonds are formed between them. A polypeptide molecule is formed.

By these steps, the DNA code is translated into an amino acid sequence in a polypeptide. There are different ways of presenting the genetic code and this is the significance of the term '**genetic dictionary**'. The genetic dictionary is a list of the particular base sequences that correspond with particular amino acids. It depends on whether it is the code in DNA, messenger RNA (or transfer RNA) that is quoted.

Can you say why?

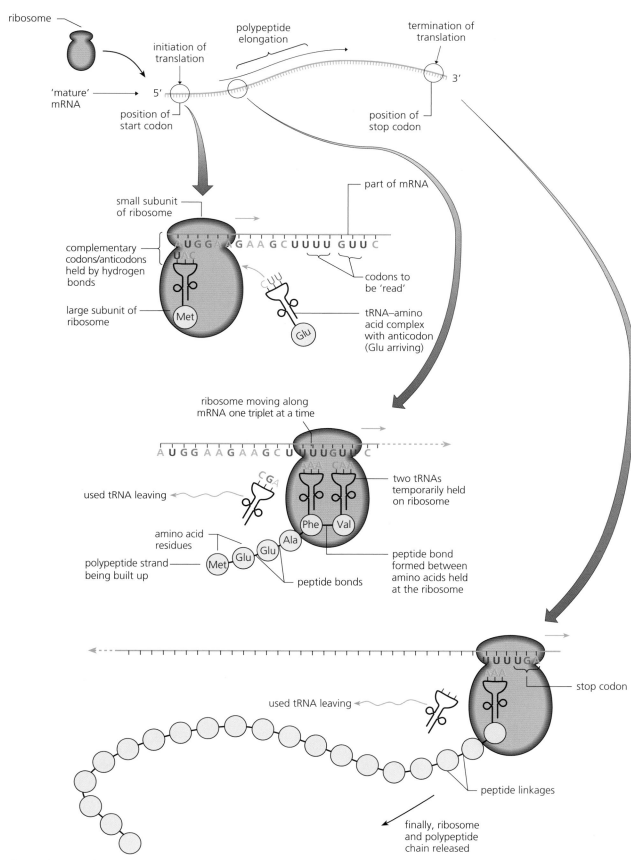

Figure 6.15 Translation

Amino acid	Abbreviation
alanine	Ala
arginine	Arg
asparagine	Asn
aspartic acid	Asp
cysteine	Cyc
glutamine	Gln
glutamic acid	Glu
glycine	Gly
histidine	His
isoleucine	Ile
leucine	Leu
lysine	Lys
methionine	Met
phenylalanine	Phe
proline	Pro
serine	Ser
threonine	Thr
tryptophan	Trp
tyrosine	Tyr
valine	Val

The DNA triplet codes are recorded in Figure 6.10 whilst those of messenger RNA are listed in Figure 6.16. These codes do not need memorising but their relationship is important and you do need to be able to use them.

		Second base			
		U	**C**	**A**	**G**
First base	**U**	UUU Phe UUC Phe UUA Leu UUC Leu	UCU Ser UCC Ser UCA Ser UCG Ser	UAU Tyr UAC Tyr UAA Stop UAG Stop	UGU Cys UGC Cys UGA Stop UGG Trp
	C	CUU Leu CUC Leu CUA Leu CUG Leu	CCU Pro CCC Pro CCA Pro CCG Pro	CAU His CAC His CAA Gln CAG Gln	CGU Arg CGC Arg CGA Arg CGG Arg
	A	AUU Ile AUC Ile AUA Ile AUG Met	ACU Thr ACC Thr ACA Thr ACG Thr	AAU Asn AAC Asn AAA Lys AAG Lys	AGU Ser AGC Ser AGA Arg AGG Arg
	G	GUU Val GUC Val GUA Val GUG Val	GCU Ala GCC Ala GCA Ala GCG Ala	GAU Asp GAC Asp GAA Glu GAG Glu	GGU Gly GGC Gly GGA Gly GGG Gly

Figure 6.16 The messenger RNA genetic dictionary

Extension

DNA also codes for the RNA of cells

The DNA of the chromosomes codes for all the RNA molecules that a cell contains, as well as the proteins. There is a different enzyme system that 'reads' the DNA that specifically codes for these RNA molecules, additional to the enzymes that catalyse the formation of messenger RNA.

Question

7 A sequence of bases in a sample of messenger RNA was found to be:

GGU, AAU, CCU, UUU, GUU, ACU, CAU, UGU.

a What sequence of amino acids does this code for?

b What was the sequence of bases in the coding strand of DNA from which this messenger RNA was transcribed?

c Within a cell, where are triplet codes, codons and anticodons found?

DNA can change

We have seen that a gene is a sequence of nucleotide pairs which codes for a sequence of amino acids. Normally, the sequence of nucleotides in the DNA is maintained without changing but very occasionally it does. If a change occurs we say a mutation has occurred.

A **gene mutation** involves a change in the sequence of bases of a particular gene. At certain times in the cell cycle mutations are more likely than at other times. One such occasion is when the DNA molecule is replicating. We have noted that the enzyme DNA polymerase that brings about the building of a complementary DNA strand also 'proof reads' and corrects most errors. However, gene mutations can and do occur spontaneously during this step. Also, certain conditions or chemicals may cause change to the DNA sequence of bases (Table 5.2, page 105).

The **mutation** that produces sickle cell haemoglobin (**Hg^S**) is in the gene for β-haemoglobin. It results from the substitution of a single base in the sequence of bases that make up all the codons for β-haemoglobin.

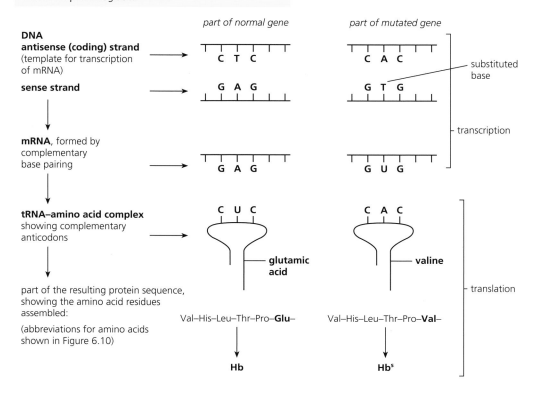

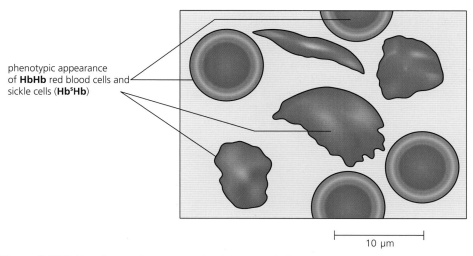

drawing based on a photomicrograph of a blood smear, showing blood of a patient with sickle cells present among healthy red blood cells

Figure 6.17 Sickle cell anaemia – an example of a gene mutation

An example of a gene mutation occurs in the condition known as **sickle cell anaemia**. Here the gene that codes for the amino acid sequence of a part of the respiratory pigment haemoglobin of red blood cells occasionally mutates at one base pair (Figure 6.17). The result of this change is that the base adenine is replaced by thymine at one position along the coding strand. The messenger RNA produced from this DNA contains the triplet code GUG in place of GAG. This causes the amino acid valine (a non-polar amino acid) to appear at that point, instead of polar glutamic acid. The presence of non-polar valine in the beta chain of haemoglobin gives a hydrophobic spot in the otherwise hydrophilic outer section of the protein. This tends to attract other haemoglobin molecules to bind to it.

When a person who is a sickle cell 'carrier' undertakes sudden physical exercise (or moves to a high altitude) the oxygen content of the blood is lowered. In these conditions the sickle cell haemoglobin molecules readily clump together into long fibres. These fibres distort the red blood cells into sickle shapes. In this condition the red blood cells cannot transport oxygen. Also, sickle cells get stuck together, blocking smaller capillaries and preventing the circulation of normal red blood cells. The result is that people with sickle cell trait suffer from anaemia – a condition of inadequate delivery of oxygen to cells.

Summary

- **Nucleic acids** are **polymers**. They are polynucleotides composed of long chains of nucleotide **monomers** (a pentose sugar, a phosphate molecule and a nitrogenous base) combined together. There are two forms of nucleic acid, **DNA** and **RNA**. These forms differ in the sugar and in the bases they contain, and in their **roles in the cell**. DNA is found in the chromosomes and RNA in both the nucleus and the cytoplasm. DNA also occurs in mitochondria and chloroplasts.

- **ATP** is the universal energy currency molecule by which energy is transferred to do useful work. ATP is a soluble molecule, formed in the mitochondria but able to move into the cytosol by facilitated diffusion. It diffuses freely about cells.

- The **sequence of four bases**, adenine (**A**), thymine (**T**), guanine (**G**) and cytosine (**C**), along a DNA molecule is the **genetic code**. This is a three-letter code – three of the four bases code for a particular amino acid. The sequence of bases in the DNA molecule may be transcribed to a complementary sequence in **messenger RNA (mRNA)**. Once formed, messenger RNA passes out of the nucleus into the cytoplasm and to a **ribosome** where **polypeptide synthesis** occurs. DNA 'codes' are referred to as **triplet codes** and those in messenger RNA are called **codons**.

- **Transfer RNA (tRNA)** molecules occur in the cytoplasm. These molecules have a particular triplet of bases at one end (known as an **anticodon**) and a point of attachment for a **specific amino acid** at the other end. The role of transfer RNA is to pick up free amino acids and bring them to the messenger RNA in the ribosome. Complementary base pairing between the codons of messenger RNA and the anticodons of transfer RNA results in the original DNA code being translated into an amino acid sequence in the process of **polypeptide and protein synthesis**.

- There are **three stages** to polypeptide and protein synthesis.
  - In the nucleus, a copy of the genetic code of a gene is made in the form of a single strand of **messenger RNA** in the stage called **transcription**. The messenger RNA passes out into the cytoplasm.
  - In the cytoplasm **amino acid activation** occurs. Each of the 20 amino acids is attached to a specific **transfer RNA** molecule.
  - Finally a **new polypeptide** is assembled when the information of the messenger RNA is **translated** ('read') in a ribosome and the transfer RNA molecules with their attached amino acids are aligned in the order given in the code. Peptide bonds are formed between the amino acids.

Examination style questions

1 Fig. 1.1 shows the replication of one strand of a DNA double helix.

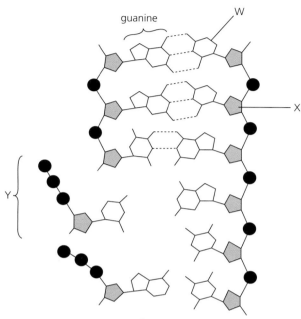

guanine

W

X

Y

Fig. 1.1

a) Name **W** to **Y**. [3]

b) Explain how the structure of DNA enables it to replicate semi-conservatively. [3]

c) Explain why it is important that an exact copy of DNA is made during replication. [2]

[Total: 8]

(Cambridge International AS and A Level Biology 9700, Paper 02 Q1 June 2005)

2 a) A gene in the form of a double-stranded length of DNA contains 12 000 nucleotides. Explain the maximum number of amino acids that could be required to transcribe it. [6]

b) What may the triplet codes of a gene represent other than the amino acids of the polypeptide that it codes for? [2]

c) The following is a short sequence of the triplet codes for part of the primary structure of a polypeptide:
AAA ATA GTA TAA CAC TGC CTC TCG
Give the sequence of codons of mRNA transcribed from this part of the gene, and the sequence of anti-codons that will base pair with them. [8]

d) How does the structure of tRNA differ from that of mRNA? [4]

[Total: 20]

3 A molecule of messenger RNA (mRNA) was produced during the transcription of a gene. Part of the template sequence of DNA was ATGC.

The diagram shows the part of the molecule of messenger RNA corresponding to that sequence of four bases.

D

G

E

F

guanine

a) Name the parts of the mRNA molecule shown in the diagram above labelled **D**, **E**, **F** and **G**. [4]

b) Copy and complete the table to show **three** ways in which mRNA differs from DNA. [3]

	mRNA	DNA
1		
2		
3		

c) Describe the role of mRNA after it leaves the nucleus and enters the cytoplasm of a eukaryotic cell. [4]

[Total: 11]

(Cambridge International AS and A Level Biology 9700, Paper 21 Q3 June 2011)

7 Transport in plants

Flowering plants do not have compact bodies like those of animals. Leaves and extensive root systems spread out to obtain the light energy, water, mineral ions and carbon dioxide that plants gain from their environment to make organic molecules, such as sugars and amino acids. Transport systems in plants move substances from where they are absorbed or produced to where they are stored or used. Plants do not have systems for transporting oxygen and carbon dioxide; instead these gases diffuse through air spaces within stems, roots and leaves.

7.1 Structure of transport tissues

Plants have two transport tissues: xylem and phloem.

By the end of this section you should be able to:

a) draw and label from prepared slides plan diagrams of transverse sections of stems, roots and leaves of herbaceous dicotyledonous plants using an eyepiece graticule to show tissues in correct proportions

b) draw and label from prepared slides the cells in the different tissues in roots, stems and leaves of herbaceous dicotyledonous plants using transverse and longitudinal sections

c) draw and label from prepared slides the structure of xylem vessel elements, phloem sieve tube elements and companion cells and be able to recognise these using the light microscope

d) relate the structure of xylem vessel elements, phloem sieve tube elements and companion cells to their functions

Internal transport – the issues

The cells of organisms need a constant supply of **water** and **organic nutrients** such as glucose and amino acids, and most need **oxygen**. The **waste products** of cellular metabolism have to be removed, too. In single-celled organisms and small multicellular organisms, internal distances (the length of diffusion pathways) are small. Here, movements of nutrients and other molecules can occur efficiently by **diffusion** (page 79). However, some substances must be transported across membranes, such as the cell surface membrane, by **active transport** (page 90).

In larger organisms diffusion alone is not sufficient. This is because the amount of gas an organism needs to exchange is largely proportional to its volume (the bulk of respiring cells). At the same time, the amount of exchange that can occur is proportional to the surface area over which diffusion takes place. The surface area to volume ratio decreases as size increases, as we have seen. Consequently, an **internal transport system** is essential. This alone can provide enough of what is required by the cells in larger and more bulky organisms. Also, the more active an organism is the more nutrients are likely to be required in cells and so the greater is the need for an efficient system. Question 2 on page 132 explores these issues of size, shape and mechanisms of internal transport.

Transport in multicellular plants

The flowering plants are the dominant land plants. Some are trees and shrubs with woody stems, but many are non-woody (herbacious) plants. Whether woody or herbaceous, a flowering plant consists of stem, leaves and root.

Internal transport occurs by **mass flow** but here there is no pumping organ. Two separate tissues are involved: water and many ions travel in the **xylem**, whilst manufactured foods (mainly sugar and amino acids) are carried in the **phloem**. Xylem and phloem are collectively known as the vascular tissue. They occur together in the centre of roots and in the **vascular bundles** of stems and leaves of the plant.

The stem

The role of the stem is to support the leaves in the sunlight and to transport organic materials (mainly sugar and amino acids), ions and water between the roots and leaves. This transport occurs in the **vascular bundles**. The vascular bundles of the stem occur in a ring just below the epidermis. To the outside of the vascular bundles lies the cortex and to the inside is the pith. Both cortex and pith mostly consist of living, relatively unspecialised cells with thin cellulose walls (Figure 7.6, page 133).

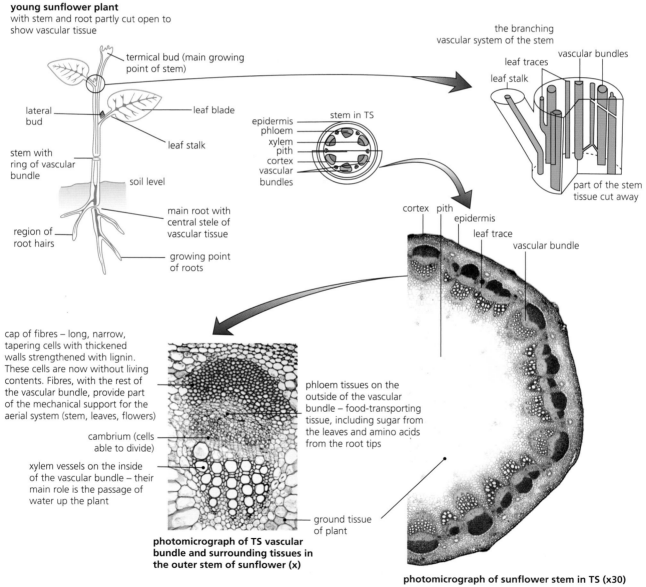

young sunflower plant
with stem and root partly cut open to show vascular tissue

termical bud (main growing point of stem)

lateral bud

leaf blade

leaf stalk

stem with ring of vascular bundle

soil level

region of root hairs

main root with central stele of vascular tissue

growing point of roots

the branching vascular system of the stem

leaf traces

vascular bundles

leaf stalk

part of the stem tissue cut away

stem in TS

epidermis
phloem
xylem
pith
cortex
vascular bundles

cortex   pith

epidermis

leaf trace

vascular bundle

cap of fibres – long, narrow, tapering cells with thickened walls strengthened with lignin. These cells are now without living contents. Fibres, with the rest of the vascular bundle, provide part of the mechanical support for the aerial system (stem, leaves, flowers)

cambrium (cells able to divide)

xylem vessels on the inside of the vascular bundle – their main role is the passage of water up the plant

phloem tissues on the outside of the vascular bundle – food-transporting tissue, including sugar from the leaves and amino acids from the root tips

ground tissue of plant

photomicrograph of TS vascular bundle and surrounding tissues in the outer stem of sunflower (x)

photomicrograph of sunflower stem in TS (x30)

Figure 7.1 The distribution and structure of the vascular bundles of the stem

The leaves

The leaf is an organ specialised for **photosynthesis** – the process in which light energy is used to make sugar. Carbon dioxide from the air and water from the soil are the raw materials used, oxygen is the waste product.

The leaf consists of a leaf blade connected to the stem by a leaf stalk. The blade is a thin structure with a large surface area. Within the leaf blade are **mesophyll** cells that are packed with **chloroplasts** – the organelles in which photosynthesis occurs. There is a thin film of water on the outer surface of these cells. The **epidermis** is a tough, transparent, single layer of cells surrounding the mesophyll tissue. There is an external waxy **cuticle** to the epidermis which is impervious. This reduces water vapour loss from the leaf surface (Figure 7.2).

The epidermis has many tiny pores, called **stomata** – the sites of gas exchange. Within the leaf's mesophyll tissue are continuous **air spaces**. Throughout the leaf blade is a **network of vascular bundles**. These provide mechanical support but the leaf is also supported by the turgidity of all its cells, surrounded as they are by the tough epidermis. Leaves are relatively delicate structures and, at times, they have to withstand destructive forces such as wind and rain.

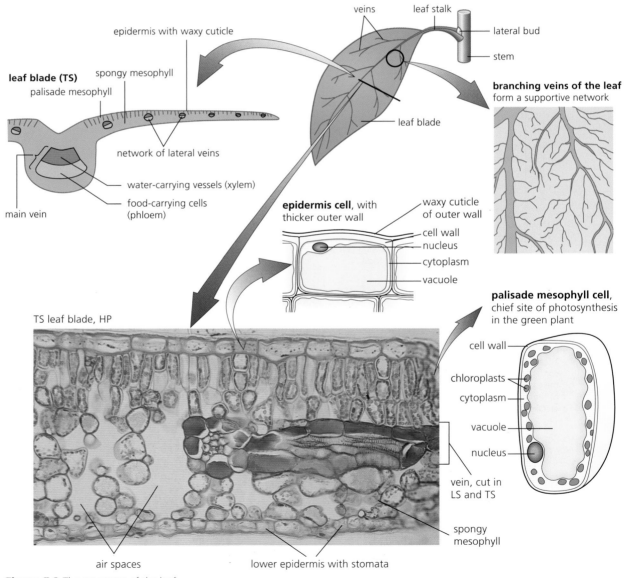

Figure 7.2 The structures of the leaf

The root

The **root** anchors the plant in the ground and is the site of absorption of water and ions from the soil. The main tap root forms lateral roots, and growth continues at each root tip. The growing plant is continuously making contact with fresh soil. This is important because the thin layer of dilute soil solution found around soil particles is where the plant obtains the huge volume of water it requires. Minerals salts as ions are also absorbed from here.

The vascular tissue of the root occurs in a single, central **stele**. Around the stele is a single layer of cells, the **endodermis**. These cells have a waxy strip in their radial walls, known as the **Casparian strip**. Immediately within the endodermis is another single layer of living cells, known as the pericycle (Figure 7.3). Only at the base of the stem is the vascular tissue of the root reorganised into several vascular bundles arranged around the outside of the stem.

Question

1 Outline the properties of cellulose that make it an ideal material for plant cell walls.

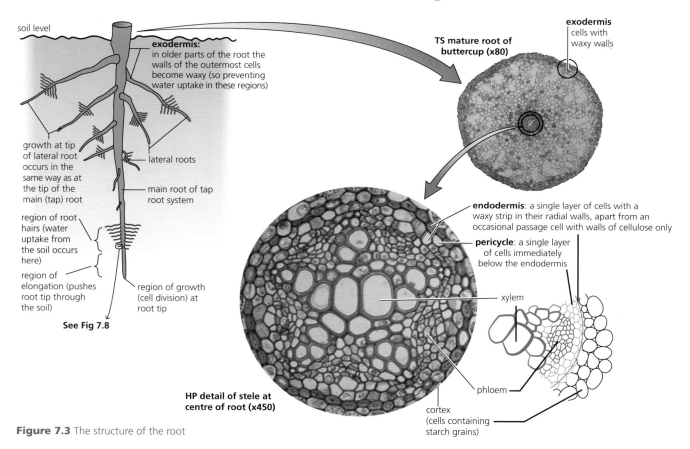

Figure 7.3 The structure of the root

Question

2 The external dimensions, volume and surface area of six geometrically shaped objects – three of a compact shape and three that are flat and thin – are listed in the table below.

	Compact			Flat and thin		
	Small	**Medium**	**Large**	**Small**	**Medium**	**Large**
Dimensions/mm	1 × 1 × 1	2 × 2 × 2	4 × 4 × 4	2 × 1 × 0.5	8 × 2 × 0.5	16 × 8 × 0.5
Volume/mm³	1	8	64	1	8	64
Surface area/mm²	6	24	96	7	42	280

a Calculate the surface area to volume ratio of the six objects.

b Comment on the effect of shape on the surface area of an object as the size of the object increases.

c Comment on the effect of shape on the surface area to volume ratio of an object as the size increases.

d If these objects were organisms, what are the implications of their size and shape on the movement of nutrients or waste products by diffusion?

The structure of xylem and phloem tissue

Xylem begins as elongated cells with cellulose walls and living contents, connected end to end.

photomicrograph of xylem tissue in LS

drawing of xylem vessels in TS and LS

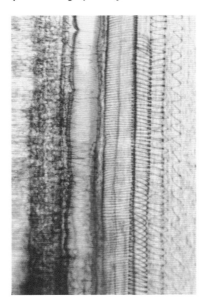

all vessels have lignin-free 'pit' areas in their walls

TS

fibre

xylem parenchyma (the only living cells of xylem tissue)

LS

lignin thickening as spirals, rings, network or solid blocks – deposited on inside of vessel, strengthening the cellulose layers

Figure 7.4 The structure of xylem tissue

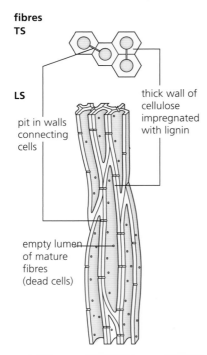

fibres TS

LS

pit in walls connecting cells

thick wall of cellulose impregnated with lignin

empty lumen of mature fibres (dead cells)

fibres are a thick-walled tissue, with walls of cellulose strengthened by lignin – when mature the cell contents die; fibres are long, narrow, pointed, empty cells with pits in their walls where the living contents once connected

Figure 7.5 The structure of fibres

During development, the end walls are dissolved away so that mature xylem vessels are **long, hollow tubes**. The living contents of a developing xylem vessel are used up in the process of depositing cellulose thickening to the inside of the lateral walls of the vessel. This is hardened by the deposition of a chemical substance, **lignin**. Lignin is a substance that makes cross links with the hemicelluloses that fill the gaps between cellulose fibres. This greatly strengthens the cellulose and makes it impermeable to water. Consequently, xylem is an extremely tough tissue. Furthermore, it is strengthened internally, which means it is able to resist negative pressure (suction) without collapsing in on itself. When you examine xylem vessels in longitudinal sections by light microscopy you will see that they may have differently deposited thickening; many have rings of thickening, for example. Other xylem vessels have more massive thickenings. However, all xylem vessels have areas in their lateral walls (including the pits where the cellulose wall is especially thin) in which there is a layer of cellulose but no lignin. Here the walls are entirely permeable and permit lateral movements of water to surrounding tissues. This is role of the xylem – to supply water to the living cells of the plant.

Fibres also occur between xylem vessels, thereby strengthening the whole bundle. Interestingly, in ferns and cone-bearing trees, xylem tissue is absent. Here the water-conducting tissue consists of fibre-like **tracheids** with large pits in their lateral walls through which water passes from tracheid to tracheid (Figure 7.4).

Phloem tissue consists of **sieve tubes** and **companion cells,** and is served by transfer cells in the leaves. Sieve tubes are narrow, elongated elements, connected end to end to form tubes. The end walls, known as sieve plates, are perforated by pores. The cytoplasm of a mature sieve tube has no nucleus, nor many of the other organelles of a cell. However, each sieve tube is connected to a companion cell by strands of cytoplasm called plasmodesmata, that pass through narrow gaps (called pits) in the walls. The companion cells service and maintain the cytoplasm of the sieve tube, which has lost its nucleus. Phloem is a living tissue, and has a relatively high rate of aerobic respiration during transport. In fact, transport of manufactured food in the phloem is an active process, using energy from metabolism (Figure 7.6).

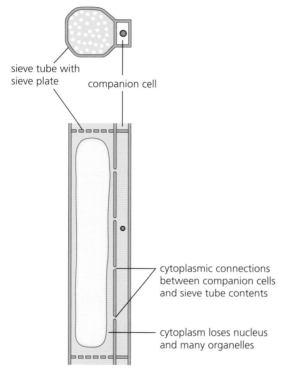

sieve tube with sieve plate

companion cell

cytoplasmic connections between companion cells and sieve tube contents

cytoplasm loses nucleus and many organelles

Figure 7.6 The structure of phloem tissue

photomicrograph of LS phloem in HP (×200) phloem is a tissue that may become crushed and damaged as stem tissue is sectioned for microscopy

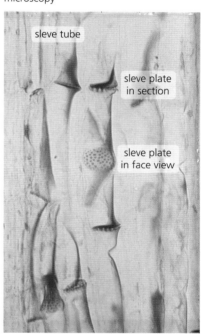

sieve tube

sieve plate in section

sieve plate in face view

phloem tubes are the site of movement of sugar and amino acids.

Observing and recording high power detail of cells

The cells of the ground tissues make up the bulk of stem (and root) of herbaceous (non-woody) plants, around the vascular tissue. These are **collenchyma** and **parenchyma** cells, and they show relatively few structural adaptations. They are concerned with support due to the turgidity of all the living cells contained within the stem. Starch storage also occurs here. **Fibres** are different. Here there are no living contents, but their thickened walls are impregnated with lignin, and are extremely tough (Figure 7.5).

The outermost layer of the stems (and leaves) of herbaceous plants is a continuous layer of compact, tough cells, one cell thick, called the **epidermis** (Figure 7.2). These cells have a layer of wax deposited on the external walls, and their strength and continuity provides mechanical support by counteracting the internal pressure due to turgidity.

You will be examining prepared slides of sections through plant organs in the laboratory, using the light microscope. With guidance, the different types of xylem vessels, sieve tubes and companion cells, fibres, parenchyma and collenchyma ground tissue can be identified in longitudinal and transverse sections and their structures recorded. Follow the guidance on recording observations in drawings on page 8.

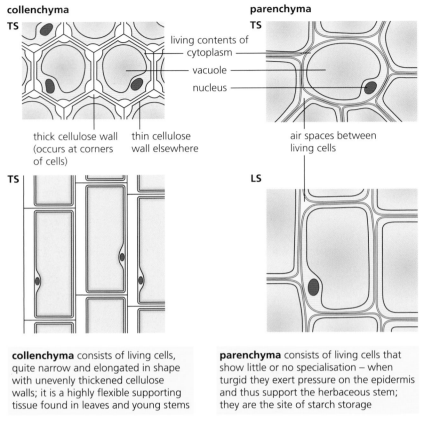

collenchyma consists of living cells, quite narrow and elongated in shape with unevenly thickened cellulose walls; it is a highly flexible supporting tissue found in leaves and young stems

parenchyma consists of living cells that show little or no specialisation – when turgid they exert pressure on the epidermis and thus support the herbaceous stem; they are the site of starch storage

Figure 7.7 The structure of parenchyma, collenchyma and fibres

Observing and recording plant structure using low-power plan diagrams

You will be examining prepared slides of sections through plant organs in the laboratory, using the light microscope. The name given to the detailed study of the structure of living tissues is **Histology**. Slides of thin sections were commonly used to discover the structure of multicellular organisms. You may notice that the material they show has been selectively stained to highlight structures.

With guidance, the different tissues a slide shows can be identified and their precise distribution represented in a **plan diagram** (also called a tissue map). This will show their relative positions to scale. Individual cells are not shown. Then the size of the specimen can be calculated and the magnification represented by a scale bar.

Figure 7.1 has the necessary information to create a plan diagram of part of a plant stem. However, it is important you create your own plan diagrams by examining prepared slides, using an eyepiece graticule (page 9) to show tissues in correct proportions. Follow the guidance on recording observations in drawings on page 8.

Figure 7.2 has the necessary information to create a plan diagram of part of a plant leaf. However, it is important you create your own plan diagrams by examining prepared slides, using an eyepiece graticule to show tissues in correct proportions.

Figure 7.3 has the necessary information to create a plan diagram of part of a plant root. However, it is important you create your own plan diagrams by examining prepared slides, using an eyepiece graticule to show tissues in correct proportions.

Question

3 Draw and label a plan diagram of a representative part of the section of the stem in Figure 7.1 to accurately record the distribution of the tissues.
Using the magnification data given, add an appropriate scale bar line to your drawing.

7.2 Transport mechanisms

Movement of xylem sap and phloem sap is by mass flow. Movement in the xylem is passive as it is driven by evaporation from the leaves; plants use energy to move substances in the phloem.

Xylem sap moves in one direction from the roots to the rest of the plant. The phloem sap in a phloem sieve tube moves in one direction from the location where it is made to the location where it is used or stored. At any one time phloem sap can be moving in different directions in different sieve tubes.

By the end of this section you should be able to:

a) explain the movement of water between plant cells, and between them and their environment, in terms of water potential

b) explain how hydrogen bonding of water molecules is involved with movement in the xylem by cohesion-tension in transpiration pull and adhesion to cellulose cell walls

c) describe the pathways and explain the mechanisms by which water and mineral ions are transported from soil to xylem and from roots to leaves

d) define the term transpiration and explain that it is an inevitable consequence of gas exchange in plants

e) investigate experimentally and explain the factors that affect transpiration rate using simple potometers, leaf impressions, epidermal peels, and grids for determining surface area

f) make annotated drawings, using prepared slides of cross-sections, to show how leaves of xerophytic plants are adapted to reduce water loss by transpiration

g) state that assimilates, such as sucrose and amino acids, move between sources (e.g. leaves and storage organs) and sinks (e.g. buds, flowers, fruits, roots and storage organs) in phloem sieve tubes

h) explain how sucrose is loaded into phloem sieve tubes by companion cells using proton pumping and the co-transporter mechanism in their cell surface membranes

i) explain mass flow in phloem sap down a hydrostatic pressure gradient from source to sink

The movement of water through the plant

We have seen that:

- xylem is the water conducting tissue, found in the vascular bundles of the stem and leaves, and in the central stele of the root
- the main tap root and lateral roots grow through the soil, continuously making contact with fresh soil
- the thin layer of dilute soil solution found around soil particles is where the plant obtains the huge volume of water it requires
- minerals salts as ions are also absorbed from the soil solution.

Contact between root and soil is also vastly increased by structures called **root hairs**. The root hairs are formed at a short distance behind the growing tip of each root. They are extensions of individual epidermal cells and are relatively short-lived (Figure 7.8). Elsewhere the outer cells of the root have walls that are waxy and therefore waterproof. Consequently, it is only in the region that the root hairs occur that the root is able to absorb water and ions.

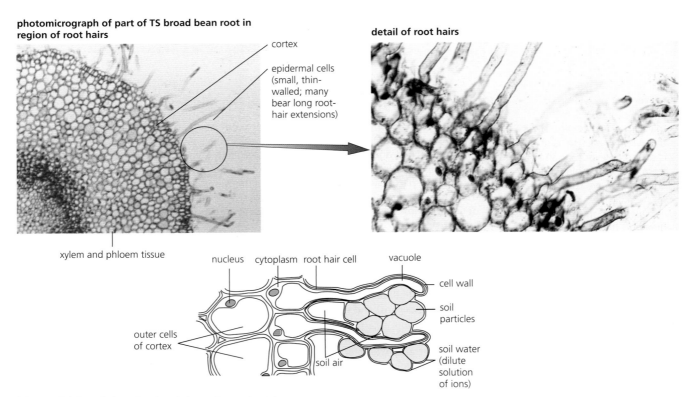

photomicrograph of part of TS broad bean root in region of root hairs

cortex

epidermal cells (small, thin-walled; many bear long root-hair extensions)

detail of root hairs

xylem and phloem tissue

nucleus   cytoplasm   root hair cell   vacuole

cell wall

soil particles

outer cells of cortex

soil air

soil water (dilute solution of ions)

Figure 7.8 Root hairs – the site of absorption and uptake

Uptake of water – the roles of the root hairs

Root hairs are the point of entry of water on its passage through the green plant.

soil → root → stem → leaf → air

The very large numbers of delicate root hairs form behind each root tip (Figure 7.3). These are in close contact with the soil solution which has a very high water potential. (Remember, **water potential**, the name we give to the tendency of water molecules to enter or leave solutions, was introduced in Topic 4, page 83.) At the same time the cells of the root cortex have a water potential much lower than that of the soil solution. Water enters the root from the soil solution, passing down a water potential gradient (Figure 7.9a).

Another process occurring in the root hairs is the active **uptake of soluble ions** from the soil solution. These ions include nitrate ions, magnesium ions and several others, all essential to the metabolism of the plant. Once absorbed, the ions are transported across the cortex of the root. They converge on the central stele and there they are actively secreted into the water column flowing up the xylem. The continuing movement of ions across the cortex of the root is partly responsible for the gradient in water potential between soil solution and the cells of the stele. We return to the issue of ion uptake later in this topic.

Question

4 What features of root hairs facilitate absorption from the soil?

From root hairs to xylem – apoplast and symplast

Look at Figure 7.9b. It shows the two pathways that water can take through plant tissues.

The alternative **pathways of water movement** shown are:

- the **apoplast**. This is the interconnected 'free spaces' that occur between the cellulose fibres in the plant cell walls. These free spaces make up about 50 per cent of the volume of the cell wall. The apoplast also includes the water-filled spaces of dead cells and of the xylem vessels. Note that this route entirely avoids the living contents of cells. We shall see that most of the water passing through plant tissue travels by this pathway, most of the time – a case of transport by mass flow.

- the **symplast**. This is the living contents of cells. Water enters cells through the cell surface membrane by osmosis and diffuses through the cytoplasm of cells and the cytoplasmic connections between cells (called **plasmodesmata**). You will remember that the cells are packed with many organelles. These tend to slow the flow of water, so this pathway is not the major one for water movement.

So, most water travels across the root via the apoplast pathways, until that is, the endodermis, is reached.

a)

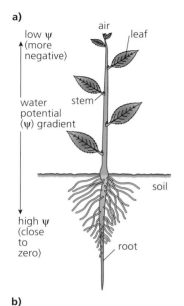

low ψ
(more negative)

air   leaf

stem

water potential (ψ) gradient

soil

high ψ (close to zero)

root

b)

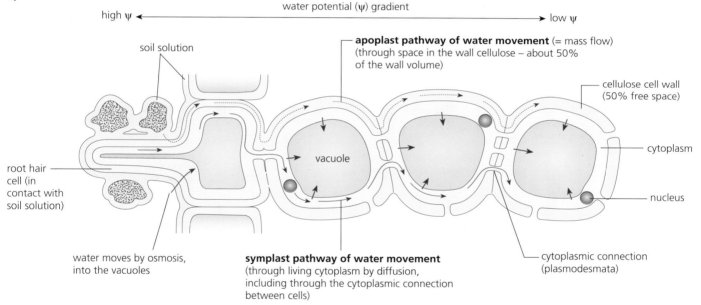

water potential (ψ) gradient

high ψ ⟵   ⟶ low ψ

soil solution

apoplast pathway of water movement (= mass flow)
(through space in the wall cellulose – about 50% of the wall volume)

cellulose cell wall (50% free space)

cytoplasm

vacuole

root hair cell (in contact with soil solution)

nucleus

water moves by osmosis, into the vacuoles

symplast pathway of water movement
(through living cytoplasm by diffusion, including through the cytoplasmic connection between cells)

cytoplasmic connection (plasmodesmata)

Figure 7.9 Water movement – force and pathways
a) The water potential (ψ) gradient of the whole plant
b) The water potential (ψ) gradient across cells of the root cortex

The role of the endodermis

This is a single layer of cells that surrounds the central stele of the root. These cells have a waxy strip in their radial walls, known as the **Casparian strip**. The wax is so strongly attached to the cellulose that it stops the movement of water by the apoplast pathway – but only temporarily. All water passes into the cytoplasm of the endodermis cells and then travels by the symplast pathway. You can see the position of the Casparian strip and its effect on water movement in Figure 7.10.

After that, water can and mostly does return to the apoplast pathway and travels the remaining short distance to reach the xylem vessels. This movement, too, is down a water potential gradient. However, the next steps – entry into the xylem and movement of water up to the leaves – depend on events in the leaves. Next, we need to investigate the process of loss of water vapour from the cells of the leaves.

Question

5 What is the consequence of the Casparian strip for the apoplast pathway of water movement?

137

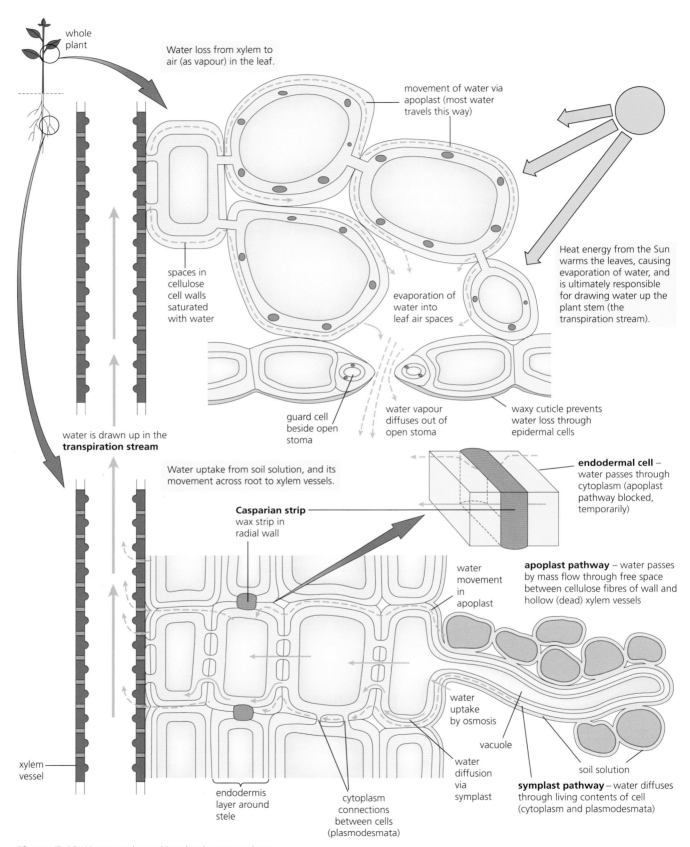

whole plant

Water loss from xylem to air (as vapour) in the leaf.

movement of water via apoplast (most water travels this way)

spaces in cellulose cell walls saturated with water

evaporation of water into leaf air spaces

Heat energy from the Sun warms the leaves, causing evaporation of water, and is ultimately responsible for drawing water up the plant stem (the transpiration stream).

water is drawn up in the **transpiration stream**

guard cell beside open stoma

water vapour diffuses out of open stoma

waxy cuticle prevents water loss through epidermal cells

Water uptake from soil solution, and its movement across root to xylem vessels.

endodermal cell – water passes through cytoplasm (apoplast pathway blocked, temporarily)

Casparian strip wax strip in radial wall

water movement in apoplast

apoplast pathway – water passes by mass flow through free space between cellulose fibres of wall and hollow (dead) xylem vessels

water uptake by osmosis

vacuole

water diffusion via symplast

soil solution

symplast pathway – water diffuses through living contents of cell (cytoplasm and plasmodesmata)

xylem vessel

endodermis layer around stele

cytoplasm connections between cells (plasmodesmata)

Figure 7.10 Water uptake and loss by the green plant

Transpiration and the role of stomata

Transpiration: the process through which water vapour is lost from the aerial parts of plants. It occurs as the result of evaporation of water at the surface of mesophyll cells into the airspaces within the leaf, followed by diffusion of water vapour out of the leaf, mainly through stomata, down a water potential gradient from the surface of spongy mesophyll cells via airspaces in the leaf to the atmosphere.

Transpiration is the name given to the loss of water vapour from the aerial parts of plants. The mesophyll cells in the leaf are moist and their walls contain water between all the cellulose fibres there. Consequently, there is a film of water around the outside of the cell walls, too. So water is freely available to evaporate from the cell walls and it does so. The air inside the leaf is usually saturated with water vapour.

Stomata are pores in the epidermis of leaves. Each stoma consists of two elongated **guard cells**. These are attached to the surrounding epidermal cells and joined together at each end but they are free to separate to form a pore between them (Figure 7.11). The epidermal cell besides a guard cell is called a subsidiary cell.

The water vapour that accumulates in the air spaces between the mesophyll cells diffuses out through the pores of open stomata, provided there is a concentration gradient between the leaf spaces and the air outside the leaf. This difference in concentration of water vapour between the leaf and the external atmosphere is highest in hot, dry, windy weather. Transpiration is greatest in these conditions, as we shall shortly demonstrate. The site of transpiration is shown in Figure 7.12.

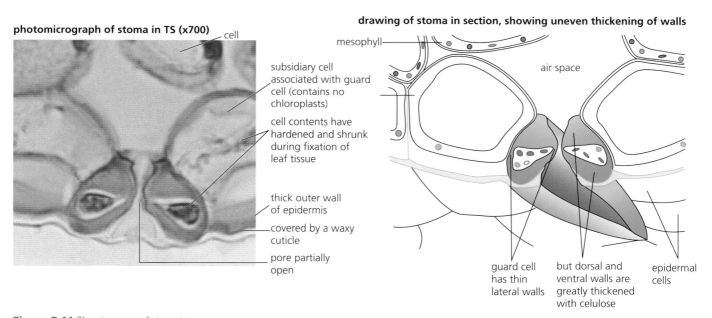

photomicrograph of stoma in TS (x700)

cell

subsidiary cell associated with guard cell (contains no chloroplasts)

cell contents have hardened and shrunk during fixation of leaf tissue

thick outer wall of epidermis

covered by a waxy cuticle

pore partially open

drawing of stoma in section, showing uneven thickening of walls

mesophyll

air space

guard cell has thin lateral walls

but dorsal and ventral walls are greatly thickened with celulose

epidermal cells

Figure 7.11 The structure of stomata

Stomatal aperture and water loss – an experimental approach

Water vapour diffusing out of a leaf through open stomata can be detected by cobalt chloride paper (blue when dry, pink when it absorbs water). The time this takes to happen will indicate the number of open stomata (Figure 7.13). This experiment can be adapted to compare water vapour loss from the leaves of different plants.

The results can be related to stomatal frequency in the same plant leaves by spreading a film of nail varnish on the leaf surfaces. When the film has dried it can be peeled off with fine forceps, placed on a microscope slide with a drop of water and a cover slip added. The number of open stomata observed in the microscope field of view can be counted. The mean of five counts for each leaf will provide a significant result.

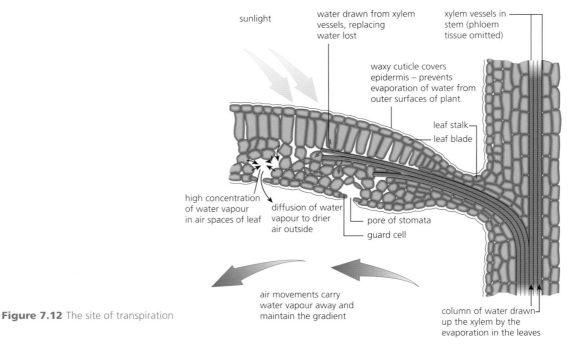

Figure 7.12 The site of transpiration

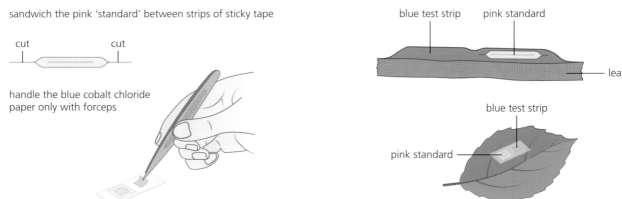

Figure 7.13 Cobalt chloride paper used to measure water vapour loss from a leaf surface

Question

6 Explain precisely why it is that a rise in temperature around a green plant normally leads to an increase in the rate of transpiration.

The transpiration stream

The water that evaporates from the walls of the mesophyll cells of the leaf is continuously replaced. It comes, in part, from the cell cytoplasm (the symplast pathway), but mostly it comes from the water in the spaces in cell walls in nearby cells and then from the xylem vessels in the network of vascular bundles nearby (the apoplast pathway).

These xylem vessels are full of water. As water leaves the xylem vessels in the leaf a tension is set up on the entire water column in the xylem tissue of the plant. This tension is transmitted down the stem to the roots because of the **cohesion** of water molecules. Cohesion is the force by which individual molecules stick together. Water molecules stick together as a result of hydrogen bonding. These bonds continually break and reform with other adjacent water molecules, but at any one moment a large number are held together by their hydrogen bonds, and so are strongly attracted to each other.

Adhesion is the force by which individual molecules cling to surrounding material and surfaces. Materials with an affinity for water are described as **hydrophilic** (see 'The solvent properties of water' on page 53). Water adheres strongly to most surfaces and can be drawn up in long columns, through narrow tubes like the xylem vessels of plant stems, without danger of the water column

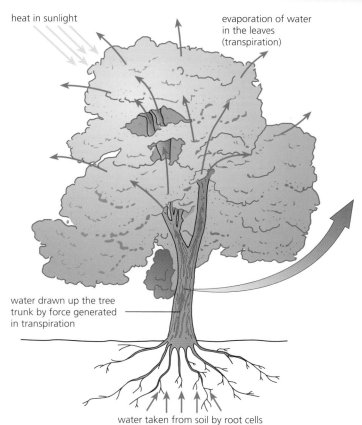

heat in sunlight

evaporation of water in the leaves (transpiration)

water drawn up the tree trunk by force generated in transpiration

water taken from soil by root cells

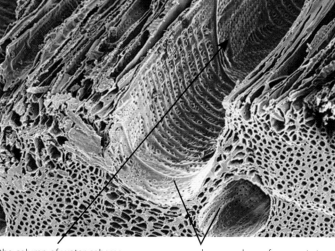

water can be drawn up to a great height without the column breaking off or pulling apart

the column of water coheres, adheres to the walls of the xylem vessels and flows smoothly through them (because its viscosity is low)

xylem vessels run from roots to leaves, as continuous narrow tubes

Figure 7.14 Water is drawn up a tree trunk

breaking (Figure 7.14). Compared with other liquids, water has extremely strong adhesive and cohesive properties that prevent it breaking under tension.

Consequently, under tension the water column does not break or pull away from the sides of the xylem vessels. The result is that water is drawn (literally *pulled*) up the stem. So water flow in the xylem is always upwards. We call this flow of water the **transpiration stream** and the explanation of water transport up the stem, the **cohesion–tension theory**.

The tension on the water column in xylem is demonstrated experimentally when a xylem vessel is pierced by a fine needle. Immediately, a bubble of air enters the column (and will interrupt water flow). If the contents of the xylem vessel had been under pressure, then a jet of water would have been released from the broken vessel.

Further evidence of the cohesion–tension theory comes from measurement of the diameter of a tree trunk over a 24-hour period. In a large tree there is an easily detectable shrinkage in the diameter of the trunk during the day. This is explored in the data handling activity on the CD. Under tension, xylem vessels get narrower in diameter, although their collapse is prevented by the lignified thickening of their walls. Notice that these are on the inside of the xylem vessels (Figure 7.15).

The diameter of a tree trunk recovers during the night when transpiration virtually stops and water uptake replaces the earlier losses of water vapour from the aerial parts of the plant.

Investigating transpiration

The conditions that affect transpiration may be studied using a **potometer**. You will find that these are also the conditions which effect evaporation of water from any moist surface. They include the humidity of the air, temperature and air movements. In Figure 7.16 the ways in which these factors may be investigated is suggested.

Actually, the potometer measures water uptake by the shoot. In the potometer the shoot has an unlimited supply of water. By contrast, in the intact plant the supply of water from the roots may slow down (as in drought conditions, for example), so the water supply is also a factor in the rate of transpiration in the intact plant. This point will help you answer Question 7.

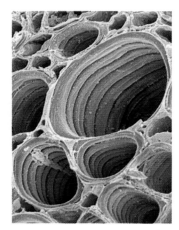

Figure 7.15 Xylems vessel with internal rings of lignin thickening their lateral walls

Table 7.1 Factors that influence transpiration and how to investigate them

The factors that influence transpiration:	How to investigate factor using the potometer:
humidity – at low humidity more water evaporates	compare transpiration by shoot contained in polythene bag with that by shoot in moving air
temperature – heat energy drives evaporation from surface of mesophyll cells	compare transpiration at different air temperatures
air movement – wind carries away saturated air from around leaves, maintaining a concentration gradient between leaf interior and air outside stomata	compare transpiration at different fan speeds
water supply – if leaves become flaccid the stomata close	(not applicable – potometer supplies unlimited water to shoot)

Question

7 Explain the significance of the fact that in large trees there is an easily detectable shrinkage in the diameter of the trunk during the daytime in hot weather that recovers in the night.

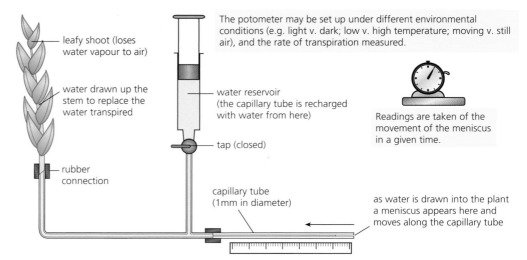

Figure 7.16 Investigating transpiration using a potometer

Does transpiration have a role?

We have seen that transpiration is a direct result of plant structure, plant nutrition and the mechanism of gas exchange in leaves, rather than being a valuable process. In effect, the living plant is a 'wick' that steadily dries the soil around it.

However, transpiration confers advantages, too.

- Evaporation of water from the cells of the leaf in the light has a strong cooling effect (page 52).
- The stream of water travelling up from the roots passively carries the dissolved ions that have been actively absorbed from the soil solution in the root hairs. These are required in the leaves and growing points of the plant (see below).
- All the cells of a plant receive water by lateral movements of water down a water potential gradient from xylem vessels, via pits in their walls. This allows living cells to be fully hydrated. It is the turgor pressure of these cells that provides support to the whole leaf, enabling the leaf blade to receive maximum exposure to light. In fact, the entire aerial system of non-woody plants is supported by this turgor pressure (Figure 4.17, page 88).

So, transpiration does have significant roles in the life of the plant.

Xerophytes: plants of permanently dry and arid conditions

Most native plants of temperate and tropical zones and many crop plants grow best in habitats with plenty of rain and well-drained soils. These sorts of plants are known as **mesophytes**. It is the structure of these plants that has been described in this topic so far. Typically, their leaves are exposed to moderately dry air. Much water is lost from their leaves, particularly in drier periods. This is especially the case in the early part of the day. Later, if excessive loss of water continues, it is prevented by the responses of the stomata – they close (Figure 7.11). After that, the water lost can be replaced by the water uptake that continues, day and night.

On the other hand, some plants are well able to survive and often grow well in habitats where water is permanently scarce. These plants are known as **xerophytes**. They show features that directly or indirectly help to reduce the water loss due to transpiration to a minimum. The adaptations they have that enable them to survive where water is short are referred to as **xeromorphic features**. These are summarised in Table 7.2.

Table 7.2 Typical structural features of xerophytes

Structural feature	Effect
Exceptionally thick cuticle to leaf (and stem) epidermis	Prevents water loss through the external wall of the epidermal cells
Layer of hairs on the epidermis	Traps moist air over the leaf and reduces diffusion
Reduction in the number of stomata	Reduces the outlets through which moist air can diffuse
Stomata in pits or groves	Moist air is trapped outside the stomata, reducing diffusion
Leaf rolled or folded when short of water (cells flaccid)	Reduces area from which transpiration can occur
Leaf area severely reduced – possibly to 'needles'	Stems take over a photosynthetic function and support much reduced leaves
Have two types of roots: a) superficial roots b) deep and extensive roots	Obtain water by both: exploiting overnight condensation at soil surface exploiting a deep water table in the soil

Question

8 Suggest why it is that, of all the environmental factors which affect plant growth, the issue of water supply is so critical.

Sorghum, a grain plant that thrives in extremely arid conditions, shows marked adaptations in its leaf structure and root system. Marram grass is a pioneer plant of sand dunes.

Examine Figures 7.17 and 7.18 carefully. Which of the features listed in Table 7.2 does each plant exhibit?

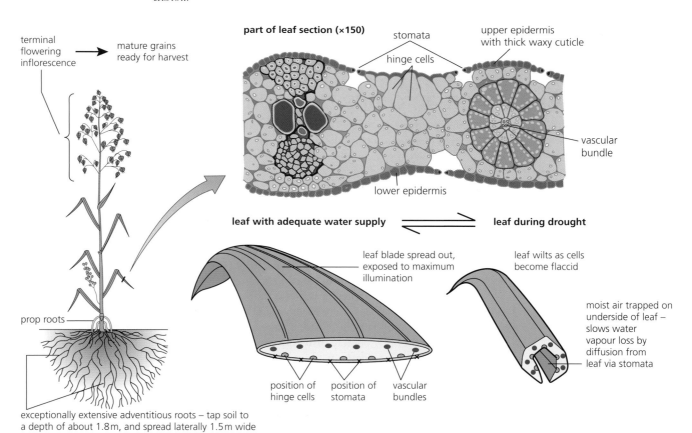

Figure 7.17 *Sorghum*: a drought-resistant tropical plant of economic importance

marram grass has the ability to grow in the extremely arid environment of sand dunes, accelerating the build-up of sand

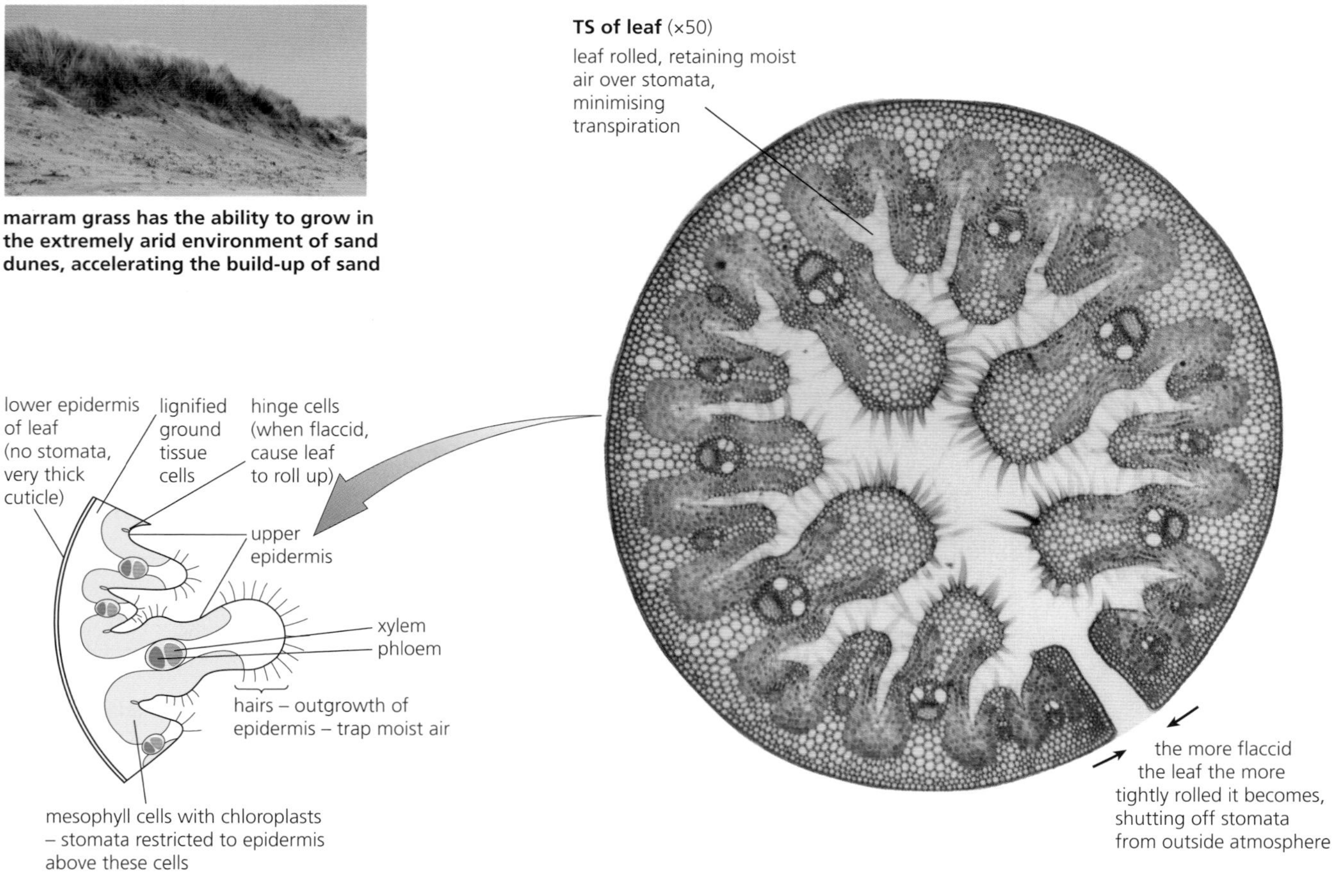

TS of leaf (×50)
leaf rolled, retaining moist air over stomata, minimising transpiration

lower epidermis of leaf (no stomata, very thick cuticle)

lignified ground tissue cells

hinge cells (when flaccid, cause leaf to roll up)

upper epidermis

xylem
phloem

hairs – outgrowth of epidermis – trap moist air

mesophyll cells with chloroplasts – stomata restricted to epidermis above these cells

the more flaccid the leaf the more tightly rolled it becomes, shutting off stomata from outside atmosphere

Figure 7.18 Marran grass (*Ammophila*): a part of sand dunes

Question

9 Explain the significance of:
  a root hair cells being able to take up nitrate ions from the soil solution even though their concentration in the cell is already higher than in the soil
  b plants often failing in soil that is permanently waterlogged. (Think about the importance of oxygen for respiration.)

Inorganic ions and plants

'Minerals' are inorganic elements, some of which are needed by organisms. They include substances like sodium chloride, in which ionic bonding occurs. In ionic bonding one or more electrons is transferred completely from one atom to another, to produce a stable arrangement of electrons. Ionic bonding transforms atoms into stable ions. The example of ionic bonding in sodium chloride is shown in Figure A3 of Appendix 1 (on the CD). Ionic bonding and the formation of ions contrasts with covalent bonding, which is shown in water (Figure 2.27, page 52) and in methane, for example.

Living things need several different inorganic ions but only a handful are needed in relatively large quantities. These are known as the **major mineral elements**. Major mineral elements required by plants include anions, like phosphate ($H_2PO_4^-$), **nitrate (NO^{3-})**, and cations like calcium (Ca^{2+}), and **magnesium (Mg^{2+})**.

The source of mineral ions for plants is the soil solution – the layer of water that occurs around the individual soil particles. Here ions occur in quite low concentrations but plants can take them up through their root hairs. So, water uptake and the absorption of mineral ions occur at the root hairs but they occur by entirely **different mechanisms**.

Ions are taken up **selectively**, by an **active process**, using energy from respiration. This means that the ions that are useful are selected and the ions not required by the plant are ignored. Uptake occurs by means of protein pumps (page 91) in the cell surface membrane of the root cells including root hair cells. Protein pumps are specific for particular ions so, if an ion is required, the particular transport protein is produced by the cells and built into the cell membranes. If the required ion is available in the soil solution, it can be pumped into the cell, despite there being a higher concentration in the cell already.

Signs and symptoms of mineral deficiency

In the soil, the ions dissolved in the soil solution come from the mineral skeleton of the soil by chemical erosion. They also come from the decay of the dead organic matter that is constantly being added to the soil. Among these are nitrate and magnesium ions, both required for particular biochemical pathways (Table 7.3). Consequently, plants that are deficient in these (and other) essential nutrients normally show characteristic signs and symptoms. Notice that the signs of deficiency of nitrate and magnesium ions are very similar.

Can you think why this is? Remember, the green colour of plants is due to the presence of chlorophyll. Are nitrate and magnesium ions required for synthesis of new chlorophyll molecules?

When deficiency symptoms appear in cultivated commercial crops, the mineral nutrients available in the soil may be checked by chemical analysis, to confirm the deficiency. Then, artificial fertiliser can be applied to counter the deficiency and restore normal growth patterns.

Question

10 Why and how is the supply of essential ions maintained in the soil in agriculture?

Table 7.3 Nitrate and magnesium: symptoms of deficiency and roles in metabolism

Symptoms of deficiency	Roles in metabolism
Nitrate deficiency	**Key ingredient in amino acid (and protein) synthesis**
upper leaves light green, lower leaves yellow	1 Nitrate ions are reduced to ammonium ions: NO_3 reduction, using reducing power (NADH) from respiration → NH_4^+ 2 Ammonium ions are combined with an organic acid to form the amino acids required for protein synthesis: NH_4^+ organic acids → amino acids 3 From the pool of amino acids are formed the polypeptides and proteins the cell requires: pool of amino acids in cells where protein is synthesised → formation of peptide bonds → protein NH_3 amino acids peptide bonds COOH Nucleic acids also contain nitrogen

(continued)

Extension   (continued)

Symptoms of deficiency	Roles in metabolism
Magnesium deficiency	**Component of chlorophyll**
lower leaves yellow from margins inwards, veins remain green	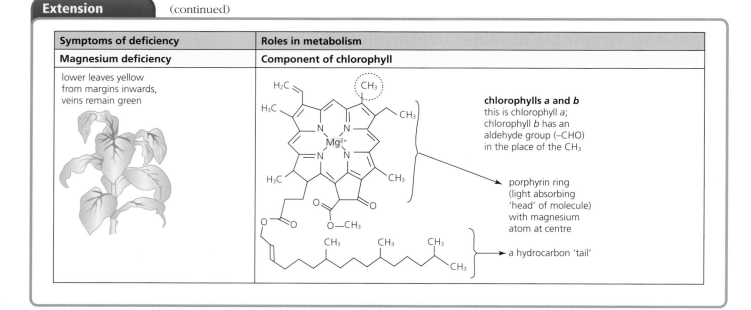

chlorophylls *a* and *b*
this is chlorophyll *a*; chlorophyll *b* has an aldehyde group (–CHO) in the place of the CH₃

porphyrin ring (light absorbing 'head' of molecule) with magnesium atom at centre

a hydrocarbon 'tail'

Translocation

Translocation is the process by which manufactured food (sugars and amino acids – sometimes called **assimilates**) are moved around the plant. This occurs in the **phloem tissue** of the vascular bundles (Figure 7.1). The sugars made in the leaves in the light are moved around the plant as sucrose. For example, young leaves export sugars to sites where new stems, new leaves or new roots are being formed. Then, in older plants, sucrose is sent to sites of storage, such as the cortex of roots or stems, and to seeds and fruits.

Meanwhile, nitrates are absorbed from the soil solution by the root hairs. Nitrates are one of the raw materials used in the synthesis of amino acids and also nucleic acids. Amino acid synthesis takes place in the roots. Amino acids are then moved to sites in the plant where protein synthesis is occurring, such as in buds, young leaves, young roots and developing fruits. So you can see that the contents of phloem (known as sap) may vary. Also, it may flow in either direction in the phloem.

Phloem structure – a reminder

Phloem tissue consists of living cells – sieve tube elements and companion cells. Both have cellulose cell walls. These walls do not become lignified. Translocation is dependent on living cells and there is a close relationship between sieve tube elements and their companion cells.

Sieve tubes are built from narrow, elongated cells, connected end to end to form continuous tubes. Their touching end walls are pressed together and become perforated by large pores which remain open. The whole structure, known as **sieve plate**, allows free flow of liquid through the tube system. The cytoplasm of a mature sieve tube element has lost its nucleus, and neither does it have many of the other organelles of a cell. A thin layer of cytoplasm lines the walls.

At least one **companion cell** accompanies each sieve tube element. Sieve tube elements and companion cells are connected by strands of cytoplasm, the plasmodesmata, that pass through gaps in the walls. The contents of the companion cells differ from those of sieve tube elements. They retain their nucleus and the other organelles that are common to cells that are metabolically active – particularly mitochondria and ribosomes. We can assume that companion cells service and maintain the sieve tube elements in various ways.

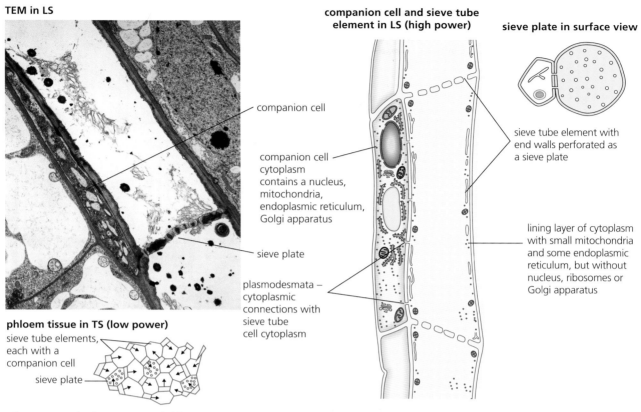

TEM in LS

companion cell

companion cell cytoplasm contains a nucleus, mitochondria, endoplasmic reticulum, Golgi apparatus

sieve plate

plasmodesmata – cytoplasmic connections with sieve tube cell cytoplasm

companion cell and sieve tube element in LS (high power)

sieve plate in surface view

sieve tube element with end walls perforated as a sieve plate

lining layer of cytoplasm with small mitochondria and some endoplasmic reticulum, but without nucleus, ribosomes or Golgi apparatus

phloem tissue in TS (low power)

sieve tube elements, each with a companion cell

sieve plate

Figure 7.19 The fine structure of phloem

Phloem transport

Phloem transport occurs by **mass flow** but the pressure difference that drives it is produced by a different mechanism from that in xylem transport. Most significantly, translocation requires living cells (Figure 7.20). Cell respiration is the process by which energy is transferred to drive metabolic reactions and processes. Much of the respiration occurs in the mitochondria. ATP is the molecule produced in mitochondria to transfer energy. We shall see that ATP is involved in the mechanism of phloem transport.

Translocation can be illustrated by the movement of sugar from the leaves. The story starts at the point where sugars are made and accumulate within the mesophyll in the leaf. This is the **source area**.

Sugars are loaded into the phloem sieve tubes through transfer cells (Figure 7.21). From here, sucrose is pumped into the companion cells by the combined action of **primary** and **secondary pumps**. These pumps are special proteins in the cell surface membrane. The primary pumps remove hydrogen ions (protons) *from* the cytoplasm of the companion cell, so setting up a gradient in concentration of hydrogen ions between the exterior and interior of the companion cell. This movement requires ATP. Hydrogen ions then flow back into the companion cell down their concentration gradient. This occurs at specific sites called secondary pumps where their flow is linked to the transport of sucrose molecules in the same direction.

As sucrose solution accumulates in the companion cells it moves on by diffusion into the sieve tubes, passing along the plasmodesmata (Figure 7.19). The accumulation of sugar in the phloem tissue lowers the water potential and water follows the sucrose, diffusing down a water potential gradient. This creates a high hydrostatic pressure in the sieve tubes of the source area.

Meanwhile, in living cells elsewhere in the plant – often, but not necessarily in the roots, sucrose may be converted into insoluble starch deposits. This is a **sink area**. As sucrose flows out of the sieve tubes here, the water potential is raised. Water then diffuses out down a water potential gradient and the hydrostatic pressure is lowered. These processes create the difference in hydrostatic pressures in the source and sink areas that drives mass flow in the phloem.

Questions

11 What does the presence of a large number of mitochondria in the companion cells suggest to you about their role in the movement of sap in the phloem?

12 Examine Figure 7.20 carefully.

a What sequence of events would you anticipate in a leaf stalk as the content of a water-jacket is raised to 50°C?

b How would you expect the phloem sap sampled from a sieve tube near leaves in the light and at the base of the same stem to differ?

Figure 7.20 Experiment to show that translocation requires living cells

That translocation requires living cells is shown by investigation of the effect of temperature on phloem transport

Note: in neither experiment did the leaf blade wilt – xylem transport is not heat sensitive at this range of temperatures (because xylem vessels are dead, empty tubes)

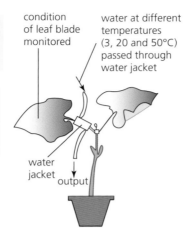

condition of leaf blade monitored

water at different temperatures (3, 20 and 50°C) passed through water jacket

water jacket output

(a) at 50°C, translocation of sugar from the leaf blade stopped – this is above the thermal death point of cytoplasm

conclusion: living cells are essential for translocation

(b) at 3°C, compared with 20°C translocation of sugar from leaf blade was reduced by almost 10% of leaf dry weight over a given time

conclusion: rate of metabolic activity of phloem cells affects rate of translocation

Figure 7.21 Transfer cells and the loading of sieve tubes

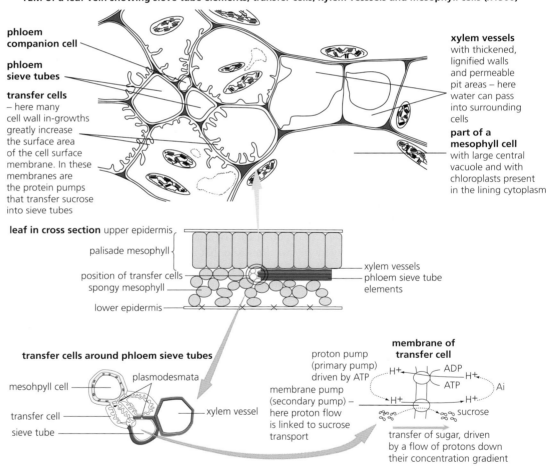

TEM of a leaf vein showing sieve tube elements, transfer cells, xylem vesssels and mesophyll cells (x1500)

phloem companion cell

phloem sieve tubes

transfer cells – here many cell wall in-growths greatly increase the surface area of the cell surface membrane. In these membranes are the protein pumps that transfer sucrose into sieve tubes

xylem vessels with thickened, lignified walls and permeable pit areas – here water can pass into surrounding cells

part of a mesophyll cell with large central vacuole and with chloroplasts present in the lining cytoplasm

leaf in cross section upper epidermis

palisade mesophyll

position of transfer cells

spongy mesophyll

lower epidermis

xylem vessels

phloem sieve tube elements

transfer cells around phloem sieve tubes

mesohpyll cell

plasmodesmata

transfer cell

sieve tube

xylem vessel

proton pump (primary pump) driven by ATP

membrane pump (secondary pump) – here proton flow is linked to sucrose transport

membrane of transfer cell

H+ ADP
H+
ATP Ai
H+ H+
sucrose

transfer of sugar, driven by a flow of protons down their concentration gradient

The pressure flow hypothesis

The principle of the pressure flow hypothesis is that the sugar solution flows down a hydrostatic pressure gradient. There is a high hydrostatic pressure in sieve elements near mesophyll cells in the light (source area) but low hydrostatic pressure in elements near starch storage cells of the stem or root (sink area). Mass flow is illustrated in Figure 7.22 and the annotations explain the steps.

In this hypothesis, the role of the companion cells (living cells, with a full range of organelles in the cytoplasm) is to maintain conditions in the sieve tube elements favourable to mass flow of solutes. Companion cells use metabolic energy (ATP) to do this.

Question

13 What conditions maintain
 a the high water potential of the cell of the root cortex
 b the low water potential of the mesophyll cells of a green leaf?

Table 7.4 Evidence for the pressure flow hypothesis

For	Against
The contents of the sieve tubes are under pressure and sugar solution exudes if phloem is cut.	Phloem tissue carries manufactured food to various destinations (in different sieve tubes), rather than to the greatest sink.
Appropriate gradients between 'source' and 'sink' tissue do exist.	Sieve plates are a barrier to mass flow and might be expected to have been 'lost' in the course of evolution if mass flow is the mechanism of transport.

model demonstrating pressure flow
(A = mesophyll cell, B = starch storage cell)

In this model, the pressure flow of solution would continue until the concentration in A and B is the same.

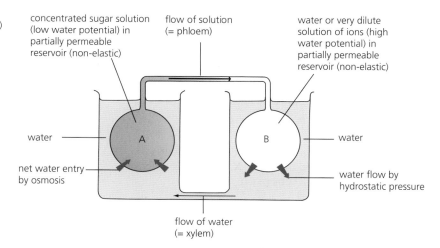

pressure flow in the plant

In the plant, a concentration difference between A and B is maintained by conversion of sugar to starch in cell B, while light causes production of sugar by photosynthesis in A.

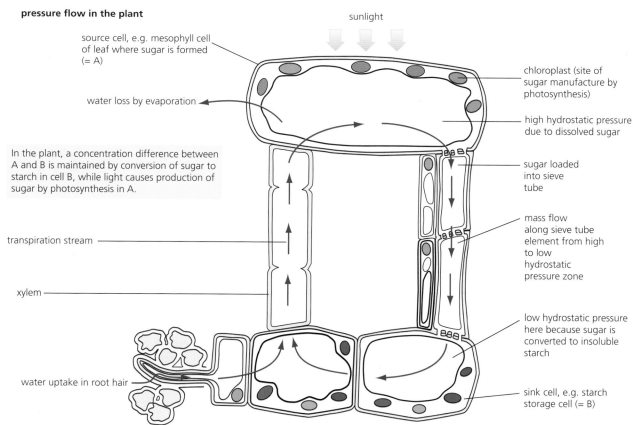

Figure 7.22 The pressure flow hypothesis of phloem transport

Summary

- Internal transport within plants occurs by **mass flow** but there is no pumping organ. Two separate tissues are involved in transport. Water and ions travel in the **xylem**, a system of vessels connected end to end to form non-living tubes. Manufactured foods are carried in a living tissue, the **phloem**, consisting of sieve tubes and companion cells. The xylem and phloem make up the **vascular tissue** that branches throughout the plant body and serves roots, stems, leaves and growing points (buds).

- Water moves from the soil to the aerial system by a **gradient in water potential**. Water is drawn up the stem by a force generated in the leaves by the evaporation of water vapour from the aerial system (**transpiration**).

- Hydrogen bonding between water molecules, and between these molecules and hydrophilic surfaces, is responsible for

the **cohesive** and **adhesive** properties of water, and makes possible the upward transport of water in the xylem by cohesion–tension.

- Transport of manufactured food in the phloem is by **active transport** requiring living phloem cells. According to the **mass flow hypothesis**, solutes flow through the phloem from a region of high hydrostatic pressure to a region of low hydrostatic pressure. Hydrostatic pressure is high in cells where sugar is formed, the **source area**, but low in **sink areas**, where sugar is converted to starch.

- Plants that survive permanently arid conditions (xerophytes), including the crop plant *Sorghum* and the sand-dune plant Marram grass, have leaves adapted to reduce water loss by transpiration.

Examination style questions

1 Fig. 1.1 is a drawing of a transverse section of a leaf.
  a) i) Use label lines and the letters **X**, **S**, **E** and **D** to indicate the following on a copy of Fig. 1.1:
      X   a xylem vessel
      S   a phloem sieve tube
      E   a lower epidermal cell
      D   a palisade mesophyll cell   [4]
   ii) Calculate the magnification of Fig. 1.1. Show your working and express your answer to the nearest whole number.   [2]

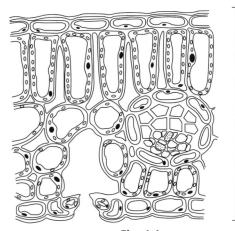

0.5 mm

Fig. 1.1

  b) Name **two** assimilates that move from the palisade mesophyll cells to the vascular tissue to be exported from the leaf.   [2]
  c) Explain, using the term **water potential**, how water moves from the vascular tissue to the atmosphere.   [4]

[Total: 12]

(Cambridge International AS and A Level Biology 9700, Paper 02 Q2 June 2005)

2 a) Draw and annotate a diagram of a potometer in use in the measurement of transpiration.   [8]
  b) Name three external factors that may influence the rate of transpiration in a plant in the light and explain why they have an effect on the rate of loss of water vapour.   [6]
  c) What is a xerophyte?   [1]
  d) Tabulate the chief xeromorphic features that *Sorghum* exhibits and explain how and why they are effective.   [5]

[Total: 20]

3 The diagram shows a vascular bundle from the stem of *Peperomia dahlstedtii*, a plant from Brazil. The vascular bundles in the stems of *P. dahlstedtii* are unusual because they are surrounded by an endodermis with a Casparian strip.

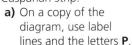

  a) On a copy of the diagram, use label lines and the letters **P**, **Q** and **R** to identify the following in the vascular bundle.
      P an endodermal cell with a Casparian strip
      Q a cell wall strengthened with lignin
      R a tissue that transports assimilates   [3]
  b) Vascular tissue in roots is surrounded by an endodermis. Describe the function of the endodermis in roots.   [3]
  c) State and explain two ways in which the **structure** of a phloem sieve tube is adapted for the transport of assimilates.   [4]

[Total: 10]

(Cambridge International AS and A Level Biology 9700, Paper 22 Q5 June 2009)

8 Transport in mammals

As animals become larger, more complex and more active, transport systems become essential to supply nutrients to, and remove waste from, individual cells. Mammals are far more active than plants and require much greater supplies of oxygen. This is transported by haemoglobin inside red blood cells.

 ## 8.1 The circulatory system

The mammalian circulatory system consists of a pump, many blood vessels and blood, which is a suspension of red blood cells and white blood cells in plasma.

By the end of this section you should be able to:

a) state that the mammalian circulatory system is a closed double circulation consisting of a heart, blood vessels and blood

b) observe and make plan diagrams of the structure of arteries, veins and capillaries using prepared slides and be able to recognise these vessels using the light microscope

c) explain the relationship between the structure and function of arteries, veins and capillaries

d) observe and draw the structure of red blood cells, monocytes, neutrophils and lymphocytes using prepared slides and photomicrographs

e) state and explain the differences between blood, tissue fluid and lymph

f) describe the role of haemoglobin in carrying oxygen and carbon dioxide with reference to the role of carbonic anhydrase, the formation of haemoglobinic acid and carbaminohaemoglobin

g) describe and explain the significance of the oxygen dissociation curves of adult oxyhaemoglobin at different carbon dioxide concentrations (the Bohr effect)

h) describe and explain the significance of the increase in the red blood cell count of humans at high altitude

Efficient internal transport in animals

The cells of organisms need a constant supply of water and organic nutrients such as glucose and amino acids; most need oxygen, and the waste products of cellular metabolism have to be removed. Larger animals have evolved a blood circulatory system for efficient internal transport. This links the parts of the body and makes these resources available where they are required.

Mammals have a closed circulation in which blood is pumped by a powerful, muscular heart and circulated in a continuous system of tubes – the arteries, veins and capillaries – under pressure. The heart has four chambers and is divided into right and left sides. Blood flows from the right side of the heart to the lungs, then back to the left side of the heart. From here it is pumped around the rest of the body and back to the right side of the heart. As the blood passes twice through the heart in every single circulation of the body this is called a double circulation. The circulatory system of mammals is shown in Figure 8.1, alongside alternative systems.

Look at these systems carefully.

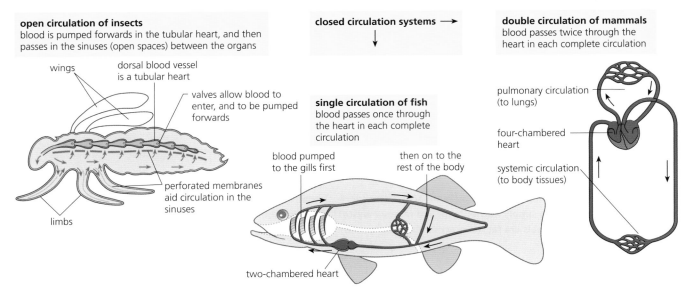

open circulation of insects
blood is pumped forwards in the tubular heart, and then passes in the sinuses (open spaces) between the organs

closed circulation systems ⟶

double circulation of mammals
blood passes twice through the heart in each complete circulation

wings

dorsal blood vessel is a tubular heart

valves allow blood to enter, and to be pumped forwards

single circulation of fish
blood passes once through the heart in each complete circulation

perforated membranes aid circulation in the sinuses

limbs

blood pumped to the gills first

then on to the rest of the body

two-chambered heart

pulmonary circulation (to lungs)

four-chambered heart

systemic circulation (to body tissues)

Figure 8.1 Open and closed circulations

Question

1 In an open circulation there is 'little control over circulation'. Suggest what this means?

Looking at these other systems helps us to understand the features of the mammalian circulation. It becomes clear that the major advantages of the mammalian circulation are:
- simultaneous high pressure delivery of oxygenated blood to all regions of the body
- oxygenated blood reaches the respiring tissues, undiluted by deoxygenated blood.

Blood – the transport medium

Blood is a special tissue of several different types of cell suspended in a liquid medium called plasma (Figure 8.2).

Plasma is a straw-coloured, very slightly alkaline liquid consisting mainly of **water**. It is the medium for the continual exchange of substances by cells and tissues throughout the body. Dissolved in the plasma are **nutrients** in transit from the gut or liver to all the cells. Excretory products are also transported in solution, mainly **urea** from the liver to the kidneys. So too are **hormones**, from the ductless (endocrine glands), where they are formed and released, to the tissues and organs.

The plasma also contains **dissolved proteins**. The principle blood protein is **albumin**, which has the role of regulating the water potential of the blood. The presence of dissolved albumin stops too much water leaving at the capillaries by osmosis and also helps in the return of fluids (Figure 8.8, page 159). Albumin and other proteins also assist in the transport of lipids and iron in the plasma. Other blood proteins present are **antibodies** (page 222) and components of the **blood clotting mechanism**.

Suspended in the plasma are the **blood cells**. The majority of these are **red blood cells** (erythrocytes) – about 5 million in every cubic millimetre of blood. The red blood cells are formed in the bone marrow tissue. Initially each red blood cell has a nucleus but early in development the nucleus is lost and these tiny cells ($7\,\mu m$ in diameter) adopt the shape of a biconcave disc. Mitochondria and endoplasmic reticulum are also lost and a great deal of a protein, **haemoglobin**, takes their place, together with the enzyme **carbonic anhydrase**. The role of mature red blood cells is exclusively the transport of respiratory gases in the blood. The lifespan of a working red blood cell is only about 120 days, after which they are broken down and replaced. Most of their components are retrieved and reused.

Question

2 Assuming the body contains 5 litres of blood, that there are 5 million (5×10^6) red blood cells per mm^3 and that a red blood cell lasts for 120 days, how many red blood cells must be replaced per day?

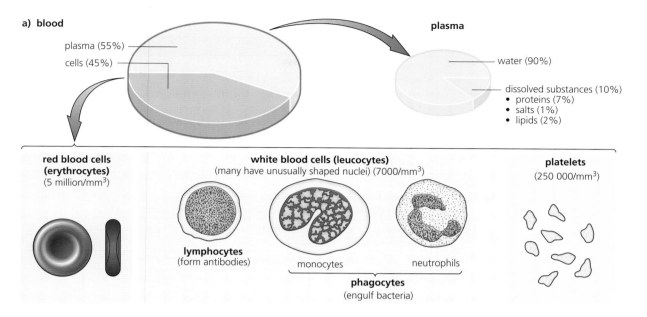

a) **blood**

plasma (55%)

cells (45%)

plasma

water (90%)

dissolved substances (10%)
- proteins (7%)
- salts (1%)
- lipids (2%)

**red blood cells
(erythrocytes)**
(5 million/mm³)

white blood cells (leucocytes)
(many have unusually shaped nuclei) (7000/mm³)

lymphocytes
(form antibodies)

monocytes

neutrophils

phagocytes
(engulf bacteria)

platelets
(250 000/mm³)

b)

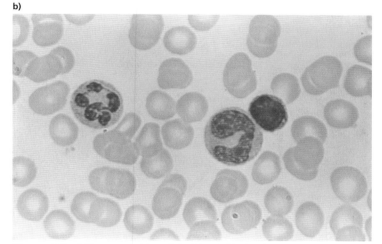

Figure 8.2 a) The composition of the blood; **b)** Blood smear stained to show the few white blood cells present

Also present are some **white blood cells** (leucocytes) – about 7000 per cubic millimetre of blood. These also originate in the bone marrow but they retain their nucleus. In the growth and development stage, white blood cells may migrate to the thymus gland, lymph nodes or the skin. The white blood cells do not necessarily function in the blood stream but all use the blood circulation as a means of reaching specific body organs or tissues. Many leave the bloodstream by migrating between the cells of the capillary wall. There are several different types of white blood cells, all of which play a part in the body's defences (Topic 11) and are mostly relatively short-lived cells. The white blood cells that are particularly important are:

- **lymphocytes** – these form antibodies
- **phagocytes** – these ingest bacteria or cell fragments.

Finally the blood contains **platelets**, best described as cell fragments. These are tiny packages of cytoplasm containing vesicles of substances that, when released, have a role in the blood clotting mechanism.

Observing and drawing the structure of red blood cells, monocytes, neutrophyls and lymphocytes

You need to observe and draw the structure of red cells, monocytes, neutrophyls and lymphocytes, using photomicrographs and prepared slides and a light microscope under medium and high power magnification, so that you will be able to recognise them.

Table 8.1 A summary of the roles of blood components

Component	Role
Plasma	Transport of: • **nutrients** from the gut or liver to all the cells • excretory products such as **urea** from the liver to the kidneys • **hormones** from the endocrine glands to all tissues and organs • **dissolved proteins** which have roles including regulating the osmotic concentration (water potential) of the blood • dissolved proteins that are **antibodies** **Heat** distribution to all tissues
Red blood cells (erythrocytes)	Transport of: • **oxygen** from the lungs to respiring cells • **carbon dioxide** from respiring cells to the lungs (also carried in plasma)
White blood cells (leucocytes)	**Lymphocytes**: major roles in the **immune system**, including forming antibodies (Topic 7) **Phagocytes: ingest bacteria** and cell fragments
Platelets	Part of the **blood clotting mechanism**

The blood circulation in mammals

Mammals have a closed circulation system in which blood is pumped by a powerful, muscular **heart** and circulated under pressure in a continuous system of tubes – the **arteries**, **veins** and **capillaries**. The heart has four chambers, and is divided into right and left sides. This is shown in Figure 8.3, in the diagram of the layout of the human blood circulation.

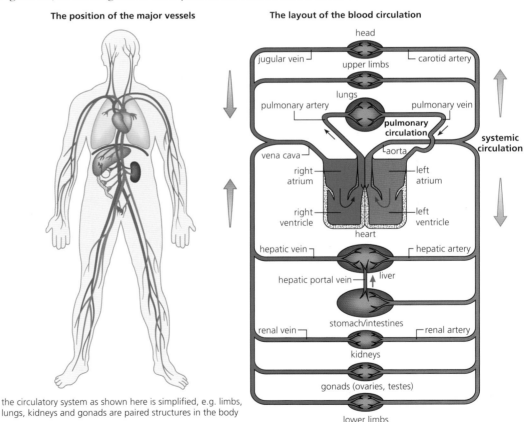

The position of the major vessels

The layout of the blood circulation

the circulatory system as shown here is simplified, e.g. limbs, lungs, kidneys and gonads are paired structures in the body

Figure 8.3 The human blood circulation

Notice that blood is pumped from the right side of the heart to the **lungs**. From the lungs, blood returns to the left side of the heart. This part of the system is called the **pulmonary circulation**.

From the left side of the heart, blood is pumped around the **rest of the body** before returning to the right side of the heart. This is called the **systemic circulation**.

So you can see that blood passes *twice* through the heart in every single circulation of the body. This arrangement is a called a **double circulation**. This means there is no mixing of oxygenated and deoxygenated blood in the heart, so the blood circulating around the body has a higher concentration of oxygen than would otherwise be possible.

Arteries, veins and capillaries

There are three types of vessel in the circulation system, each with different roles.
- **Arteries** carry blood away from the heart.
- **Veins** carry blood back to the heart.
- **Capillaries** are fine networks of tiny tubes linking arteries and veins.

In Figure 8.3 you can see how the main arteries and veins serve the organs of the body. (Notice that arteries and veins are often named after the organs they serve.)

For example, in the systemic circulation, each organ is supplied with blood by a separate **artery** that branches from the main **aorta**. Inside organs, the artery branches into numerous **arterioles** (smaller arteries). The smallest of these arterioles supply numerous and very extensive **capillary networks**. These capillaries then drain into **venules** (smaller veins) and these venules join to form **veins**. Finally, the veins join the **vena cava** carrying blood back to the right side of the heart. The branching sequence of blood vessels that make up the systemic circulation is, therefore:

aorta → artery → arteriole → capillary → venule → vein → vena cava

The structure of arteries, capillaries and veins

Blood leaving the heart is under very high pressure and travels in waves or pulses following each heart beat. By the time the blood has reached the capillaries it is under very much lower pressure, without a pulse. These differences in blood pressure account for the differences in the wall structure of arteries and veins. There are many subtle changes on the way between a main artery and the corresponding vein.

Arteries and veins have three distinct layers to their walls (Figure 8.4 and Table 8.2). Both arteries and veins have strong, elastic walls but the walls of the arteries are very much thicker and stronger than those of the veins. The strength of the walls comes from a combination of collagen and elastic fibres. Their elasticity is due to the elastic fibres and to the involuntary (smooth) muscle fibres.

Capillaries have an altogether different structure – they consist of a single layer of cells and they are tiny in cross-section when compared with the main arteries and veins.

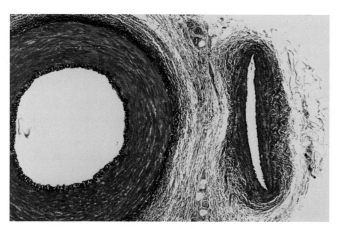

TS artery and vein, LP (×20) – in sectioned material (as here), veins are more likely to appear squashed, whereas arteries are circular in section

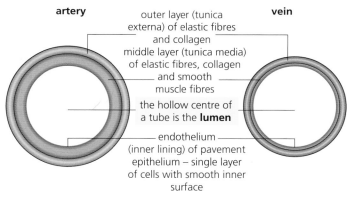

artery

outer layer (tunica externa) of elastic fibres and collagen

middle layer (tunica media) of elastic fibres, collagen and smooth muscle fibres

the hollow centre of a tube is the **lumen**

endothelium (inner lining) of pavement epithelium – single layer of cells with smooth inner surface

vein

capillary wall (enlarged)

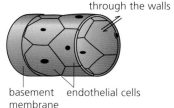

substances leave (and enter) capillaries through the walls

basement membrane   endothelial cells

Figure 8.4 The structure of the wall of arteries, veins and capillaries

155

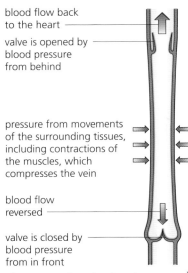

blood flow back to the heart

valve is opened by blood pressure from behind

pressure from movements of the surrounding tissues, including contractions of the muscles, which compresses the vein

blood flow reversed

valve is closed by blood pressure from in front

Figure 8.5 The valves in veins

Observing and drawing the structure of arteries, veins and capillaries

You need to observe and draw the structure of arteries, veins and capillaries, using prepared slides and a light microscope under medium and high power magnification, so that you will be able to recognise them.

Table 8.2 Differences between arteries, capillaries and veins

	Artery	Capillary	Vein
Outer layer (tunica externa) of elastic fibres and collagen	Present – thick layer	Absent	Present – thin layer
Middle layer (tunica media) of elastic fibres, collagen and involuntary (smooth) muscle fibres	Present – thick layer	Absent	Present – thin layer
Endothelium (inner lining) of pavement epithelium – a single layer of cells fitting together like jigsaw pieces, with a smooth inner surface that minimises friction	Present	Present	Present
Valves	Absent	Absent	Present

Now look carefully at the photomicrograph of an artery and vein in section and the details the wall structure of these three vessels (Figure 8.4).

If you examine a prepared slide of an artery and vein under the high power of a microscope you may see the layers of the walls in detail and you may be able to make out the muscle fibres. Because of the low pressure in veins there is a possibility of back-flow here. Veins have **valves** at intervals that prevent this (Figure 8.5).

We have noted that, as blood is transported around the body the pressure of blood progressively falls. This fact and the roles of the arteries, capillaries and veins are reflected in specific changes in their structure (Table 8.3).

Table 8.3 The changing structure of blood vessels in relation to function

Component and role	Structure in relation to function
Arteries: have the thickest and strongest walls; the tunica media is the thickest layer.	
Aorta: receives blood pumped from the heart in a 'pulse' (pressure about 120 mmHg).	• The walls stretch to accommodate the huge surge of blood when the ventricles contract. The elastic and collagen fibres of the tunica externa prevent rupture as blood surges from the heart.
Main arteries: distribute blood under high pressure to regions of the body. • Arteries become wider, so lowering the pressure. • Supply the smaller **branch arteries** that distribute blood to the main organs.	• The high proportion of elastic fibres are first stretched and then recoil, keeping the blood flowing and propelling it forwards after each pulse passes. • With increasing distance from the heart the tunica media progressively contains more smooth muscle fibres and fewer elastic fibres. By varying the constriction and dilation of the arteries, blood flow is maintained. Muscle fibres stretch and recoil, tending to even out the pressure but a pulse can still be detected.
Arterioles: deliver blood to the tissues.	• The walls have a high proportion of smooth muscle fibres and so are able to regulate blood flow from the arteries into the capillaries.
Capillaries	
Capillaries: serve the tissues and cells of the body. The blood is under lower pressure (about 35 mmHg).	• Narrow tubes, about the diameter of a single red blood cell (about 7 μm), reduce the flow rate to increase exchange between blood and tissue. • Thin walls of a single layer of endothelial cells. • The walls have gaps between cells sufficient to allow some components of the blood to escape and contribute to tissue fluid (page 146).
Veins: have thin walls; the tunica externa is the thickest layer.	
Venules: collect blood from the tissues. They are formed by the union of several capillaries (pressure about 15 mmHg).	• The walls consist of endothelium and a very thin tunica media of a few scattered smooth muscle fibres.
Veins: receive blood from the tissues under low pressure (pressure about 5 mmHg). • The veins become wider, so lowering pressure and increasing the flow rate.	• The tunica externa, containing elastic and collagen fibres, is present. • The tunica media contains a few elastic fibres and muscle fibres. • The presence of valves prevents the back-flow of blood.

Pressure: the **pascal (Pa)** and its multiple the **kilopascal (kPa)** are generally used by scientists to measure pressure. However, in medicine the older unit, **millimetres of mercury (mmHg)** is still used (1 mmHg = 0.13 kPa).

Questions

3 List the differences between the pulmonary circulation and the systemic circulation of blood.

4 The graph in Figure 8.6 shows the changing blood velocity, blood pressure and cross-sectional area of the vessels as blood circulates in the body. Use the information in the graph and in Table 8.2 where appropriate to answer the questions below.

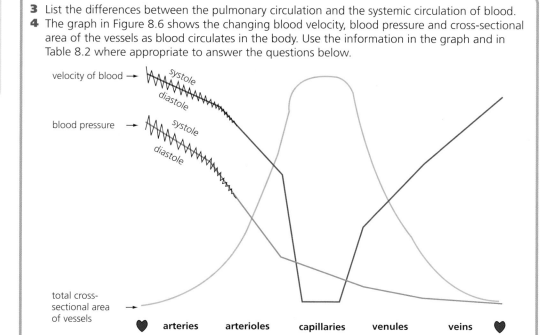

Figure 8.6 The changing blood velocity, blood pressure and cross-sectional area of the vessels as blood circulates in the body

a Comment on the velocity and pressure of the blood as it enters the aorta compared with when it is about to re-enter the heart.

b What aspects of the structure of the walls of arteries may be said to be in response to the condition of blood flow through them?

c What factors may cause the velocity of blood flow to be at the level it is in the capillaries and why is this advantageous?

The roles of the blood circulation system

The blood circulation has a role in the body's defence against diseases (Topic 11), in internal communications and in the maintenance of a constant internal environment (Topic 14), as well as being the transport system of the body. It is the transport roles of the blood circulation (Table 8.4), a key feature of which is the formation and fate of **tissue fluid**, that we consider next.

Table 8.4 Transport roles of the blood circulation

Function in organism	Transport role of circulation
Tissue respiration	Transport of oxygen to all tissues and carbon dioxide back to the lungs
Hydration	Transport of water to all the tissues
Nutrition	Transport of nutrients (sugars, amino acids, lipids, vitamins) and inorganic ions to all cells
Excretion	Transport of waste products of metabolism to kidneys, lungs and sweat glands
Temperature regulation	Distribution of heat
Development and co-ordination	Transport of hormones from endocrine glands to target organs

Exchange in the tissues and the formation of tissue fluid

The blood circulation delivers essential resources (nutrients and oxygen, for example) to the tissues of the body. This occurs as the blood flows through the capillaries, between the cells of the body (Figure 8.7). There are tiny gaps in the capillary walls, found to vary in size in different parts of the body. Through these gaps passes a watery liquid, very similar in composition to plasma. This is **tissue fluid**. However, red blood cells, platelets and most blood proteins are not present in tissue fluid, instead these are retained in the capillaries. Tissue fluid bathes the living cells of the body. Nutrients are supplied from this fluid and carbon dioxide and waste products of metabolism are carried away by it.

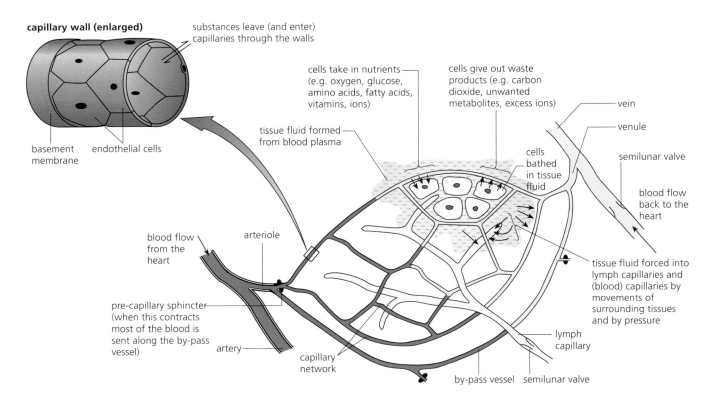

capillary wall (enlarged)

substances leave (and enter) capillaries through the walls

basement membrane   endothelial cells

cells take in nutrients (e.g. oxygen, glucose, amino acids, fatty acids, vitamins, ions)

cells give out waste products (e.g. carbon dioxide, unwanted metabolites, excess ions)

tissue fluid formed from blood plasma

cells bathed in tissue fluid

vein

venule

semilunar valve

blood flow back to the heart

tissue fluid forced into lymph capillaries and (blood) capillaries by movements of surrounding tissues and by pressure

blood flow from the heart

arteriole

pre-capillary sphincter (when this contracts most of the blood is sent along the by-pass vessel)

artery

capillary network

lymph capillary

by-pass vessel   semilunar valve

The role of the by-pass vessel
Which tissues of the body are being served by the blood circulation is regulated. When tiny pre-capillary sphincter muscles are contracted, blood flow to that capillary bed is restricted to a minimum. Most blood is then diverted through by-pass vessels – it is shunted to other tissues and organs in the body.

Figure 8.7 Exchange between blood and cells via tissue fluid

Given the quantity of dissolved solutes in the plasma (including and especially all the blood plasma proteins), we would expect the water potential of the blood to limit the loss of water by osmosis. In fact, we might expect water to be passing back into the capillaries from the tissue fluid, due to osmosis. However, the force applied to the blood by the heart creates enough hydrostatic pressure to overcome osmotic water uptake, at least at the arteriole end of the capillary bed. Here the blood pressure is significantly higher than at the venule ends. Then, as the blood flows through the capillary bed there is progressive loss of water and of hydrostatic pressure, too. As a result, much of the tissue fluid is able to eventually return to the plasma – about 90 per cent returns by this route (Figure 8.8).

So, further along the capillary bed there is a net inflow of tissue fluid to the capillary. However, not all tissue fluid is returned to the blood circulation by this route – some enters the **lymph capillaries** (Figure 8.9). **Lymph** is the tissue fluid that drains into the lymphatic vessels, rather than directly back into a blood capillary.

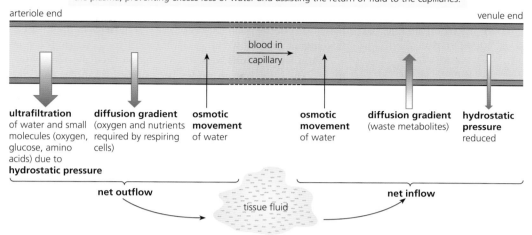

Blood proteins, particularly the albumins, cannot escape; they maintain the water potential of the plasma, preventing excess loss of water and assisting the return of fluid to the capillaries.

arteriole end

venule end

blood in capillary

ultrafiltration of water and small molecules (oxygen, glucose, amino acids) due to **hydrostatic pressure**

diffusion gradient (oxygen and nutrients required by respiring cells)

osmotic movement of water

osmotic movement of water

diffusion gradient (waste metabolites)

hydrostatic pressure reduced

net outflow

tissue fluid

net inflow

Figure 8.8 Forces for exchange in capillaries

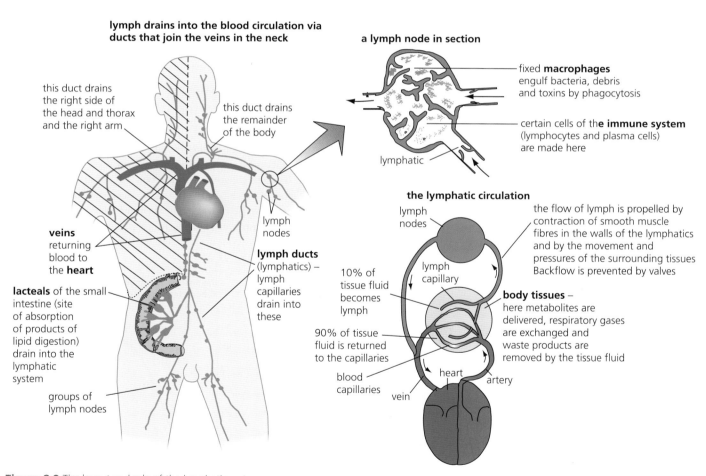

lymph drains into the blood circulation via ducts that join the veins in the neck

a lymph node in section

this duct drains the right side of the head and thorax and the right arm

this duct drains the remainder of the body

fixed **macrophages** engulf bacteria, debris and toxins by phagocytosis

certain cells of th**e immune system** (lymphocytes and plasma cells) are made here

lymphatic

veins returning blood to the **heart**

lymph nodes

lymph ducts (lymphatics) – lymph capillaries drain into these

the lymphatic circulation

lymph nodes

the flow of lymph is propelled by contraction of smooth muscle fibres in the walls of the lymphatics and by the movement and pressures of the surrounding tissues Backflow is prevented by valves

lacteals of the small intestine (site of absorption of products of lipid digestion) drain into the lymphatic system

10% of tissue fluid becomes lymph

lymph capillary

body tissues – here metabolites are delivered, respiratory gases are exchanged and waste products are removed by the tissue fluid

90% of tissue fluid is returned to the capillaries

groups of lymph nodes

blood capillaries

vein

heart

artery

Figure 8.9 The layout and role of the lymphatic system

Molecules too large to enter capillaries are removed from the tissues via the lymph system. There are tiny valves in the walls of the lymph capillaries that permit this. The network of lymph capillaries drains into larger lymph ducts (known as **lymphatics**). Then, lymph is moved along the lymphatics by contractions of smooth muscles in their walls and by compression from body movements. Back-flow is prevented by valves, as it is in the veins.

Lymph finally drains back into the blood circulation in veins close to the heart. At intervals along the way there are **lymph nodes** present in the lymphatics. Lymph passes through these nodes before it is returned to the blood circulation. In the nodes phagocytic macrophages engulf bacteria and any cell detritus present. Also the nodes are a site where certain cells of the immune system are found. We return to this issue in Topic 11.

Transport of respiratory gases between the lungs and respiring tissues

Oxygen

A single red blood cell contains about 280 million molecules of **haemoglobin**. We have seen that each haemoglobin molecule is built of four, interlocking subunits composed of a large, globular protein with a non-protein haem group, containing iron, attached (Figure 2.28, page 53). At the concentration of oxygen that occurs in our lungs, one molecule of oxygen will combine with each haem group. This means each haemoglobin molecule is able to transport four molecules of oxygen.

$$\text{haemoglobin} + \text{oxygen} \rightarrow \text{oxyhaemoglobin}$$

$$\text{Hb} \qquad 4O_2 \qquad HbO_8$$

Figure 8.10 shows the relationship between haemoglobin and oxygen as the concentration of oxygen around the haemoglobin molecule changes. (Oxygen concentration is expressed as its partial pressure, in kPa.) This is called an **oxygen dissociation curve**. The curve is S-shaped. This tells us that the first oxygen molecule attaches with difficulty. This is because the addition of the first oxygen molecule has to slightly distort the shape of the haemoglobin molecule. However, once this has occurred, the second oxygen molecule combines slightly more easily and then the third and forth combine progressively more easily. This sequence of changes accounts for the S-shape of the curve.

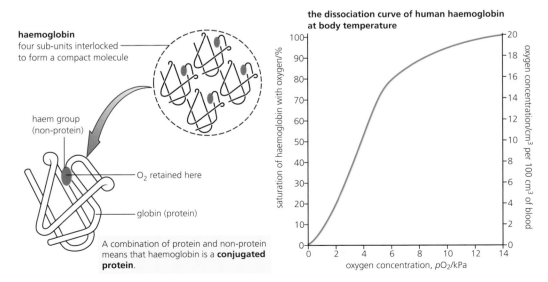

Figure 8.10 The structure of haemoglobin and its affinity for oxygen

Extension

A note on partial pressure

We live at the bottom of a 'sea of air' – hence air pressure. The air we breathe is a mixture of gases and, in a gas mixture, each gas exerts a pressure. In fact, the pressure of a mixture of gases is the sum of the pressure of the component gases. Consequently, the pressure of a specific gas in a mixture of gases is called its **partial pressure**. The symbol for partial pressure is p and for the partial pressure of a gas X is $p\mathbf{X}$ (so, for example, $p\mathrm{O}_2$ denotes the partial pressure of oxygen). The unit of pressure is the **Pascal (Pa)** and its multiple the **kilopascal (kPa)**.

Atmospheric pressure at sea level is approximately 100 kPa and oxygen makes up about 21 per cent of the air. *What is $p\mathrm{O}_2$ at sea level?*

Extension

Carbon monoxide poisoning

Carbon monoxide is found in vehicle exhaust gas, **cigarette smoke**, and as a product of incomplete combustion of natural gas where there is restricted access for air.

Carbon monoxide combines irreversibly with the iron of haemoglobin at the site that oxygen would otherwise occupy, forming **carboxyhaemoglobin**. The affinity of carbon monoxide for haemoglobin is about 300 times greater than oxygen's and at low partial pressures of this poison the blood's ability to transport oxygen may be fatally reduced.

(Incidentally, a catalytic converter, when fitted to a vehicle's exhaust, can significantly reduce the quantity of the gas by oxidising it to carbon dioxide.)

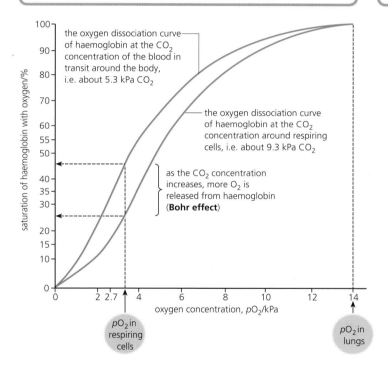

Figure 8.11 How carbon dioxide favours the release of oxygen in respiring tissues

To return to haemoglobin, most haemoglobin molecules are fully saturated at a partial pressure of oxygen of only 8 kPa so oxygen is efficiently loaded under the conditions in our ventilated lungs. However, in respiring tissue the situation is quite different. Here the partial pressure of oxygen is much lower because oxygen is used in aerobic respiration. Oxyhaemoglobin will dissociate if the partial pressure is less than 8 kPa, making oxygen molecules readily available for use in the cells of respiring tissues.

So oxygen is scarce in respiring cells. On the other hand, the concentration of carbon dioxide is high. In this environment it is interesting to see what effect the concentration of carbon dioxide has on the loading and unloading of oxygen by haemoglobin.

In fact, carbon dioxide has a marked effect. An increased carbon dioxide concentration shifts the oxygen dissociation curve to the right. That is, where the carbon dioxide concentration is high (obviously in the actively respiring cells), oxygen is released from oxyhaemoglobin even more readily. This very useful outcome is known as the **Bohr effect**.

Question

6 Use Figure 8.11 to deduce the change in the percentage saturation of haemoglobin if the partial pressure of oxygen drops from 4 to 2.7 kPa when the partial pressure of carbon dioxide is 5.3 kPa.

Carbon dioxide

Carbon dioxide is transported in the blood, in both the plasma and in red blood cells, mainly as hydrogencarbonate ions.

$$CO_2 + H_2O \xrightarrow{\text{carbonic anhydrase}} HCO_3^2 + H^+$$

The enzyme **carbonic anhydrase**, which catalyses this reaction, is present in the red blood cells. You can see that hydrogen ions are one of the products. The hydrogen ions released become associated with haemoglobin to form **haemoglobinic acid**. In effect, the haemoglobin molecules act as a buffer for hydrogen ions, preventing the blood from becoming acidic (Figure 8.12).

In addition, about 5 per cent of the carbon dioxide remains as carbon dioxide in the blood, dissolved in the plasma. A further 10 per cent combines with haemoglobin, reacting with amine groups (–NH₂) to form **carbaminohaemoglobin**. In the lungs, the partial pressure of carbon dioxide ($p\mathrm{CO}_2$) there allows these forms of carbon dioxide to be released, also.

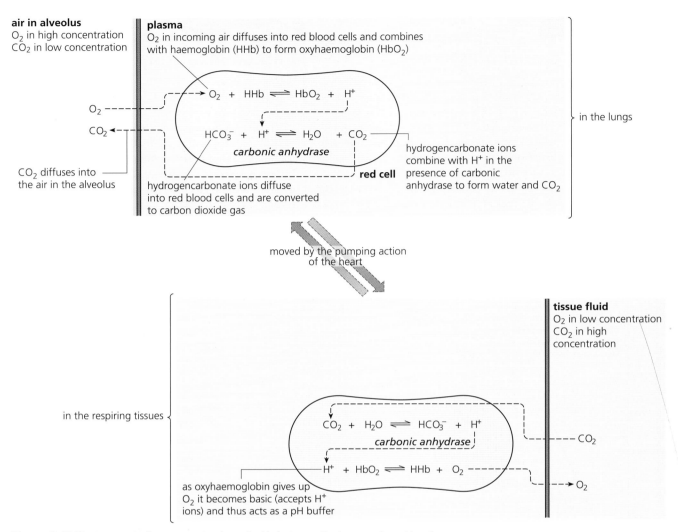

air in alveolus
O_2 in high concentration
CO_2 in low concentration

plasma
O_2 in incoming air diffuses into red blood cells and combines with haemoglobin (HHb) to form oxyhaemoglobin (HbO_2)

O_2

CO_2

$$O_2 + HHb \rightleftharpoons HbO_2 + H^+$$

$$HCO_3^- + H^+ \rightleftharpoons H_2O + CO_2$$
carbonic anhydrase

in the lungs

hydrogencarbonate ions combine with H^+ in the presence of carbonic anhydrase to form water and CO_2

CO_2 diffuses into the air in the alveolus

hydrogencarbonate ions diffuse into red blood cells and are converted to carbon dioxide gas

red cell

moved by the pumping action of the heart

tissue fluid
O_2 in low concentration
CO_2 in high concentration

in the respiring tissues

$$CO_2 + H_2O \rightleftharpoons HCO_3^- + H^+$$
carbonic anhydrase

$$H^+ + HbO_2 \rightleftharpoons HHb + O_2$$

CO_2

O_2

as oxyhaemoglobin gives up O_2 it becomes basic (accepts H^+ ions) and thus acts as a pH buffer

Figure 8.12 The transport of oxygen and carbon dioxide between the lungs and respiring tissues

Humans at high altitude

It is estimated that more than 40 million people live and work at altitudes between 3000 and 5500 m, for example in the Andes and Himalayas. The problems of gaseous exchange at high altitude do not arise because the percentage of oxygen is lower up there – it isn't. The percentage of oxygen does not vary significantly between sea level and high altitudes. However, with increasing altitude it is the atmospheric pressure that falls, and so the partial pressure of oxygen falls also (Table 8.5).

Table 8.5 Change in the partial pressure of oxygen with altitude

Altitude/m above sea level	Atmospheric pressure/kPa	Oxygen content/%	Partial pressure of oxygen/kPa
0	101.3	20.9	21.2
2 500	74.7	20.9	15.7
5 000	54.0	20.9	11.3
7 000	38.5	20.9	8.1
10 000	26.4	20.9	5.5

The result of these changes is that as altitude increases it becomes increasingly hard for haemoglobin in red cells in the lungs to load oxygen. Once the percentage saturation with oxygen is lowered, this is detected by chemoreceptors. The response of the respiratory centre is to cause the taking of extra deep breathes. As a result, more carbon dioxide is lost from the body, which causes a small but significant rise in the pH of the blood. Now the chemoreceptors become ineffective. Ventilation regulation is hampered.

The body cannot adapt to high altitude immediately; sudden, prolonged exposure at very high altitudes by people without prior experience of these conditions can be fatal. However, progressively the following changes take place:

- A more alkaline urine is secreted by the kidney tubules via the collecting ducts, and the pH of the blood returns to normal. As a result, the carbon dioxide chemoreceptors become sensitive again, and normal ventilation is maintained.
- Bone marrow tissue, the site of red cell formation, produces and releases more red cells, thereby enhancing the oxygen-carrying capacity of the blood (Table 8.6).

Table 8.6 The change in red blood cell count in humans at different altitudes

Altitude/m above sea level	Red cell count/$\times 10^{12}d^3$
0 (sea level)	5.0
5000$^+$ as a temporary visitor	5.95
5000$^+$ as a resident	7.37

These changes are called 'acclimatisation'. Other mammal species have evolved at high altitudes, such as the llama of the mountains of South America. In this case the mammal has a form of haemoglobin that loads more readily at lower partial pressures of oxygen.

8.2 The heart

The mammalian heart is a double pump: the right side pumps blood at low pressure to the lungs and the left side pumps blood at high pressure to the rest of the body.

By the end of this section you should be able to:

a) describe the external and internal structure of the mammalian heart
b) explain the differences in the thickness of the walls of the different chambers in terms of their functions with reference to resistance to flow
c) describe the cardiac cycle (including blood pressure changes during systole and diastole)
d) explain how heart action is initiated and controlled

The heart as a pump

The human heart is the size of a clenched fist. It lies in the thorax between the lungs and below the breast bone (sternum). The heart is a hollow organ with a muscular wall. It is contained in a tightly fitting membrane, the **pericardium** – a strong, non-elastic sac which anchors the heart within the thorax and prevents the overfilling of the heart with blood. Within the pericardium is fluid which reduces friction between the beating heart muscle and the surrounding tissues.

The wall of the heart is supplied with oxygenated blood from the aorta via **coronary arteries**. All muscle tissue consists of cells called fibres that are able to shorten by a half to a third of their length. **Cardiac muscle** consists of cylindrical branching columns of fibres, uniquely forming a three-dimensional network and allowing contractions in three dimensions. The fibres have a single nucleus, are striped or striated and are surrounded by **sarcolemma** (muscle cell surface membrane) (Figure 8.13). They are very well supplied by mitochondria and capillaries. The fibres are connected by special junctions called **intercalated discs**. These discs transmit the impulse to contract to all cells simultaneously. The impulse to contract is generated within the heart muscle itself (known as a **myogenic origin**), not by nervous stimulation (a **neurogenic origin**). Heart muscle fibres contract rhythmically from their formation until they die.

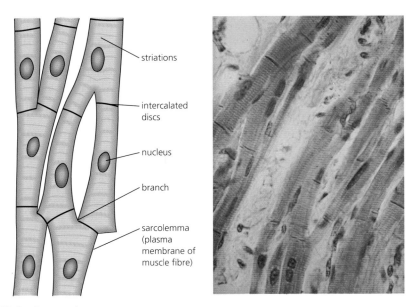

Figure 8.13 Cardiac muscle

Structure of the heart chamber walls in relation to function

The cavity of the heart is divided into four **chambers**. The chambers of the right side of the heart are separated from those of the left side by a wall of muscle called the **septum**.

The upper chambers are thin-walled and are called **atria** (singular: **atrium**). These chambers receive blood into the heart under relatively low pressure. Later, when the walls of the atria contract, the blood they now contain is pumped into the ventricles. At this stage the ventricles are in a relaxed condition – there is little resistance to this inward flow of blood.

The two lower chambers are thick-walled and called **ventricles**. They eventually pump the blood out of the heart – the left ventricle pumps blood through the entire systemic circulation, whereas the right ventricle pumps blood through the pulmonary circulation. Consequently, the muscular wall of the left ventricle is much thicker than that of the right ventricle, for it has to overcome the greater resistance of the capillaries throughout the body tissues in comparison with a lower resistance in the lungs. However, the volumes of the right and left sides (the quantities of blood they contain) are identical.

The functions of the heart valves

The direction of blood flow through the heart is maintained by valves, because it is the valves that prevent back-flow of the blood when the muscles of the chamber walls contract. Figure 8.15 shows the action of all the valves of the heart.

The **atrioventricular valves** are large valves that separate the upper and lower chambers. The edges of these valves are supported by tendons anchored to the muscle walls of the ventricles below and which prevent the valves from folding back due to pressure from the ventricles. The valves separating the right and left sides of the heart are individually named – on the right side, the **tricuspid valve** and on the left, the **bicuspid** or mitral valve.

Valves of a different type separate the ventricles from pulmonary artery (right side) and aorta (left side). These are pocket-like structures called **semilunar valves**, rather similar to the valves seen in veins.

heart viewed from the front of the body with pericardium removed

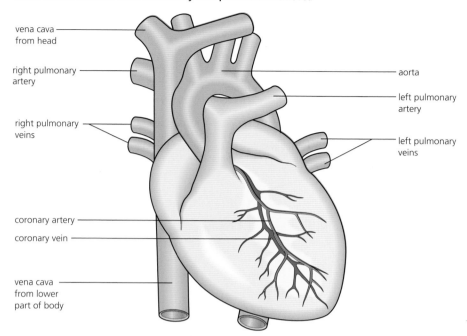

vena cava from head

right pulmonary artery

right pulmonary veins

coronary artery

coronary vein

vena cava from lower part of body

aorta

left pulmonary artery

left pulmonary veins

heart in LS

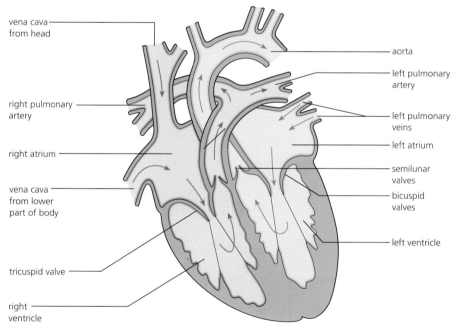

vena cava from head

right pulmonary artery

right atrium

vena cava from lower part of body

tricuspid valve

right ventricle

aorta

left pulmonary artery

left pulmonary veins

left atrium

semilunar valves

bicuspid valves

left ventricle

Figure 8.14 The structure of the heart

What causes the valves of the heart to open and close?

The heart valves are opened and closed by differences in the blood pressure caused by alternating contraction of heart muscle, first in the atrial walls (atrial systole), then in the ventricles (ventricular systole).

● Valves open when pressure on the input side (caused by muscle contraction) exceeds that on the output side.
● Valves close when pressure on the output side (caused by muscle contraction) exceeds that on the input side – typically caused by relaxation of the muscles on the input side.

Heart valves allow blood flow in one direction only. They are opened and closed by differences in blood pressure.
- They are opened when the pressure on the input side (caused by muscle contraction) exceeds that on the output side.
- They are closed when the pressure on the output side exceeds that on the input side. Typically caused by the relaxation of muscle on the input side.

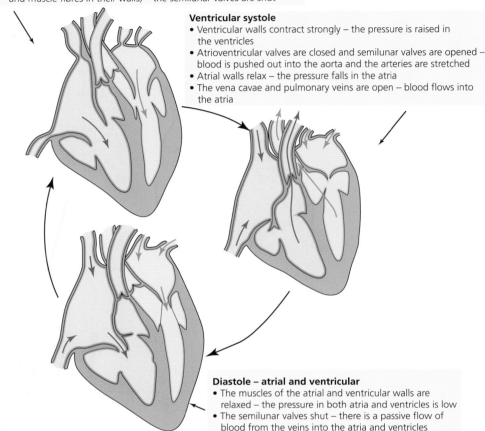

Atrial systole
- Atrial walls contract – the pressure is raised in the atria
- Vena cavae and pulmonary veins are closed
- Atrioventricular valves pushed open – blood flows into the ventricles
- Ventricular walls relaxed – the pressure is low in the ventricles
- High pressure in the aorta and pulmonary arteries (due to the elastic and muscle fibres in their walls) – the semilunar valves are shut

Ventricular systole
- Ventricular walls contract strongly – the pressure is raised in the ventricles
- Atrioventricular valves are closed and semilunar valves are opened – blood is pushed out into the aorta and the arteries are stretched
- Atrial walls relax – the pressure falls in the atria
- The vena cavae and pulmonary veins are open – blood flows into the atria

Diastole – atrial and ventricular
- The muscles of the atrial and ventricular walls are relaxed – the pressure in both atria and ventricles is low
- The semilunar valves shut – there is a passive flow of blood from the veins into the atria and ventricles

Figure 8.15 The action of the heart valves

Question

7 The edges of the atrioventricular valves have non-elastic strands attached which are anchored to the ventricle walls (Figure 8.14). What is the role of these strands?

The cardiac cycle

The cardiac cycle is the sequence of events of a heart beat, by which blood is pumped all over the body. The heart beats at a rate of about 75 times per minute, so each cardiac cycle is about 0.8 seconds long. This period of 'heart beat' is divided into two stages which are called **systole** and **diastole**. In the systole stage heart muscle contracts and during the diastole stage heart muscle relaxes. When the muscular walls of the chambers of the heart contract, the volume of the chambers decreases. This increases the pressure on the blood contained there. This forces the blood to a region where pressure is lower. Valves prevent blood flowing backwards to a region of low pressure, so blood always flows on through the heart.

Look at the steps in Figures 8.15 and 8.16. You will see that Figure 8.16 illustrates the cycle on the left side of the heart only, but both sides function together, in exactly the same way, as Figure 8.15 makes clear.

We can start with contraction of the atrium (**atrial systole**, about 0.1 second). As the walls of the atrium contract, blood pushes past the atrioventricular valve, into the ventricles where the contents are under low pressure. At this time, any back-flow of blood from the aorta back into the ventricle chamber is prevented by the semilunar valves. Notice that back-flow from the atria into the vena cavae and the pulmonary veins is prevented because contraction of the atrial walls seals off these veins. Veins also contain semilunar valves which prevent back-flow here, too.

The atrium then relaxes for the remainder of the cycle (**atrial diastole**, about 0.7 seconds).

Next the ventricle contracts (**ventricular systole**, about 0.3 seconds). The high pressure this generates slams shut the atrioventricular valve and opens the semilunar valves, forcing blood into the aorta. A 'pulse', detectable in arteries all over the body, is generated (see below).

This is followed by relaxation of the ventricles (**ventricular diastole**, about 0.5 seconds) until the next contraction of the ventricles.

Each contraction of cardiac muscle is followed by relaxation and elastic recoil. The changing pressure of the blood in the atria, ventricles, pulmonary artery and aorta (shown in the graph in Figure 8.16) automatically opens and closes the valves.

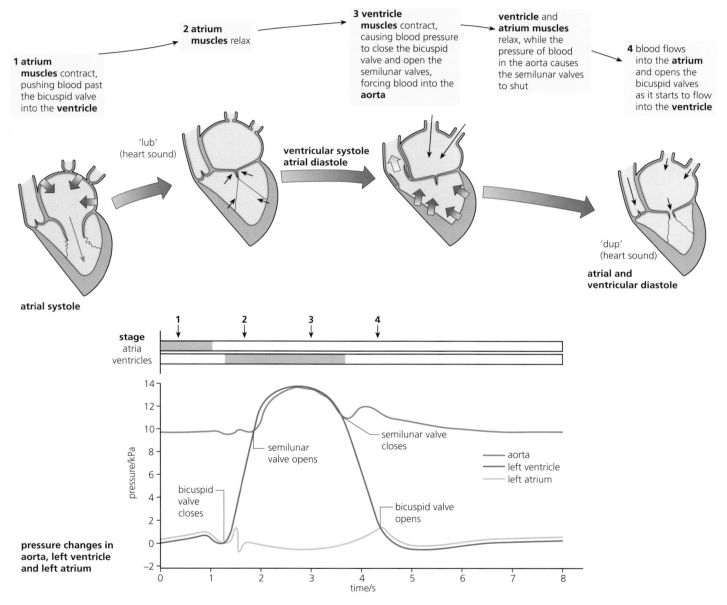

Figure 8.16 The cardiac cycle

8 Examine the data on pressure change during the cardiac cycle in Figure 8.15. Determine or suggest why:

a pressure in the aorta is always significantly higher than that in the atria

b pressure falls *most* abruptly in the atrium once ventricular systole is underway

c the semilunar valve in the aorta does not open immediately that ventricular systole commences

d when ventricular diastole commences, there is a significant delay before the bicuspid valve opens, despite rising pressure in the atrium

e it is significant that about 50 per cent of the cardiac cycle is given over to diastole.

Heart rate and the pulse

The contraction of the ventricle walls forces a surge of blood into the aorta and the pulmonary arteries come under great pressure. This volume of blood is called the **stroke volume**. Each surge stretches the elastic fibres in the artery walls (Table 8.3, page 156). The artery walls are distended as the surge passes, before the subsequent elastic recoil occurs. This is known as the **pulse**. Each contraction of the ventricles generates a pulse, so when we measure our pulse rate we are measuring our heart rate. We can measure heart rate in the carotid artery in the neck or at the wrist, where an artery passes over a bone. Incidentally, the amount of blood flowing from the heart is known as the **cardiac output**. At rest, our cardiac output is typically about 5 litres of blood per minute.

$$\text{cardiac output} = \text{stroke volume} \times \text{pulse rate}$$

Pulsation of the blood flow has entirely disappeared by the time it reaches the capillaries. This is due to the extensive nature of the capillary networks and to the resistance the blood experiences as it flows through the capillary networks.

Origin and control of the heart beat

The heart beats rhythmically throughout life, without rest, apart from the momentary relaxation that occurs between each beat. Even more remarkably, heart beats occur naturally, without the cardiac muscle needing to be stimulated by an external nerve. Since the origin of each beat is within the heart itself, we say that heart beat is 'myogenic' in origin. However, the alternating contractions of the cardiac muscle of the atria and ventricles are controlled and co-ordinated precisely. Only in this way can the heart act as an efficient pump. The positions of the structures within the heart that bring this about are shown in Figure 8.17.

The steps in the control of the cardiac cycle are as follows.

- The heart beat originates in a tiny part of the muscle of the wall of the right atrium, called the **sinoatrial node (SA node)** or **pacemaker**.
- From here a wave of excitation (electrical impulses) spreads out across both atria.
- In response, the muscle of both atrial walls contracts simultaneously (**atrial systole**).
- This stimulus does not spread to the ventricles immediately because of the presence of a narrow band of non-conducting fibres at the base of the atria. These block the excitation wave, preventing its conduction across to the ventricles. Instead, the stimulus is picked up by the **atrioventricular node (AV node)**, situated at the base of the right atrium.

- After a delay of 0.1–0.2 seconds, the excitation is passed from the atrioventricular node to the base of both ventricles by tiny bundles of conducting fibres known as the **Purkyne tissue**. These are collectively called the **bundles of His**.
- On receiving this stimulation from the bundles of His, the ventricle muscles start to contract from the base of the heart upwards (**ventricular systole**).
- The delay that occurs before the atrioventricular node acts as a 'relay station' for the impulse permits the emptying of the atria into the ventricles to go to completion and prevents the atria and ventricles from contracting simultaneously.
- After every contraction, cardiac muscle has a period of insensitivity to stimulation, the **refractory period** (a period of enforced non-contraction – **diastole**). In this phase, the heart begins to refill with blood passively. This period is a relatively long one in heart muscle and enables the heart to beat throughout life.

The heart's own rhythm, set by the sinoatrial node, is about 50–60 beats per minute. Conditions in the body can and do override this basic rate and increase heart performance. The action of the pacemaker is modified according to the needs of the body. For example, it may be increased during physical activity. This occurs because increased muscle contraction causes an increased volume of blood passing back to the heart. The response may be more powerful contractions without an increase in the rate of contractions. Alternatively, the rate of 75 beats per minute of the heart 'at rest' may be increased to up to 200 beats a minute during very strenuous exercise.

stimulation of heart beat originates in the muscle

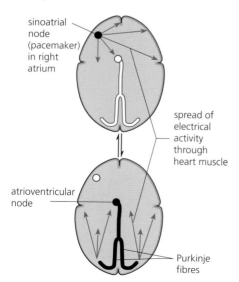

sinoatrial node (pacemaker) in right atrium

spread of electrical activity through heart muscle

atrioventricular node

Purkinje fibres

Figure 8.17 Myogenic origin of the heart beat

Extension

Electrocardiography

The impulses (action potentials) that originate in the sinoatrial node (pacemaker) of the heart during the cardiac cycle produce electrical currents that are also conducted through the fluids of the body as a whole and can be detected at the body surface by electrocardiography. Here, electrodes are attached to the patient's chest and the electrical activity detected is displayed as an electrocardiogram (ECG) by means of a chart recorder (Figure 8.18).

Electrocardiography has clinical applications; it is an aid in the diagnosis of **cardiovascular disease** (**CVD**, page 198). Some heart conditions that can be detected by the analysis of electrocardiograms are listed in Table 8.7.

normal electrocardiogram (ECG), analysed

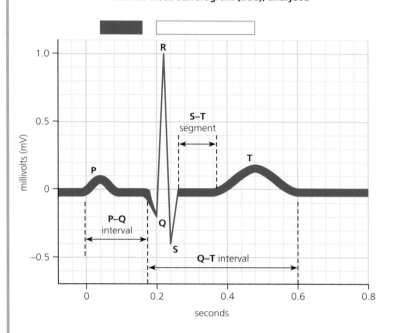

Key

▉	atrial contraction (atrial systole)
☐	ventricular contraction (systole)
P wave	atrial depolarisation – leads to atrial contraction
P–R interval	time for impulse to be conducted from SA node to ventricles, via AV node
QRS complex	onset of ventricular depolarisation – leads to ventricular contraction
T wave	ventricular repolarisation – relaxation phase

abnormal traces showing

1 tachycardia
  heart rate is over 100 beats/minute

2 ventricular fibrillation
  uncontrolled contraction of the ventricles – little blood is pumped

3 heart block
  ventricles not always stimulated

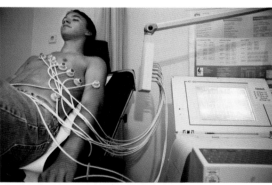

electrical activity detected through electrodes attached to the patient's chest is displayed on the chart recorder as an electrocardiogram

Figure 8.18 Electrocardiography

(Continued)

Extension (Continued)

Table 8.7 Heart conditions detected by ECG analysis

Heart conditions detected by abnormal ECG traces	
Tachycardia	A normal adult heart beats between 60 and 100 times a minute; a heart rate over 100 beats a minute is called tachycardia. Tachycardia may be relatively harmless and need no treatment but some forms can be life-threatening.
Ventricular fibrillation	Asynchronous contraction of the ventricle muscle fibres results in a failure of the heart to pump sufficient blood because some muscle fibres are contracting whilst others are relaxing.
Heart block	The most common site of blockage is at the atrioventricular node.
Arrhythmia	Arrhythmia is a condition of irregularity in heart rhythm due to a defect in the conduction system of the heart. It may be due to: • drugs, such as nicotine or alcohol • anxiety, hypothyroidism or potassium deficiency.

Summary

- Mammals have a **closed circulation** in which blood is pumped by a muscular **heart** through a continuous series of tubes – the **arteries**, **veins** and **capillaries**. It is a **double circulation** in that blood goes twice through the heart in every complete circulation. The **pulmonary circulation** is to the lungs, supplied by the right side of the heart. The **systemic circulation** is to the rest of the body, supplied by the left side of the heart.

- **Blood** consists of a straw-coloured fluid, the **plasma**, and **blood cells**. In humans, **red blood cells (erythrocytes)** are without a nucleus, contain **haemoglobin** and transport oxygen. **White blood cells (leucocytes)** combat disease. They are circulated by the blood to locations in the body where they act.

- **Tissue fluid** is the liquid that bathes all the body cells, formed from the blood plasma that has escaped from the capillaries. This fluid differs from plasma in that blood cells and plasma proteins are absent. **Lymph** is the tissue fluid that drains back into the blood circulation via the lymphatic system (lymph vessels and nodes).

- The main **role of the blood circulation** is the **transport** of respiratory gases, water, nutrients, waste products, heat and hormones. The transfer of substances in solution or suspension is essential since diffusion is insufficient.

- **Haemoglobin**, a conjugated protein, and the enzyme **carbonic anhydrase** occur in red blood cells and together affect the transport of **respiratory gases**. Oxygen is transported as **oxyhaemoglobin**. Haemoglobin buffers the H^+ ions from the reaction of carbon dioxide and water in which carbon dioxide is converted to **hydrogencarbonate ions**. These are transported mainly in the plasma.

- The **heart** is a hollow, muscular organ of **four chambers**. The right and left halves of the heart are completely separate. The upper chambers, the **atria**, have relatively thin walls. The lower chambers, the **ventricles**, have thick walls. The thickness of the wall of the left ventricle is much greater than the right because the left ventricle has to pump blood all around the rest of the body whereas the right ventricle pumps blood only to the lungs which are very close to the heart. Direction of blood flow through the heart is maintained by **valves**.

- The cycle of changes during a heart beat, known as the **cardiac cycle**, lasts about 0.8 seconds when the body is at rest. It consists of alternate contraction (**systole**) and relaxation (**diastole**). Atrial systole precedes ventricular systole and both are followed by periods of diastole that partly overlap. Then atrial systole commences again.

- The heart beat originates in the heart itself (**myogenic**), at the **sinoatrial node** (SAN or pacemaker) in the upper wall of the right atrium. This activates the atrial muscle to contract and triggers the atrioventricular node (AVN), which carries the signal to contract on to the ventricles. The excitation passes from the atrioventricular node to the base of both ventricles by tiny bundles of conducting fibres, known as the **Purkyne tissue**. These are collectively called the **bundles of His**. Impulses from the cerebral hemispheres or involuntary reflexes from stretch receptors in arteries outside the heart, or hormone action can all alter the rate at which the heart beats.

Examination style questions

1 Fig. 1.1 shows the changes in blood pressure in the left atrium, left ventricle and aorta during one complete contraction of the heart. It also shows a recording of the electrical activity of the heart.

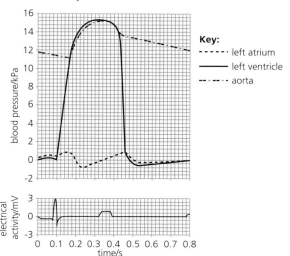

Key:
- - - - - left atrium
——— left ventricle
- - · - - · · aorta

Fig. 1.1

a) Name the source of the electrical activity in the heart. [1]

b) Explain how the heart is co-ordinated so that the ventricle contracts after the atrium has contracted. [4]

c) With reference to Fig. 1.1, calculate the heart rate in beats per minute. Show your working and express your answer to the nearest whole number. [2]

d) The pressure in the **right ventricle** is rarely higher than 4.0 kPa.

Explain why the pressure in the right ventricle is much lower than that in the left ventricle. [2]

[Total: 9]

(Cambridge International AS and A Level Biology 9700, Paper 02 Q4 November 2007)

2 a) What tissue is found in all types of blood vessel? How it is adapted to its functions? [4]

b) In a table, give the differences between the structure of an artery and a vein as seen in transverse section. [4]

c) Explain the difference between a single and a double circulation, and the benefits gained by respiring tissues served by the latter. [2]

d) In a table, give the differences in composition between plasma and tissue fluid. [4]

e) List in order the pathway of vessels and chambers taken by a red blood cell, from leaving the liver until it reaches cells of the muscles of the arm. [6]

[Total: 20]

3 Both red blood cells in mammals and companion cells in flowering plants play a part in internal transport. Describe the mature structure of both these cells, and how those structures are adapted to deliver their precise roles.

9 Gaseous exchange and smoking

The gas exchange system is responsible for the uptake of oxygen into the blood and excreting carbon dioxide. An understanding of this system shows how cells, tissues and organs function together to exchange these gases between the blood and the environment. The health of this system and of the cardiovascular system is put at risk by smoking.

9.1 The gas exchange system

The gas exchange surface in the lungs is extensive, very thin, well supplied with blood and well ventilated. The trachea and bronchi provide little resistance to the movement of air to and from the alveoli.

By the end of this section you should be able to:

a) describe the gross structure of the human gas exchange system
b) observe and draw plan diagrams of the structure of the walls of the trachea, bronchi, bronchioles and alveoli indicating the distribution of cartilage, ciliated epithelium, goblet cells, smooth muscle, squamous epithelium and blood vessels
c) describe the functions of cartilage, cilia, goblet cells, mucous glands, smooth muscle and elastic fibres and recognise these cells and tissues in prepared slides, photomicrographs and electron micrographs of the gas exchange system
d) describe the process of gas exchange between air in the alveoli and the blood

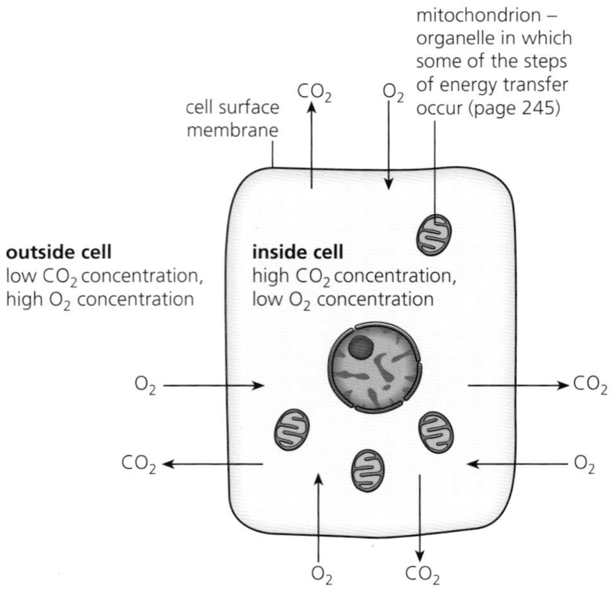

Figure 9.1 Gas exchange in an animal cell

Gaseous exchange – meeting the needs of respiration

Living things need energy to build, maintain and repair body structures and also for activities such as metabolism, excretion, and movement. Respiration is the process that transfers that energy. Respiration occurs continuously in living cells, largely in the mitochondria (Topic 1). Most respiration is aerobic and requires oxygen to transfer energy from carbohydrates and lipids. Respiration results in the formation of ATP which is the energy-transferring molecule used in metabolic processes in cells. Carbon dioxide is released as a waste product. This gas exchange in cells occurs by **diffusion**. For example, in a cell respiring aerobically there is a higher concentration of oxygen outside the cells than inside and so there is a net inward diffusion of oxygen (Figure 9.1).

Respiratory systems in large animals

In larger animals most if not all of their cells are too far from the surface of the body to receive enough oxygen by diffusion alone. Remember, the surface area to volume ratio decreases as the size of an organism increases. In addition, many of these animals have developed an external surface that provides protection for the body. This is true of water-tight (impervious) outer coverings and tough or hardened skins. These

1 List three factors affecting the rate of diffusion across a surface and explain why each is significant. (You may need to refer to Table 4.1, page 79.)

2 Due to the presence of haemoglobin, about 20 cm³ of oxygen is carried per 100 cm³ of blood. By contrast, the solubility of oxygen in water is only 0.025 cm³ per cm³ under the same conditions. How much more oxygen is carried by a litre (1000 cm³) of blood compared with the same quantity of water?

outer surfaces are no longer suitable for gas exchange and the organism requires an alternative **respiratory surface**. Active organisms have an increased metabolic rate, too, and the demand for oxygen in their cells is higher than in slow-moving or inactive organisms. So we find that, for many reasons, larger active animals have specialised organs for gas exchange.

Efficient respiratory surfaces in animals take various forms, such as the gills of fish, the lungs of mammals, the tubular system of many insects that carries air to the most actively respiring organs. All these systems provide a large thin surface area, suitable for gas exchange. In addition, conditions for diffusion are often improved by three refinements.

1 **A ventilation mechanism** – a pumping mechanism that moves the respiratory medium (water or air) over the gills or into and out of the lungs or tubes. This maintains the concentration gradient for diffusion.

2 **A blood circulation system** – a means of speeding up the transport of dissolved oxygen from the respiratory surface as soon as it has diffused in. This, too, maintains the concentration gradient.

3 **A haem protein**, such as haemoglobin, that can associate with oxygen and so increase the gas-carrying ability of the blood. For example, we have seen that our blood contains red blood cells packed with the **respiratory pigment** haemoglobin (Figure 8.10, page 160).

The lungs of mammals

We can now consider the human lungs as structures adapted for rapid gas exchange. The structure of the human thorax is shown in Figure 9.2.

The lungs are housed in the **thorax**, an air-tight dome-shaped chamber formed by the **rib cage** and its muscles (**intercostal muscles**) and with a domed floor, the **diaphragm**. The diaphragm is a sheet of muscle attached to the body wall at the base of the rib cage, separating the thorax from the abdomen. The internal surfaces of the thorax are lined by the **pleural membrane**, which secretes and maintains pleural fluid. Pleural fluid is a lubricating liquid made from the blood plasma. This fluid provides the surface tension that holds the lungs to the rib cage and protects the lungs from friction during breathing movements.

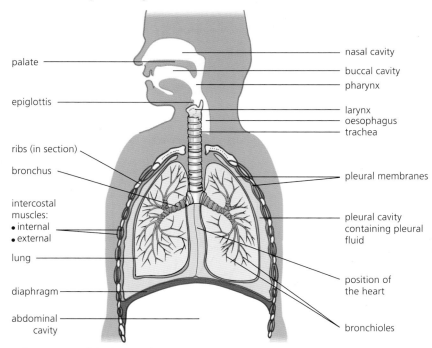

Figure 9.2 The structure of the human thorax

The lungs connect with the pharynx at the rear of the mouth by the **trachea**. Air reaches the trachea from the mouth and nostrils, passing through the larynx ('voice box'). Entry into the larynx is via a slit-like opening, the glottis. Above it is a cartilaginous flap, the **epiglottis**. Glottis and epiglottis work to prevent the entry of food into the trachea.

The trachea then divides into two **primary bronchi**, one to each lung. Within the lungs the primary bronchi divide into **secondary bronchi** and these continue to divide into smaller bronchi, a branching system referred to as the 'bronchial tree'. The smallest bronchi themselves divide into **bronchioles**. The finest bronchioles, the **terminal bronchioles**, end in air sacs – the **alveoli**.

Trachea, bronchi and bronchioles – structure in relation to function

To protect the delicate lining of the alveoli, the air reaching them needs to be warmed (preferably to body temperature), to be moist and to be as free of dust particles and other foreign bodies as possible. In the nostrils, hairs trap and filter out large dust particles from the incoming air stream. Superficial blood vessels in the nostrils start to warm the incoming air. The trachea and bronchi are lined with a **ciliated epithelium** with numerous **goblet cells** (Figure 9.3). The sticky mucus produced here moistens the incoming air and traps finer dust particles. The ciliated epithelium beats the mucus stream up into the buccal cavity, where it is swallowed.

SEM showing that the lining of the bronchioles is of simple, ciliated columnar epithelium, with mucus secreted by goblet cells (x6 000)

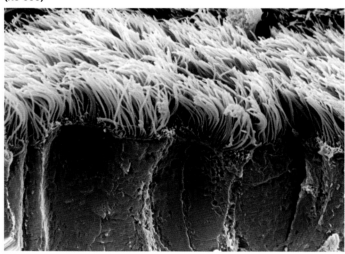

a drawing of the ciliated epithelium cells with goblet cells (HP)

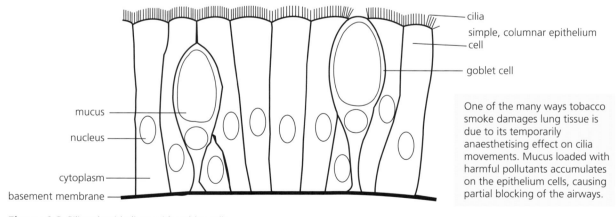

cilia

simple, columnar epithelium cell

goblet cell

mucus

nucleus

cytoplasm

basement membrane

One of the many ways tobacco smoke damages lung tissue is due to its temporarily anaesthetising effect on cilia movements. Mucus loaded with harmful pollutants accumulates on the epithelium cells, causing partial blocking of the airways.

Figure 9.3 Ciliated epithelium with goblet cells

The trachea lies beside and in front of the oesophagus (the 'food tube'). Any hard mass of food passing down the oesophagus might interrupt the air supply to the lungs. Incomplete **rings of cartilage** in the trachea wall prevent collapse under pressure from a large bolus passing down the oesophagus (Figure 9.4).

The walls of the bronchi and larger bronchioles, in addition to rings or tiny plates of cartilage, also contain **smooth muscle**. At each division of the bronchial tree the amount of cartilage decreases and the amount of smooth muscle increases. Together they prevent collapse of these tubes – collapse might be triggered by the sudden reduction in pressure that occurs with powerful inspirations of air, for example. The smooth muscles also regulate the size of the smaller airways as the muscle fibres contract or relax.

The walls of the smaller **bronchioles** are without cartilage (Figure 9.5). These tubes, the narrowest of the airways serving the alveoli, branch repeatedly and, as they do so, become progressively narrower. Here, smooth muscle is the major component of their walls and the lining of columnar epithelium contains only very occasional goblet cells.

So the air sacs are supplied with warm moist clean air for gaseous exchange, but of course the lungs cannot prevent some water loss during breathing – a significant issue for most terrestrial organisms (Figure 9.6).

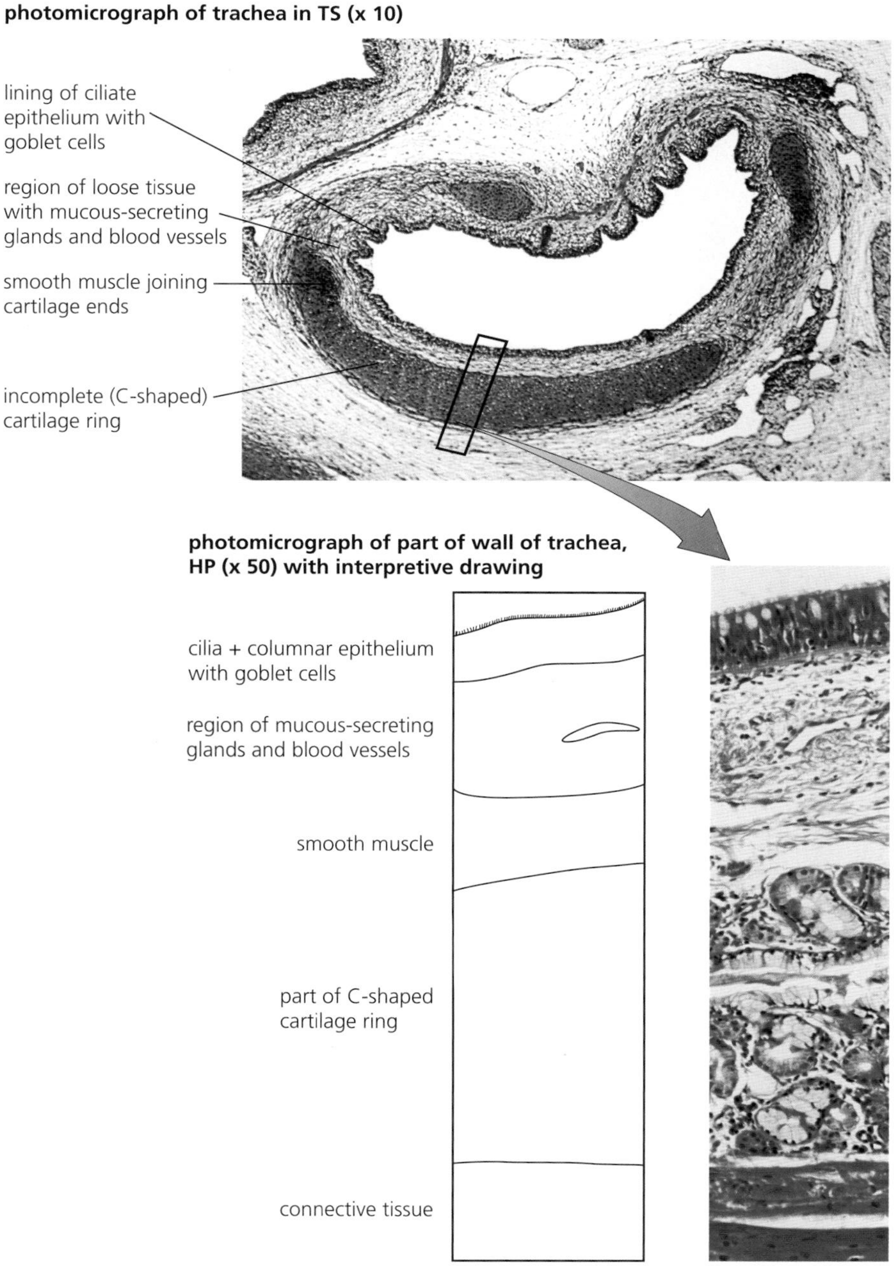

photomicrograph of trachea in TS (x 10)

lining of ciliate epithelium with goblet cells

region of loose tissue with mucous-secreting glands and blood vessels

smooth muscle joining cartilage ends

incomplete (C-shaped) cartilage ring

photomicrograph of part of wall of trachea, HP (x 50) with interpretive drawing

cilia + columnar epithelium with goblet cells

region of mucous-secreting glands and blood vessels

smooth muscle

part of C-shaped cartilage ring

connective tissue

Figure 9.4 The structure of the trachea

photomicrograph of a trachiole HP (x 000)

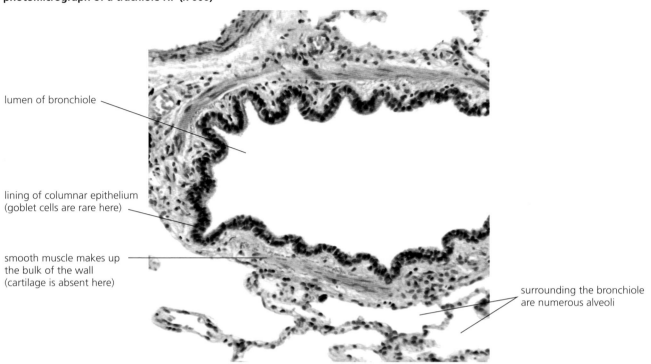

lumen of bronchiole

lining of columnar epithelium
(goblet cells are rare here)

smooth muscle makes up
the bulk of the wall
(cartilage is absent here)

surrounding the bronchiole
are numerous alveoli

Figure 9.5 The structure of a small bronchiole

Questions

3 Draw and label a low-power plan of a representative part of the wall of the bronchiole, as see under high power magnification in Figure 9.5.

4 What percentage of our total daily water loss occurs from the lungs? Explain why this happens.

Exhaled air is saturated with water vapour, an invisible component except in freezing weather or when breathed onto a very cold surface.

The water balance of the body:
• about 3 litres is taken in and lost daily
• about 10–15% of this total is lost from the lungs.

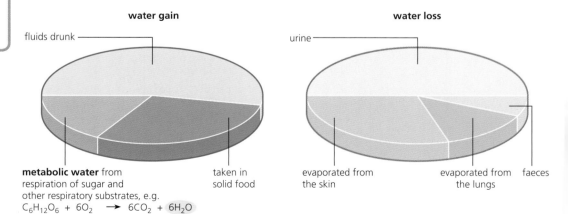

water gain

water loss

fluids drunk

urine

metabolic water from respiration of sugar and other respiratory substrates, e.g.
$C_6H_{12}O_6 + 6O_2 \rightarrow 6CO_2 + 6H_2O$

taken in solid food

evaporated from the skin

evaporated from the lungs

faeces

Figure 9.6 Water loss during gas exchange in humans

Extension

Bronchiole structure and asthma

Figure 9.7 Asthma patient using an inhaler

Asthma is a disease of the airways of the lungs. In an asthma attack, the bronchioles are narrowed by excessive contraction of the smooth muscle in their walls. Immediately, getting air into and out of the lungs becomes difficult. Also, extra mucus is produced, exaggerating the symptoms. Breathing can become very difficult indeed.

Asthma attacks may be started by the arrival of irritants like pollen, dust from pets or droppings from house dust mites. Certain viruses and the oxides of nitrogen present in vehicle exhaust fumes may also be triggers.

During times of physical activity our involuntary nervous system naturally releases hormones that relax the smooth muscles of the bronchioles, dilating them so that air reaches the alveoli more quickly. Medical treatments for asthma sufferers seek to trigger this relaxation, too (Figure 9.7).

Ventilation of the lungs

Air is drawn into the alveoli when the air pressure in the lungs is lower than atmospheric pressure and it is forced out when the pressure is higher than atmospheric pressure. Since the thorax is an air-tight chamber, pressure changes in the lungs occur when the volume of the thorax changes (Figure 9.8).

The volume of the thorax is increased when the ribs are moved upwards and outwards, and the diaphragm dome is lowered. These movements are brought about by contraction of the diaphragm and external intercostal muscles.

The volume of the thorax is decreased by the diaphragm muscles relaxing and the diaphragm becoming more dome shaped by pressure from below (the natural elasticity of the stomach and liver mean that they can be displaced and stretched at inspiration). The ribs move down and inwards, largely brought about by the elastic recoil of the lungs and the weight of the rib cage, as the external intercostal muscles relax.

Question

5 Explain the difference between gas exchange and cellular respiration.

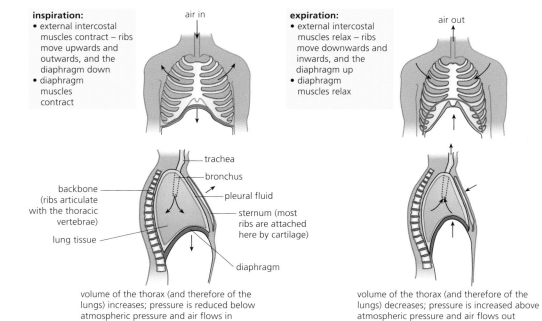

Figure 9.8 Ventilation of the lungs in quiet breathing

Alveolar structure and gaseous exchange

The lung tissue consists of the alveoli, arranged in clusters, each served by a tiny terminal bronchiole. Alveoli and the terminal bronchioles have **elastic connective tissue** as an integral part of their walls. These fibres are stretched during inspiration, when the alveoli are caused to expand, but recoil during expiration, aiding the expulsion of air as the alveoli return to their resting condition (Figure 9.9).

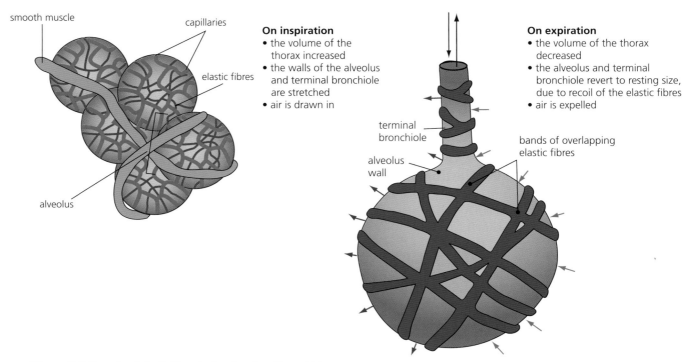

smooth muscle

capillaries

elastic fibres

alveolus

On inspiration
- the volume of the thorax increased
- the walls of the alveolus and terminal bronchiole are stretched
- air is drawn in

On expiration
- the volume of the thorax decreased
- the alveolus and terminal bronchiole revert to resting size, due to recoil of the elastic fibres
- air is expelled

terminal bronchiole

alveolus wall

bands of overlapping elastic fibres

Figure 9.9 The role of elastic fibres in the alveoli and bronchioles

A **capillary system** wraps around the clusters of alveoli. Each capillary is connected to a branch of the **pulmonary artery** and is drained by a branch of the **pulmonary vein** (Figure 9.10). The pulmonary circulation is supplied with deoxygenated blood from the right side of the heart and oxygenated blood is returned to the left side of the heart to be pumped to the rest of the body (page 165).

There are some 700 million alveoli present in our lungs, providing a surface area of about 70 m^2 in total. This is an area 30–40 times greater than that of the body's external skin. The wall of an alveolus is formed by pavement epithelium, one cell thick. An epithelium is a sheet of cells bound strongly together, covering internal or external surfaces of multicellular organisms. Lying very close is a capillary, its wall also just a single, flattened (endothelium) cell thick. The combined thickness of walls separating air and blood is typically 2–4 µm thick. The capillaries are extremely narrow, just wide enough for red blood cells to squeeze through, so red blood cells are close to or in contact with the capillary walls.

The extremely delicate structure of the alveoli is protected by two types of cell, present in abundance in the surface film of moisture (Figure 9.10).

- **Macrophages** (dust cells) are the main detritus-collecting cells of the body. They originate from bone marrow stem cells and are then dispersed about the body in the blood circulation. These amoeboid cells migrate into the alveoli from the capillaries. Here these phagocytic white blood cells ingest any debris, fine dust particles, bacteria and fungal spores present. They also occur lining the surfaces of the airways leading to the alveoli.
- **Surfactant cells** produce a detergent-like mixture of lipoproteins and phospholipid-rich secretion that lines the inner surface of the alveoli. This lung surfactant lowers surface tension, permitting the alveoli to flex easily as the pressure of the thorax falls and rises. It reduces a tendency of alveoli to collapse on expiration.

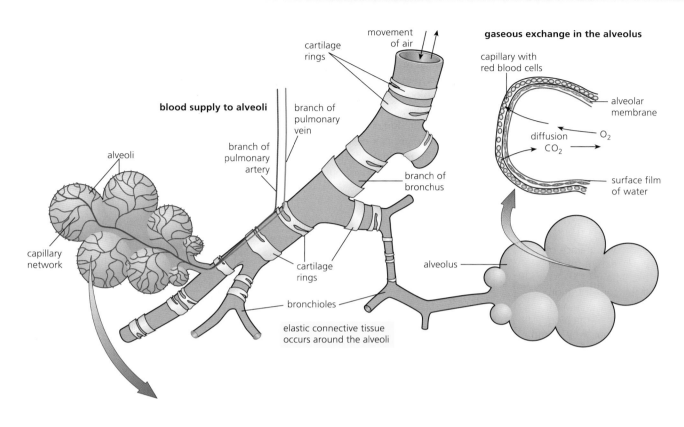

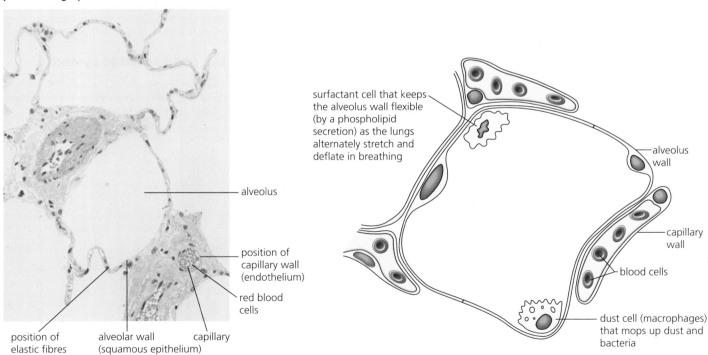

photomicrograph of TS alveoli, HP

Figure 9.10 Gas exchange in the alveoli

Questions

6 Explain how the macrophages remove foreign matter around them. (The discussion on page 222 in Topic 11 may help in the answering of this question.)

7 Create a table listing the main structural features of the air sacs (alveoli) and identifying the effects and consequence of each feature for gas exchange.

8 If the concentration of carbon dioxide were to build up in the blood of a mammal, why would this be harmful?

Blood arriving in the lungs is low in oxygen (it has a lower **partial pressure** of oxygen than the alveolar air – see Table 9.1) but high in carbon dioxide (it has a higher partial pressure of carbon dioxide than the alveolar air). As blood flows past the alveoli, gaseous exchange occurs by diffusion. Oxygen dissolves in the surface film of water, diffuses across into the blood plasma and into the red blood cells where it combines with haemoglobin to form oxyhaemoglobin. At the same time, carbon dioxide diffuses from the blood into the alveolus.

Air flow in the lungs of mammals is tidal in that air enters and leaves by the same route. Consequently there is a **residual volume** of air that cannot be expelled. Incoming air mixes with and dilutes the residual air. The effect of this is that air in the alveoli contains significantly less oxygen than the atmosphere outside (Table 9.1). Nevertheless, the lungs are efficient organs of gaseous exchange.

Table 9.1 The composition of air in the lungs

	Inspired air	Alveolar air	Expired air
Oxygen	20%	14%	16%
Carbon dioxide	0.04%	5.5%	4.0%
Nitrogen	79%	81%	79%
Water vapour	variable	saturated	saturated

We have already noted that, in a mixture of gases, each component gas exerts a partial pressure in proportion to how much is present. The partial pressure of oxygen is written as pO_2 and of carbon dioxide as pCO_2.

At sea level, the atmospheric pressure is 101.3 kPa so, for example, the partial pressure of oxygen in the air at sea level is:

$$pO_2 = \frac{101.3}{100} \times 20 = 20.3 \text{ kPa}.$$

but in the alveolus the partial pressure of oxygen is only:

$$pO_2 = \frac{101.3}{100} \times 14 = 14.2 \text{ kPa}.$$

Observing and drawing plan diagrams of the gas exchange system

You will be required to observe and draw plan diagrams of the structure of the walls of the trachea, bronchi, bronchioles and alveoli in which the distribution and roles of cartilage, ciliated epithelium, and blood vessels are identified by your annotations. Sources should include prepared slides observed by medium and high power, photomicrographs and electron micrographs.

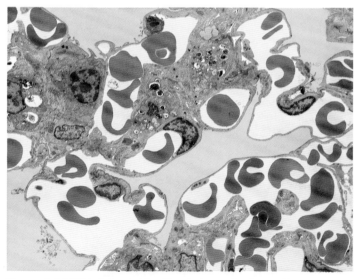

Figure 9.11 TEM of the air–blood barrier in an alveolus/capillary section

9.2 Smoking

Smoking is one of the major avoidable risk factors of chronic, life-threatening diseases of the gas exchange and circulatory systems.

By the end of this section you should be able to:

a) describe the effects of tar and carcinogens in tobacco smoke on the gas exchange system with reference to lung cancer and chronic obstructive pulmonary disease (COPD)
b) describe the short-term effects of nicotine and carbon monoxide on the cardiovascular system

Smoking and health

The cultivation and use of the tobacco plant (*Nicotiana tobacum*) has origins in Central America. By the fifteenth century the tobacco leaf had reached Europe, mostly to be smoked in pipes or as cigars, but increasingly considered a 'cure-all' for many common conditions! Only with the mass production of cigarettes and the invention of the match in the 1800s did this form of smoking become available to many – it had become easy and affordable. In the twentieth century, cigarette smoking in the developed world was advanced hugely by the availability of cigarettes to the troops of two World Wars. Subsequent bold and aggressive advertising campaigns, persuasive product placement in films and the generous sponsorships of sporting and cultural events by cigarette manufacturers, all encouraged greater smoking by men, and persuaded women to take it up, too. Very slowly, the dangers of smoking became known but many people doubted the evidence. However, by the 1950s it started to be recognised just how dangerous cigarette smoking and the inhalation of cigarette smoke was.

The composition of cigarette smoke

Analysis of cigarette smoke shows it contains a cocktail of harmful substances – more than 4000, in fact. These include acetone, ammonia, arsenic, butane, cadmium, hydrogen cyanide, methanol, naphthalene, toluene and vinyl chloride. However, to understand the danger to health that cigarette smoke poses we shall focus on the following components.

- **Carcinogens in the 'tar' component**, of which there are at least twenty different types. Particularly harmful are certain polycyclic aromatic hydrocarbons and nitrosamines. Remember, a carcinogen is any agent that may cause cancer by damage to the DNA molecules of chromosomes. Such 'mistakes' or mutations of different types may build up in the DNA of body cells exposed to these substances.

- **Nicotine**, a stimulating and relaxing drug which, on entering the blood stream, is able to cross the blood–brain barrier. In the brain it triggers the release of dopamine, the natural neurotransmitter substance (page 319) associated with our experience of pleasure. Long-term exposure to nicotine eventually comes to have the reverse effect, actually depressing the ability to experience pleasure. So more nicotine is needed to 'satisfy', and cigarettes become addictive; it is as addictive as heroin and cocaine, in fact. Smokers find it increasingly hard to quit the habit.

  Nicotine also increases the heart rate and blood pressure. It decreases the blood flow, particularly in the hands and feet, and it makes blood clotting more likely. It can make platelets stick together.

- **Carbon monoxide**, a gas that diffuses into the red blood cells and combines irreversibly with haemoglobin (page 160). In smokers, the blood is able to transport less oxygen. The strain this puts on the heart and circulation is most apparent during physical activity.

  Carbon monoxide also promotes the release of damaging free radicals and makes platelets and neutrophils stick together.

Cigarette smoke reaches the smoker's lungs when it is drawn down the cigarette and inhaled, but it reaches other people, too, when it escapes from the glowing tip into the surrounding air. These latter fumes normally have a higher concentration of the toxic ingredients and it is this mixture that others inhale. '**Passive smoking**' has itself been shown to be dangerous, too.

Two of the major diseases that are directly induced by cigarette smoke are **chronic obstructive pulmonary disease (COHD)** and **lung cancer**. We will review each in turn, focusing on the effects of tar and carcinogens in tobacco smoke as the causative agents.

Chronic obstructive pulmonary disease (COPD)

Cigarette smoke reaching the lungs stimulates the secretion of viscous mucus by the goblet cells and, at the same time, inhibits the movements of the cilia (Figure 9.3, page 174) of the epithelium lining the airways. The result is that mucus, in which dust and carcinogenic chemicals are trapped, accumulates in the bronchioles. The smallest bronchioles may be blocked off. Also, irritation of the airways by the smoke results in inflammation and scarring in the lungs. The elasticity that permits the air sacs to expand and contract is progressively lost. The outcome may be chronic obstructive pulmonary disease, which takes two main forms.

Chronic bronchitis, the symptoms of which are a long-term cough with phlegm (mucus), has a gradual onset but is of long duration. The bronchi are inflamed and produce excess mucus which ultimately has to be coughed up, leading to recurring attacks of coughing. As the disease progresses the bronchioles narrow and breathing becomes difficult. The airways are narrowed by the mucus but, unlike asthma where the narrowing of the airways can be reversed, here the narrowing is progressive and not fully reversible. Cigarette smoking is the most important cause of chronic bronchitis but other forms of air pollution and some respiratory infections may be the cause in some cases.

Emphysema is a disease in which the walls of the alveoli lose their elasticity. It results in the destruction of the lung tissue with time. This is because such lungs now contain large numbers of macrophages, accumulated from the blood circulation (Figure 9.10). These phagocytic cells release a high level of the natural hydrolytic enzymes that break down elastic fibres of the alveolar walls. With failing elastic fibres, the air sacs are left over-inflated when air becomes trapped in them and they fail to recoil and expire air properly. Small holes also develop in the walls of the alveoli. These begin to merge, forming huge air spaces with drastically lowered surface area for gas exchange. The patient becomes permanently breathless. The destruction of air sacs can be halted by stopping smoking but any damage done to the lungs cannot be reversed.

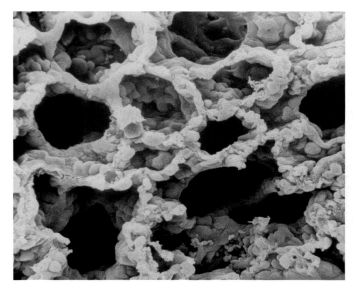 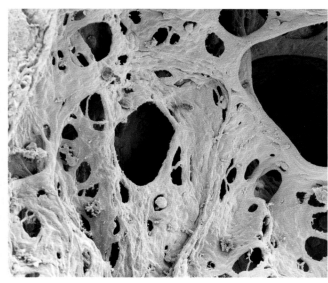

Figure 9.12 SEMs of Healthy human lung tissue (left) and human lung tissue showing advanced emphysema (right)

Lung cancer

Persistent exposure of the bronchi to cigarette smoke results in damage to the goblet cells. These are then replaced by an abnormally thickened epithelium. With prolonged exposure to the carcinogens, permanent mutations may be triggered in the DNA of some of the cells that have replaced goblet cells. If this occurs in their proto-oncogenes or tumour-suppressing genes the result is loss of control over normal cell growth (page 104).

A single mutation is unlikely to be responsible for triggering lung cancer; it is the accumulation of mutations with time that causes a group of cells to divide by mitosis repeatedly, without control or regulation, forming an irregular mass of cells – the **tumour**. Tumour cells then emit signals promoting the development of new blood vessels to deliver oxygen and nutrients, all at the expense of the surrounding healthy tissues. Sometimes tumour cells break away and are carried to other parts of the body, forming a secondary tumour (a process called **metastasis**). Unchecked, cancerous cells ultimately take over the body, leading to malfunction and death.

Signs and symptoms of lung cancer

Lung cancer is dangerous because the symptoms are vague and only become severe when the disease is in its late stages. Lung cancer can be cured but only if it is detected early enough. The symptoms of the disease are:

- frequent and severe coughing that persists for more than three weeks
- the coughing up of blood
- shortness of breath and a feeling of weakness or tiredness
- an unexpected and unexplained loss of weight
- pain in the rib cage and/or the shoulder areas
- chest infections that will not go away, even with antibiotics
- hoarseness of the voice and swelling of the face and neck.

SEM of human lung tissue with cancer

ciliated epithelium lining the bronchial tree

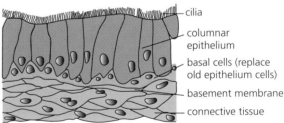

in response to continuing exposure to cigarette smoke

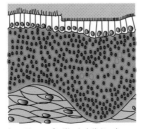

beating of cilia inhibited; excessive, abnormal proliferation of basal cells

ciliated epithelium replaced by squamous epithelium

mass of basal cells showing abnormal growth

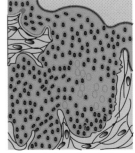

accumulation of mutations in mass of abnormal cells; forms a tumour

Figure 9.13 The development of a lung tumour

Diagnosis and treatment of lung cancer

From the initial critical mutations until symptoms of the disease are evident typically takes 20–30 years. A diagnosis may be confirmed by chest X-ray, CT scan (a three-dimensional image of part of the body produced using X-rays) and ultimately, by bronchoscopy (where a device is used to see inside the lungs). The latter is combined with biopsy (the removal of a small piece of tissue for laboratory examination).

Treatment of lung cancer largely depends upon the general health of the patient, on what type of lung cancer has been contracted, and how far it has spread. Many of the tumours detected by early diagnosis may be removed by surgery or suppressed by radiotherapy (with X-rays or another form of radiation) – possibly followed in either case by chemotherapy using anti-cancer drugs. However, more than half of patients diagnosed currently die within a year because lung cancer is a deadly disease that is rarely diagnosed early enough. Incidentally, a person who gives up smoking may eventually achieve a life expectancy very similar to that of a non-smoker (Figure 9.14).

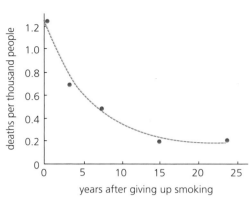

Figure 9.14 Death rates from cancer after giving up smoking

Cigarette smoking causes lung diseases – the evidence

It was **epidemiology** (the study of the incidence and distribution of diseases and of their control and prevention) that first identified a likely causal link between smoking and disease. For example, a 1938 study of longevity in people who smoked lightly and heavily compared with those who did not smoke established that heavy smoking most shortened life expectancy (Table 9.2).

In 1950, an American study of over 600 smokers compared with a similar group of non-smokers found **lung cancer** was 40 times higher among the smokers. The risk of contracting cancer increased with the numbers of cigarettes smoked (Table 9.3).

Table 9.2 Longevity in smokers and non-smokers

Age	Number of people still alive at each age		
	Non-smokers	**Light smokers**	**Heavy smokers**
30	100 000	100 000	100 000
40	91 546	90 883	81 191
50	81 160	78 436	62 699
60	66 564	61 911	46 226
70	45 919	41 431	30 393
80	21 737	19 945	14 494

Table 9.3 Cancer rates and the number of cigarettes smoked per day

Number of cigarettes smoked per day	Incidence of cancer per 100 000 men
0	15
10	35
15	60
20	135
30	285
40	400

A survey of smoking in the UK was commenced in 1948, at which time 82 per cent of the male population smoked and of whom 65 per cent smoked cigarettes. This had fallen to 55 per cent by 1970 and continued to decrease. In same period the numbers of females who smoked remained just above 40 per cent until 1970, after which numbers also declined (Figure 9.15). Look carefully at the changing pattern in the incidences of lung cancer – figures are available since 1975.

Concerning smoking and **respiratory diseases**, the National Institute for Health and Clinical Excellence (NICE) reports that chronic obstructive pulmonary disease (COPD) accounts for 30 000 deaths per year in the UK, of which 85 per cent could be attributed to smoking – a figure approximately double the European average. The World Health Organisation (WHO) provides leadership on health matters for the United Nations by monitoring and reporting on the health of peoples and assessing health trends worldwide. The World Health Organisation predicts that by 2030, chronic obstructive pulmonary disease will be the third most common cause of death worldwide.

Question

9 Look at Figure 9.15. Comment on the incidence of lung cancer in men and women between 1975 and 2007 in relation to the changing pattern of smoking since UK records began.

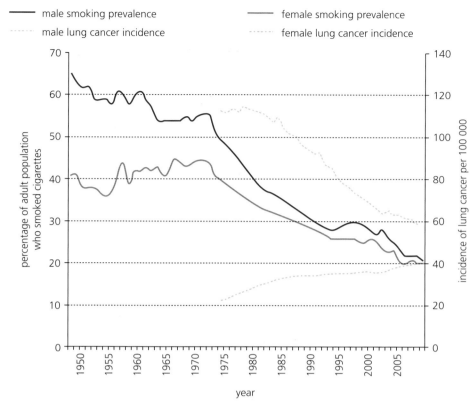

Figure legend:
— male smoking prevalence
— female smoking prevalence
···· male lung cancer incidence
···· female lung cancer incidence

y-axis left: percentage of adult population who smoked cigarettes

y-axis right: incidence of lung cancer per 100 000

x-axis: year

Figure 9.15 Lung cancer incidence and smoking trends in the UK, 1948–2007

Figure 9.16 An animal experiment investigating the harmful effects of cigarette smoke

Question

10 Explain what is meant by a 'control group' in a medical investigation.

Meanwhile, experimental **laboratory-based investigations** have demonstrated *how* cigarette smoke causes disease. For example, experiments in which laboratory animals (mice, rats and dogs) were exposed to cigarette smoke (passive smoking) for an extended period resulted in damage and disease in their lungs identical to that often observed in the lungs of deceased human smokers. Additionally, cancers have been induced when components of cigarette smoke, mainly polycyclic aromatic hydrocarbons (PAHs), were applied in concentrated form to the exposed external skin of experimental mammals.

Most significantly, specific damage to DNA, induced by exposure to polycyclic aromatic hydrocarbons, has been identified. Firstly, labelled polycyclic aromatic hydrocarbons were added to cigarettes that were then smoked by human volunteers taking part in a clinical experiment. These labelled polycyclic aromatic hydrocarbons were shown to be absorbed into the bloodstream and quickly metabolised into substances that damaged DNA – within about 15–30 minutes of the cigarettes being smoked, in fact. Secondly, it was found that these substances actually reacted with specific points on the p53 gene. Remember, this gene in an un-mutated state restricts unlimited cell division (a feature of cancerous cells) and it stimulates repair of damaged DNA. Further, the p53 gene has been shown to be inactivated in the majority of people suffering from lung cancers. The inescapable conclusion of these studies is that specific chemicals in cigarette smoke directly trigger lung cancer.

Experiments have been supported by **clinical investigations** of the biochemical and structural changes that cigarette smoke induces in patients. Further, **intervention studies** with vulnerable groups of patients who were persuaded to change their habits established how smoking-related diseases were reduced in them compared with rates in other patients who did not change their habits. These others were, in effect, a **control group**.

Short-term effects of nicotine on the cardiovascular system

The short-term effects that the nicotine in tobacco smoke has on the cardiovascular system are all profoundly harmful. Examining these in turn, they are:

1 As an addictive, psychoactive drug.

Smoking a single cigarette introduces about 1 mg of nicotine into the bloodstream almost immediately. The first outcomes are the release of sugar from the liver and of adrenalin from the adrenal medulla – with stimulating effects on the body. Long before the nicotine circulating in the bloodstream can been expelled from the body, however, nicotine-rich blood reaches the brain, and nicotine crosses the blood–brain barrier. When present among the neurones of the brain, nicotine stimulates the release of chemical messengers in synapses there (page 320). For example, the release of the neurotransmitter dopamine is enhanced, so a sense of well-being is generated, and pain and anxiety are suppressed. Further, nicotine is similar in structure to the neurotransmitter acetylcholine and fits acetylcholine receptors on post-synaptic membranes. However, it is not broken down by enzymes that inactivate acetylcholine, so it remains attached, prolonging the effects. Initially, these are enhanced concentration and memory, so alertness appears to be enhanced. It is these combined psychoactive effects that generate the condition of dependence – leading to a state of addiction that is one of the hardest to break. Moreover, with increasing dependence on nicotine, the impact of the drug switches to a sedative effect.

2 As a vasoconstrictor.

The impacts of this effect on the vascular system are raised blood pressure and a reduced oxygen supply to the major organs and limbs. At the same time, the enhanced release of the hormone adrenalin causes an increased heart rate, which in these circumstances, contributes to raised blood pressure, too. A condition of permanently raised blood pressure is known as hypertension (see 'Blood pressure and its measurement' below). The epidemiological evidence is of an increased risk of coronary artery disease at an earlier age in smokers than in non-smokers. In fact the first evidence of the link between smoking and cardiovascular disease (CVD) established this relationship.

It was in 1950 that Dr Richard Doll and colleagues, working at St Thomas's Hospital, London, published the results of their investigation of the causes of death due to cardiovascular disease in a large sample of working doctors (among whom the habit of smoking was widespread *at that time*). These implicated smoking as the culprit (Table 9.4). Today, few doctors smoke.

Table 9.4 Mortality from cardiovascular disease among a sample of doctors

Cause of death in a sample of 40 000 doctors	Non-smokers	Cigarette smokers
Coronary heart disease	606	2067
Stroke	245	802
Aneurysm	14	136
Arteriosclerosis	23	111
Total	888	3116

Blood pressure and its measurement

By blood pressure we mean the pressure of the blood flowing through the arteries. We have noted that, initially, flow is a surge or **pulse** (page 168). So it is arterial blood pressure that is measured, in a part of our body relatively close to the heart. Blood pressure is quoted as two values (typically, one over the other). The higher pressure is produced by ventricular systole (systolic pressure) and is followed by a lower pressure at the end of ventricular diastole (diastolic pressure). Normally, systolic and diastolic pressures are about 15.8 and 10.5 kPa respectively. (The medical profession give these values as 120 and 70–80 mmHg. The unit 'mmHg' was recognised as a unit of pressure before the SI System was introduced.)

To measure these two values an inflatable cuff called a **sphygmomanometer** is used with a **stethoscope**, as shown in Figure 9.17. The three steps are as follows.

1 The cuff is inflated and blood flow is monitored in the artery in the arm (the brachial artery) at the elbow. Inflation is continued until there is no sound (indicating there is no flow of blood).

2 Air is now allowed to escape from the cuff, slowly, until blood can just be heard spurting through the constriction point in the artery. This pressure at the cuff is recorded, as it is equal to the maximum pressure created by the heart (the **systolic pressure**).

3 Pressure in the cuff is allowed to drop until the blood can be heard flowing constantly. This is the lowest pressure the blood falls to between beats (the **diastolic pressure**).

Table 9.5 Screening of blood pressure in adults

Systolic pressure/mmHg	Diastolic pressure/mmHg	Condition	Response
120	80	Optimum	
120–129	80–84	Normal	Biennial checks
130–139	85–89	High-normal	Annual checks
140–159	90–99	Stage 1 hypertension	Check in two months
160–179	100–109	Moderate (stage 2) hypertension	Treatment essential if these conditions persist.
180–209	110–119	Severe (stage 3) hypertension	
210+	120+	Very severe (stage 4) hypertension	

The pascal (Pa) and its multiple the kilopascal (kPa) are generally used by scientists to measure pressure, but in medicine the older unit of pressure, 'millimetre of mercury' (mmHg) is still used (1 mmHg = 0.13 kPa).

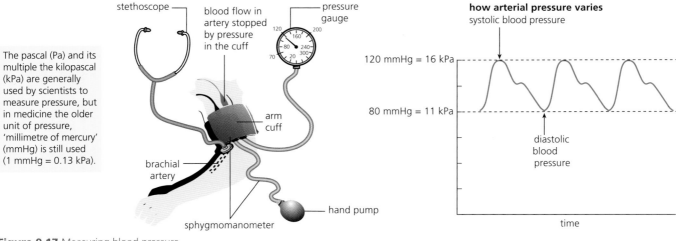

Figure 9.17 Measuring blood pressure

Short-term effects of carbon monoxide on the cardiovascular system

The iron of haemoglobin (Figure 8.10, page 160) is not oxidised to iron(III) during oxyhaemoglobin formation in the lungs. This oxidation can occur, however – with serious consequences. For example, carbon monoxide converts this iron(II) to iron(III) when it reacts with haemoglobin, in the formation of carboxyhaemoglobin. In this form, the haemoglobin molecule cannot carry oxygen. In the blood of smokers about 10 per cent of their haemoglobin occurs in the form of carboxyhaemoglobin, due to exposure to the carbon monoxide in cigarette smoke. Carboxyhaemoglobin does not dissociate, for the reaction between carbon monoxide and haemoglobin is irreversible. Oxygen transportation in the blood is hampered, and so too is the supply of oxygen to body organs and the brain.

Extension

Prevention – a challenging but more sustainable response?

Figure 9.18 Persuasive advertising, it seems!

Smoking is clearly a health risk with life-threatening consequences. Perhaps the best response is to prevent the problems arising by persuasion, backed up by legislation where this is thought effective. One developed country's current response to the issue is to restrict the advertising and display of tobacco products, enforce bold, unavoidable health warnings on cigarette packets and outlaw smoking in public buildings and workplaces. At the same time, the authority responsible for public health is assisting those addicted to nicotine through smoking to discard the habit if they can. This sustainable 'carrot and stick' approach is having some significant positive outcomes but there remain many who continue to smoke.

Meanwhile, throughout the world there continues to be huge investments by the tobacco industry in the growing of the crop, the manufacturing of 'attractive' packaging for tobacco products and the placing of advertisements in countries and places that are open to them.

Question

11 Cardiovascular disease was not an issue for people before the twentieth century. Today it is a major health issue. Make a concise, annotated list of all the factors that you think may lie behind this change to use in a group discussion with your peers.

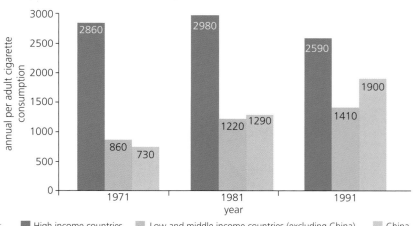

Figure 9.19 World trends in per capita cigarette consumption

(Continues)

Whilst the number of people smoking cigarettes is actually falling in developed countries, smoking is on the rise in the developing world (Figure 9.19). The global health issues generated by cigarette smoking are of epidemic proportions – about one-third of the global adult male population smokes. Among the World Health Organisation areas, the East Asia and the Pacific Region has the highest smoking rate, with nearly two-thirds of men smoking. Today, apparently, one in every three cigarettes is smoked in China.

Globally, the consumption of manufactured cigarettes continues to rise steadily. Consequently, the death toll from the use of tobacco, currently estimated to be 6 million per year worldwide, is predicted to rise to steeply (Figure 9.20b). This is inevitable because of the continuing popularity of cigarettes and of the time lag between the taking up of smoking and the appearance of the symptoms of lung cancer (Figure 9.20a).

How has such a detrimental personal habit gained such a foothold worldwide? Why does it continue to attract people? The answer must lie partly in the impact of the drug nicotine on the central nervous system. Another is that people, particularly the young, have been (and are still being) persuaded that the use of tobacco is linked with highly desirable, personal qualities, such as adult sophistication, self-fulfilment, athletic prowess and sexual attractiveness.

- Studies show that teenagers are strongly influenced by tobacco advertising.
- Today, the multinational conglomerates that form the bulk of the tobacco industry have huge budgets – in Asia, they are among the top ten advertisers.

In attempting to deal with this health issue, the scale of industrial investment focused on this product is a huge problem. There is a sound economic and moral case for continuous, powerful advertising campaigns that more effectively discourage the use of tobacco – despite the cost of such initiatives. This is an ethical issue for all nations.

Question

12 List the effects of tobacco smoke on the cardiovascular system.

a) The 20-year time lag between smoking and the diagnosis of lung cancer

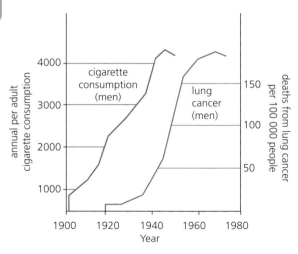

b) Premature death from tobacco use, projections for 2000–2004 and 2025–2049

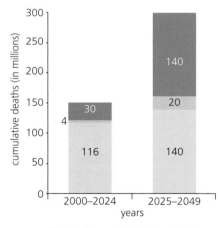

- preventable if adults quit (halving global cigarette consumption by 2020)
- preventable if young adults do not start (halving global uptake of smoking by 2020)
- other premature deaths from tobacco-related causes

Figure 9.20 World smoking habits and the consequent premature deaths

Summary

- **Gaseous exchange**, the exchange of gases between an organism and its environment, makes possible aerobic respiration in the cells of the body. Gaseous exchange between individual cells and their immediate environment occurs by **diffusion**.

- **Mammals**, with their compact bodies and protected external surface, require specialised respiratory organs, the **lungs**. These efficient organs are housed in the air-tight thorax which moves rhythmically to ventilate the lungs. The **blood circulation system** serves the lungs and transports oxygen all over the body, facilitated by the respiratory pigment **haemoglobin**, present in the red blood cells. Carbon dioxide is returned to the lungs.

- The **alveoli** (the air sacs where gaseous exchange occurs) are reached via the trachea, bronchi and bronchioles, a branching system referred to as the 'bronchial tree'. The walls of these tubes are variously supported by **cartilage and smooth muscle** and their surface is lined by a **ciliated epithelium** with **goblet cells** that secrete mucus.

- Air is warmed and made moist as it is drawn into the thorax. Dust and foreign bodies are trapped in the **mucus** that is continually swept up and out of the bronchial tree by the beating movements of the **cilia**.

- The millions of individual alveoli are served by terminal bronchioles and both are thin walled and surrounded by fine capillaries. Here, a huge, thin surface area is provided for **efficient exchange of gases** between the air and the blood circulation. Detritus and bacteria are removed by **macrophages** and the walls are kept flexible by the secretions of **surfactant cells**.

- **Smoking** generates major health problems because of carcinogenic aromatic hydrocarbons in the 'tar', the addictive nature of the nicotine and the presence of carbon monoxide – all ingredients released by the burning of tobacco. Many of them reach delicate body cells when cigarette smoke is inhaled.

- Chronic obstructive pulmonary disease (mainly **bronchitis and emphysema**) and **lung cancer** are caused by frequent and regular exposure to cigarette smoke. Cardiovascular disease (CVD) is also caused by smoking, leading to likely hypertension, coronary heart disease, strokes and aneurisms.

- Cardiovascular disease (and cancers) may be treated by various drugs and surgical procedures but **prevention** of these conditions arising by stopping smoking and wiser lifestyle choices would give superior results. However, **nicotine** is so strongly addictive and advertising campaigns by cigarette manufacturers are so influential, there is currently a growing and **global menace to human health** from the smoking of cigarettes.

Examination style questions

1 Fig. 1.1 shows the heart and associated blood vessels.

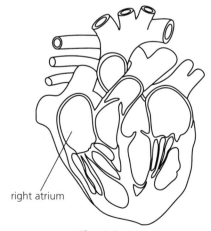

right atrium

Fig. 1.1

a) On a copy of Fig. 1.1, draw label lines and use the letters **P**, **Q** and **R** to indicate the following structures:

P a blood vessel that carries deoxygenated blood

Q a structure that prevents backflow into a ventricle

R a blood vessel that carries blood at high pressure   [3]

b) The changes in blood pressure in the right atrium are the same as those in the left atrium. The changes in blood pressure in the right ventricle are different from those in the left ventricle.
Explain why this is so.   [4]

c) Some components of tobacco smoke are absorbed into the blood stream and affect the cardiovascular system.

Describe the effects of nicotine and carbon monoxide on the cardiovascular system.   [4]
[Total: 11]

(Cambridge International AS and A Level Biology 9700, Paper 02 Q1 June 2008)

2 The global consumption of manufactured cigarettes continues to rise steadily and the health issues generated by cigarette smoking are of epidemic proportions worldwide.

a) Suggest why young people today may still choose to smoke cigarettes.

b) Explain the chief ways that cigarette smoke is directly harmful to the body.

c) Identify why cigarette smoking, once begun, is addictive.

3 Scientists at the Tibet Institute of Medical Sciences in Lhasa investigated differences between adult Tibetans who had lived in Lhasa (altitude 3658 m) all their lives and adult Han Chinese residents who had lived there for about 8 years. The Tibetans and the Han Chinese exercised at maximum effort and various aspects of their breathing were measured. Some of the results are shown in Table 3.1.

Table 3.1

Feature	Tibetans	Han Chinese
minute volume / dm³ min⁻¹	149	126
oxygen uptake / cm³ kg⁻¹ min⁻¹	51.0	46.0

- Minute volume. This is the volume of air breathed in during one minute.
- Oxygen uptake. This is the volume of oxygen absorbed into the blood during one minute. It is expressed per kg of body mass.

The researchers observed that

- the greater minute volume of the native Tibetans resulted from a greater tidal volume
- the tidal volumes of the Tibetans showed a positive correlation with their vital capacity measurements
- the Han Chinese had lower values for both tidal volume and vital capacity.

a) State what is meant by the term *tidal volume*. [1]

b) Suggest why the researchers also measured the *vital capacity* of the people in the study. [2]

c) Explain how the minute volume **at rest** would be determined. [2]

d) Suggest two differences in the **structure** of the lungs that may account for the greater oxygen uptake by the Tibetans shown in Table 3.1. [2]

e) When people who have lived all their lives at low altitude go to a place at high altitude, such as Lhasa, they are often breathless, lack energy and suffer from altitude sickness. However, with time, they often acclimatise to the high altitude.

   In another study, researchers found that the red blood cell count increases in such people by about 30% over several weeks.

   Explain why the red blood cell count increases so much when people visit places at high altitude. [2]

[Total: 9]

(Cambridge International AS and A Level Biology 9700, Paper 02 Q4 June 2008)

4 Fig 5.1 shows a drawing made from an electron micrograph of a cell from the ciliated epithelium of the bronchus.

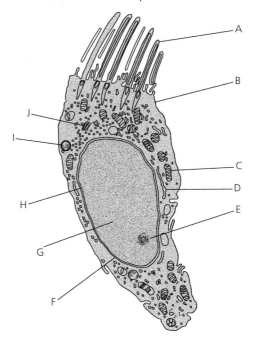

Fig. 5.1

a) Copy and complete the table below by writing the appropriate letter from the diagram to indicate the structure that carries out each of the functions listed. The first one has been completed for you. [5]

Function	Structure
facilitated diffusion of glucose	B
creates a current to move mucus	
aerobic respiration	
makes ribosomes	
a site of transcription	
packages proteins into lysosomes	

b) The alveoli in the lungs are lined by a squamous epithelium. Explain why gas exchange occurs in alveoli and not in the bronchus. [3]

c) Describe the likely appearance of the lining of the bronchus in a person who has been a heavy smoker for many years. [3]

[Total: 11]

(Cambridge International AS and A Level Biology 9700, Paper 02 Q1 June 2007)

10 Infectious disease

The infectious diseases studied in this topic are caused by pathogens that are transmitted from one human host to another. Some, like *Plasmodium* that causes malaria, are transmitted by vectors; others are transmitted through water and food or during sexual activity. An understanding of the biology of the pathogen and its mode of transmission is essential if the disease is to be controlled and ultimately prevented.

 ## 10.1 Infectious diseases

While many infectious diseases have been successfully controlled in some parts of the world, many people worldwide are still at risk of these diseases.

By the end of this section you should be able to:

a) define the term disease and explain the difference between an infectious disease and a non-infectious disease (sickle cell anaemia and lung cancer)

b) state the name and type of causative organism (pathogen) of each of the following diseases: cholera, malaria, tuberculosis (TB), HIV/AIDS, smallpox and measles

c) explain how cholera, measles, malaria, TB and HIV/AIDS are transmitted

d) discuss the biological, social and economic factors that need to be considered in the prevention and control of cholera, measles, malaria, TB and HIV/AIDS

e) discuss the factors that influence the global patterns of distribution of malaria, TB and HIV/AIDS and assess the importance of these diseases worldwide

Disease: an abnormal condition affecting an organism, which reduces the effectiveness of the functions of the organism.

Non-infectious disease: a disease with a cause other than a pathogen, including genetic disorders (such as sickle cell anaemia) and lung cancer (linked to smoking and other environmental factors).

What are 'health' and 'disease'?

By disease we mean 'an unhealthy condition of the body'. There are many different diseases but good health is more than the absence of the harmful effects of disease-causing organisms. We have already seen that illness may be caused by poor lifestyle choices or unfavourable environmental conditions, such as lung cancer due to the smoking of cigarettes (page 184). Diseases of this type are called non-communicable or **non-infectious diseases**. They include conditions such as chronic obstructive pulmonary diseases (COPD) and cardiovascular disease (CVD), discussed in Topic 9. Non-infectious disease may also be caused by malnutrition. Inherited or genetic disorders such as sickle cell anaemia are included here, too.

Infectious disease

In this topic the focus is on diseases that can be 'caught' – sometimes described as contagious or communicable. These are **infectious diseases**. They are caused by invading organisms living parasitically on or in the body. These disease-causing organisms are known as **pathogens**. Pathogens may be viruses, some types of bacteria or certain other organisms that are passed from one organism to another. The range of pathogens is introduced in Figure 10.1.

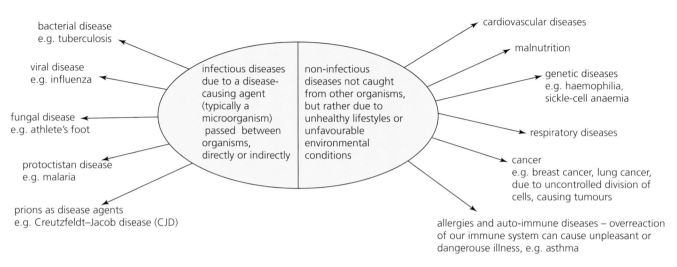

Figure 10.1 Types of human disease

Infectious disease:
a disease caused by a pathogen that can be transmitted from one host organism to another.

Pathogen: a biological agent (such as a virus, bacterium, fungus or protoctist) that causes disease. A pathogen causing human diseases will have, as part of its structure, proteins that are different from those of the human host and are therefore antigens.

The pathogens that cause some infectious diseases exist permanently in a particular region or population. Such a disease is said to be **endemic** for it has a constant presence. **Malaria** is an example – a serious threat to humans living in parts of Africa and other tropical and sub-tropical regions in the world.

On the other hand, **epidemics** are diseases that appear suddenly and then spread rapidly in a specific area or within a particular population group for a time. Cases of food poisoning are examples.

A **pandemic disease** is an epidemic that spreads far more widely – typically throughout a whole country, a whole continent or throughout the whole world. It affects a huge number of people. In the twentieth century several pandemics of influenza ('flu) occurred. For example, the 'flu pandemic that occurred in 1918 was responsible for 10 million deaths worldwide.

In the **prevention of infectious disease**, knowledge of the biology of the pathogen is essential. So the fight against disease involves the work of microbiologists as well as the vigilance of the health professionals. In this topic, six specific infectious diseases of humans are examined, most of which continue to present major threats to life. There are reasons why this is so in the case of particular diseases but four general points are significant.

- We live in a 'global village' world with speedy, worldwide travel facilities available to some. A person, infected on one continent, may travel to another within hours – before any symptoms have appeared.
- The human population is in a phase in which it is increasing more and more rapidly (Figure 10.2). In ever-increasing numbers, people congregate together – many in mega-cities. Pathogens do not have far to travel!
- The work of health professionals is most effective where there is social stability, the absence of war and an equitable division of wealth. Unfortunately, these conditions are absent from many human societies.
- As natural resistance to pathogens develops and drugs to overcome them are found, so pathogens themselves evolve greater infectivity – there is a continuing 'arms race' between pathogens and **hosts**.

Accidents

Not included in any of the types of ill-health listed in Figure 10.1 is damage due to accidents, such as broken bones. A bone fracture is not a disease but it is certainly an unhealthy condition of the body which takes time to repair. Clearly, there is a borderline between health and disease that is not easy to define.

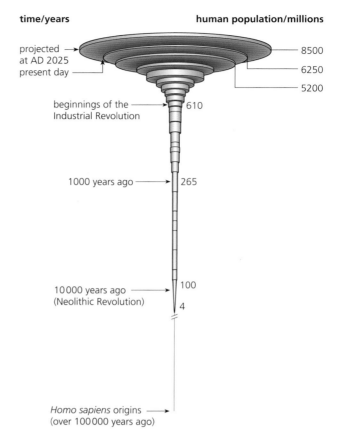

Figure 10.2 The changing pattern of the estimated world human population

Case studies in infectious disease

Cholera

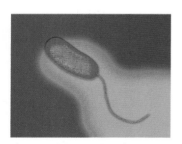

Figure 10.3 Scanning electron micrograph of *Vibrio cholerae* (×60 000)

Cholera is a dangerous disease, caused by the curved rod-shaped bacterium, *Vibrio cholerae* (Figure 10.3). This pathogen is typically acquired from drinking water that has become heavily contaminated by the faeces of patients (or from human 'carriers' of the pathogen). Alternatively, it may be picked up from food contaminated by flies that previously fed on human faeces. The consumption of raw shellfish taken from waters polluted with untreated sewage is another common source.

Cholera is endemic in many parts of Africa, India, Pakistan, Bangladesh, and Central and South America. A large number of the cholera bacteria must normally be ingested for the disease to develop; fewer than 10^8–10^9 organisms is ineffective (unless anti-indigestion tablets that neutralise stomach acid were taken beforehand). Once the bacteria has survived the stomach acid and reached the intestine, the incubation period is from 2 hours to 5 days. However, 75 per cent of those infected develop no symptoms at all – the bacteria merely remaining in their faeces for up to 2 weeks – whilst others are made very ill. This difference may arise because there are different strains of the pathogen, some mild, others virulent in their impact on the patient.

In the intestine, the pathogen increases in numbers and attaches itself to the epithelium membrane. The release of a toxin follows. This generates symptoms in susceptible patients that are mild to moderate in many people but extremely severe in about 20 per cent of people who become infected.

The effect of this toxin is to trigger a loss of ions from the cells of the epithelium. Outflow of water follows and this is constantly replaced from tissue fluid. The patient rapidly loses a massive amount of body fluid. Typically 15–20 litres may drain from the body as watery diarrhoea containing mucus and epithelium cells ('rice-water stool'), as well as vast numbers of the bacterium. At the same time, the patient has severe abdominal muscle cramp, vomiting and fever. Death may easily result from this dehydration because the severely reduced level of body fluids causes the blood circulation to collapse.

1 Digestion of the proteins in the gut begins in the stomach. What is the source and role of the 'strong acid' present in the stomach when food is ingested?

How the cholera toxin works

The cholera toxin consists of a two-protein complex composed of one 'A' subunit and five 'B' subunits. The B part is the binding protein which attaches the toxin complex to a particular site, a glycolipid, on the cell surface membranes of intestine epithelium cells. The A part is an enzyme that activates other enzymes of the cell surface membrane of the epithelium cell to which it is attached. These cause the secretion of chloride ions into the gut lumen and inhibit any uptake of sodium ions. Hypersecretion of chloride ions results and is followed by water loss (Figure 10.4).

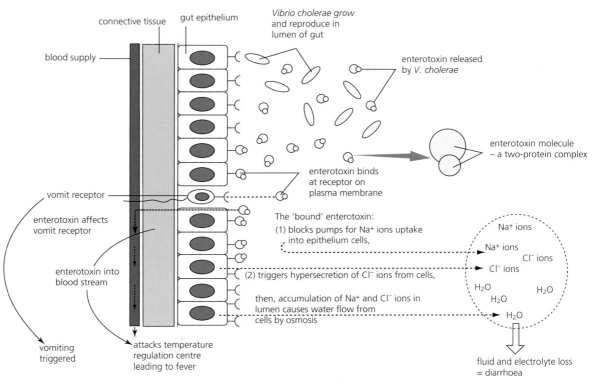

Figure 10.4 The action of the cholera toxin

Question

2 In cholera, toxic protein triggers the exit of chloride ions from the epithelium cells into the lumen of the gut. How does this movement of ions differ from the movement of water that follows ion transport?

Treatment of cholera

About 50 per cent of untreated cases of cholera are fatal but, when properly treated, generally less than 1 per cent of people are likely to die from this very unpleasant disease. A cholera patient requires immediate oral (by mouth) administration of a dilute solution of ions and glucose, present in a rehydration pack (Figure 10.5), in order to replace the fluids and ions lost from the body. The presence of glucose aids uptake of ions and compensates for energy loss. This solution prevents dehydration and restores the osmotic balance of the blood and tissue fluid. This might seem a simple process to people in the developed world but, in those places where cholera is endemic, conditions may be very different. Many who contract the disease are weakened by shortage of food. Their drinking water may be of dubious purity and have to be carried some distance. The boiling of water makes it safe to use in the rehydration fluid but this requires fuel that may be scarce.

Very severely dehydrated patients require the administration of intravenous fluids. An additional treatment, the taking of the antibiotics streptomycin or tetracycline, will assist in ridding the gut of the cholera cells. However, these drugs are of little use unless fluid, ions and glucose replacement has succeeded in restoring normal body fluid composition. An oral vaccine against cholera now exists but it is not effective against all strains.

Prevention of cholera

Cholera has been eradicated where there is effective sanitation. In the longer term, the disease can be virtually eliminated by satisfactory processing of sewage and the purification of drinking water, including a chlorination phase. In endemic areas, the spreading of infection between households may be prevented by the boiling of drinking water. The faecal contamination of foods by flies must also be prevented.

However, where populations have been displaced into inadequate overcrowded refugee camps or have migrated to crowded urban slums where clean water and sanitation are not available, cholera remains a threat to life (Figure 10.6). Malnourished children and those living with HIV are at greatest risk.

Figure 10.5 An oral rehydration pack for the treatment of cholera

Figure 10.6 A refugee camp – ideal conditions for cholera

Table 10.1 Cholera – the profile

Pathogen	Pathogen taxonomy	A rod-shaped bacterium, *Vibrio cholerae*
	Transmission	Via polluted water and contaminated foods
Disease parameters	**Incubation and venue**	Once in contact with the wall of the small intestine, a short incubation period of between 2 hours and 5 days
	Signs, symptoms and diagnosis	Diarrhoea, vomiting, abdominal muscle cramp, fever Confirmation by microscopic examination of faeces
	Treatment	Replacement of lost water and ions by oral rehydration with dilute solution of salts and glucose
	Prognosis	Typically 50% of untreated cases are fatal but with prompt correct treatment almost all patients recover
	Prevention and control	Boiling of drinking water, correct treatment of sewage and the prevention of food contamination by flies

Malaria

Malaria is the most important of all insect-borne diseases, posing a threat to 2400 million humans in 90 countries – 40 per cent of the world's population, in fact (Figure 10.7). About 80 per cent of the world's malaria cases are found in Africa south of the Sahara. Here, some 90 per cent of the fatalities due to the disease occur. It is estimated that around 400 million people are infected, of which 1.5 million (mostly children under 5 years) die each year. Malaria kills more people than any other communicable disease, with the exception of tuberculosis.

Malaria is caused by **Plasmodium**, a protoctist, which is transmitted from an infected person to another by blood-sucking mosquitoes of the genus **Anopheles**. The mosquito is described as the **vector** – an organism that transfers a disease or parasite from one host organism to another. Of the four species of *Plasmodium* that cause malaria in humans, only one (*P. falciparum*) causes severe illness. The others species trigger milder infections that are rarely fatal. Insecticides, particularly **DDT**, had been effective against mosquitoes in the past but there has been a resurgence of malaria since the 1970s.

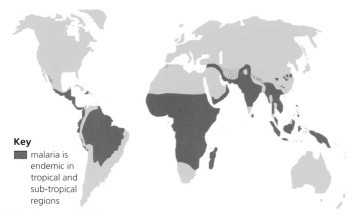

Key
malaria is endemic in tropical and sub-tropical regions

Figure 10.7 World distribution of malaria

Until 100 years ago, malaria was endemic in Europe and the USA, too. In Britain it was called 'fen ague' because it was common in communities living near marshes and because of the periodic shivering fits patients experienced. Marsh and fen drainage for agriculture changed the environment so that it was no longer favourable to mosquitoes and their transmission of the parasite. This is one reason for its disappearance in some areas.

Transmission of *Plasmodium* by the mosquito

Some pathogens need a vector to reach a new host. Stages in the life cycle of the pathogen may occur in the vector as well as the host. The female mosquito is the vector for malaria. (The male mosquito feeds on plant juices.) The mosquito is a fluid-feeding heterotroph, not a parasite, as sometimes stated. The female detects its human host, lands and inserts a long thin tube (called a proboscis) into a blood vessel below the skin surface. A 'meal' of blood is taken quickly; there is a danger that an active, alert human will 'swat' the intruder. For a blood-sucking insect, the mammalian's blood-clotting mechanism presents a problem that has to be overcome. This occurs when the mosquito's proboscis penetrates the vein and a secretion from its salivary glands passes into the victim's blood, inhibiting it from clotting. At this point *Plasmodium* may enter its human host (if the mosquito carries an infective stage). Meanwhile, the mosquito loads up with a blood meal (Figure 10.8).

Mosquitoes tend to feed at night, on sleeping victims. Those that alight on patients already ill with malaria are likely to be able to feed unhindered. The life cycle of the malarial parasite is complex. It consists of a sexual stage beginning in the human host (the primary host) and completed in the mosquito (the secondary host or vector). There is an asexual multiplication phase in the mosquito, too, and a phase of growth and multiplication in the liver cells and red blood cells of the human host. Details of this life cycle are not required here.

Question

3 The insecticide DDT is harmless to humans at concentrations that are toxic to mosquitoes. This insecticide is a stable molecule that remains active long after it has been applied and it is stored in the body rather than being excreted. Nevertheless, the use of this compound has been discontinued. Suggest why this is so?

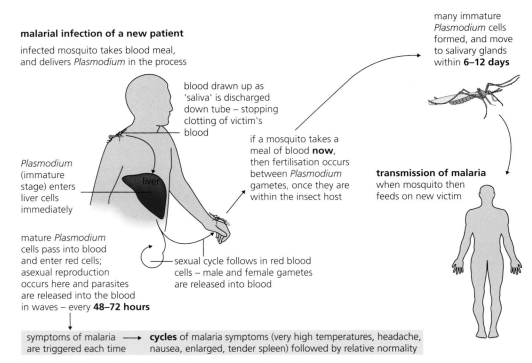

malarial infection of a new patient

infected mosquito takes blood meal, and delivers *Plasmodium* in the process

blood drawn up as 'saliva' is discharged down tube – stopping clotting of victim's blood

many immature *Plasmodium* cells formed, and move to salivary glands within **6–12 days**

Plasmodium (immature stage) enters liver cells immediately

if a mosquito takes a meal of blood **now**, then fertilisation occurs between *Plasmodium* gametes, once they are within the insect host

transmission of malaria when mosquito then feeds on new victim

mature *Plasmodium* cells pass into blood and enter red cells; asexual reproduction occurs here and parasites are released into the blood in waves – every **48–72 hours**

sexual cycle follows in red blood cells – male and female gametes are released into blood

symptoms of malaria are triggered each time → **cycles** of malaria symptoms (very high temperatures, headache, nausea, enlarged, tender spleen) followed by relative normality

Figure 10.8 Mosquito feeding and the transmission of malaria

Control measures

Control measures currently applied include the following.

- Interruption of the mosquito life cycle by attacking the larval stage which is an air-breathing aquatic animal. Swamps are drained and open water sprayed with oil (which forms a surface film, blocking the larva's air tube), and with insecticide.
- Use of insecticides to kill adult mosquitoes on and around the buildings that humans occupy.
- Protection from mosquito bites by sleeping under insecticide-treated mosquito nets and by the burning of mosquito coils.
- Use of drugs to kill the stages of *Plasmodium* found in the blood and liver of infected people.
- Use of drugs to kill *Plasmodium* as soon as it is introduced into a healthy person's blood by an infected female mosquito whilst feeding.

The effect of the sickle cell trait on death from malaria

The malarial parasite *Plasmodium* completes its life cycle in red blood cells but it cannot do so in red blood cells containing haemoglobin-S (Hb^S) (Figure 6.17, page 125). People with **sickle cell trait** are heterozygous for the sickle cell allele ($Hb^A Hb^S$). Their red blood cells contain some sickle haemoglobin so they are protected to a significant extent from malaria – they are much less likely to die from the disease. Where malaria is endemic in Africa, possession of one mutant allele is advantageous. Consequently, people with sickle cell trait are more likely to survive to pass on this allele to the next generation (Figure 10.9).

How will malaria be eradiated?

After many soldiers in the wars of the first half of the twentieth century became casualties of malaria, money was found for research into its control. Powerful insecticides (first DDT, then gamma BHC and dieldrin) became available. By the 1960s mosquito-eradication programmes had been successful in many countries. However, by the 1970s, resistant strains of mosquito had appeared (and *Plasmodium* resistant to anti-malaria drugs, too).

distribution of haemoglobin-S is virtually the same as that of malaria

distribution of malaria caused by *Plasmodium falciparum* or *P. vivax* (the forms of malaria that are most frequently fatal, especially in childhood)

distribution of sickle cell gene in the population

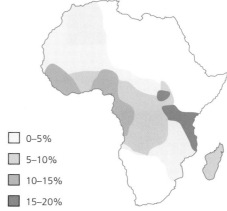

☐ 0–5%

☐ 5–10%

☐ 10–15%

☐ 15–20%

Figure 10.9 The distribution of sickle cell trait in Africa confers an advantage

A determined search for a vaccine, begun in the 1980s, led to unsuccessful trials. Now the situation is more hopeful. A huge injection of funds from the Gates Foundation into this project occurred in 1994 and recently more has been pledged. The subsequent, energetically pursued search for a vaccine that will confer lifetime active immunity for treated humans has yielded several promising 'candidates'. One is undergoing advanced trials. The malaria research community remains cautious; the complex parasitic lifestyle of *Plasmodium* makes it a difficult to eradicate and there have been many earlier set-backs. In the longer term, control of mosquitoes by the careful use of selective insecticides will be necessary, as well as the development of powerful vaccines, before this very well-adapted malarial parasite is overcome.

Table 10.2 Malaria – the profile

Active immunity: immunity resulting from exposure to an antigen. During the subsequent immune response, antibodies are produced by plasma cells and the body makes memory cells that provide ongoing long-term immunity. There is a delay before the immune response is complete, so immunity takes some days to build up.

Pathogen	Pathogen taxonomy	*Plasmodium*, a protoctist, with four species, of which *P. falciparum* is the most dangerous
	Transmission	Via the blood meals of the female *Anopheles* mosquito
Disease parameters	**Incubation and venue**	*Plasmodium* enters liver cells, reproduces asexually, then parasites are released into the patient's blood stream and enter red blood cells immediately. A cycle of red blood cell invasion, reproduction and release (with toxins) is then repeated.
	Signs, symptoms and diagnosis	A cycle of very high temperatures, headache, nausea and enlarged tender spleen every 48–72 hours, followed by normality
	Treatment	A combination of anti-malarial drugs
	Prognosis	Most patients improve within 48 hours after initiation of treatment and become fever-free after 96 hours. Infection by *P falciparum* causes a form of malaria with a high mortality rate if untreated.
	Prevention and control	Use of insecticides around human habitations Protection from insect bites, particularly by sleeping under insecticide-treated mosquito netting Destruction of the mosquito larva's habitat by draining swamps and spraying open water with oil to prevent larva's access to oxygen via surface air tube
WHO status	**Distribution**	Endemic in tropical and sub-tropical regions
	Global incidence and mortality	About 247 million cases per year with about 90% of the cases occurring in sub-Saharan Africa About 800 000 deaths per year with the majority being young children in remote, rural areas of Africa

Figure 10.10 Electron micrograph of a colony of *Mycobacterium tuberculosis* (×25 000)

Figure 10.11 Droplet infection

Tuberculosis (TB)

Tuberculosis (TB), a major, worldwide public health problem of long standing, is caused by a **bacillus** (rod-shaped bacterium) called *Mycobacterium tuberculosis* (Figure 10.10). Viewed under the microscope after staining by the Ziehl–Nielsen technique, these cells appear bright red. This red colouration is unique to bacteria of the genus *Mycobacterium* and is due to wax and other lipids in their walls. This is how the pathogen is identified in infected patients.

How tuberculosis spreads

People with pulmonary TB have a persistent cough. The droplets they inevitably spread in the air are infected with live *Mycobacterium* (Figure 10.11). Tuberculosis is chiefly spread by droplet infection in this way. Because of their lipid-rich cell walls, the bacilli are protected from drying out so the pathogen may survive for many months in the air and the dust of homes. This is another source of infection. Overcrowded and ill-ventilated living conditions are especially favourable for the transmission of the infection. However, it still requires quite prolonged contact with a viable source before people will succumb, because the bacterium is not strongly infectious.

A bovine form of TB occurs in cattle and the bacillus can enter the milk. Unpasteurised milk from infected cows is another potential source of infection. In some rural communities especially, this type of milk may be consumed by both adults and children. Today, in some countries, milk is supplied from 'tuberculin-tested' cows that are certified free of *Mycobacterium*. However, TB may still be found in dairy herds. This is because of a 'reservoir' of the bacterium that exists in wild animals that have chance contacts with the herds. Treatment of milk by pasteurisation is a major control mechanism as this heat treatment kills the bacteria without affecting the quality of the milk.

How TB develops

Once inside the lungs, the bacteria are engulfed by macrophages in the alveoli and bronchioles. If the person is in good health these white blood cells (with the help of the T-lymphocytes which migrate in from lymph nodes – page 159) kill the pathogen. Alternatively, and particularly with strains of *Mycobacterium* that are more infectious, the pathogens may remain alive within the macrophage, although localised and effectively controlled by the immune system. This explains why it is that not all infections with *M. tuberculosis* result in TB – many people show no symptoms. In fact, about one-third of the world's population is infected with *M. tuberculosis*, but only 5–10 per cent becomes sick or infectious at some stage during their life.

However, if the patient is malnourished or is in inferior health with a weakened immune system, a chronic infection may develop, typically within the lungs. Cavities appear as bacteria destroy the lung tissues. Blood vessels are broken down and fluid collects. The structural damage to the lungs can be seen X-ray examination (Figure 10.12).

If the condition is not treated promptly, the pathogen may be carried in the bloodstream and lead to TB in almost any part of the body. For example, the meninges of the brain, bone tissue, the lymph glands, the liver, the central nervous system, the kidneys or the genital organs can all be attacked. Generally, patients show loss of appetite, loss of weight, excessive sweating and a decline in physical activity.

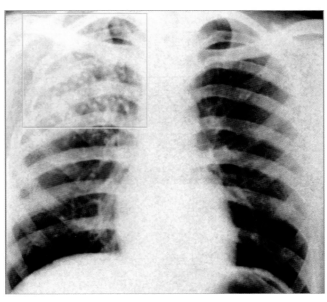

the heart is visible as a **white bulge**, with **lung tissue** on each side, enclosed in the **rib cage**
the white patches contain live *Mycobacterium tuberculosis* where lung structure and function are permanently destroyed

Figure 10.12 Chest X-ray showing TB in the lungs

Treatment of TB

Today, TB is treatable but it is still a 'killer' disease if not diagnosed early on in the infection. Amongst those with the disease of HIV/AIDS, TB is the leading cause of death.

Since TB is infectious, on confirmation of the diagnosis of a case of active tuberculosis, all contacts of the patient need to be traced and screened by a community public health team. Infectious patients require isolation and treatment with specific antibiotics until they cease to cough up viable bacilli. Then, antibiotics continue to be administered to the patient, now back in the community, until there is evidence the infection has been eradicated from the body.

Unfortunately, microorganisms develop resistance to antibiotics and other drugs used against them, with time. In the case of TB, this has already happened. Patients are now treated with several antibiotics simultaneously because of the emergence of multidrug-resistant TB (MDR-TB). This is particularly in the case of AIDs patients. MDR-TB is especially common in the USA and South-East Asia (see Table 10.3).

Prevention of TB

A vaccine against tuberculosis, the Bacillus Calmette–Guérin (**BCG**) vaccine, is prepared from a strain of attenuated (weakened) live bovine tuberculosis bacillus, *Mycobacterium bovis*. It is most often used to prevent the spread of TB among children.

A tiny quantity of inoculum is injected under the skin of the upper arm. Before vaccination takes place, people are tested for existing immunity, since these people will react unfavourably to the vaccine (the Haef test, Figure 10.13). Only those with no reaction may receive the vaccine. Immunity typically lasts for at least 15 years – longer if the individual is re-exposed to TB bacilli, for example, by accidental encounter with it. At best, the BCG vaccine is effective in preventing tuberculosis in a majority of people for a period of 15 years. However, immunity wanes with time and disappears entirely from the elderly.

1 instrument with six needles by which the skin is punctured and a small amount of protein, extracted from TB bacilli, is introduced below the skin (**haef test**)

2 after 2 or 3 days, if this test has proved positive, i.e. the patient has TB or has already acquired immunity

3 in a TB vaccination, the needle introduces the BCG vaccine just below the skin – a blister forrms and eventually subsides, leaving the patient protected against TB

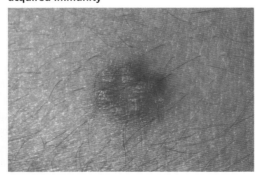

Figure 10.13 Haef test and a TB vaccination

How effective the vaccine is varies according to geography. Trials of BCG in two States in the USA were only 14 per cent effective. Another trial showed no protective effect. In the UK the vaccine is 80 per cent effective. How BCG vaccination is applied in different countries is reviewed in Table 10.3.

Table 10.3 BCG vaccination policy recommended in various countries

Country	Vaccination policy/recommended for
United Kingdom	Babies and children under 16 years with parents or grandparents born in counties with high rates of TB, or where born in one themselves Those who have been in close contact with a TB patient Those under the age of 35 whose occupation puts them at risk of exposure to TB
USA	Not used
India and Pakistan	Introduced in 1948 – the first countries outside Europe to do so
Brazil	Introduced in 1967
Malaysia	Given once only, at birth, from 2001
Singapore	Given once only, at birth, from 2001
Taiwan	Given at birth and repeated at the age of 12
South Korea	Recently discontinued altogether
USSR	In the former USSR (including Russia) it was given regularly, throughout life

Deaths from TB in England and Wales, 1900–2000

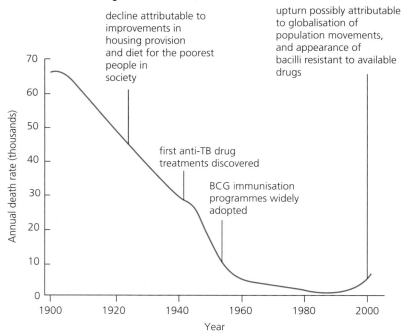

Figure 10.14 How social, economic and medical factors have influenced the occurrence of TB in developed countries

The changing pattern of the disease

There is evidence that TB was present in some of the earliest human communities and it has persisted as a major threat to health where vulnerable people lived in crowded conditions. In the nineteenth century, when the TB bacillus was first identified, one in seven of all deaths among Europeans was due to TB.

Today, in most developed countries this disease is relatively rare – in the UK for example, in 2007 the national average was about 15 cases per 100 000 of the population. In these countries, the reduction in TB throughout most of the past century has been due to steadily improving living conditions, particularly in housing and diet (Figure 10.14). However, today there is evidence of a resurgence of the disease due to inward migration of new citizens, to the globalisation of travel and to the emergence of multidrug-resistant TB (MDR-TB).

Globally, almost 14 million people have active TB, with approaching 2 million deaths due to the bacillus annually. In developing countries with a high incidence of this disease it primarily affects young adults. In developed countries where TB had ceased to be a major health threat, the rising incidence of the disease is largely among those described as immunocompromised (page 205), particularly those with HIV/AIDS, and among new migrants from developing countries – specially where they are now living in poorer housing in large cities, such as New York and London.

The World Health Organisation's Stop TB strategy – DOTS

In their vision for a TB-free world, the World Health Organisation (WHO) goal is to dramatically reduce the global burden of TB by 2015 by:

- achieving universal access to high-quality care for all people with TB, so reducing the human suffering and socioeconomic burden associated with TB
- protecting vulnerable populations from TB, co-infection with TB and HIV, and multidrug-resistant TB by supporting the development of new tools and enabling their timely and effective use
- protecting and promoting human rights in TB prevention care and control.

Direct Observation, Treatment, Short course (**DOTS**) is an approach aimed at health workers and family members who are responsible for patients taking prescribed medicines, to see that they do so regularly. The drugs now used (isoniazid and rifampicin, typically with a combination of others) cure the majority of patients with multidrug-resistant strains of TB, provided they are taken for 6–8 months.

Question

4 Examine the data in Table 10.4 concerning the disease TB in selected WHO Regions. Then suggest what factors are mostly likely to account for these contrasting figures.

Table 10.4 Estimated prevalence and mortality from TB in 2009

Selected WHO region	Prevalence/rate per 100 000 of the population	Mortality/rate per 100 000 of the population
Africa	450	50
South-East Asia	280	27
Europe	63	7

Table 10.5 Tuberculosis – the profile

Pathogen	Pathogen taxonomy	A rod-shaped bacterium, *Mycobacterium tuberculosis*
	Transmission	Droplet transmission from lung-infected patients and from contaminated dust of homes Prolonged contact is required – the bacterium is not strongly infectious (Also *M. bovis* from infected cows via unpasteurised milk)
Disease parameters	Incubation and venue	Within weeks or months, typically in the lungs of patients already vulnerable due to malnutrition or inferior health
	Signs, symptoms and diagnosis	A persistent cough, producing sputum (phlegm), with fatigue, loss of appetite, weight loss and fever Confirmation by a skin test, a chest X-ray and bacteriological test of sputum
	Treatment	A combination of antibiotics, particularly isoniazid and rifampicin, for at least 6 months
	Prognosis	Improvement within 2–3 weeks of early diagnosis and treatment, full recovery in 6–12 months An ignored infection that spreads to other organs is likely to be fatal For patients with HIV/AIDS, TB is likely to be the cause of death
	Prevention and control	BCG vaccination often protects younger people and is typically given to children and young adults at risk
WHO status	Distribution	Worldwide; strains resistant to anti-TB drugs have emerged in all countries
	Global incidence and mortality	About 8.8 million, for 1.4 million per year of whom the condition proves fatal Within the UK there are over 8000 cases per year, most non-UK born, from Sub-Saharan Africa (37%) and South Asia (47%)

HIV/AIDS

Auto-immune deficiency syndrome (AIDS) is a disease caused by the human immunodeficiency virus (**HIV**, Figure 10.15). **Viruses** are disease-causing agents, rather than 'organisms'.

glycoprotein

lipid membrane

RNA

reverse transcriptase (enzyme)

protease (enzyme)

capsid (protein)

Figure 10.15 The structure of the human immunodeficiency virus (HIV)

Antibiotics are not effective against viruses. What antibiotics are and why this is so is explained on pages 210–16.

Antigen: a protein (normally – some carbohydrates and other macromolecules can act as antigens) that is recognised by the body as foreign (non-self) and that stimulates an immune response. The specificity of antigens (which is a result of the variety of amino acid sequences that are possible) allows for responses that are customised to specific pathogens.

Infection with HIV is only possible through contact with blood or body fluids of infected people, such as may occur during sexual intercourse, sharing of hypodermic needles by intravenous drug users, and during pregnancy, labour, delivery and breast feeding of a newborn baby. Also, blood transfusions and organ transplants will transmit HIV but donors are now screened for HIV infection in most countries. HIV is *not* transferred by contact with saliva on a drinking glass, nor by sharing a towel, for example. Nor does the female mosquito transmit HIV when feeding on human blood.

The spread of HIV and the eventual onset of AIDS in patients are outpacing the current efforts of scientists and doctors to prevent them. The World Health Organisation WHO records the current state of this pandemic (Figure 10.16).

T-lymphocytes (CD4 helper cells) and the development of AIDS

T-lymphocytes are a particular type of white blood cell (page 153). An **antigen** is, normally, a protein that is recognised by the body as foreign (**non-self**, see page 221) and that stimulates an immune response (page 231). HIV attacks T-lymphocytes.

HIV is one of a minority of viruses that has an additional external envelope – a membrane of lipids and proteins. (This membrane is actually 'borrowed' from the last cell that it infected.) Once in the bloodstream, specific proteins on the outer membrane of the virus are antigens. These proteins attach the virus to protein receptors of the surface of T-lymphocytes (CD4 helper cells of the immune system, page 221) and the core of the virus penetrates to the cytoplasm (Figure 10.17).

Estimated number of adults and children newly infected with HIV during 1998 – Total: 2.7 million

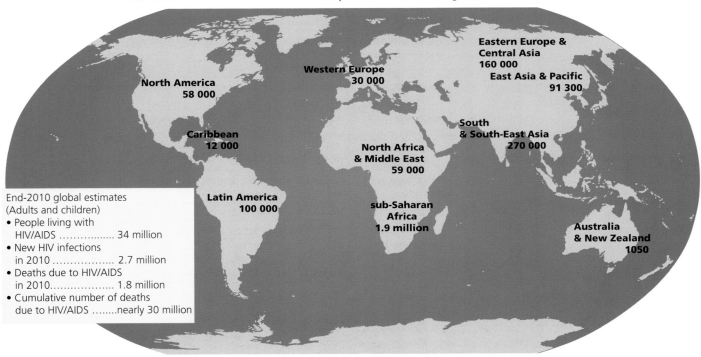

North America
58 000

Western Europe
30 000

Eastern Europe &
Central Asia
160 000

East Asia & Pacific
91 300

Caribbean
12 000

South
& South-East Asia
270 000

North Africa
& Middle East
59 000

Latin America
100 000

sub-Saharan
Africa
1.9 million

Australia
& New Zealand
1050

End-2010 global estimates
(Adults and children)
• People living with
  HIV/AIDS .... 34 million
• New HIV infections
  in 2010 .... 2.7 million
• Deaths due to HIV/AIDS
  in 2010................... 1.8 million
• Cumulative number of deaths
  due to HIV/AIDS ....nearly 30 million

Figure 10.16 HIV/AIDs – global estimates (WHO data)

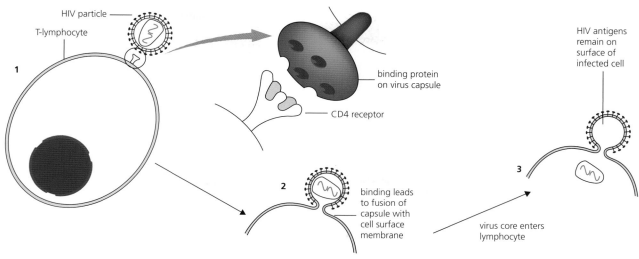

Figure 10.17 HIV infection of a white blood cell

HIV is a virus belonging to a group known as retroviruses. **Retroviruses** are RNA viruses with a unique 'infection process'. On entering the cytoplasm of a host cell the virus RNA is translated into DNA which then attaches to that of a chromosome in the host's nucleus (Figure 10.18). The steps to this are as follows.

Inside the lymphocyte, the RNA strands and an enzyme called reverse transcriptase are released from the core of the virus. Then, using the viral RNA as the template, a DNA copy is formed by the action of the reverse transcriptase. (Remember, RNA is single stranded but DNA is a double-stranded molecule, page 112).

This DNA enters the nucleus of the lymphocyte and attaches itself to a chromosome. It becomes a permanent part of the host cell's genome. It is known as a provirus. After an initial, mild form of AIDS (a brief 'flu-like illness within a few weeks, or possibly the development of a rash or swollen glands) the provirus remains dormant (called the latency period).

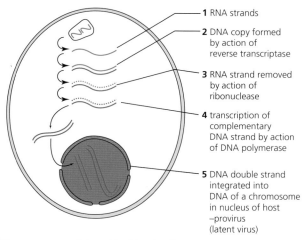

Figure 10.18 How HIV becomes part of the white blood cell's genome

However, at some later time the viral DNA replicates, leading to the production of new viruses which then invade and kill other lymphocytes (Figure 10.19). It is not clear whether infected T-lymphocytes are killed by the virus they harbour or by the actions of the patient's immune system against these infected cells. However, AIDS-related symptoms follow, caused by pathogens that are around us all the time. These are normally resisted by a healthy immune system but, without it, the body cannot effectively resist (Figure 10.20). We say the patient has now become **immunocompromised**.

SEM of a white cell from which HIV are budding off

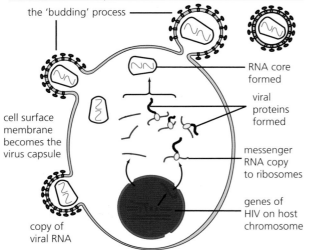

viral proteins and RNA being synthesised and new HIV 'cores' assembled then budded off with part of the host cell surface membrane as the capsules

the 'budding' process

cell surface membrane becomes the virus capsule

copy of viral RNA

RNA core formed

viral proteins formed

messenger RNA copy to ribosomes

genes of HIV on host chromosome

Figure 10.19 Activation of the HIV genome and the production of new HIV

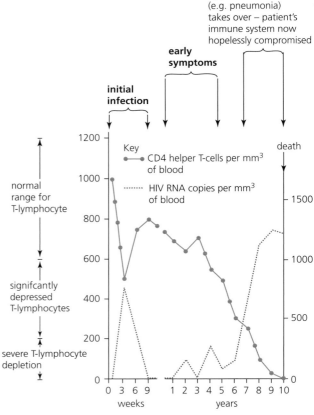

Figure 10.20 The profile of an AIDS infection

Social, economic and biological factors in the prevention and control of HIV/AIDS

AIDS is difficult to treat because viruses are not controlled by antibiotics. AIDS patients are offered **drugs** that slow down the progress of the infection. The three most popular drugs (AZT and two protease inhibitors) interrupt the reverse transcription of nucleic acid. A combination of drugs is used to prevent HIV from rapidly developing resistance to any one drug. This also helps to avoid dangerous or unpleasant side effects that some patients experience.

However, there is no cure for HIV/AIDS. Ideally, a **vaccine** against HIV would be the best solution – one designed to wipe out infected T-lymphocytes and free virus particles in the patient's bloodstream. The work of several laboratories is dedicated to this solution. The problem is that infected T4 cells in the latent state frequently change the membrane marker proteins they carry due to the HIV genome within. Effectively, HIV can hide from the body's immune response by frequently changing its identity.

Current effective measures to prevent the spread of HIV include **male circumcision**. Recent studies have shown that the incidence of AIDS is significantly higher in uncircumcised males. Circumcised males were seven times less likely to transmit to or receive HIV from their partner. The reason for this is that the mucous membrane of the inner surface of the foreskin has cells with receptors that HIV can exploit. Consequently, it is possible that infection of future generations with HIV might be reduced if male circumcision was practised more widely. However, if this were to encourage males to participate in unsafe sex with numerous partners, it would defeat the object.

Table 10.6 The prevention and control of HIV/AIDS

Factors	Prevention	Control
Social	Effective education programmes so that the vulnerable understand the cause and effects of HIV infection, and the best steps to remain healthy, whether or not they are literate Encouraging faithfulness to one partner	Contact tracing, where a person diagnosed as HIV-positive is willing and able to identify those who have been put at risk by contact
Economic	Condoms being made freely available to the sexually active population Sterile needles being made freely available to intravenous drug users	Funding of testing of the sexually-active population to identify HIV-positive people (if people will accept this invasion of their privacy) Funding and supply of the drugs that prevent the replication of the virus inside host cells
Biological	The practice of 'safe sex' by the use of condoms to prevent the transmission through infected blood or semen Screening of blood donors for HIV, so that infected blood is not taken and used	Drug therapy Research on the development of a vaccine Replacement of breast feeding by bottle feeding of infants of HIV-positive mothers, and other WHO strategies for helping HIV-infected mothers

Question

5 Suggest what features of the HIV virus and the disease it causes create particular problems for the people and the economies of less-developed countries such as Zimbabwe or Zambia?

Table 10.7 HIV/AIDS – the profile

Pathogen	Pathogen taxonomy	A single-stranded RNA retrovirus, human immunodeficiency virus (HIV)
	Transmission	Contact with blood or body fluids of infected people, via sexual intercourse, the sharing of hypodermic needles by intravenous drug users or during breast feeding of infants
Disease parameters	Incubation and venue	Once in the blood, the virus immediately enters T-lymphocytes (CD4 cells) – components of the immune system
	Signs, symptoms and diagnosis	Early symptoms typical of mild 'flu occur in a few weeks Presence of HIV can only be confirmed by HIV blood test The patient's terminal vulnerability to any opportunistic infections typically delayed months or years
	Treatment	A combination of drugs that interrupt reverse transcription of nucleic acid (AZT and two protease inhibitors)
	Prognosis	There is no cure for HIV/AIDS but an effective vaccine is sought Patients ultimately succumb to other infection, e.g. TB or measles
	Prevention and control	Safe sex using condoms Needle exchange schemes for intravenous drug users Male circumcision
WHO status	Distribution	Worldwide, after the first cases of the disease came to light in the 1980s in the USA
	Global incidence and mortality	Estimated annual total of newly-infected individuals of the order of 6 million, two-thirds are in sub-Saharan Africa Estimated annual death rate approaching 3 million

Measles

Measles is a highly infectious disease that people of any age can get. It is most commonly contracted by children aged between one and four years. Unlike other childhood diseases, such as mumps and rubella, it can generate severe complications. Measles is a virus disease entirely specific to humans and with a long history. For example, there is a tenth-century record of it being '*more dreaded than smallpox*'. In fact, in the past most people contracted measles. The incidence of this infection and the dangers it posed were abruptly diminished only as recently as the 1960s. A cheap and reliable vaccine became available then but it was effective only when taken up by 95 per cent of the population. This is often the case in the developed world (but not always – see below). Today, over 95 per cent of fatalities due to measles occur in less-developed countries where health infrastructures are poor. Measles is also a major killer of those in an advanced stage of HIV/AIDS.

The disease and its transmission

Measles is caused by an RNA virus belonging to the *Morbilivirus* genus. It is closely related to the canine distemper virus but both viruses are host-specific. The disease is spread by droplet transmission when infected individuals cough or sneeze. It infects the upper respiratory tract or conjunctiva of the eyes. There is an incubation period of 10–14 days before the symptoms appear. Initially these are a cold-like condition, red eyes that are sensitive to light, fever, and grey or white spots appearing in the throat and mouth. Subsequently, a red-coloured rash spreads from the head to the rest of the body, starting behind the ears. The patient is most infectious for about four days before the appearance of the rash and then remains mildly infectious for a further four days after its appearance. The result is that measles is often spread before the condition is recognised or confirmed.

In susceptible patients the complications that arise include pneumonia, hepatitis, and encephalitis (inflammation of the brain). The latter occurs in only 1 in 5000 cases but it can be fatal.

Prevention and control of measles

The first measles vaccines became available in 1963 and were then replaced by a superior version in 1968. The effect of the widespread use of these vaccines was dramatic, including in the USA (Figure 10.21).

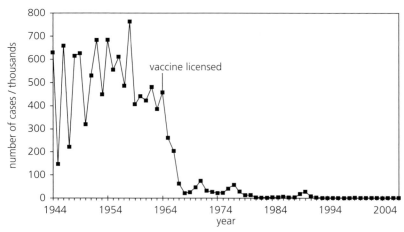

Figure 10.21 Number of measles cases in the USA, 1944–2007

A combined vaccine against three common childhood illness, referred to as the **MMR vaccine**, became available in 1971. The MMR vaccine is made by using live, attenuated (weakened) strains of the measles, mumps and rubella viruses. Infants in the UK, for example, are now offered a first MMR injection aged 15–24 months and then receive a booster injection between 3 and 5 years' of age. (A baby under 6 months is protected by antibodies received from the mother, via the placenta, whilst *in utero*; conferring what is known as '**passive immunity**'.) The success of vaccines has made measles a rare disease, at least where the vaccination regime has been adhered to.

> **Question**
>
> **6** What is meant by a disease-causing agent being 'host-specific'?

> **Passive immunity**: immunity involving the transfer of antibodies (already made in the body of another organism or *in vitro*) into the body where they will bind to their specific antigen if it is present. This gives instant immunity but does not lead to the development of memory cells, so the immunity only lasts for a few weeks.

Table 10.8 Measles – the profile

Pathogen	Pathogen taxonomy	A single-stranded RNA virus of the *Morbilivirus* genus
	Transmission	Droplet transmission when an infected person coughs or sneezes The virus has a short survival time outside its host
Disease parameters	Incubation and venue	10–14 days Respiratory mucous membranes, lymph glands and other body tissues
	Signs, symptoms and diagnosis	Cold-like symptoms, red eyes that are sensitive to light, fever, grey or white spots in the mouth and throat, followed by red skin rash
	Treatment	Pain and fever-reducing medications, e.g. paracetamol Anti-viral drugs in the (rare) case of complications
	Prognosis	In the absence of complications, recovery occurs within 7 days of the appearance of the rash
	Prevention and control	Measles can kill or leave severe complications Avoided by prior vaccination The MMR vaccine has made measles rare in developed countries
WHO status	Distribution	Worldwide
	Global incidence and mortality	Estimated 900 000 deaths from measles (2008), in countries with low incomes and limited health infrastructures

Extension

The MMR–Autism controversy

Recently, a controversy over the MMR vaccine has influenced some parents' choices.

Dr Andrew Wakefield, a gastroenterologist employed at the Royal Free Hospital in London, studied 12 children who had all apparently developed autism within 14 days of being given MMR. He published his opinion (in the medical journal *The Lancet* in 1998) that for some children, inoculation with the MMR vaccine triggered an inflammatory bowel condition. In this state the gut might become leaky, allowing 'rogue peptides' into the blood circulation, quickly leading to brain damage – in particular, to **autism**. (Autism is a life-long disorder in which children withdraw into themselves, reject human contact and often become difficult to handle. To the families in which it occurs, autism is most distressing. The film '*Rain Man*', in which Dustin Hoffman starred, helped raise the profile of this disease so it is more widely recognised now.)

Dr Wakefield noted '*We did not prove an association*', meaning this was a suggestion that needed following up (and it was). However, his ideas immediately received widespread publicity and created public anxiety sufficient to cause a marked fall in the take-up of MMR vaccine.

Why did some parents confidently blame MMR for their child's condition?

Autism, when it arises, does so at about 18 months. When this occurs soon after the MMR inoculation, it is inevitable that there may *appear* to be a connection. In fact, no study has produced evidence that MMR causes autism. Major investigations and follow-up studies have been undertaken with large samples of children in Australia, in Finland, and in Japan. In communities where autism appears to be on the increase, the increase began before MMR was introduced, and the incidence of autism continued to rise even when the numbers of children receiving MMR reached a plateau. There is no doubt in the minds of health authorities that vaccination is the safest option for children and their parents.

The measles vaccine is not universally available; there are several less developed countries where many people are not vaccinated. This is despite the fact the disease is targeted by the WHO in an expanded programme of immunisation. It is estimated there may be up to 900 000 measles-related deaths annually – many HIV/AIDs-related. Unfortunately, some children who recover are left with serious, permanent complications.

Question

7 The coincidence of autism onset and MMR administration may be described as circumstantial evidence. What part, if any, can it play in scientific discovery?

Those patients that survived smallpox were typically disfigured by scabs. These resulted from red spots that had formed on their skin and became filled with thick pus before drying out. Eyelids often swelled up and became glued together, leaving many patients blind.

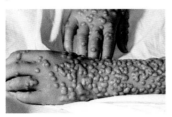

Figure 10.22 Skin vesicles on a smallpox patient

Question

8 Outline how it has been possible to eradicate smallpox but not other infectious diseases like malaria.

Smallpox

Smallpox, a highly contagious disease, was once endemic throughout the world. It killed or disfigured all those who contracted it (Figure 10.22). Smallpox was caused by a DNA *Variola* virus. The virus was stable, meaning it did not mutate or change its surface antigens.

Eventually, a suitable vaccine was identified – made from a harmless but related virus – *Vaccinia*. This was used in a 'live' state and could be freeze dried for transport and storage. Consequently, the virus vaccine was relatively easy to handle and stable for long periods in tropical climates.

How eradication came about

This disease has been eradicated (the last case of smallpox occurred in Somalia in 1977). The development of a vaccine played an important part in this achievement, which was the outcome of a determined World Health Organisation programme, begun in 1956. This involved careful surveillance of cases in isolated communities and within countries sometimes scarred by wars – altogether a most remarkable achievement. How was smallpox eradicated when so many other diseases continue? The reasons include:

- patients with the disease were easily identified as they had obvious clinical features
- transmission was by direct contact only
- on diagnosis, patients were isolated, all their contacts traced and all were vaccinated
- it had a short period of infectivity, about 3–4 weeks
- patients who recovered did not retain any virus in the body ('carriers' did not exist)
- there were no animals that acted as a vectors or 'reservoirs' of the infection (and so be able to pass on the virus to humans).

Other reasons, relating to the virus, its antigens and the vaccine that was used in the eradication campaign are identified on page 231.

10.2 Antibiotics

The 'age of antibiotics' began in the 1940s with the availability of penicillin. With an increase in antibiotic resistance is this age about to come to an end?

By the end of this section you should be able to:

a) outline how penicillin acts on bacteria and why antibiotics do not affect viruses

b) explain in outline how bacteria become resistant to antibiotics with reference to mutation and selection

c) discuss the consequences of antibiotic resistance and the steps that can be taken to reduce its impact

Introducing antibiotics

Antibiotics were discovered relatively recently – with a chance observation in 1928. By the 1950s, it became possible to treat patients with courses of antibiotics that cured many bacterial infections. More recently, the excessive use of antibiotics has generated problems, as we shall see later.

Antibiotics are naturally occurring chemical substances obtained mainly from certain fungi and bacteria commonly found in the soil. When antibiotics are present in low concentrations they inhibit the growth of other microorganisms or cause their outright death, as demonstrated in Figure 10.23.

The discovery, isolation and development of the first antibiotic, penicillin, was not an easy task. (It took from 1929 until 1944.) Since then, over four hundred different antibiotics have been isolated and tested. Of these, only about 50 have proved to be non-toxic to patients. These antibiotics have achieved wide usage. Those effective against a wide range of pathogenic bacteria are called **broad-spectrum antibiotics**. These include chloramphenicol. Others, including streptomycin, are effective against a limited range of bacteria.

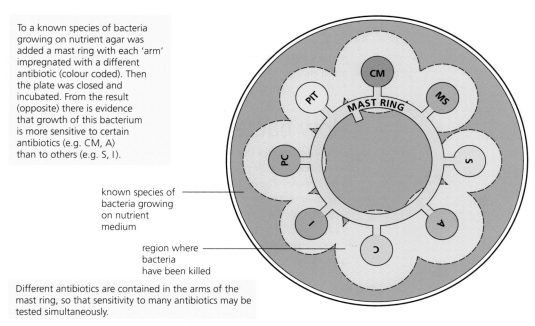

To a known species of bacteria growing on nutrient agar was added a mast ring with each 'arm' impregnated with a different antibiotic (colour coded). Then the plate was closed and incubated. From the result (opposite) there is evidence that growth of this bacterium is more sensitive to certain antibiotics (e.g. CM, A) than to others (e.g. S, I).

known species of bacteria growing on nutrient medium

region where bacteria have been killed

Different antibiotics are contained in the arms of the mast ring, so that sensitivity to many antibiotics may be tested simultaneously.

Figure 10.23 Investigating sensitivity to antibiotics

How antibiotics work

Antibiotics work by disrupting the metabolism of prokaryotic cells. It is in these division and growth phases that bacteria are vulnerable to antibiotic action – an effective antibiotic specifically disrupts one of three major aspects of the growth or metabolism of bacteria. Since the particular components, metabolites or enzymes concerned are not found in eukaryotic cells in the same form, the antibiotic is not toxic to the mammalian host tissues.

1 Cell wall synthesis inhibition

The most effective antibiotics work by interfering with the synthesis of bacteria's cell walls. Once the cell wall is destroyed, the delicate plasma membrane of the bacterium is exposed to the destructive force generated by excess uptake of water by osmosis, and possibly, as well, to attack by antibodies and phagocytic macrophages.

Several antibiotics, including **penicillin**, ampicillin, and bacitracin, bind to and inactivate specific wall-building enzymes. These are the enzymes required to make essential cross-links between the linear polymers of the walls in particular species. In the presence of the antibiotic, wall polymers continue to be synthesised by the pathogens, but the individual strands are not linked and bound together. The walls fall apart.

2 Protein synthesis inhibition

Other antibiotics inhibit protein synthesis by binding with ribosomal RNA. The ribosomes of prokaryotes (known as 70S) are made of particular RNA sub-units, together with many polypeptides (which mainly function as enzymes). The ribosomes of eukaryotic cells are larger (80S), and are built with different RNA molecules and polypeptides. Antibiotics like streptomycin, chloramphenicol, tetracyclines and erythromycin all bind to prokaryotic ribosomal RNA subunits unique to bacteria. Here their presence causes protein synthesis to be terminated.

3 Nucleic acid synthesis

A few antibiotics interfere with DNA replication or transcription, or they block mRNA synthesis. These antibiotics, for example the quinolones, are not as selectively toxic as other antibiotics. This is because the processes of replication and transcription do not differ so greatly between prokaryotes and eukaryotes as wall synthesis and protein synthesis do.

Why antibiotics do not affect viruses

We have seen that viruses are non-living particles (page 26). A virus lacks cell structure and has no metabolism of its own to be interfered with or disrupted. Instead, viruses reproduce using metabolic pathways in their host cell that are not affected by antibiotics. Antibiotics cannot be used to prevent viral diseases.

How bacteria become resistant to antibiotics

Before antibiotics became available to treat bacterial infections the typical hospital ward was filled with patients with pneumonia, typhoid fever, tuberculosis, meningitis, syphilis and rheumatic fever, for example. These diseases, all caused by bacteria, claimed many lives, sometimes very quickly. The discovery of antibiotics brought a significant change. Patients with bacterial infections were treated with an antibiotic, and quickly overcame their infection. In fact, antibiotics have been very widely used – to great success.

However, in any large population of a species of pathogenic bacteria, some of the bacteria may acquire or carry a gene for resistance to the antibiotic in question. Sometimes this gene has arisen by a spontaneous mutation. Sometimes the gene has been acquired in a form of sexual reproduction between bacteria of different populations, or when the bacterium itself has been parasitised by a virus that brought the resistance gene within its nucleic acid (Figure 10.24).

When an individual bacterium has acquired resistance it confers no selective advantage at all *in the absence of the antibiotic*. The resistant bacterium must compete for resources with non-resistant bacteria. However, when an antibiotic is administered, most bacteria of the population are killed off. Only at this stage will the bacteria with resistance to the antibiotic flourish, and then come to create the future population, all of which now carry the gene for resistance to the antibiotic. At this point, the genome has been changed abruptly, and the future effectiveness of that antibiotic is compromised.

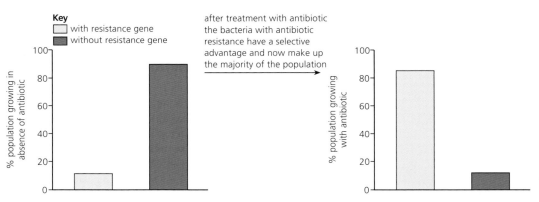

Figure 10.24 Development of antibiotic resistance in bacteria

Today, the issue of bacterial resistance to antibiotics in healthcare has become a major problem. As pathogenic bacteria developed genes for resistance to antibiotic actions, different antibiotics were used in response. The inevitable outcome of this is that pathogenic bacteria slowly acquire **multiple resistance** (Figure 10.25). For example, a strain of *Staphylococcus aureus* has acquired resistance to a range of antibiotics including methicillin. Now, so-called methicillin-resistant *Staphylococcus aureus* (MRSA) is referred to as a 'hospital superbug' because of the harm its presence has inflicted in these places. (Actually, 'superbugs' are found everywhere in the community, not just in hospitals.) MRSA poses the greatest threat to patients who have undergone surgery. With cases of MRSA to treat, the intravenous antibiotic called vanomycin is prescribed, but recently there have been cases of partial resistance to this drug, too (Figure 10.26).

Similarly, a strain of the bacterium *Clostridium difficile* is now resistant to all but two antibiotics. This bacterium is a natural component of our gut 'microflora'. It is only when *C. difficile's* activities are no longer suppressed by the surrounding, huge, beneficial (i.e. 'friendly') gut flora that it may multiply to life-threatening numbers and then may trigger toxic damage to the colon.

Suppression of beneficial gut bacteria is a typical consequence of heavy doses of broad-spectrum antibiotics, administered to overcome infections by superbugs.

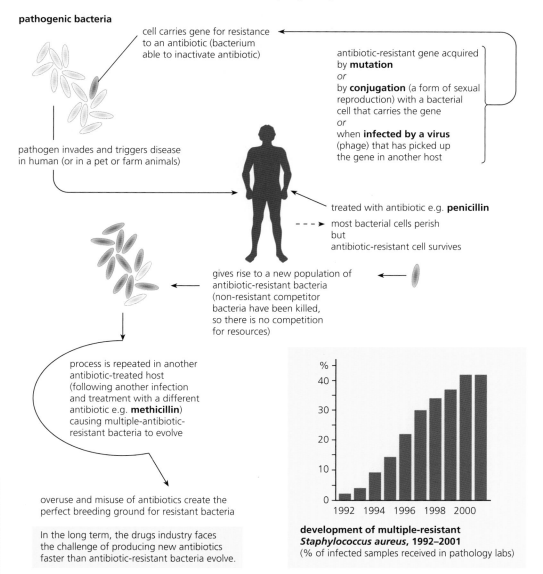

Figure 10.25 Multiple antibiotic resistance in bacteria

Question

9 Explain:

a why doctors ask patients to ensure that they complete their course of antibiotics fully

b why the medical profession tries to combat resistance by regularly alternating the type of antibiotic used against an infection.

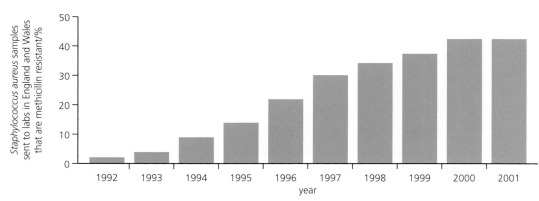

Figure 10.26 The increasing incidence of MRSA

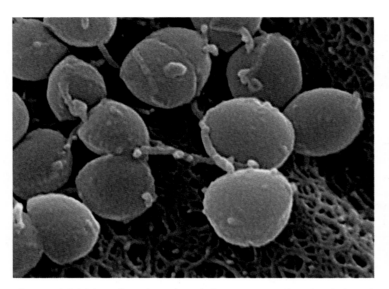
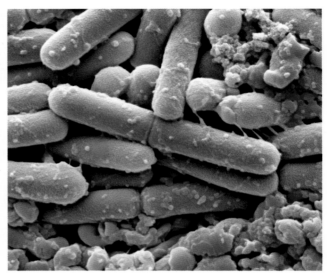

Figure 10.27 False-colour SEMs of *Staphylococcus aureus* (x000) and *Clostridium difficile* (x10 000)

THE TIMES I Thursday May 1 2014

Antibiotics crisis means scratch could kill

Common infections will once again kill as the rise of antibiotic-resistant super-bugs has a "devastating" effect on modern medicine, the World Health Organisation said yesterday.

Resistance to commonly used drugs is increasing across the planet and is now a problem that could affect "anyone, of any age, in any country", the WHO said in a report tracking the rise of super-bugs in 114 countries.

Patients are already suffering because those infected by superbugs are twice as likely to die as those with non-resistant infections. Routine operations, minor scratches and common infections could all become fatal, the report warns.

Doctors said the problem demanded an international effort modelled on the fight against Aids, including a clampdown on the use of antibiotics, the end of their use in agriculture and the urgent development of new drugs. Patients must only use antibiotics prescribed by a doctor and complete a full course even if they are feeling better, the WHO said.

Chris Smyth Health Correspondent

The consequences of antibiotic resistance

A natural response to this challenging situation is to search for new and more effective antibiotics to replace those that have become ineffective. This has been the response of pharmaceutical firms – and continues to be so. However, this is proving increasingly difficult – the number of new antibiotics being developed each year has fallen dramatically (Figure 10.28). Pathogens tend to develop resistance faster than new antibiotics can be found.

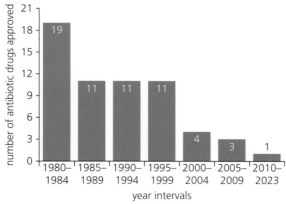

Figure 10.28 The number of new antibiotics developed and approved in the past three decades

In this situation the strategy to combat the problem includes:

● **Prudent antibiotic use.**

Avoidance of the unnecessary and inappropriate exposure of microorganisms to antibiotics in medical clinics, veterinary clinics, animal husbandry and agriculture. These drugs must be used sparingly, and only in situations where there is no alternative appropriate treatment. When a course of antibiotics is prescribed, the antibiotics should be taken to completion, and not abandoned as soon as the patient feels some improvement. The widespread prophylactic use of antibiotics in animal husbandry should be minimalised and preferably discontinued. Animals reared intensively have been fed antibiotics as a component of their diet, following the discovery that they grew faster and reached marketable weights more quickly. However, antibiotic residues have accumulated in the food chain, and many more bacteria have been exposed to situations where resistance may evolve and emerge.

● **Surveillance.**

The monitoring of the appearance of new cases of resistance to antibiotics, and prompt circulation of the data to healthcare professionals and institutions.

● **Infection control.**

A renewed focus on the reduction in the spread of infection in general, so that the need for the use of antibiotics is reduced – on the principle that prevention is better than cure.

Question

10 Antibiotics are widely used as prophylactics in animal husbandry. What does this mean, why does this happen, and what possible dangers arise from this use of antibiotics?

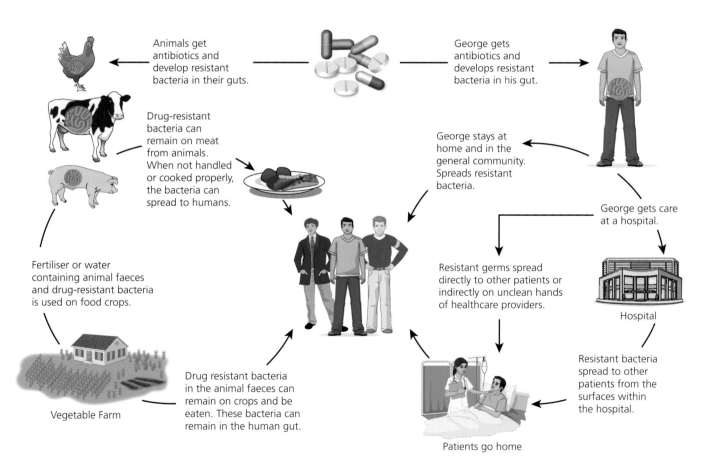

Figure 10.29 How antibiotic resistance spreads

Summary

- **Health** is a state of physical, mental and social well-being, not merely the absence of disease. Diseases caused by disease-causing agents are **infectious diseases** since they may be transferred. **Non-infectious diseases**, such as lung cancer and sickle cell anaemia, are not 'caught' from other organisms.

- **Viruses** consist of DNA or RNA, wrapped in a protein coat. Viruses take over host cells and convert the cellular machinery to produce more viruses which escape and may repeat the process. Human viral diseases include HIV/AIDS, measles and smallpox.

- **HIV** is transmitted by contact with blood or body fluids. It enters lymphocytes of the immune system and ultimately destroys them and the body's resistance to other diseases. Treatment with drugs that interfere with the replication of the virus delays the onset of AIDS but there is no vaccine as yet and therefore no cure. Transmission may be controlled by barrier methods of contraception but HIV/AIDS especially harms societies when young, economically active members and their children are infected. The distribution of the disease is global.

- The **measles** virus is transmitted by droplets discharged via coughs and sneezes, and attacks the respiratory mucous membranes and other tissues. Distinctive symptoms appear within 10 days or so, including a body rash and grey or white spots in the mouth. Contracting measles can be avoided by prior vaccination, now via the MMR vaccine, but untreated infections can kill or leave severe complications.

- **Bacterial diseases** are caused by pathogenic bacteria. They are vulnerable to antibiotics, unlike viruses. Examples of bacterial diseases include cholera and tuberculosis (TB) but most bacteria are not disease-causing at all.

- The **cholera** bacterium is principally transmitted in water or foods contaminated by human faeces. The bacterium releases a toxin that triggers the loss of water and salts from the intestines, causing diarrhoea and severe dehydration. Cholera has been eradicated where there is effective processing of sewage and the purification of drinking water. Vaccines are available but prompt treatment by rehydration with a dilute solution of glucose and salts is essential once the disease is diagnosed. About 50 per cent of untreated cases are fatal.

- The **TB** bacterium is transmitted by droplets from infected lungs during persistent coughing. In the lungs of people weakened by malnutrition or ill-health the bacterium overcomes resistance and attacks the alveoli. Prolonged treatment with a combination of antibiotics leads to recovery but, untreated, TB will attack other body systems and cause death. A vaccine is available and is given to people in some countries, particularly children, but how effective the vaccine is varies according to geography. Distribution of the disease is worldwide but it is a particular threat to HIV/AIDS patients. Strains resistant to anti-TB drugs have emerged in all countries.

- Diseases caused by eukaryotic organisms include **malaria**, due to the protoctist *Plasmodium*, transferred by the mosquito. The life cycle of *Plasmodium* is acted out in the liver and red blood cells and results in periodic release of toxins and high fever. Protection from mosquito bites is critical but a combination of anti-malarial drugs permits a patient with malaria to improve within 48 hours. Malaria is endemic in tropical and sub-tropical regions and kills thousands of children in remote areas of sub-Saharan Africa. A well-funded drive to find a vaccine continues.

- **Antibiotics** are developed from naturally-occurring chemicals obtained from certain bacteria and fungi and that inhibit the growth of other microorganisms. They work by destroying the pathogenic bacteria or by interfering in their metabolism so that the body's own defences can work effectively. Unfortunately, pathogens develop resistance so new antibiotics are constantly sought.

Examination style questions

1 **a) i)** Name the organism that causes tuberculosis (TB). [1]

ii) Explain how TB is transmitted from an infected person to an uninfected person. [2]

The World Health Organisation (WHO) collects data on TB from its six different regions as shown in Table 1.1.

In 2003, it used these figures to estimate
- the total number of people with the disease in each region
- the number of deaths from TB.

Many of those who died from TB were also infected with HIV.

Table 1.1

WHO region	Number of cases per 100 000 population	Number of deaths from TB (including TB deaths in people infected with HIV) per 100 000 population
Africa	345	78
The Americas	43	6
Eastern Mediterranean	122	28
Europe	50	8
South-East Asia	190	38
Western Pacific	112	19

b) Explain the advantage of expressing the number of cases and the number of deaths as 'per 100 000 population'. [2]

c) Using the information in Table 1.1, outline the reasons why TB has a greater impact on the health of people in some regions rather than others. [3]

d) The number of cases of TB decreased considerably in many countries during the 20th century. Over the past 20 years, the number of cases worldwide has increased very steeply. A vaccine against TB has existed since 1921.

Explain why TB has not been eradicated even though a vaccine has existed since 1921. [3]

[Total: 11]

(Cambridge International AS and A Level Biology 9700, Paper 02 Q5 November 2007)

2 **a)** Moves to prevent the spread of infectious disease within a local community have included:

A an improved supply of water, including sewage treatment

B insect control

C milk pasteurisation.

For each of these preventative measures, describe a disease they are effective against and explain why. [12]

b) What features of the lifecylce of the pathogen that causes malaria make the development of an effective vaccine especially difficult? [4]

c) Explain why malaria is endemic in tropical and sub-tropical regions but absent or rare elsewhere in the world. [2]

d) What is the significant difference between a disease outbreak which is classified as an epidemic and as a pandemic? [2]

[Total: 20]

3 **a)** The human immunodeficiency virus (HIV) is a retrovirus that is responsible for the disease of AIDs. Explain what 'retrovirus' means.

b) With the onset of AIDs, the immune system is said to be 'hopelessly compromised'. Explain this, and outline how HIV brings this about.

c) Ideally, a vaccine is sought against HIV. Why is the development of an effective vaccine proving difficult in this case?

4 **a)** Explain the mechanism by which antibiotics may be effective against pathogenic bacteria but that results in them having no effect on viruses.

b) Antibiotics have been thought of as 'wonder drugs' but today are used selectively and sometimes with restraint. What are the reasons for this change in attitudes and practice?

11 Immunity

An understanding of the immune system shows how cells and molecules function together to protect the body against infectious diseases and how the body is protected from further infection by the same pathogen. Phagocytosis is a more immediate non-specific part of the immune system, while the actions of lymphocytes provide effective defence against specific pathogens.

 ## 11.1 The immune system

The immune system has non-specific and specific responses to pathogens. Auto-immune diseases are the result of failures in the system to distinguish between self and non-self.

By the end of this section you should be able to:

a) state that phagocytes (macrophages and neutrophils) have their origin in bone marrow and describe their mode of action
b) describe the modes of action of B-lymphocytes and T-lymphocytes
c) describe and explain the significance of the increase in white blood cell count in humans with infectious diseases and leukaemias
d) explain the meaning of the term immune response, making reference to the terms antigen, self and non-self
e) explain the role of memory cells in long-term immunity
f) explain, with reference to myasthenia gravis, that the immune system sometimes fails to distinguish between self and non-self

Responses to infection

Pathogens do not gain easy entry to the body because:
- externally, the keratinised protein of the dead cells of the epidermis are tough and impervious unless broken, cut or deeply scratched
- internal surfaces, in particular the trachea, bronchi and the bronchioles of the breathing apparatus, and the gut are all lined by moist epithelial tissue. These internal barriers are protected by mucus, by the actions of cilia removing the mucus, and some by digestive enzymes or strong acid (as in the stomach).

However, these barriers are sometimes crossed by pathogens. In response, there are internal 'lines of defence'. Firstly, the body responds to localised damage (cuts and abrasions, for example) by **inflammation**. If a blood vessel is ruptured then the blood clotting mechanism is activated. In the blood and tissue fluid, the **immune system** is triggered, which is covered in more detail shortly.

Inflammation

Inflammation is the initial, rapid, localised response the tissues make to damage, whether due to a cut, scratch, bruising, or a deep wound. We are quickly aware that the site of a cut or knock (contusion) has become swollen, warm and painful. Inflammation is triggered by the damaged cells themselves, which release 'alarm' chemicals, including **histamine** and prostaglandins.

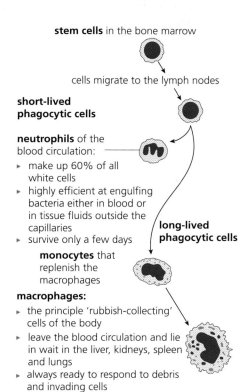

stem cells in the bone marrow

cells migrate to the lymph nodes

short-lived phagocytic cells

neutrophils of the blood circulation:
▸ make up 60% of all white cells
▸ highly efficient at engulfing bacteria either in blood or in tissue fluids outside the capillaries
▸ survive only a few days

long-lived phagocytic cells

monocytes that replenish the macrophages

macrophages:
▸ the principle 'rubbish-collecting' cells of the body
▸ leave the blood circulation and lie in wait in the liver, kidneys, spleen and lungs
▸ always ready to respond to debris and invading cells

Figure 11.1 The phagocytic cells of the body's defence mechanism

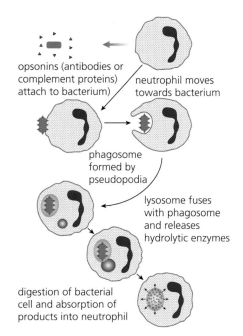

opsonins (antibodies or complement proteins) attach to bacterium)

neutrophil moves towards bacterium

phagosome formed by pseudopodia

lysosome fuses with phagosome and releases hydrolytic enzymes

digestion of bacterial cell and absorption of products into neutrophil

Figure 11.2 Phagocytosis by a white cell – a neutrophil

The initial outcome is that the volume of blood in the damaged area is increased, and **white blood cells** and **plasma** accumulate outside the enlarged capillaries. Ultimately, tissue repair is initiated, also.

The blood circulation plays a complex part in the resistance to infection that will result if microorganisms enter the body. The **increased blood flow removes toxic products** released by invading microorganisms and by damaged cells (Figure 11.3).

The white blood cells that accumulate at the site of tissue damage or invasion of pathogens fall into two functional groupings depending on their roles in the defence against disease.

General **phagocytic white cells** engulf 'foreign' material. These are the **neutrophils** and the **macrophages**. Neutrophils make up 60 per cent of all white blood cells in the blood, but they are short-lived. Macrophages are the principal 'rubbish-collecting cells' found throughout the body tissues (Figure 11.1). These responses to infection destroy any invading pathogen.

Other white blood cells called **lymphocytes**, produce the **antibody reaction** to infection or invasion of foreign matter. This, the immune response, is triggered by and directed towards *specific* pathogens.

You can see some phagocytic white cells and a lymphocyte in the photomicrograph of a blood smear, in Figure 8.2. You will probably have already seen them in a prepared microscope slide of mammalian blood.

All these white blood cells originate by division of stem cells in the bone marrow and in certain cases they mature in the lymph nodes. On average, they spend only about 10 per cent of their short lives in the blood, but all are concerned with the body's protection against infection.

The blood also delivers special proteins (**complement proteins**), which are activated by the presence of infection. Complement proteins enhance the work of white blood cells in overcoming infections. Some trigger the lysis of invading microorganisms, and others, the opsonins, bind to pathogenic bacteria and so increase phagocytosis.

Cells infected with viruses produce proteins called **interferons**. These bind to neighbouring, healthy cells and trigger synthesis of antiviral proteins. Viral replication is halted.

Non-specific responses to infection

The above responses to infection are referred to as **non-specific responses** because they help to destroy any invading pathogen. By contrast, the immune response we discuss next is triggered by and directed towards *specific* pathogens.

The immune response

The immune response is our main defence once invasion of the body by harmful microorganisms or 'foreign' materials has occurred. It is particular leucocytes (white blood cells) called **lymphocytes** that are responsible for the immune response. They make up 20 per cent of the leucocytes circulating in the blood plasma (or found in the tissue fluid – remember, white blood cells also move freely through the walls of blood vessels).

Lymphocytes detect any 'foreign' matter entering from outside our bodies, including macromolecules as well as microorganisms, as different from out body cells and our own proteins. Any molecule the body recognises as a foreign or '**non-self**' substance is known as an **antigen**.

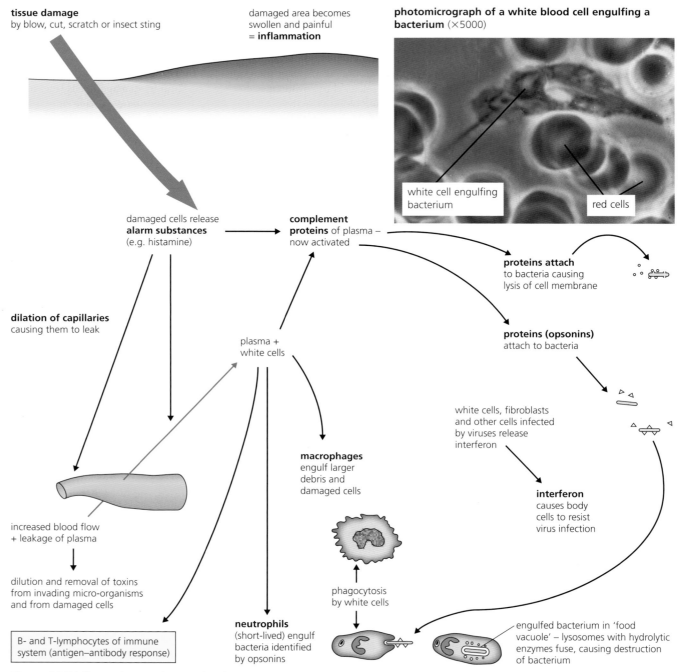

tissue damage
by blow, cut, scratch or insect sting

damaged area becomes swollen and painful
= inflammation

photomicrograph of a white blood cell engulfing a bacterium (×5000)

white cell engulfing bacterium

red cells

damaged cells release **alarm substances** (e.g. histamine)

complement proteins of plasma – now activated

proteins attach
to bacteria causing lysis of cell membrane

dilation of capillaries
causing them to leak

plasma + white cells

proteins (opsonins) attach to bacteria

white cells, fibroblasts and other cells infected by viruses release interferon

interferon
causes body cells to resist virus infection

increased blood flow + leakage of plasma

macrophages
engulf larger debris and damaged cells

dilution and removal of toxins from invading micro-organisms and from damaged cells

phagocytosis by white cells

B- and T-lymphocytes of immune system (antigen–antibody response)

neutrophils
(short-lived) engulf bacteria identified by opsonins

engulfed bacterium in 'food vacuole' – lysosomes with hydrolytic enzymes fuse, causing destruction of bacterium

Figure 11.3 Inflammation at a damage site, the processes

What recognition of 'self' entails

All cells are identified by specific molecules – markers, if you like – that are lodged in the outermost surface of the cell. In fact, these molecules include the highly variable **glycoproteins** on the cell surface membrane. Remember, glycoproteins occur attached to proteins there. Look at Figure 4.3, page 76 and remind yourself of the 'fluid mosaic' structure of this membrane.

The glycoproteins that identify cells are also known as the **major histocompatibility complex** antigens – but we can refer to the major histocompatibility complex as **MHC** from now on. There are genes on one of our chromosomes (chromosome 6, actually) that code for MHC antigens. Each individual organism's MHC is genetically determined – it is a feature we inherit. As with all inherited characteristics that are products of sexual reproduction, variation occurs. So each of us has distinctive MHC antigens present on the cell surface membrane of most of our body cells. Unless you have an identical twin, your MHC antigens are unique.

Lymphocytes of our immune system have **antigen receptors** that recognise our own MHC antigens and can tell them apart from any 'foreign' antigens detected in the body. After all, it is critically important that our own cells are not attacked by our immune system.

> **Self**: the products of the body's own genotype, which contain proteins (normally – some carbohydrates and other macromolecules can act as antigens) that do not trigger an immune response in the body's own immune system. Inside the body that produced them, self proteins do not act as antigens (and so do not stimulate an immune response) but, if introduced into another body, they become non-self.

Lymphocytes and the antigen–antibody reaction

We have two distinct types of lymphocytes (Figure 11.4), based on the ways they function. Both cell types originate in the bone marrow where they are formed from **stem cells**. As they mature they undergo different development processes in preparation for their distinctive roles.
- **B-lymphocytes** secrete antibodies.
- **T-lymphocytes**, some of which assist B-lymphocytes, and others which may attack infected cells.
How do these distinctive roles develop?

T-cells leave the bone marrow early during development and they undergo differentiation in the **thymus gland**. Here they mature. The thymus gland is found in the chest, just below the breast bone (sternum). It is an active and enlarged gland from birth to puberty. The gland then shrinks in size, its task completed.

It is whilst T-lymphocytes are present in the thymus gland that the body selects out all the lymphocytes that would otherwise react to the body's own cells. The surviving T-lymphocytes then circulate in the plasma and are stored in lymph nodes.

Question

1 Explain the significance of the role of the thymus gland in destroying T-lymphocytes that would otherwise react to body proteins.

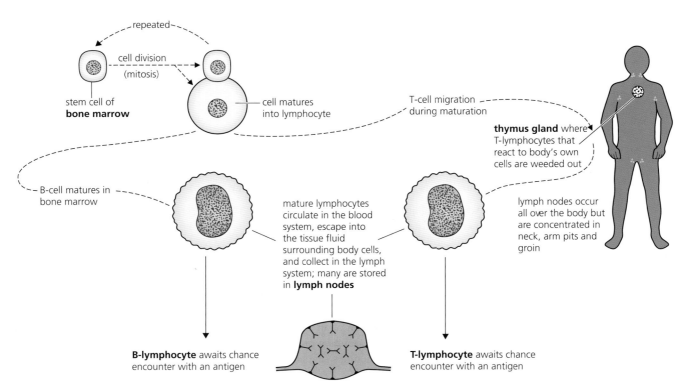

Figure 11.4 T- and B-lymphocytes

The T-lymphocytes do not secrete antibodies. The role of T-lymphocytes is **cell-mediated immunity** – meaning their role in the immune response does not directly involve antibodies, although some have a role in the activation of B-lymphocytes, as we shall see shortly.

B-lymphocytes complete their maturation in the bone marrow, prior to circulating in the blood and being stored in lymph nodes. The role of the majority of B-lymphocytes, after recognition and binding to a specific antigen, is to proliferate into cells (called **plasma cells**) that secrete antibodies into the blood system. This is known as **humoral immunity**.

Now whilst both T- and B-cells have molecules on the outer surface of their cell surface membrane that enable them to recognise antigens, each T- and B-lymphocyte has **only one type** of surface receptor. Consequently, **each lymphocyte can recognise only one type of antigen**.

How do B- and T-lymphocytes respond to an infection?

When an infection occurs the leucocyte population increases enormously and many collect at the site of the invasion. The complex response to infection is begun. The special roles of T- and B-lymphocytes in this response are given below.

Refer to Figure 11.6 as you follow these steps.

1 On the arrival of a specific antigen in the body, **B-lymphocytes** with surface receptors (antibodies) that recognise that particular antigen, **bind** to it.

Antibodies initially occur attached to the cell surface membrane of B-lymphocytes but later are also mass produced and secreted by cells derived from the B-lymphocyte, but only after that B-lymphocyte has undergone an **activation step** (step 4, below).

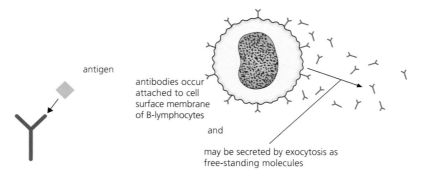

antigen

antibodies occur attached to cell surface membrane of B-lymphocytes

and

may be secreted by exocytosis as free-standing molecules

2 On binding to the B-lymphocyte, the antigen is taken into the cytoplasm by endocytosis, before being expressed on the cell surface membrane of the B-lymphocyte.

3 Meanwhile, **T-lymphocytes** can only respond to antigens when presented on the surface of other cells. Phagocytic cells of the body, including **macrophages**, engulf antigens they encounter. This may occur in the plasma, lymph or tissue fluid. Once these antigens are taken up, the macrophage presents them externally by attaching the antigen to their surface membrane proteins, **MHC** antigens. This is called **antigen presentation** by a macrophage.

4 T-cells come in contact with these macrophages and briefly bind to them. The T-lymphocyte is immediately activated. They become **activated helper T-lymphocytes**.

5 Activated helper T-lymphocytes now bind to B-lymphocytes with the same antigen expressed on their cell surface membrane (step 2 above). The B-lymphocyte is activated by the binding of this chemical. It is now an **activated B-lymphocyte**.

6 Activated B-lymphocytes immediately divide very rapidly by mitosis forming a clone of cells called **plasma cells**. Plasma cells are packed with rough endoplasmic reticulum (RER). It is in these organelles that the antibody is mass produced. The antibody is then exported from the plasma cell by exocytosis. The antibodies are normally produced in such numbers that the antigen is overcome. The production of an activated B-lymphocyte, its rapid cell division to produce a clone of plasma cells and the resulting production of antibodies that react with the antigen is called **clonal selection**. Sometimes several different antibodies react with the one antigen – this is **polyclonal selection**.

Antibody: A glycoprotein secreted by a plasma cell. An antibody binds to the specific antigen that triggered the immune response, leading to destruction of the antigen (and any pathogen or other cell to which the antigen is attached). Antibodies have regions that vary in shape (variable regions) that are complementary to the shape of the antigen. Some antibodies are called antitoxins and prevent the activity of toxins.

Figure 11.5 The structure of an antibody

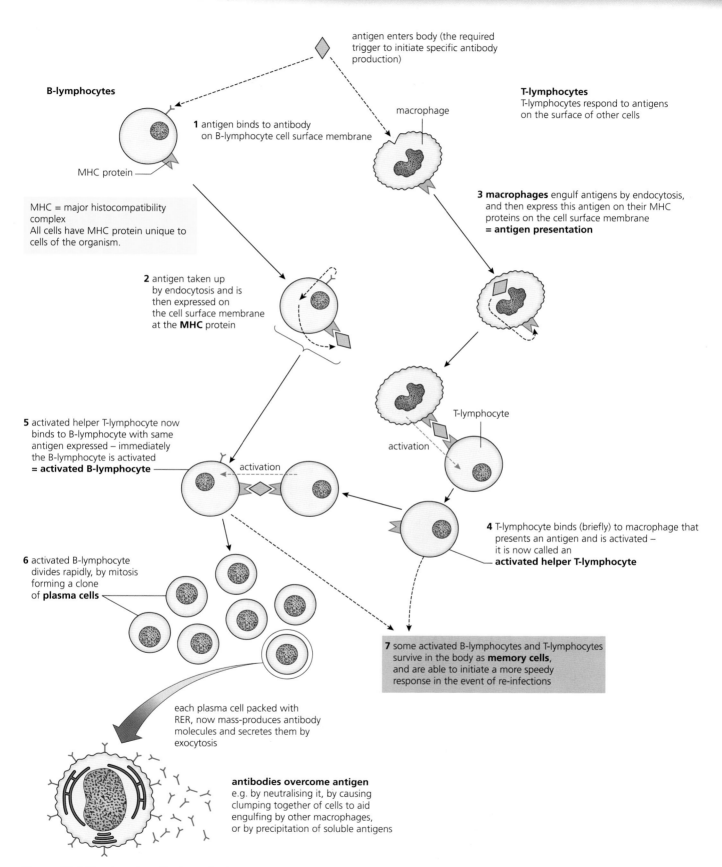

Figure 11.6 The stages in antibody production

7 After these antibodies have tackled the foreign matter and the disease threat is overcome, the antibodies disappear from the blood and tissue fluid. So too do the bulk of the specific B-lymphocytes and T-lymphocytes responsible for their formation.

However, certain of these specifically activated B- and T-lymphocytes are retained in the body as **memory cells**. These are long-lived cells, in contrast to plasma cells and activated B-lymphocytes. Memory cells make possible an early and effective response in the event of a re-infection of the body by the same antigen (Figure 11.7). This is the basis of natural immunity.

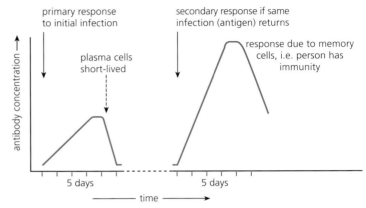

Figure 11.7 The profile of antibody production on infection and re-infection

It is now helpful to summarise the complex roles of B- and T-lymphocytes in the immune system (Figure 11.8).

The significance of white blood cell counts

A normal white blood cell count is between 4500 to 10 000 cells per µl, but during active infections in the body the number rises and may be considerably higher. Given the roles of white blood cells when in the presence of pathogens within the body, this relationship is of no surprise. Another cause of raised white blood cell count is leukaemia – a cancer of the bone marrow cells. The stem cells from which white blood cells are normally formed proliferate but fail to differentiate. Here the numbers can be very high indeed, but with a variety of abnormal cells present.

Myasthenia gravis – a failure to distinguish 'self' from 'non-self'

Myasthenia gravis is an usual auto-immune disorder in which specific antibodies attack and destroy the acetylcholine receptors at the post-synaptic junctions, thereby inhibiting the effect of the neurotransmitter acetylcholine (page 319). This disease is one of the less common auto-immune disorders, affecting about two people in every 10 000 of the population. It causes muscle weakness. The effects of the disease may be localised in the body, involving just the eye muscles for example. Alternatively, many muscle systems may be affected, leaving the sufferer susceptible to general fatigue. In the presence of this condition, muscles weaken during periods of activity but improve with resting. Some patients with myasthenia gravis are found to have a tumour of the thymus gland. These are often benign, but the gland is then removed as a precaution, and typically symptoms of myasthenia gravis improve.

Questions

2 Identify where antigens and antibodies may be found in the body.

3 Remind yourself of the appearance of the types of white blood cells observed in a blood smear preparation. See for example, the photomicrograph in Figure 8.2 (page 153). Compare this image with a prepared slide of a human blood smear so that you are familiar with the appearance of phagocytic white blood cells and lymphocytes. Make a fully annotated drawing of representative white blood cells that makes clear the differences between them.

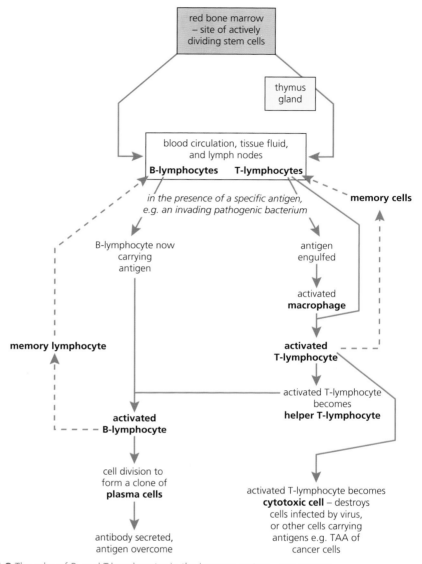

Figure 11.8 The roles of B- and T-lymphocytes in the immune system – a summary

Immune response: the complex series of reactions of the body to an antigen, such as a molecule on the outside of a bacterium, virus, parasite, allergen or tumour cell.

- The immune response begins with an innate first response, carried out by phagocytic white blood cells, which can destroy and engulf (by phagocytosis/endocytosis) many different foreign organisms.
- At the same time, the primary phase of the adaptive immune system response begins, in which specific clones of B-lymphocytes and T-lymphocytes divide and differentiate to form antibody-secreting plasma cells (from B-lymphocytes) and T helper cells and T killer cells (from T-lymphocytes) that are specific to the antigen, contributing to its destruction or preventing its activity.
- This leads into the secondary phase of the adaptive immune system response, where memory cells retain the capability to secrete antibodies or act as T helper or T killer cells as soon as the specific antigen is detected again.

11.2 Antibodies and vaccination

By the end of this section you should be able to:

a) relate the molecular structure of antibodies to their functions
b) outline the hybridoma method for the production of monoclonal antibodies
c) outline the use of monoclonal antibodies in the diagnosis of disease and in the treatment of disease
d) distinguish between active and passive, natural and artificial immunity and explain how vaccination can control disease
e) discuss the reasons why vaccination programmes have eradicated smallpox, but not measles, tuberculosis (TB), malaria or cholera

The antibody molecules – structure in relation to function

Antibodies are proteins called **immunoglobulins**. Each antibody consists of four polypeptide chains held together by disulfide bridges (–S–S–), forming a molecule in the shape of a **Y**. The arrangement of amino acids in the polypeptides that form the tips of the arms of the Ys in this molecule are unique to that antibody. It is these regions that hold the highly specific **binding site** for the antigen. Antibodies initially occur attached to the cell surface membrane of B-lymphocytes, but later are also mass-produced and secreted by cells derived from the B-lymphocyte (Figure 11.9), but only after that B-lymphocyte has undergone an **activation step**.

The structure of the antibody molecule demonstrates the four levels of protein structure, discussed earlier (page 45):

The **primary structure** of a protein is the long chain of amino acids in its molecule – amino acid residues joined by peptide linkages. You can see that there are four polypeptide chains that make up each antibody, two described as 'heavy chains' and two as 'light chains'. The polypeptides differ in the variety, number and order of their constituent amino acids.

The **secondary structure** of a protein develops when parts of the polypeptide chain take up a particular shape, immediately after formation at the ribosome. Parts of the chain become folded or twisted, or both, in various ways. The most common shapes are formed either by coiling to produce an α-**helix** or folding into β-**sheets**, and both shapes are permanent, held in place by hydrogen bonds.

The **tertiary structure** of a protein is the precise, compact structure, unique to that protein that arises when the molecule is further folded and held in a particular complex shape. This shape is made permanent by four different types of bonding, established between adjacent parts of the chain (Figure 2.22, page 47).

The **quaternary structure** of a protein arises when two or more proteins are bound together, forming a complex, biologically active molecule. In an antibody, four polypeptide chains are combined, held together in a characteristic shape by sulfide bridges (Figure 11.9).

Monoclonal antibodies

We have seen that white blood cells in the blood provide our main defence against the invasion of the body by harmful microorganisms. Special cells among the white blood cells, called **lymphocytes**, are responsible for the immune response. Among these it is the **B-lymphocytes** that secrete antibodies. Antibodies are very effective in the destruction of antigens within the body (Figure 11.7). **Monoclonal antibodies** are produced by a single clone of B-lymphocytes. They are identical antibodies effective against a single, specific antigen.

Figure 11.9 The antibody – a quaternary protein

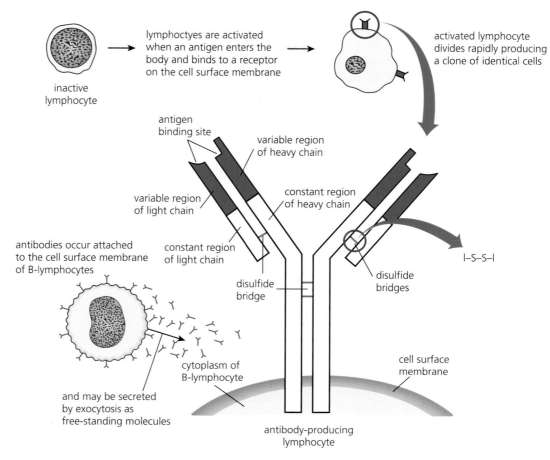

inactive lymphocyte

lymphoctyes are activated when an antigen enters the body and binds to a receptor on the cell surface membrane

activated lymphocyte divides rapidly producing a clone of identical cells

antigen binding site

variable region of heavy chain

variable region of light chain

constant region of heavy chain

antibodies occur attached to the cell surface membrane of B-lymphocytes

constant region of light chain

disulfide bridges

l–S–S–l

disulfide bridge

cytoplasm of B-lymphocyte

cell surface membrane

and may be secreted by exocytosis as free-standing molecules

antibody-producing lymphocyte

Figure 11.10 Computer-generated image of an antibody molecule

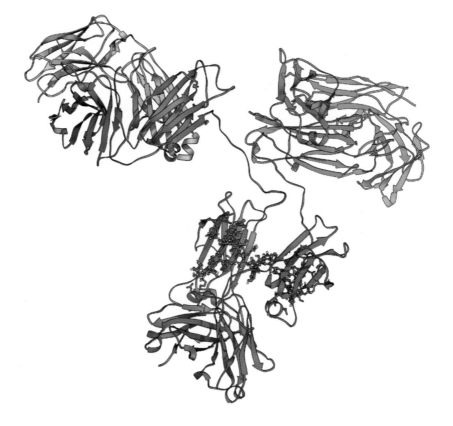

The production of monoclonal antibodies

Antibodies have great potential in modern medicine, if they can be made available. The problem has been that B-lymphocytes are short-lived. This issue of the normally brief existence of a B-lymphocyte is overcome by fusing the specific lymphocyte with a **cancer cell** (a malignant myeloma of a B-lymphocyte) which, unlike non-cancerous body cells, goes on dividing indefinitely. The fused cells, called **hybridoma** cells, are separated and each hybridoma is cultured individually. Each cell divides to form a clone of cells which persists and which conveniently goes on secreting the antibody in significant quantities. The outcome is a single antibody that is stable and that can be used over a period of time. How a specific monoclonal antibody is made is illustrated in Figure 11.11.

The use of monoclonal antibodies in medicine is already established and is under further development.

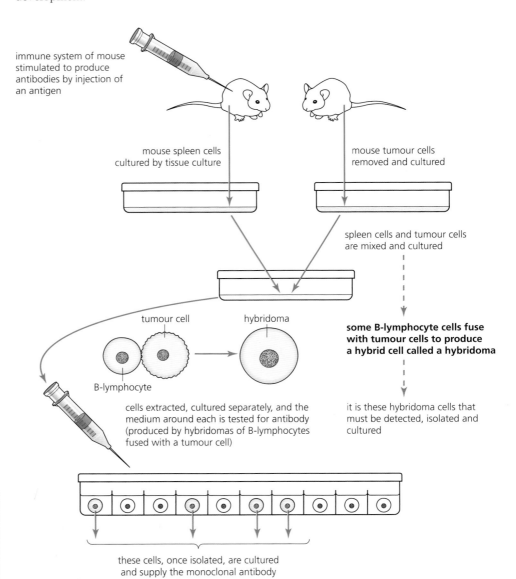

immune system of mouse stimulated to produce antibodies by injection of an antigen

mouse spleen cells cultured by tissue culture

mouse tumour cells removed and cultured

spleen cells and tumour cells are mixed and cultured

tumour cell

hybridoma

B-lymphocyte

some B-lymphocyte cells fuse with tumour cells to produce a hybrid cell called a hybridoma

it is these hybridoma cells that must be detected, isolated and cultured

cells extracted, cultured separately, and the medium around each is tested for antibody (produced by hybridomas of B-lymphocytes fused with a tumour cell)

these cells, once isolated, are cultured and supply the monoclonal antibody

Figure 11.11 The formation of monoclonal antibodies

The use of monoclonal antibodies in the treatment of disease

It is known that cancer cells carry specific **tumour-associated antigens (TAA)** on their cell surface membrane. Monoclonal antibodies to TAA have been produced. Then, drugs to kills cells or inhibitors to block key tumour proteins have been attached. The monoclonal antibodies specifically target and kill the cancer cells. The advantage is that whereas many drugs and treatments effective against cancer are also harmful to other, healthy cells, the specificity of antibodies avoids this problem. These developments are sometimes referred to as 'magic bullets'.

Monoclonal antibodies have already been developed to treat certain cancers, as described in Table 11.1. Now, monoclonal antibodies are under investigation in clinical trials for nearly every type of cancer. Perhaps this is the most important way that monoclonal antibodies are currently being used in the treatment of disease.

Table 11.1 How monoclonal antibodies are used to treat cancers

Making cancer cells 'visible' to the immune system	Cancer cells are 'self' cells – they have not invaded the body from outside. The immune system only targets non-self cells. Monoclonal antibodies that recognise and attach to cell surface membrane proteins that have resulted from the specific cancer condition present have been produced. Once introduced into the body they attach, making the cell visible to the immune system and therefore vulnerable to destruction by it.
Blocking growth signals of the cell surface membrane of cancer cells	Monoclonal antibodies have been produced that attach to and block the growth signal molecules on the cell surface membrane of cancer cells. Unblocked, these growth factors signal the cancer cell to grow. Cancer cells produce additional growth factors, enhancing their growth rate. Such 'blocking' monoclonal antibodies have been produced to treat colon cancer, for example.
Stopping the formation of new blood vessels	The growth signals manufactured by cancer cells which stimulate the growth of fresh blood vessels to supply the developing malignant tumour can be blocked by specific monoclonal antibodies. One such is in use in the treatment of breast cancer.
Delivering radiation to cancer cells	Radiation sufficient to destroy a cancer cell can be delivered specifically to the cells of a malignant tumour by combining radioactive particles with monoclonal antibodies that attach exclusively to cancer cells. These monoclonal antibodies deliver a low level of radiation over a long period of time, without damage to the surrounding cells. Such monoclonal antibodies are in use against non-Hodgkin's lymphoma.

Another application of monoclonal antibodies in the treatment of disease is as 'passive vaccines'. A monoclonal antibody specific to a particular pathogen may be produced and injected into a patient's blood circulation to good effect.

The use of monoclonal antibodies for pregnancy testing

Another application of monoclonal antibodies is in pregnancy testing. A pregnant woman has a significant concentration of **HCG hormone** in her urine whereas a non-pregnant woman has a negligible amount. Monoclonal antibodies to HCG have been engineered which also have coloured granules attached. In a simple test kit, the appearance of a coloured strip in one compartment provides immediate, visual confirmation of pregnancy. How this works is illustrated in Figure 11.12.

Because all the monoclonal antibodies produced by a clone of a B-lymphocyte are identical they can be used to identify other specific macromolecules, too.

Other examples of the use of monoclonal antibodies in diagnosis

- Since a B-lymphocyte recognises a specific antigen, monoclonal antibodies can be used to distinguish between different strains of a disease-causing microorganism
- Detection of HIV in a patient's blood sample
- Blood typing before a blood transfusion, for example to determine A, B, AB or O and Rhesus negative or Rhesus positive groupings
- Tissue typing before a transplant operation
- Identification of the particular type of leukaemia (a cancer) that a patient has contracted to determine the most appropriate treatment
- Location of tumours.

Extension

An issue to note is that, because the original monoclonal antibodies were developed from mice cells (Figure 11.11), a patient may develop an adverse reaction to an antibody. This is because the monoclonal antibody is itself a foreign protein to the patient's immune system. Genetically engineered antibodies that are compatible with the human immune system are sought to avoid the triggering of the immune response.

pregnancy testing kit

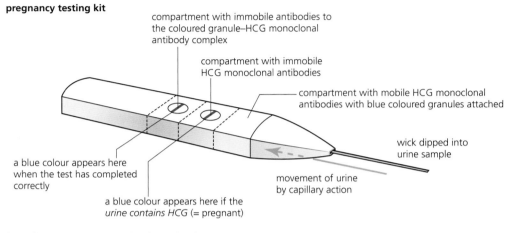

compartment with immobile antibodies to the coloured granule–HCG monoclonal antibody complex

compartment with immobile HCG monoclonal antibodies

compartment with mobile HCG monoclonal antibodies with blue coloured granules attached

wick dipped into urine sample

movement of urine by capillary action

a blue colour appears here when the test has completed correctly

a blue colour appears here if the *urine contains HCG (= pregnant)*

how the positive test result is brought about

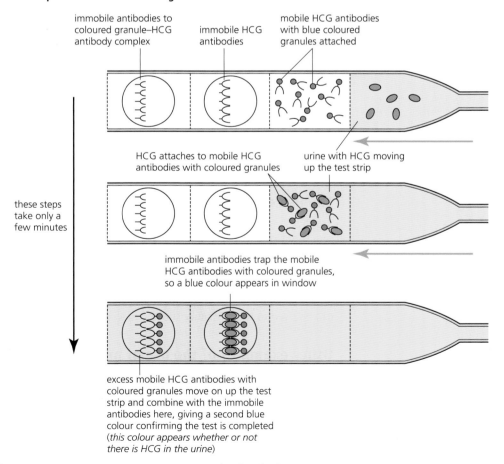

immobile antibodies to coloured granule–HCG antibody complex

immobile HCG antibodies

mobile HCG antibodies with blue coloured granules attached

HCG attaches to mobile HCG antibodies with coloured granules

urine with HCG moving up the test strip

these steps take only a few minutes

immobile antibodies trap the mobile HCG antibodies with coloured granules, so a blue colour appears in window

excess mobile HCG antibodies with coloured granules move on up the test strip and combine with the immobile antibodies here, giving a second blue colour confirming the test is completed (*this colour appears whether or not there is HCG in the urine*)

Figure 11.12 Detecting pregnancy using monoclonal antibodies

Vaccination

Vaccination is the deliberate administration of antigens that have been made harmless, after they are obtained from disease-causing organisms, in order to confer immunity in future. The practice of vaccination has made important contributions to public health (see TB, page 200; measles, page 208; smallpox, page 210).

Vaccines are administered either by injection or by mouth. They cause the body's immune system to briefly make antibodies against the disease (*without becoming infected*), and then **to retain the appropriate memory cells**. Active artificial immunity is established in this way (Table 11.2). The profile of the body's response in terms of antibody production, if later it is exposed to the antigen again, is exactly the same as if the immunity was acquired after the body overcame an earlier infection.

Vaccines are manufactured from dead or attenuated (weakened) bacteria, or from inactivated viruses, or purified polysaccharides from bacterial walls, or toxoids, or even recombinant DNA produced by genetic engineering.

In communities and countries where vaccines are widely available and have been taken up by 85–90 per cent of the relevant population, vaccination has reduced some previously common and dangerous diseases to very uncommon occurrences. As a result, in these places, the public there has sometimes become casual about the threat such diseases still represent.

A summary of the types of immunity

Our immune system may provide protection to the body from the worst effects of many of the pathogens that may invade. This immunity may be acquired **actively**, as when our body responds to invasion by a pathogen, or **passively**, as when ready-made antibodies are injected into our body. Also, immunity may be acquired **naturally**, as when our bodies respond to a natural pathogen invasion, or **artificially**, as follows vaccination.

In Table 11.2 the differences between active and passive immunity, and natural and artificial immunity are identified and illustrated.

Table 11.2 Types of immunity that humans display

Types of immunity	Natural	Artificial	. . . and their longevity
Active immunity • Results from exposure to an antigen. During the subsequent immune response, antibodies are produced by plasma cells. • There is a delay before the immune response is complete. (Immunity takes some days to build up.)	e.g.: Immunity that follows natural infection by a pathogen involving the production of memory cells (for example, natural infection with chicken pox, giving long-term protection from this virus).	e.g.: Immunity that results from the injection of antigens (such as those attached to killed or weakened pathogens) where memory cells are made (for example, after vaccination).	Active immunity persists for a prolonged period, and possibly throughout life, after it has been acquired because the body makes memory cells.
Passive immunity • Involves the transfer of antibodies (already made in the body of another organism or *in vitro*) into the body where they will bind to their specific antigen if it is present. • Gives instant immunity.	e.g.: Immunity following transfer of maternal antibodies into a fetus through the placenta, and into a newborn infant in the first milk (colostrum).	e.g.: Immunity following injection of antibodies (for example, monoclonal antibodies, to treat acute life-threatening infections, such as rabies).	Passive immunity fades with time because the recipient does not acquire memory cells and so is unable to make the antibodies in future.

Vaccination: the medical giving of material containing antigens, but with reduced or no ability to be pathogens, in order to give long-term active immunity as a result of the production of memory cells.

Artificial immunity: immunity that is acquired by a person as a result of medical intervention. This includes artificial passive immunity following injection of antibodies (for example monoclonal antibodies, to treat acute life-threatening infections, such as tetanus or rabies). It also includes the long-term immunity that results from the injection of antigens (such as those attached to killed or weakened pathogens) where memory cells are made.

Questions

5 a Identify the steps to plasma cell formation that the existence of memory cells avoids in the event of re-infection.
  b Why is it quicker to respond on re-infection in these cases?
6 State what we mean by immunity.

Question

7 **a** By means of a table such as Table 11.2, show the ways that immunity may arise according to whether it is actively or passively obtained, and by natural or artificial means. Place these terms in your table:

polio vaccine

measles infection

antibodies received via the placenta

monoclonal antibodies to treat tetanus

b Underline those terms in which memory cells are *not* formed (and therefore the effects are short term).

Issues with vaccines and vaccination

The many uses that the body has for proteins that it builds from the pool of amino acids are identified in Figure 2.26 (page 51). The antibodies (immunoglobulins) secreted by B-lymphocytes are only one of these demands. So, if our diet is deficient in proteins, and especially if we are generally malnourished, our amino acid pool is diminished and the body has difficulty in meeting all of these demands. So underfed and starving people typically have a defective immune system and they are at risk of contracting infectious diseases and of passing them on to others.

Nevertheless, *one* infectious disease, once common among people throughout the world has been eradicated. This is smallpox. This is partly for reasons that relate to the virus and the vaccine produced, including:

- the virus was stable – it did not mutate or change its surface antigens. (Minor changes in antigens (**antigenic shift**) will cause memory cells to fail to recognise a pathogen.)
- transmission was by direct contact only, so the people possibly infected by an isolated outbreak were limited and were easily identified. (The vaccine could be quickly given to all contacts to achieve **herd immunity**. In this way, transmission within any community is interrupted.)
- the vaccine was freeze dried and stocks could be distributed and held at high temperatures (as in the tropics) and remain effective.
- the made-up vaccine could be administered with a reusable (stainless steel) needle.

Other reasons why smallpox has been eradicated relate to the impact of the disease on patients and their contacts and to the ways the eradication campaign was administered and delivered, for example. They are identified earlier in the topic, on page 210.

Other diseases we have discussed in this topic have not been eradicated. The reasons are due to particular features of these diseases and the pathogens or factors that cause them. These have been identified and discussed earlier in this topic. They are summarised in Table 11.3.

Table 11.3 Why vaccination has not eradicated measles, TB, malaria, sickle cell anaemia or cholera

Disease	Impact (or other otherwise) of vaccination – and why
Cholera – vaccines exist	The cholera-causing bacterium and its toxin are active in the intestine – a region of the body where the vaccine administered by injection cannot operate. The oral vaccine that has been developed offers protection from some but not all strains of the cholera bacterium. (Standard hygiene precautions – clean water and efficient sewage disposal – if they can be achieved, are the most effective protection. Cholera is endemic only in countries where these conditions cannot be maintained.)
Malaria – no vaccine exists as yet	No effective vaccine has been produced as yet. The difficulties are: • the pathogen is a protoctist, with many more genes than bacteria and viruses. The alleles of these genes result in vast numbers of antigens being present on the surface of the pathogen – including different antigens on the separate stages in the life cycle. • the pathogen spends a very short time in the blood plasma, before 'hiding' from the immune system in liver or red blood cells in the human host. This **antigenic concealment** leaves little time for an effective immune response.
Sickle cell anaemia	This is a genetic disease expressed in people who are homozygous for the sickle cell haemoglobin allele (Hb^s). Vaccines and the immune response cannot effect changes in the genetic constitution (the genome).
TB – a vaccine exists	The effectiveness of the vaccine against tuberculosis (BCG vaccine) varies according to geography – 80% in the UK, 14% in the USA and in some places it shows no protective effect. (Strains of the pathogenic bacterium resistant to anti-TB drugs have emerged in all countries.)
Measles – a vaccine exits	The disease persists because the virus is highly infectious and the vaccine is about 95% effective, so a whole infant population needs to be treated within 8 months of birth in order to prevent transmission of the virus. Some children require one or more 'booster' injections before full immunity is achieved, too. Many countries now have about 80% of the population vaccinated but in African countries it is currently less than 50% that receive the vaccine. For most communities, **herd immunity** has not been established. In less-developed countries with a high birth rate the living conditions of many people are crowded – ideal conditions for transmission of the virus. Also, many victims are malnourished and the diets of infants are low in protein (and deficient in vitamin A). Any natural resistance to infection by the measles virus is minimal.

Summary

- Defence against disease that enters when the body is wounded and then invaded by a pathogen is provided by **phagocytic white blood cells**. These are able to engulf and destroy foreign matter.

- The **immune system** is provided by **B- and T-lymphocytes** which, when sensitised by foreign material (**antigens**), respond in complex ways to overcome the invasion, including by the production of specific **antibodies** that destroy or inactivate the foreign matter. Memory cells are formed and retain the ability to respond again in the event of re-infection.

- **Vaccination** is the deliberate administration of antigens that have been rendered harmless but nevertheless are able to

stimulate antibody production and the retention of appropriate memory cells.

- **Monoclonal antibodies** are a source of a single antibody made by a particular type of B-lymphocyte fused with a cancer cell to form a **hybridoma**. This divides indefinitely and so secretes the antibody in significant quantities. Monoclonal antibodies are used in the treatment of cancer by targeting specific **damaged** or **diseased cells** and killing them by the delivery of a 'magic bullet'. They are used in diagnosis to identify the presence of **specific macromolecules**, as in pregnancy testing, and in several other medical applications.

Examination style questions

1 During an immune response, plasma cells secrete antibody molecules. Fig.1.1 is a diagram of an antibody molecule. The diagram is **not** complete.

Fig. 1.1

a) On a copy of Fig. 1.1:

  i) Draw a circle around a variable region. [1]

  ii) Draw in and label the position of the disulfide bonds in the molecule. [1]

  iii) Explain the importance of disulfide bonds in protein molecules, such as antibodies. [3]

b) Describe how antibodies provide protection against pathogens. [4]

c) Other proteins are found in cell surface membranes.

Describe three roles of the proteins in cell surface membranes. [3]

[Total: 12]

(Cambridge International AS and A Level Biology 9700, Paper 21 Q1 November 2009)

2 a) Plasma cells secrete specific antibodies but are short-lived in the body. Explain, with use of a diagram, how a plasma cell is made 'immortal' in the production of a monoclonal antibody. [8]

b) Detail the use of monoclonal antibodies in one application of their use in diagnosis and one application in the treatment of diseases. [8]

[Total: 16]

3 Phagocytes and lymphocytes are part of the body's cellular response to infection by pathogens.

The diagram shows the origin and maturation of phagocytes and lymphocytes.

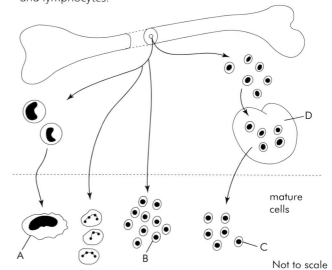

Not to scale

a) Name the site of origin of phagocytes and lymphocytes. [1]

b) Name:

  i) cells **A**, **B** and **C** [3]

  ii) organ **D**. [1]

c) Explain the roles of the cells, **A**, **B** and **C** in an immune response.

In your answer use the terms *antigen* and *non-self*. [5]

[Total: 10]

(Cambridge International AS and A Level Biology 9700, Paper 21 Q6 November 2011)

12 Energy and respiration

Energy is a fundamental concept in biology. All living things require a source of cellular energy to drive their various activities. ATP is the universal energy currency as its molecules are small, soluble and easily hydrolysed to release energy for cellular activities. All organisms respire to release energy from energy-rich molecules such as glucose and fatty acids and transfer that energy to ATP. Respiration is a series of enzyme-catalysed reactions that release energy in small 'packets'. In eukaryotes, aerobic respiration occurs in mitochondria.

12.1 Energy

ATP is the universal energy currency as it provides the immediate source of energy for cellular processes.

By the end of this section you should be able to:

a) outline the need for energy in living organisms, as illustrated by anabolic reactions, such as DNA replication and protein synthesis, active transport, movement and the maintenance of body temperature
b) describe the features of ATP that make it suitable as the universal energy currency
c) explain that ATP is synthesised in substrate-linked reactions in glycolysis and in the Krebs cycle
d) outline the roles of the coenzymes NAD, FAD and coenzyme A in respiration
e) explain that the synthesis of ATP is associated with the electron transport chain on the membranes of mitochondria and chloroplasts
f) explain the relative energy values of carbohydrate, lipid and protein as respiratory substrates and explain why lipids are particularly energy-rich
g) define the term respiratory quotient (RQ) and determine RQs from equations for respiration
h) carry out investigations, using simple respirometers, to determine the RQ of germinating seeds or small invertebrates (e.g. blowfly larvae)

Cell respiration – the controlled transfer of energy

Living things need a **continuous transfer of energy** to keep them alive and active. Energy is used to build, maintain and repair body structures, and for all the activities of life. Energy is transferred in cells by the breakdown of **nutrients**. Typically, it is carbohydrates like glucose from which energy is transferred by the process of **cell respiration**. It is respiration that is the subject of this topic.

So organisms need a supply of certain nutrients for respiration, but animals and plants obtain these in very different ways.

Green plants make sugars from simpler substances by photosynthesis. In photosynthesis they transfer light energy into the chemical **potential energy** of sugars. Using these sugars and a supply of ions from the soil, they make their proteins, lipids and all other requirements. This type of nutrition is called **autotrophic**. It means 'self-feeding'. Photosynthesis is discussed in Topic 13.

On the other hand, animals and other **heterotrophic** organism are dependent on green plants for their nutrients – directly or indirectly. Their nutrients are taken up into their body tissues following the digestion of food.

Cellular respiration: the controlled transfer of energy from organic compounds in cells to ATP.

glucose burning in air

sprinters completing a race

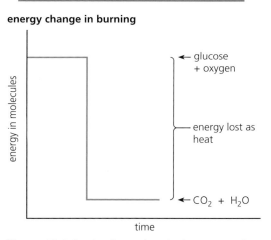

energy change in burning

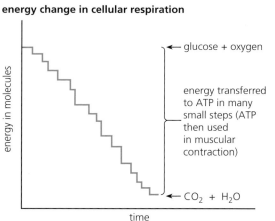

energy change in cellular respiration

Figure 12.1 Combustion and respiration compared

People sometimes talk about 'burning up food' in respiration. In fact, linking respiration to burning (combustion) is unhelpful. In combustion, energy in fuels is released, in a one-step reaction, as heat. Such a violent change would be a disaster for body tissues. In cell respiration, a large number of small steps occur. Each is catalysed by a particular enzyme. Because energy in respiration is transferred in small quantities, much of the energy is made available to the cells. It may be trapped in the energy currency molecule **adenosine triphosphate (ATP)**. However, some energy is still lost as heat in each step (Figure 12.1). In animals that regulate their body temperature, this source of heat is important (Figure 12.13).

The need for energy in living organisms

Energy is needed for all the activities of living organisms, such as nutrition, excretion, sensitivity, movement and reproduction. It is also required to build and maintain body structures, and to maintain body temperature. Specifically, all organisms transfer usable energy for:

- the reactions of **metabolism**, and especially for the steps by which complex molecules are built up, including macromolecules. All these reactions are known as **anabolism**. Protein synthesis is just one example.

- the active transport of molecules and ions across membranes, carried out by membrane pumps.
- movement by organisms, including those due to muscle contractions. Energy is also transferred for movements within cells and by cell organelles.
- formation and secretion of substances from cells, including enzymes and hormones.
- cell division, and the formation and maintenance of all the organelles of the cytoplasm.

ATP – the universal energy currency

Adenosine triphosphate (ATP) is a relatively small, soluble organic molecule. It occurs in all cells, typically at a concentration of $0.5–2.5\,mg\,cm^{-3}$. Like many organic molecules of its size, it contains a good deal of chemical energy locked up in its structure.

Chemically, ATP is a **nucleotide** (page 111), but it has a particular structural feature we have not seen in nucleotides previously – it has three terminal phosphate groups linked together in a linear sequence (Figure 12.2). These phosphate groups are involved in ATP's role in cells.

Question

1 In Figure 12.2 the structural formula of ATP is shown. In Figure 6.1 (page 111) the way the components of nucleotides are represented diagrammatically is shown. Use the symbols in Figure 6.1 to represent the structure of ATP.

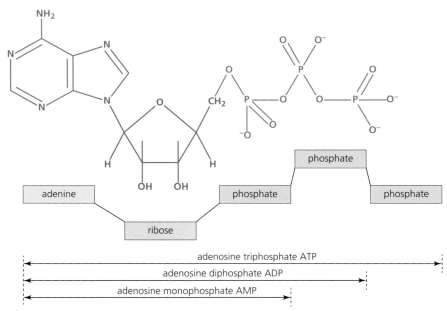

Figure 12.2 ATP, ADP and AMP

ATP is referred to as **'energy currency'** because, like money, it can be used in different contexts and it is constantly recycled in the living cell. When the outermost phosphate is removed, adenosine diphosphate (ADP) is formed and energy is transferred.

Later, it is energy transferred from respiration that reforms ATP from ADP and **phosphate (P_i)**. The reactions that yield energy and generate ATP from ADP and phosphate (P_i) are particular steps in respiration. We will return to these shortly.

All these changes make up the ATP–ADP cycle (Figure 12.3).

The **free energy** available in the step from ATP to ADP is approximately $30–34\,kJ\,mol^{-1}$. Some of this energy is lost as heat in the reaction, but much of the free energy is made available to do useful work – more than sufficient to drive a typical energy-requiring reaction of metabolism, for example. Actually, ATP may lose both of the outer phosphate groups, but normally only one is lost. It is only in a few reactions that more energy is required and in these cases both may be lost.

An important way in which ATP serves as a reservoir of usable chemical energy is its role as a common **intermediate** between energy-yielding reactions and energy-requiring reactions and processes. We saw an example of this in protein synthesis (Figure 6.12, page 121). There are other ways in which energy is transferred, including muscle contraction (page 249).

In summary, ATP is a molecule universal to all living things. It is the source of energy for chemical change in cells, tissues and organisms. The important features of ATP are that it is:
- a substance that moves easily within cells and organisms – by **facilitated diffusion**
- formed in cellular respiration and takes part in many reactions of metabolism
- able to transfer energy in relatively small amounts, sufficient to drive individual reactions.

In the course of the ATP–ADP cycle in cells:
- ATP is formed from adenosine diphosphate (ADP) and phosphate ions (P_i) by the transfer of energy in respiration
- in the presence of specific enzymes, ATP is involved in energy-requiring reactions.

We will return to the issue of just how the energy of ATP brings about useful work in cells and organisms shortly. We will focus on the process of cellular respiration first.

Question

2 Outline why ATP is an efficient energy currency molecule.

Figure 12.3 The ATP–ADP cycle

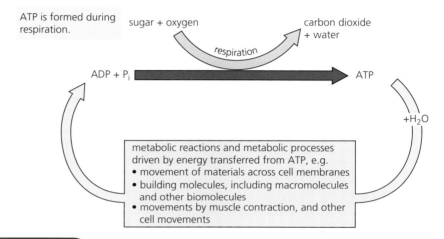

ATP is formed during respiration.

sugar + oxygen → (respiration) → carbon dioxide + water

ADP + P_i → ATP

+H_2O

metabolic reactions and metabolic processes driven by energy transferred from ATP, e.g.
- movement of materials across cell membranes
- building molecules, including macromolecules and other biomolecules
- movements by muscle contraction, and other cell movements

Extension

Respiration as a series of redox reactions

The terms 'oxidation' and 'reduction' occur frequently in respiration. First, we need to remind ourselves what is meant by these terms.

In biological oxidation, oxygen atoms may be added to a compound. Alternatively, hydrogen atoms may be removed. In respiration, all the hydrogen atoms are gradually removed from glucose. They are added to hydrogen **acceptors**, which are then reduced. In respiration, most of the hydrogen atoms are added to the hydrogen acceptor molecule known as **nicotinamide adenine dinucleotide (NAD)**. The process of removal of hydrogen from a compound is known as **dehydrogenation** and the enzyme involved in such a reaction is called a dehydrogenase.

Now a hydrogen atom consists of an electron and a proton. This means that in a reduction reaction, one or more electrons are gained. In fact, the best definition of **oxidation** is the **loss of electrons** and of **reduction** is the **gain of electrons**. So tissue respiration is a series of oxidation and reduction reactions. The shorthand name for **red**uction–**ox**idation reactions is **redox reactions**.

Redox reactions take place in biological systems because of the presence of a compound with a strong tendency to take electrons from another compound (an **oxidising agent**) or the presence of a compound with a strong tendency to donate electrons to another compound (a **reducing agent**).

Another feature of oxidation and reduction is an **energy change**. When reduction occurs, energy is absorbed (it is an **endergonic reaction**). When oxidation occurs, energy is released (an **exergonic reaction**). An example of energy release during oxidation is the burning of a fuel in air. Here, energy is given out as heat. In fact, the amount of energy in a molecule depends on its degree of oxidation. An oxidised substance has less stored energy than a reduced substance. An example of this is the fuel molecule methane (CH_4) which has more stored chemical energy than carbon dioxide (CO_2).

The role of nucleotides in cell biochemistry

In addition to being the building blocks of the nucleic acids, certain nucleotides and molecules made from nucleotides are important in the metabolism of cells. An example is **adenosine triphosphate** (**ATP**). The structure and roles of ATP were introduced in Topic 1 (page 21). In Figure 12.2 we viewed the structure of ATP as a nucleotide.

Look at the details of the structure and roles of ATP in these references again, now.

Other nucleotides involved in respiration

1 Nicotinamide adenine dinucleotide (NAD)

NAD consists of two nucleotides combined with two phosphate groups. In one of the combined nucleotides, the organic base is replaced by a substance called nicotinamide. Nicotinamide is derived from a vitamin of the B complex, called nicotinic acid. The importance of the ring molecule, nicotinamide, is that it accepts hydrogen ions and electrons. When NAD accepts electrons it is said to be reduced, but it can pass them to other acceptor molecules with relative ease. When reduced NAD passes electrons on, it has become oxidised again. The transport of hydrogen ions like this is an important aspect of respiration. NAD is a substance known as a coenzyme, which works with different enzymes of respiration. It also takes part in other processes in cells.

2 Flavine adenine dinucleotide (FAD)

FAD is a coenzyme derived from vitamin B2. Like NAD this is a hydrogen-carrying molecule involved in oxidation and reduction reactions, including in the Krebs cycle (Figure 12.9, page 245). It occurs tightly associated with protein, forming a complex known as a flavoprotein.

Figure 12.4 NAD, structure and role

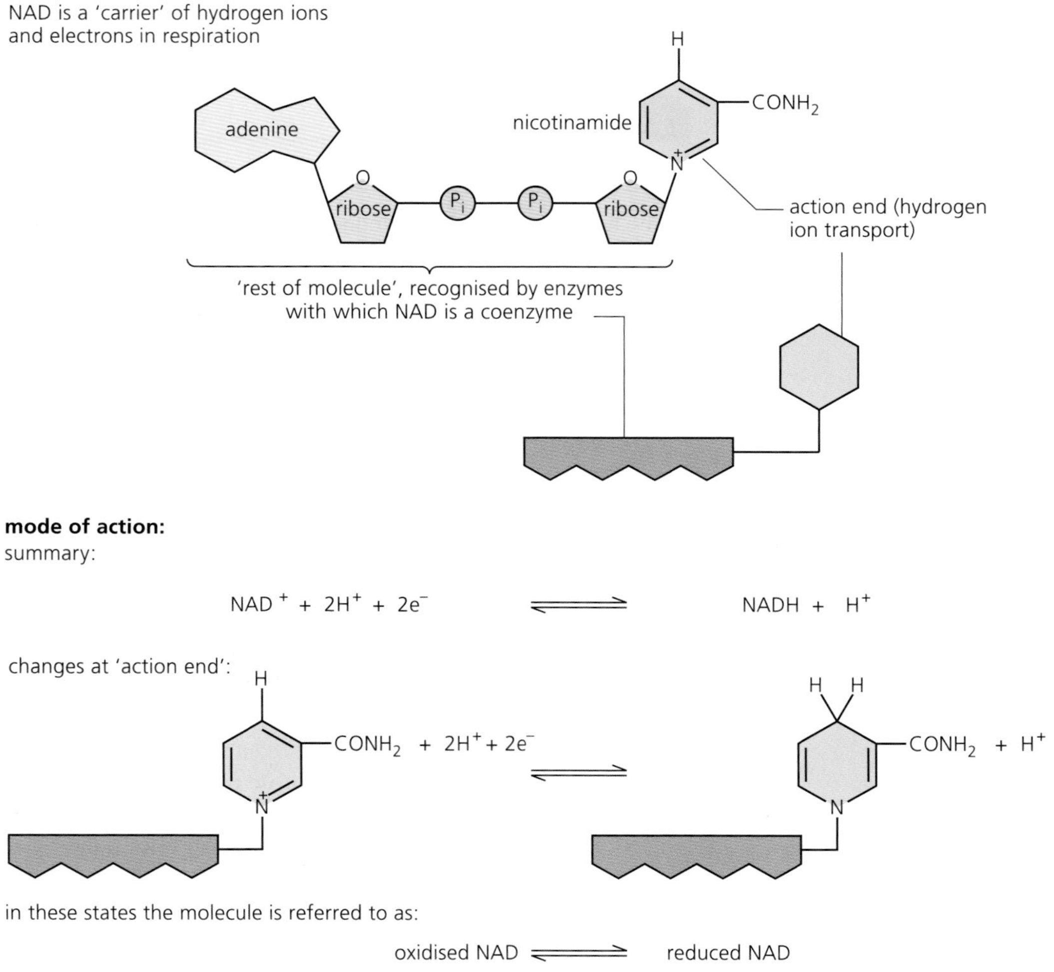

NAD is a 'carrier' of hydrogen ions and electrons in respiration

mode of action:
summary:

$$NAD^+ + 2H^+ + 2e^- \rightleftharpoons NADH + H^+$$

changes at 'action end':

$$\cdots CONH_2 + 2H^+ + 2e^- \rightleftharpoons \cdots CONH_2 + H^+$$

in these states the molecule is referred to as:

oxidised NAD $\rightleftharpoons$ reduced NAD

3 Coenzyme A (CoA)

CoA is involved in the removal of two-carbon fragments from certain carbohydrates and lipids during respiration, and also in food-storage reactions. You can see CoA 'at work' in the link reaction of cell respiration (Figure 12.9).

flavine adenine dinucleotide (FAD)   **coenzyme A (CoA)**

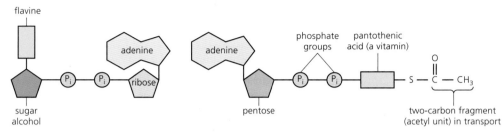

Figure 12.5 Other nucleotide derivatives involved in respiration

The ways ATP is generated in respiration – an introduction

We have seen that ATP is synthesised in cells from ADP and inorganic phosphate. This occurs in respiration in two ways.

1 **Substrate-level phosphorylation** takes place within the cytosol of cells in reactions forming part of the cell respiration pathway called glycolysis. The steps to glycolysis are listed in order, on page 243, and the reactions by which substrate-level phosphorylation is brought about (Step 3. Oxidation, and Step 4. ATP synthesis) are shown in Figure 12.8.

   Look up these references now.

   Actually, the reactions of substrate-level phosphorylation occur twice in glycolysis, and it also occurs in the Krebs cycle. However, the amount of ATP generated by this mechanism is trivial.

2 The bulk of ATP synthesis takes place in special structures **associated with the electron transport chain**, present in the membrane-bound organelles, the mitochondria (page 18) (and also in green plants in the chloroplasts (pages 19–20), as we shall shortly discuss (see 'The electron transport chain and oxidative phosphorylation', page 246).

Extension

The respiration of fats and proteins

In addition to glucose, **fats** (lipids) are also commonly used as respiratory **substrates** – first being broken down to **fatty acids** and glycerol, page 42.

Fatty acids are 'cut up', by enzyme action of course, into 2-carbon fragments. These are then fed into the Krebs cycle via **coenzyme A**. Vertebrate muscle (including our heart muscle) is well adapted to the respiration of fatty acids in this way and they are just as likely as glucose to be the respiratory substrate. This is one reason why vigorous physical exercise can help in body weight reduction.

Proteins are also used as respiratory substrates by animals, especially carnivorous animals. Proteins are first hydrolysed to amino acids. Then, individual amino acids are deaminated. (That is, the amino group, $-NH_2$, is split off and excreted, as ammonia or urea, for example.) The residual carbon compound, an organic acid, then enters the respiratory pathway as pyruvic acid and acetyl coenzyme A, or as a Krebs cycle acid.

The relative energy values of carbohydrates, lipids and proteins as respiratory substrates

The energy value of respiratory substrates is expressed in **joules** (J) and kilojoules (kJ, 1 kJ = 1000 J). We have seen that this may be estimated experimentally, using a calorimeter (Figure 12.4). The typical amounts of energy that the major food substances yield are as follows:

carbohydrates:	1600–1760 kJ per 100 g
lipids (fats and oils):	3700–4000 kJ per 100 g
protein:	1700–1720 kJ per 100 g

Why do lipids (and proteins) transfer more energy than carbohydrates?

The greater part of the energy transferred in aerobic respiration comes from the oxidation of reduced NAD and the formation of water (during oxidative phosphorylation). So, the greater the amount of hydrogen in a particular respiratory substrate, the greater is its energy value.

Triglycerides are the form in which lipids are respired. Triglycerides contain more hydrogen in each molecule than carbohydrates. Mass for mass, fats and oils release more than twice as much energy when they are respired as carbohydrates do. In effect, fats are more reduced than carbohydrates – they therefore form a more concentrated energy store than carbohydrates.

Although the primary purpose of the protein in food is to provide the amino acids for the production and maintenance of the body's proteins, we must remember that they may also be broken down to provide energy. In other words, a secondary role of proteins, when present in excess, is as a supplementary energy source. However, proteins must first be deaminated before respiration can occur safely. And you can see from the energy value above that the carbon in proteins is slightly more reduced than that in carbohydrates.

Question

3 Explain why the amino group must be removed from amino acids and be converted to urea before the remainder of the molecule can be safely respired.

Extension

Measuring the energy of nutrients

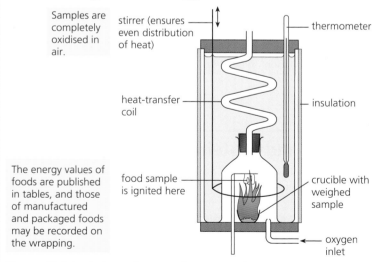

Samples are completely oxidised in air.

The energy values of foods are published in tables, and those of manufactured and packaged foods may be recorded on the wrapping.

stirrer (ensures even distribution of heat)

thermometer

heat-transfer coil

insulation

food sample is ignited here

crucible with weighed sample

oxygen inlet

Figure 12.6 A calorimeter for measuring energy values

We can estimate the total energy in a common nutrient such as glucose by means of a simple **calorimeter** (Figure 12.6). Here, a known mass of glucose is placed in the crucible in a closed environment. It is then burnt in oxygen. The energy available is released as heat and is transferred to the surrounding jacket of water. The resulting rise in temperature of the water is measured. This enables us to calculate the energy value of the glucose sample. This is because it takes 4.2 joules (J) of heat energy to raise 1 g of water by 1 °C.

One well known outcome of this technique is the energy values of manufactured foods that we may read on the packaging. In countries where food is plentiful there is often much talk about so-called 'low-calorie' items – the basis of 'slimming' diets.

Respiratory quotient

The **respiratory quotient (RQ)** is the ratio of the amount of carbon dioxide produced to the amount of oxygen taken in, in a given time, by an organism.

$$RQ = \frac{CO_2 \text{ produced}}{O_2 \text{ taken in}}$$

e.g., for a carbohydrate:

$$C_6H_{12}O_6 + 6O_2 \rightarrow 6CO_2 + 6H_2O$$

$$RQ = \frac{6}{6} = 1$$

e.g., for a lipid:

$$2C_{57}H_{110}O_6 + 163O_2 \rightarrow 114CO_2 + 110H_2O$$

$$RQ = \frac{114}{163} = 0.7$$

Calculation of the respiratory quotient is useful because it may indicate which substrates are being used. For example, when **hexose** sugars are respired aerobically:

$$C_6H_{12}O_6 + 6O_2 \rightarrow 6CO_2 + 6H_2O$$

the respiratory quotient is:

$$RQ = \frac{CO_2 \text{ produced}}{O_2 \text{ taken in}}$$

$$= 6/6 = 1$$

Whereas, when fatty acids are respired aerobically:

$$C_{18}H_{36}O_2 + 26O_2 \rightarrow 18CO_2 + 18H_2O$$

the respiratory quotient is:

$$RQ = \frac{CO_2 \text{ produced}}{O_2 \text{ taken in}}$$

$$= \frac{18}{26} = 0.7$$

The respiratory quotient due to the respiration of fat (RQ = 0.7) is significantly lower than when carbohydrates are respired (RQ = 1.0). This is because fats have a greater proportion of hydrogen relative to oxygen (they are a more highly reduced form of respiratory substrate than sugars). More oxygen is required for the respiration of fats.

The respiratory quotient when proteins and amino acids are respired is typically 0.9. Where proteins are used as the substrate, we have noted that the amino group ($-NH_2$) has first to be removed (**deamination**). In mammals the amino group is converted to urea and excreted. Plants rarely respire proteins. If they did, the ammonia gas formed would be harmful if not quickly lost from the cells of the organism.

Despite this clear difference in values for different respiratory substrates, respiratory quotient studies have only limited value. Rarely will an organism (such as germinating seeds) be respiring a single respiratory substrate. Many organisms are found to have an respiratory quotient of 0.8–0.9, but this may be due to respiration of both hexose sugars (such as glucose) and fatty acids, in varying proportions.

4 Calculate the respiratory quotient of the saturated fatty acid, palmitic acid ($C_{15}H_{31}COOH$).

5 The respiratory quotient (RQ) of two species of germinating seeds was measured at two-day intervals after germination.

Day	2	4	6	8	10
Seedling A	0.65	0.35	0.48	0.68	0.70
Seedling B	0.70	0.91	0.98	1.0	1.0

a Plot a graph to compare the changes in respiratory quotient of the two species during early germination.

b Suggest which respiratory substrates may be being respired.

12.2 Respiration

Respiration is the process whereby energy from complex organic molecules is transferred to ATP.

By the end of this section you should be able to:

a) list the four stages in aerobic respiration (glycolysis, link reaction, Krebs cycle and oxidative phosphorylation) and state where each occurs in eukaryotic cells

b) outline glycolysis as phosphorylation of glucose and the subsequent splitting of fructose 1,6-bisphosphate (6C) into two triose phosphate molecules, which are then further oxidised to pyruvate with a small yield of ATP and reduced NAD

c) explain that, when oxygen is available, pyruvate is converted into acetyl (2C) coenzyme A in the link reaction

d) outline the Krebs cycle, explaining that oxaloacetate (a 4C compound) acts as an acceptor of the 2C fragment from acetyl coenzyme A to form citrate (a 6C compound), which is reconverted to oxaloacetate in a series of small steps

e) explain that reactions in the Krebs cycle involve decarboxylation and dehydrogenation and the reduction of NAD and FAD

f) outline the process of oxidative phosphorylation including the role of oxygen as the final electron acceptor

Aerobic respiration

Cellular respiration which requires oxygen from the air is called **aerobic respiration**. Most animals and plants and very many microorganisms respire aerobically. They do so most, if not all the time.

In aerobic respiration, sugar is oxidised to carbon dioxide and water and much energy is transferred. The steps of aerobic respiration are summarised by a single equation. This equation is equivalent to a balance sheet of inputs (the raw materials) and outputs (the products). *It tells us nothing about the steps.*

$$\text{glucose} + \text{oxygen} \rightarrow \text{carbon dioxide} + \text{water} + \text{energy}$$
$$C_6H_{12}O_6 + 6O_2 \rightarrow 6CO_2 + 6H_2O + 2870\,kJ$$

So, the most significant outcome of aerobic respiration is the large quantity of energy that is transferred to ATP when glucose is oxidised. To understand how this energy is transferred from glucose, we need to look into the steps of aerobic respiration.

The stages of aerobic respiration

In cellular respiration, glucose undergoes a series of enzyme-catalysed oxidation reactions. These reactions are grouped into three major stages and a link reaction (Figure 12.7):

- **glycolysis**, in which glucose is converted to **pyruvate**
- a **link reaction**, in which pyruvate is converted to acetyl coenzyme A (acetyl CoA)
- the **Krebs cycle**, in which acetyl coenzyme A is converted to carbon dioxide
- the **electron transport chain and oxidative phosphorylation**, in which hydrogen atoms removed in the oxidation reactions of glycolysis, the link reaction and the Krebs cycle are converted to water. Most of the ATP is synthesised in this stage (oxidative phosphorylation).

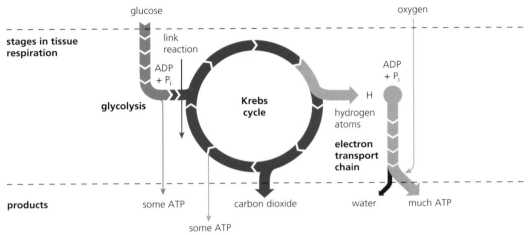

Figure 12.7 The stages of aerobic respiration

Glycolysis

Glycolysis is a linear series of reactions in which 6-carbon sugar is broken down to two molecules of the 3-carbon pyruvate ion (Figure 12.8). The enzymes of glycolysis are located in the cytoplasm outside organelles (known as the cytosol) rather than in the mitochondria. Glycolysis occurs in four steps.

- **Phosphorylation** by reactions with ATP. In this way glucose is first activated. The product is **a six-carbon sugar with two phosphate groups attached**. Note that, at this stage of glycolysis, rather than producing ATP, two molecules of ATP are consumed per molecule of glucose respired.
- **Lysis** (splitting) of 6-carbon sugar phosphate now takes place, forming two molecules of **three-carbon sugar** (called triose phosphate).
- **Oxidation** of the 3-carbon sugar molecules occurs by removal of hydrogen. The enzyme for this reaction (a dehydrogenase) works with a coenzyme, **nicotinamide adenine dinucleotide (NAD)**. NAD is a molecule that can accept hydrogen. It then becomes reduced NAD. (Later, reduced NAD will pass hydrogen ions and electrons on to other acceptor molecules. When it does, it is oxidised back to NAD).
- **ATP synthesis**. This happens twice in the reactions by which each triose phosphate molecule is converted to pyruvate. This form of ATP synthesis is known as '**at substrate level**'. It is quite different from the way the bulk of ATP synthesis occurs later in cell respiration (during the operation of the electron transport chain, see page 246).

So, at this stage, as two molecules of triose phosphate are converted to pyruvate, four molecules of ATP are synthesised. In total, there is a net gain of two molecules of ATP during glycolysis.

The link reaction and the Krebs cycle

The subsequent stages of aerobic respiration occur in the organelles known as **mitochondria**.

Remind yourself of the size and structure of the mitochondrion now (Figure 1.17, page 18). Notice the tiny gap between the inner and outer membranes.

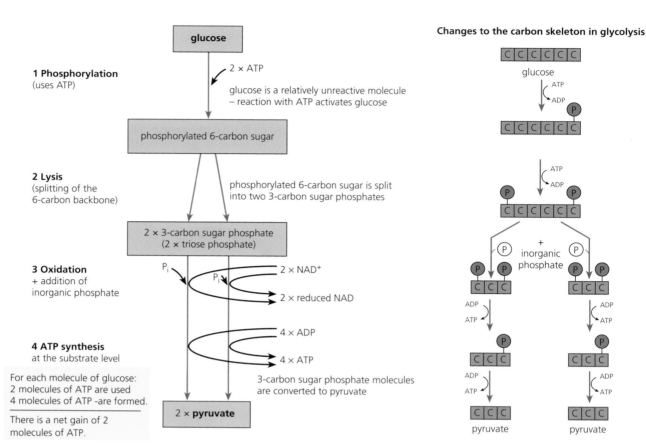

1 Phosphorylation
(uses ATP)

glucose is a relatively unreactive molecule
– reaction with ATP activates glucose

2 Lysis
(splitting of the
6-carbon backbone)

phosphorylated 6-carbon sugar is split
into two 3-carbon sugar phosphates

3 Oxidation
+ addition of
inorganic phosphate

4 ATP synthesis
at the substrate level

For each molecule of glucose:
2 molecules of ATP are used
4 molecules of ATP -are formed.

There is a net gain of 2
molecules of ATP.

3-carbon sugar phosphate molecules
are converted to pyruvate

Figure 12.8 Glycolysis: a summary

Question

6 State which of the
following are produced
during glycolysis.
a CO_2
b lactate
c reduced NAD
d glycogen
e ATP
f NAD
g pyruvate
h glucose

As it is formed during the **link reaction**, pyruvate diffuses from the cytosol into the **matrix** of
a mitochondrion. There it is metabolised. Firstly, carbon dioxide is removed from the 3-carbon
pyruvate, a reaction called **decarboxylation**. At the same time, oxidation occurs by removal of
hydrogen, a **dehydrogenation** reaction. NAD is converted to reduced NAD during this reaction.

The product of the oxidative decarboxylation reaction is an acetyl group (meaning a 2-carbon
fragment). This acetyl group is then combined with a coenzyme called **coenzyme A**, forming
acetyl coenzyme A.

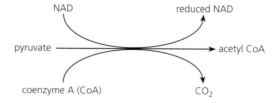

The production of acetyl coenzyme A from pyruvate is known as the **link reaction** because it
connects glycolysis to the reactions of the **Krebs cycle**, which now follow.

In the **Krebs cycle** (Figure 12.9), the acetyl coenzyme A enters by reacting with a **4-carbon
organic acid** (oxaloacetate). The products of this reaction are a **6-carbon acid** (citrate) and, of
course, coenzyme A. This latter, on release, is reused in the link reaction.

The Krebs Cycle is named after Hans Krebs who led the team that discovered it. It is also
sometimes referred to as the citric acid cycle, after the first intermediate acid formed.

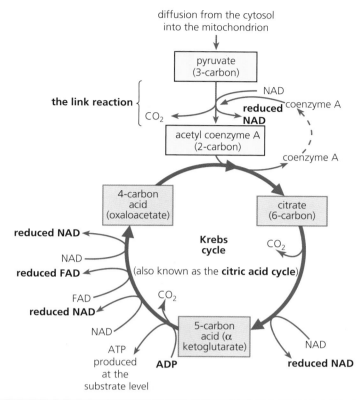

diffusion from the cytosol
into the mitochondrion

Figure 12.9 The link reaction and Krebs cycle: a summary

Next, the citrate is converted back to the 4-carbon acid (an acceptor molecule, in effect) by a series of reactions. This is why Krebs is called a 'cycle'. These are the changes.

- Two molecules of carbon dioxide are given off, in separate decarboxylation reactions.
- A molecule of ATP is formed, as part of one of the reactions of the cycle. As is the case with glycolysis, this ATP synthesis is at substrate level.
- Three molecules of reduced NAD are formed in dehydrogenation reactions.
- One molecule of another hydrogen acceptor, FAD (flavin adenine dinucleotide), is reduced. (NAD is the chief hydrogen-carrying coenzyme of respiration but FAD is another coenzyme with this role in the Krebs cycle.)

Finally, remember that glucose was converted to two molecules of pyruvate in the glycolysis phase. This means that the whole Krebs cycle sequence of reactions 'turns' twice for every molecule of glucose that is metabolised by aerobic respiration.

We are now in a position to summarise the changes to the molecule of glucose that occurred in the reactions of glycolysis, the link reaction and the Krebs cycle. The products formed during glycolysis and two turns of the Krebs cycle are shown in Table 12.1.

Question

7 Outline the types of reaction catalysed by
a dehydrogenases
b decarboxylases.

Table 12.1 Net products of aerobic respiration per molecule of glucose at the end of the Krebs cycle

Stage	Products			
	CO_2	ATP	Reduced NAD	Reduced FAD
glycolysis	0	2	2	0
link reaction	2	0	2	0
Krebs cycle	4	2	6	2
Totals	6	4	10	2

The electron transport chain and oxidative phosphorylation

We have seen that the removal of pairs of hydrogen atoms from various intermediates of the respiratory pathway is a feature of several of the steps in glycolysis, the link reaction and the Krebs cycle. *Can you locate these points in Figures 12.8 and 12.9 now?*

You can see that, at most of these events, NAD is converted to reduced NAD. However, at one point in the Krebs cycle, an alternative hydrogen-acceptor coenzyme (known as FAD) is reduced.

Now, in the final stage of aerobic respiration, the hydrogen electrons are transported along a **series of carriers**, from the reduced NAD (or FAD) produced in the earlier stages. The terminal acceptor is oxygen. Now, finally, hydrogen has been oxidised and oxygen reduced to water (Figure 12.10).

The 'carriers' (they are conjugated proteins – page 50) are located in the inner membrane of the mitochondria. These carriers occur in highly ordered sequences, each consisting of hydrogen acceptors and electron carriers next to a final enzyme, **ATP synthase** (known for short as ATPase). The location and physical structure of these systems are illustrated in Figure 12.11.

Theoretically at least, the total yield from aerobic respiration is about 38 molecules of ATP per molecule of glucose respired. In practice, it is less than that, probably about 32 molecules of ATP.

Question

8 Distinguish between the following pairs.
   a substrate and intermediate
   b glycolysis and the Krebs cycle
   c oxidation and reduction

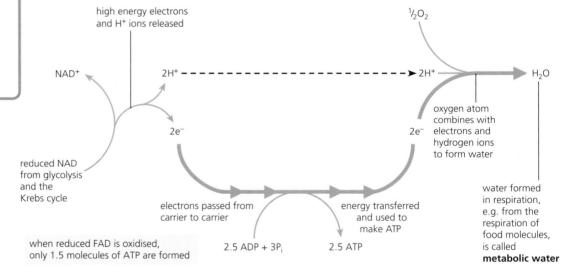

Figure 12.10 The electron transport chain and oxidative phosphorylation: a summary

12.2 *continued...* The process of ATP synthesis

This process of ATP synthesis using the energy in proton gradients is common to both respiration and photosynthesis.

By the end of this section you should be able to:

g) explain that during oxidative phosphorylation:
- energetic electrons release energy as they pass through the electron transport system
- the released energy is used to transfer protons across the inner mitochondrial membrane
- protons return to the mitochondrial matrix by facilitated diffusion through ATP synthase providing energy for ATP synthesis

h) carry out investigations to determine the effect of factors such as temperature and substrate concentration on the rate of respiration of yeast using a redox indicator (e.g. DCPIP or methylene blue)

i) describe the relationship between structure and function of the mitochondrion using diagrams and electron micrographs

ATP synthesis – the electron transport chain at work

When a small sample of hydrogen gas is burnt in air there is a small explosion. If you have seen this experiment demonstrated behind a safety screen, you will remember it! Alternatively, you might know that a car can be adapted to run on hydrogen gas as the fuel. These points illustrate that there is a lot of energy to be transferred when hydrogen is oxidised to water.

Of course, in cells, hydrogen is oxidised to water in many small steps. So, in the mitochondrion, water is formed from reduced NAD and oxygen in a highly controlled way. The energy of oxidation is conserved by the respiratory chain and much is transferred to ADP and P_i to form ATP. The energy can then be used by the cell.

How energy transfer and ATP synthesis are brought about

Firstly, we need to recognise that the inner membrane of the mitochondria (in which the electron carriers are located) is a barrier to the movement of ions and electrons. Hydrogen ions (protons) can get across but they have to be pumped, using energy. And once they have been moved from the inner matrix of the mitochondrion to the tiny space between the inner and outer membranes they are trapped there – temporarily, as it turns out.

It is the carrier proteins (a sequence of hydrogen carriers and electron carriers) that use the energy of oxidation at several steps along the way, to pump the protons into the inter-membrane space. Here, if they accumulate, they cause the pH to drop.

It is because the inner mitochondrial membrane is largely impermeable to ions that a significant **gradient in hydrogen ion concentration** builds up across the inner membrane, generating a potential difference across the membrane. This represents a store of potential energy.

Then the protons are allowed to flow back into the matrix at particular points. These points are ATP synthase molecules, each of which has a tiny channel. Note that this enzyme occurs in the inner mitochondrial membrane, right besides each electron transport chain. As the protons flow down their concentration gradient through the channel in the enzyme, the energy is transferred to the synthesis of ATP from ADP and Pi. The reaction is catalysed by the ATP synthase (Figure 12.11).

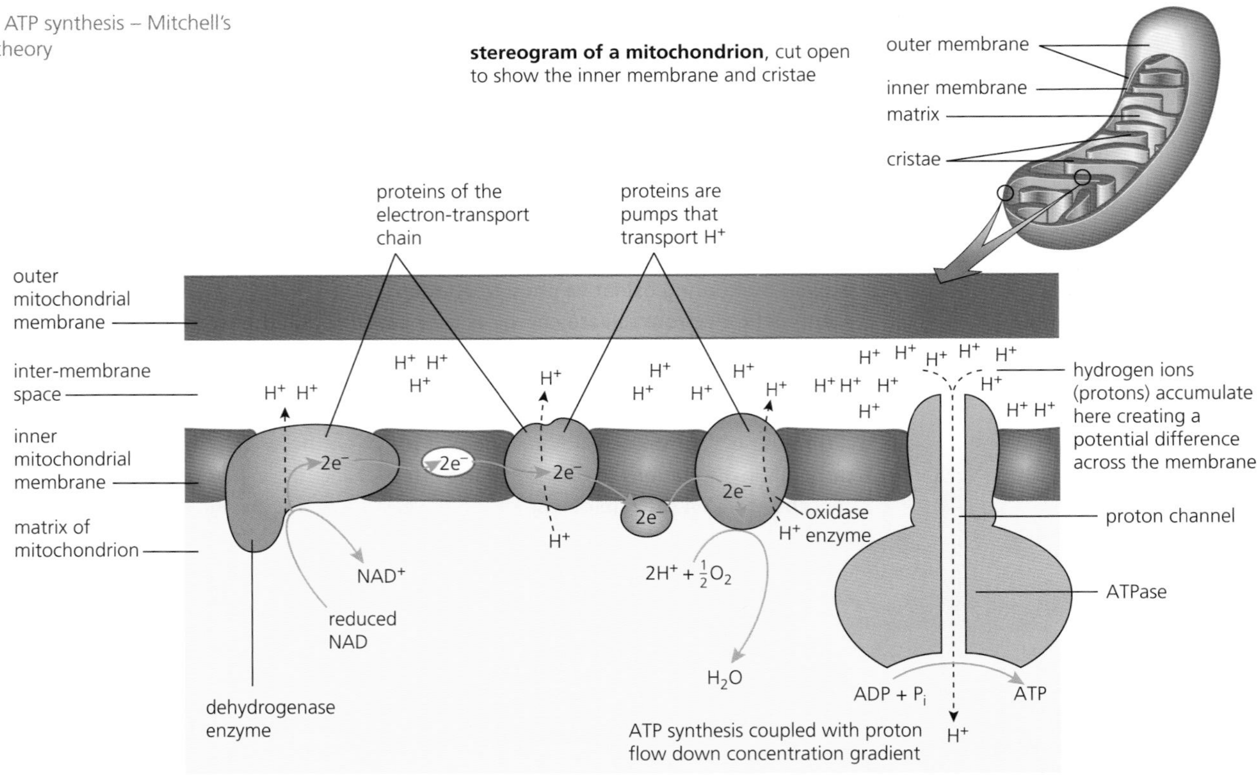

Figure 12.11 ATP synthesis – Mitchell's chemiosmotic theory

stereogram of a mitochondrion, cut open to show the inner membrane and cristae

This process by which ATP is generated has become known as **chemiosmosis**. It was a biochemist, Peter Mitchell, who suggested the chemiosmotic theory. This was in 1961. His hypothesis was not generally accepted for some years but, about two decades later, he was awarded a Nobel Prize for his discovery. Today there is a great deal of evidence for the occurrence of chemiosmosis. We can describe it as a correct hypothesis. So the word 'theory' in its title is used in its broadest sense – as an accepted concept.

How ATP supports useful work in cells and organisms

We have noted that organisms need energy for many reasons, including:
- driving anabolic reactions
- the active transport of molecules and ions across membranes
- movement
- the maintenance of body temperature, where this occurs.

ATP is the chief source of free energy sustaining the metabolism and these activities of cells and organisms (although it is not the form in which energy is stored long term).

How does the energy of ATP bring about such useful work in cells and organisms?

We have seen that ATP is formed from adenosine diphosphate (ADP) and a phosphate ion (P_i) by transfer of energy during respiration (Figure 12.3). Then, in the presence of enzymes, ATP is involved in energy-requiring reactions. The free energy made available when ATP is formed from ADP and a phosphate ion is approximately $30–34\,kJ\,mol^{-1}$. Some of this energy is lost as heat in the reaction, but much is made available – more than sufficient to drive a typical energy-requiring reaction of metabolism. ATP typically reacts with other **metabolites** and forms phosphorylated intermediates, making them more reactive in the process. The phosphate groups are released later, so both ADP and phosphate ions become available for reuse as metabolism continues.

Consequently, there is a huge turnover of ATP molecules in the cell cytoplasm including in many organelles. ATP is used up very quickly after it has been formed from respiratory substrates. Even whilst in a mainly resting state, the human body produces and consumes about $40\,kg$ of ATP in 24 hours. If we switch to very strenuous activity about $0.5\,kg$ of ATP is required every minute!

Question

9 State the most common long-term energy store
  a in human cells
  b in plant cells.

Case studies of ATP 'at work'

1 An example of an anabolic reaction of metabolism 'driven' by ATP is the manufacture of proteins in the ribosomes. Each amino acid used to build up the primary structure of the protein is first attached to a specific molecule of transfer RNA. ATP is required for this 'activation' reaction (Figure 6.12, page 121).

2 ATP also reacts directly with protein pumps in membranes, so providing the energy needed to change the shape or activate the protein and so move ions and other metabolites across cell membranes which would otherwise block their movement. The movement of a single substance is illustrated in Figure 4.21 (page 91). The sodium–potassium pump is shown in Figure 4.22.

3 Sometimes ATP reacts with water (a hydrolysis reaction) and is converted to ADP and a phosphate ion. For example, direct hydrolysis of the terminal phosphate groups like this happen in **muscle contraction**.

Muscles that cause locomotion, skeletal muscle, consist of bundles of muscle fibres. These fibres are long multinucleate cells that are able to shorten to half or even a third of the relaxed or 'resting' length. The whole muscle fibre has a 'stripy' appearance.

This is due to an interlocking arrangement of two types of protein filaments, made of the proteins **myosin** and **actin**.

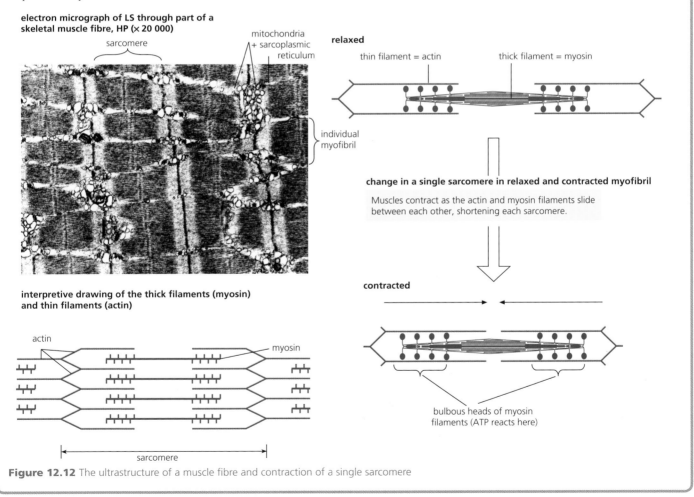

electron micrograph of LS through part of a skeletal muscle fibre, HP (× 20 000)

sarcomere

mitochondria + sarcoplasmic reticulum

individual myofibril

interpretive drawing of the thick filaments (myosin) and thin filaments (actin)

actin

myosin

sarcomere

relaxed

thin filament = actin

thick filament = myosin

change in a single sarcomere in relaxed and contracted myofibril

Muscles contract as the actin and myosin filaments slide between each other, shortening each sarcomere.

contracted

bulbous heads of myosin filaments (ATP reacts here)

Figure 12.12 The ultrastructure of a muscle fibre and contraction of a single sarcomere

(Continued)

Look at Figure 12.12 now so that you can understand where ATP operates to bring about muscle contraction.

Skeletal muscle contracts when the actin and myosin filaments slide past each other in a series of steps, sometimes described as a 'ratchet mechanism'. A great deal of ATP is used in the contraction process. This cycle is repeated many times per second, with thousands of myosin heads working along each myofibril. ATP is used up rapidly and the muscle may shorten by about 50 per cent of its relaxed length.

4   Another significant application of the energy transferred in respiration concerns body temperature. The major source of heat in mammals (and in birds – both groups of vertebrate which regulate their body temperature) is as a waste product of the biochemical reactions of metabolism. Heat is then distributed by the blood circulation.

   The organs of the body vary greatly in the amount of heat they yield. For example, the liver is extremely active metabolically, but most of its metabolic reactions require an input of energy (they are endergonic reactions) and little energy is lost as heat. Mostly, the liver is thermally neutral.

   When at rest, the bulk of our body heat comes from other abdominal organs, mainly from the heart and kidneys, but also from the lungs and brain (which, like a computer central processing unit, needs to be kept cool). Under resting conditions, the skeleton, muscles and skin, which make up over 90 per cent of the body mass, produce less than 30 per cent of the body heat (Figure 12.13). Of course, in times of intense physical activity, the skeletal muscles generate a great deal of heat as a waste product of respiration and contraction.

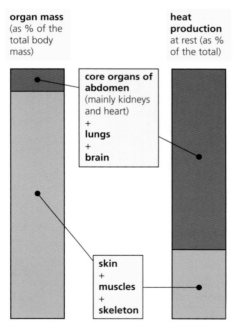

Figure 12.13 Heat production in the body at rest

Question

10 What does it mean when we say most reactions in liver cells are endergonic.

The site of cell respiration – structure in relation to function

The location of the stages of cellular respiration are summarised in Figure 12.14. You can see that, once pyruvate has been formed from glucose in the cytosol, the remainder of the pathway of aerobic respiration is located in the mitochondria. The relationship between structure and function in this organelle is summarised in Table 12.2.

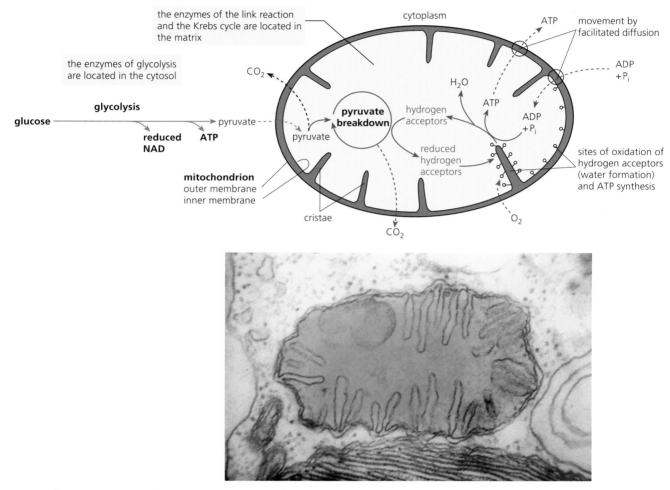

Figure 12.14 The location of the stages of cellular respiration

Table 12.2 Mitochondrial structure in relation to function

Structure	Function
external double membrane	permeable to pyruvate, CO_2, O_2, NAD and reduced NAD
matrix	site of the enzymes of the link reaction and the Krebs cycle
inner membrane	• surface area greatly increased by intucking to form **cristae** – since the electron transport chain and ATP synthase enzymes are housed here, opportunities for ATP synthesis are enhanced • impermeable to hydrogen ions (protons) – allowing the formation of a potential difference between the inter-membrane space and the matrix
inter-membrane space	relatively tiny space – allowing the accumulation of hydrogen ions (protons) there to generate a large potential difference with the matrix, making phosphorylation possible.

Question

11 State which stages in cellular respiration occur whether or not oxygen is available to cells.

Respiration as a source of intermediates

We have seen that the main role of respiration is to provide a pool of ATP, and that this is used to drive the endergonic reactions of synthesis – among other things. But the compounds of the glycolysis pathway and the Krebs cycle (**respiratory intermediates**) may also serve as the starting points of synthesis of other metabolites needed in the cells and tissues of the organism. These include polysaccharides like starch and cellulose; glycerol and fatty acids; and many others.

Investigating factors that affect the rate of respiration in yeast

The effect of temperature on the rate of respiration of a suspension of yeast cells can be investigated using a solution of methylene blue. Methylene blue is a **redox indicator** – in solution it is blue when in an oxidised state but turns colourless if exposed to a reducing agent.

Identical yeast suspensions in test tubes are required, as shown in Figure 12.15. The layer of oil prevents oxygen from the air dissolving in the yeast suspension. In these conditions, yeast respires anaerobically. The oil also prevents colourless methylene blue solution becoming re-oxidised and turning blue again. Using simple water baths at a range of appropriate temperatures, the time taken for tubes at each temperature to turn colourless is measured and recorded. Typical results of investigations of temperature and respiration rate are shown in Figure 12.21.

Can you explain the events and changes in the tubes as methylene blue becomes colourless?

Question

12 Detail what changes are required to this experiment in order to investigate the effect of substrate concentration on the rate of respiration.

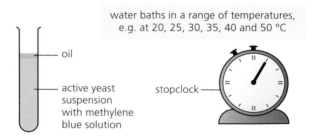

water baths in a range of temperatures, e.g. at 20, 25, 30, 35, 40 and 50 °C

oil

active yeast suspension with methylene blue solution

stopclock

Figure 12.15 An investigation of the rate of respiration in yeast

The variables in this investigation are shown in Table 12.3.

Table 12.3

controlled variables	the amount and concentration of the yeast suspension the volume of oil and methylene blue solution
independent variable	the range of temperatures
dependent variable	the time taken for decolourisation of the methylene blue

12.2 *continued...* Anaerobic respiration

Some organisms and some tissues are able to respire in both aerobic and anaerobic conditions. When yeast and plants respire under anaerobic conditions, they produce ethanol and carbon dioxide as end-products; mammalian muscle tissue produces lactate when oxygen is in short supply.

By the end of this section you should be able to:

j) distinguish between respiration in aerobic and anaerobic conditions in mammalian tissue and in yeast cells, contrasting the relative energy released by each

k) explain the production of a small yield of ATP from respiration in anaerobic conditions in yeast and in mammalian muscle tissue, including the concept of oxygen debt

l) explain how rice is adapted to grow with its roots submerged in water in terms of tolerance to ethanol from respiration in anaerobic conditions and the presence of aerenchyma

m) carry out investigations, using simple respirometers, to measure the effect of temperature on the respiration rate of germinating seeds or small invertebrates

Anaerobic respiration

Many organisms are able to respire in the absence of oxygen, at least for a short time. They do so by a process called fermentation or anaerobic respiration. Some tissues in an organism may respire anaerobically, too, when deprived of sufficient oxygen (see below).

In the absence of oxygen, flowering plants may respire by **alcoholic fermentation**, for example, the cells of roots in waterlogged soil. Many species of yeast (*Saccharomyces*) respire anaerobically (some do so even in the presence of oxygen). The products are ethanol and carbon dioxide. **Alcoholic fermentation** of yeast has been exploited by humans in wine and beer production for many thousands of years.

$$\text{glucose} \rightarrow \quad \text{ethanol} \quad + \text{carbon dioxide} + \text{energy}$$
$$C_6H_{12}O_6 \rightarrow 2CH_3CH_2OH \ + \ 2CO_2 \qquad + \text{energy}$$

Incidentally, the **respiratory quotient for anaerobic respiration** by alcoholic fermentation cannot be calculated since no oxygen is taken in. However, if some respiration in a tissue is aerobic then a small amount of oxygen will be absorbed. The respiratory quotient will be high, but less than 2. Generally, tissues with high respiratory quotient values (RQs well above 1) indicate that some anaerobic respiration is occurring.

Vertebrate muscle is an example of a tissue that can respire anaerobically, but only when the demand for energy for contractions is very great and cannot be fully met by aerobic respiration. It is **lactic acid fermentation** that occurs here. The sole waste product is lactate.

$$\text{glucose} \rightarrow \quad \text{lactic acid} \qquad + \text{energy}$$
$$C_6H_{12}O_6 \rightarrow 2CH_3CHOHCOOH \ + \text{energy}$$

Once again, these equations are **balance sheets** of the inputs (raw materials) and the output (products) of cellular respiration. They tell us nothing about the separate steps involved, each catalysed by a specific enzyme. They do not mention pyruvate, for example.

The pathways of anaerobic respiration

The respiratory pathways of alcoholic and lactic acid fermentation are shown in Figure 12.16. When oxygen is not available, glycolysis continues and pyruvate accumulates, at least initially. However, the Krebs cycle and oxidative phosphorylation cannot take place because the aerobic oxidation of reduced NAD is now blocked. In a tissue in which reserves of oxidised NAD have run out, glycolysis would cease.

How is the supply of pyruvate maintained in cells in the absence of oxygen?

In fermentation, reduced NAD is oxidised in other reactions. In effect, oxygen is replaced as the hydrogen acceptor. In alcoholic fermentation, ethanal is the hydrogen acceptor, and in lactic acid fermentation, pyruvate is the hydrogen acceptor. You can see this in Figure 12.16. Now, glycolysis can continue in the absence of oxygen. However, the total energy yield in terms of ATP generated in both alcoholic and lactic acid fermentation is limited to the net two molecules of ATP generated in glycolysis per molecule of glucose respired.

Figure 12.16 The respiratory pathways of anaerobic respiration

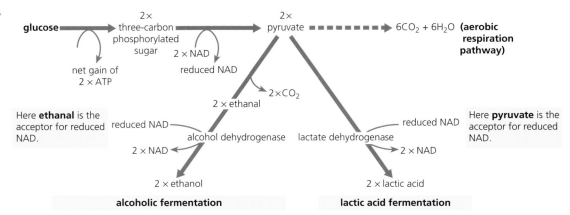

Anaerobic respiration is less efficient than aerobic respiration

Anaerobic respiration is wasteful of respiratory substrate. The useful product of fermentation is a tiny quantity of ATP. This is minimal when compared with the yield of ATP from aerobic respiration of the same quantity of respiratory substrate. Also, the waste product, ethanol or lactate, contains much unused chemical energy. In fact, both lactate and ethanol are energy-rich molecules. The energy in these molecules may be used later, however. For example, in humans, lactate is transported to the liver and later metabolised aerobically.

   Energy yields of cellular respiration are compared in Table 12.4. You do not need to memorise the details here, but give careful attention to two points:

● the limited yield in terms of molecules of ATP generated per molecule of glucose respired by anaerobic compared to the yield from aerobic respiration
● the limited yield from **substrate level phosphorylation** compared to the yield from **oxidative phosphorylation**.

Lactate and ethanol are harmful to organisms (in different ways) if they accumulate in quantity. However, organisms may be able to tap the energy locked up in the waste products of fermentation by converting them back to sugar, which can then be respired. For example, as a result of prolonged, very strenuous exercise, lactate builds up in the muscles. Eventually this lactate is carried in the blood stream to the liver. Here the lactate is converted back to glucose in the liver. This process requires energy from respiration, too, and therefore requires extra oxygen. The oxygen needed to remove the lactate is called the 'oxygen debt'.

Extension

A few organisms respire only in the absence of oxygen. One is the tetanus bacillus *Clostridium tetani* which causes the disease lockjaw. This bacterium is active in oxygen-free environments rich in nutrients. These conditions occur in deep body wounds. Otherwise the bacterium merely survives as resistant spores in the soil or among the faeces of the lower gut of many animals. During growth (but not as a dormant 'spore'), *C. tetani* produces toxins which, when released inside wounds, are carried around the body in the blood circulation, causes agonising muscular spasms. Today, tetanus can be prevented by vaccination, supported by periodic booster injections.

Table 12.4 Energy yield of aerobic and anaerobic cellular respiration compared

	Yield from each molecule of glucose respired			
	Aerobic respiration		**Anaerobic respiration**	
Stage	**Reduced NAD (or FAD)**	**ATP**	**Reduced NAD**	**ATP**
Glycolysis		2 (net production at the substrate level)	2 reduced NAD (re-oxidised during fermentation)	2 (net production at the substrate level)
	2 reduced NAD	$2 \times 2.5 = 5$ (produced by oxidative phosphorylation)		
Link reaction	2 reduced NAD	$2 \times 2.5 = 5$ (produced by oxidative phosphorylation)		
Krebs cycle		2 (net production at the substrate level)		
	6 reduced NAD	$6 \times 2.5 = 15$ (produced by oxidative phosphorylation)		
	2 reduced FAD	$2 \times 1.5 = 3$ (produced by oxidative phosphorylation)		
Totals		32		2

Many species of yeast cannot metabolise ethanol; the energy locked up in ethanol remains unavailable to the cell. This makes these species industrially useful organisms in the production of ethanol. However, other yeast species can metabolise ethanol when other respiratory substrates are used up.

Rice – a plant of waterlogged soils

Rice (*Oryza sativa*) is a semi-aquatic annual grass plant that flourishes in poorly draining, silty soils. The plant was domesticated in south-west Asia about 12 000 years ago, which is much earlier than the first records of cultivated wheat (in the Fertile Crescent). If so, rice has certainly fed more people over a longer period than any other crop. Today, rice is the staple for more than 60 per cent of the world's population.

Rice is adapted to grow with its roots submerged in water (Figure 12.17). Here the environment for root growth is deprived of oxygen. In fact, the mud here is in a permanent anaerobic state because of the respiration of the saprotrophic bacteria present. These bacteria live off the dead organic matter present and tend to consume all the available oxygen in the process. At the same time, plant root growth requires oxygen for aerobic respiration. All plants absorb ions from the soil solution in their roots by active uptake (page 90). Here, too, the plant manufactures essential amino acids, which it then transports all over the plant. So the roots are highly metabolically active and require oxygen for aerobic respiration and ATP production. Meanwhile, in waterlogged soils oxygen is virtually absent.

How is rice adapted to grow in these conditions?

Firstly, the ground tissue cells (parenchyma) that make up the bulk of the stems and roots of all plants show a special adaptation in the rice plant. Here, the cells have huge, interconnecting air spaces between them, particularly of the roots. This tissue is known as *aerenchyma tissue* – you can see it in Figure 12.17. These spaces between cells aid the diffusion of oxygen to the roots, growing deeply in the waterlogged soil, typically 5–10 cm below the water surface. In fact, the aerenchyma tissue of the rice stem and root makes possible normal growth and metabolism of the root system, in anaerobic conditions.

Secondly, the roots can and do respire anaerobically by alcoholic fermentation, when necessary. The ethanol produced in the cells may be metabolised by enzymes present, but the root cells of the rice plant have a higher tolerance than other cereals for ethanol.

paddy field culture of rice – planting out of rice seedlings

Question

15 List the features of the rice plant that enable it to survive swamp conditions. Annotate your list, so that the value of the adaptation is made clear.

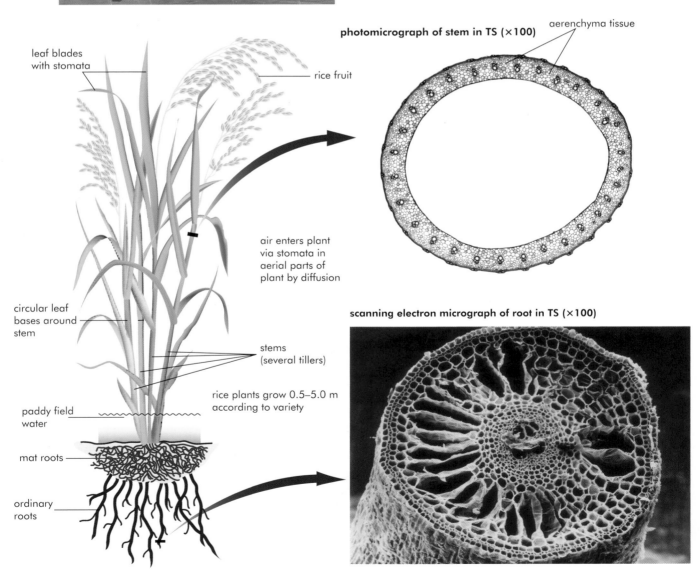

Figure 12.17 The structure of mature rice plants, as vegetative growth gives way to fruit formation

A source of combined nitrogen for rice plants of the paddy field

Azolla is a tiny, aquatic fern that grows profusely in flooded paddy fields, for example, in China and Vietnam. Living in cavities of the fern leaf is a species of a photosynthetic bacterium, *Anabaena* (known as *Anabaena azollae*). This cyanobacterium is a nitrogen-fixing organism. The ammonium ions it forms in excess are shared with the fern cells. When the fern dies and decays, the soil of the paddy field is fertilised with additional combined nitrogen which also supports the growth of the rice plants and makes possible productive, intensive cultivation of rice. Human agriculture has benefited from this relationship for centuries, long before plant nutrition or the importance of combined nitrogen was understood.

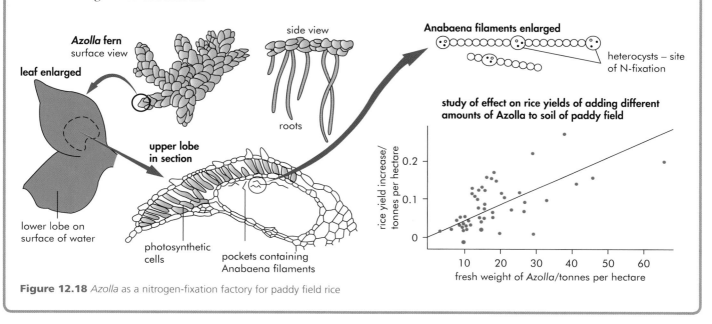

Figure 12.18 *Azolla* as a nitrogen-fixation factory for paddy field rice

Practical: Investigating the effect of temperature on respiration rate of germinating seeds or small invertebrates

The rate of respiration of an organism tells us about its demand for energy. The more active the organism, the higher its rate of respiration. By 'respiration rate' we mean the uptake of oxygen per unit time. This can be measured by means of a respirometer. A respirometer is a form of manometer because it detects change in pressure or volume of a gas. So, respiration rates are investigated by manometry.

A simple respirometer is shown in Figure 12.19. Respiration by an organism trapped in the chamber changes the composition of the gas there, once the screw clip has been closed. Soda lime, enclosed in the chamber, removes carbon dioxide gas as it is released by the respiring organism. Consequently, only oxygen uptake should cause a change in volume. The drop of coloured liquid in the attached capillary tube will move in response. The change in the volume of the apparatus, due to oxygen uptake, can be estimated from measurements of the movement in the manometric fluid during the experiment.

Can you think of problems that might arise in using this apparatus?

Figure 12.19 A simple respirometer

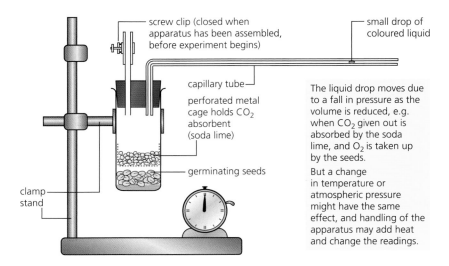

screw clip (closed when apparatus has been assembled, before experiment begins)

small drop of coloured liquid

capillary tube

perforated metal cage holds CO_2 absorbent (soda lime)

germinating seeds

clamp stand

The liquid drop moves due to a fall in pressure as the volume is reduced, e.g. when CO_2 given out is absorbed by the soda lime, and O_2 is taken up by the seeds.

But a change in temperature or atmospheric pressure might have the same effect, and handling of the apparatus may add heat and change the readings.

With a **differential respirometer** (Figure 12.20), the sources of error arising from the simple manometer are eliminated by having a control chamber connected by a U-tube manometer to the respirometer chamber. External temperature or pressure changes act equally on both sides of the manometer and cancel out. In this apparatus, the volume of gas absorbed per unit time is given by readings on the syringe.

Question

16 In the respirometer shown in Figure 12.20, explain how changes in temperature or pressure in the external environment are prevented from interfering with the measurement of oxygen uptake by the respiring organisms in the apparatus.

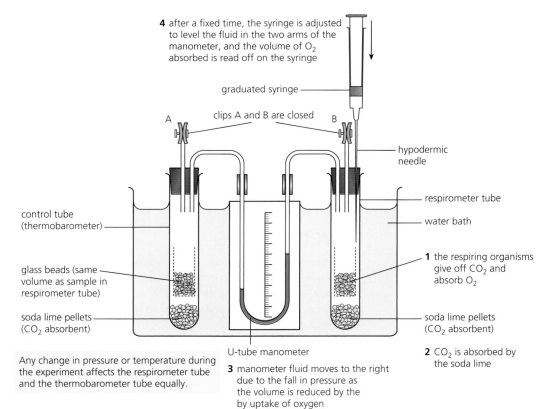

4 after a fixed time, the syringe is adjusted to level the fluid in the two arms of the manometer, and the volume of O_2 absorbed is read off on the syringe

graduated syringe

clips A and B are closed

A   B

hypodermic needle

respirometer tube

water bath

control tube (thermobarometer)

glass beads (same volume as sample in respirometer tube)

soda lime pellets (CO_2 absorbent)

1 the respiring organisms give off CO_2 and absorb O_2

soda lime pellets (CO_2 absorbent)

2 CO_2 is absorbed by the soda lime

U-tube manometer

Any change in pressure or temperature during the experiment affects the respirometer tube and the thermobarometer tube equally.

3 manometer fluid moves to the right due to the fall in pressure as the volume is reduced by the by uptake of oxygen

Figure 12.20 A differential respirometer

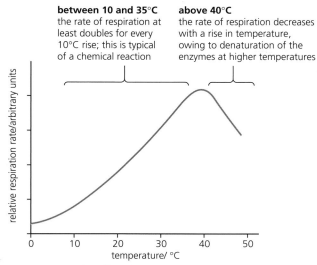

between 10 and 35°C
the rate of respiration at least doubles for every 10°C rise; this is typical of a chemical reaction

above 40°C
the rate of respiration decreases with a rise in temperature, owing to denaturation of the enzymes at higher temperatures

Figure 12.21 The effect of temperature on the rate of respiration

The differential respirometer is used with a water bath. By repeating readings at a range of different temperatures, the effect of temperature on respiration can be measured. Data such as that in Figure 12.21 can then be obtained with this equipment.

Why does temperature affect the rate of respiration?

The enzymes of respiration are proteins. All proteins are denatured by higher temperatures, although the temperature at which proteins are denatured varies widely. For example, some bacteria live successfully in hot springs where the water temperature may approach that of boiling water.

From the graph in Figure 12.21 we can work out the temperature at which significant denaturing of respiratory enzymes in the organisms used in the experiment takes place. Below this, a rise in temperature causes the rate of respiration to increase. It is a feature of chemical reactions that their rate at least doubles for every 10°C rise in temperature.

Summary

- **Respiration** is a cellular process in which energy is transferred from nutrients, such as glucose, to the cellular machinery. **Energy** is required to do useful work, such as the transport of metabolites across membranes, the driving of anabolic reactions and to cause movements in organisms.

- **Aerobic respiration** involves the complete oxidation of glucose to carbon dioxide and water with the release of a large amount of energy. In addition to hexose sugar, fats and proteins may be used as respiratory substrates. Anaerobic respiration involves the partial oxidation of glucose with the release of only a small amount of energy.

- ATP is the **universal energy currency** molecule by which energy is transferred to do useful work. ATP is a soluble molecule, formed in the mitochondria but able to move into the cytosol by facilitated diffusion. It diffuses freely about cells.

- The stages of aerobic respiration are:

  - **glycolysis,** in which glucose is converted to pyruvate

  - the **link reaction**, in which pyruvate is converted to acetyl coenzyme A

  - the **Krebs cycle**, in which acetyl coenzyme A is metabolised, carbon dioxide is given off and NAD (or FAD) is reduced

  - the **electron transport chain** and **oxidative phosphorylation**, in which reduced NAD (and reduced FAD) are oxidised and water is formed. Most of the ATP is produced during this stage.

- Glycolysis occurs in the **cytoplasm**, but the Krebs cycle is located in the matrix of **mitochondria** and oxidative phosphorylation occurs on the inner mitochondrial membrane including the **cristae**.

- In **anaerobic respiration**, the products are either lactate (in lactic acid fermentation, typically found in vertebrate muscle) or ethanol and carbon dioxide (in alcoholic fermentation, found in yeast and in plants under anaerobic conditions).

- Anaerobic respiration is **wasteful of respiratory substrate**. It yields only a tiny quantity of ATP, when compared with the yield of ATP from aerobic respiration of the same quantity of respiratory substrate. The waste products, ethanol or lactate, contain much unused chemical energy. They are both energy-rich molecules.

- The **rate of aerobic respiration** can be measured manometrically, in a respirometer. The **respiratory quotient** (RQ) is the ratio of the volume of carbon dioxide produced to the volume of oxygen used, in a given time. The respiratory quotient is an indicator of the respiratory substrate.

- Plants show adaptations to different environments. The adaptations to swamp conditions by rice include extensive **aerenchyma tissue**, with continuous, interconnecting air spaces throughout stems, leaves and roots, and physiological tolerance of ethanol by root cells.

Examination style questions

1 Fig. 1.1 shows the structure of ATP.

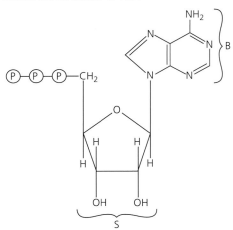

Fig. 1.1

a) i) Name the nitrogenous base labelled **B**. [1]
   ii) Name the sugar labelled **S**. [1]

b) ATP is described as having a universal role as the energy currency in all living organisms. Explain why it is described in this way. [4]

c) State **precisely** two places where ATP is synthesised in cells. [2]

[Total: 8]

(Cambridge International AS and A Level Biology 9700, Paper 04 Q6 November 2008)

2 a) For the stages of aerobic respiration listed below, identify the site of the process and the chief products of each step.

Component of process of aerobic respiration	Location of the process in the eukaryotic cell	Chief products
glycolysis		
link reaction		
Krebs cycle		
oxidative phosphorylation		

[12]

b) What do you understand by:

i) decarboxylation [2]
ii) dehydrogenation. [2]

[Total: 16]

3 a) Describe the process of oxidative phosphorylation in the mitochondrion. [9]

b) Explain the roles of NAD in anaerobic respiration in **both** plants and animals. [6]

[Total: 15]

(Cambridge International AS and A Level Biology 9700, Paper 04 Q9 June 2008)

4 Outline and explain:

a) the difference between substrate level phosphorylation and oxidative phosphorylation,

b) why aerobic respiration is described as more efficient than anaerobic respiration,

c) how reduced NAD is reoxidised in anaerobic respiration.

5 During the process of glycolysis, glucose is converted by a series of steps into two molecules of pyruvate.
The diagram below outlines glycolysis.

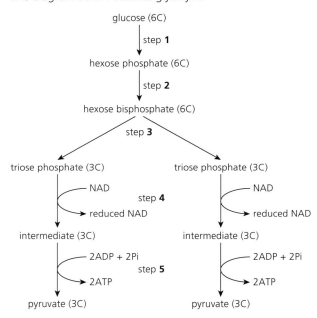

a) With reference to the diagram above, state the process occurring at:

i) steps 1 and 2 [1]
ii) step 3 [1]
iii) step 4 [1]

b) Explain why glucose needs to be converted to hexose bisphosphate. [2]

c) Pyruvate can enter a mitochondrion when oxygen is present.
Describe what happens to pyruvate in a yeast cell when oxygen is **not** present. [4]

[Total: 9]

(Cambridge International AS and A Level Biology 9700, Paper 41 Q6 November 2011)

13 Photosynthesis

Photosynthesis is the energy transfer process that is the basis of much of life on Earth. It provides the basis of most food chains providing energy directly or indirectly for all other organisms. In eukaryotes, the process occurs within chloroplasts.

Students use their knowledge of plant cells and leaf structure from the topic on Cell structure while studying photosynthesis. Various environmental factors influence the rate at which photosynthesis occurs.

13.1 Photosynthesis as an energy transfer process

Light energy absorbed by chloroplast pigments in the light dependent stage of photosynthesis is used to drive reactions of the light independent stage that produce complex organic compounds.

Chromatography is used to identify chloroplast pigments and was also used to identify the intermediates in the Calvin cycle.

By the end of this section you should be able to:

a) explain that energy transferred as ATP and reduced NADP from the light dependent stage is used during the light independent stage (Calvin cycle) of photosynthesis to produce complex organic molecules

b) state the sites of the light dependent and the light independent stages in the chloroplast

c) describe the role of chloroplast pigments (chlorophyll a, chlorophyll b, carotene and xanthophyll) in light absorption in the grana

d) interpret absorption and action spectra of chloroplast pigments

e) use chromatography to separate and identify chloroplast pigments and carry out an investigation to compare the chloroplast pigments in different plants

f) describe the light dependent stage as the photoactivation of chlorophyll resulting in the photolysis of water and the transfer of energy to ATP and reduced NADP

g) outline the three main stages of the Calvin cycle:
- fixation of carbon dioxide by combination with ribulose bisphosphate (RuBP), a 5C compound, to yield two molecules of GP (PGA), a 3C compound
- the reduction of GP to triose phosphate (TP) involving ATP and reduced NADP
- the regeneration of ribulose bisphosphate (RuBP) using ATP

h) describe, in outline, the conversion of Calvin cycle intermediates to carbohydrates, lipids and amino acids and their uses in the plant cell

Photosynthesis as energy transfer

In photosynthesis, green plants use the energy of sunlight to produce organic molecules, mostly **sugars**, from the inorganic raw materials **carbon dioxide** and **water**. **Oxygen** is the waste product. Photosynthesis occurs in green plant cells – typically, in the leaves. Here energy from light is trapped by **chlorophyll** pigments, present in **chloroplasts**. This light energy is then transferred to energy in the chemical bonds of a range of organic molecules, including sugars and **ATP**. We summarise photosynthesis in single equations:

$$\text{carbon dioxide} + \text{water} + \text{energy} \rightarrow \text{glucose} + \text{oxygen}$$
$$6CO_2 + 6H_2O + \text{energy} \rightarrow C_6H_{12}O_6 + 6O_2$$

*Compare this equation with that for **aerobic respiration** (page 243).*

At first sight it seems that the one process is simply the reverse of the other. In fact these equations are merely 'balance sheets' of the inputs and outputs. Photosynthesis and respiration occur by different pathways with many enzymes unique to one or the other pathway. And of course, photosynthesis is an anabolic process, whereas respiration is a catabolic one.

The sugars produced in photosynthesis provide the carbon 'skeletons' for the synthesis of all the **metabolites** the plant needs. We can say that green plants are 'self-feeding' (**autotrophic nutrition**). This type of nutrition contrasts with the ways animals and other organisms get their nutrients. In fact, it is green plant nutrition that supports the living world as a whole. Green plants are the source of nutrients for almost all other organisms, directly or indirectly. In this topic we look at photosynthesis as an **energy transfer process**.

Photosynthesis – inorganic to organic carbon

The bulk of living things consist of the elements hydrogen, oxygen, carbon and nitrogen. This is due to the large quantities of water they contain and because most other compounds present consist of carbon combined with hydrogen and oxygen. Compounds containing carbon combined with hydrogen are known as **organic compounds**. At least two and a half million organic compounds are known – more than the total of known compound made from all the other elements. This great diversity of organic compounds has made possible the diversity of life as we know it.

However, carbon is also abundant on earth as the **inorganic compounds** – chiefly carbon dioxide and carbonates. It is by the process of photosynthesis that inorganic carbon is converted to organic carbon. Photosynthesis is a reduction–oxidation process. In Topic 12 the idea of redox reactions was introduced (page 237).

Question

2 What colour does the pigment chlorophyll chiefly reflect or transmit rather than absorb?

$$\overset{\text{reduction}}{\overbrace{6CO_2 + 6H_2O + \text{ENERGY} \rightarrow \underset{\text{oxidation}}{\underbrace{C_6H_{12}O_6 + 6O_2}}}}$$

A feature of oxidation and reduction is an energy change, for when reduction occurs energy is absorbed. In photosynthesis it is solar light energy that is transferred to the chemical energy of organic compounds. Photosynthesis is thus simultaneously a process of energy transfer and a process by which inorganic carbon is converted to organic forms and built up into living organisms.

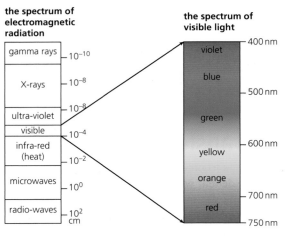

the spectrum of electromagnetic radiation

the spectrum of visible light

Light as an energy source

Light is only part of the total electromagnetic radiation from the Sun that reaches the Earth. When visible light itself is projected through a prism we see a continuous spectrum of light – a rainbow of colours, from red to violet. These colours have different wavelengths. The wavelengths of the electromagnetic radiation spectrum and of the components of light are shown in Figure 13.1. Of course, not all the colours of the spectrum present in white light are absorbed equally by chlorophyll. In fact, one colour is even transmitted (or reflected), rather than being absorbed.

Figure 13.1 The electromagnetic radiation spectrum and the spectrum of visible light

Investigating chlorophyll

Some plant pigments are soluble in water but chlorophyll is not. Whilst we cannot extract chlorophyll from leaves with water, it can be extracted by dissolving in an organic solvent like propanone (acetone). With this solvent, extraction of chlorophyll is fairly straightforward (Figure 13.2).

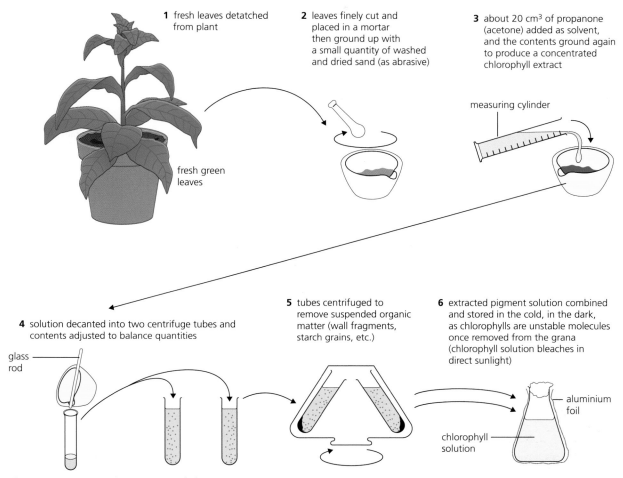

1 fresh leaves detatched from plant

2 leaves finely cut and placed in a mortar then ground up with a small quantity of washed and dried sand (as abrasive)

3 about 20 cm³ of propanone (acetone) added as solvent, and the contents ground again to produce a concentrated chlorophyll extract

measuring cylinder

fresh green leaves

4 solution decanted into two centrifuge tubes and contents adjusted to balance quantities

5 tubes centrifuged to remove suspended organic matter (wall fragments, starch grains, etc.)

6 extracted pigment solution combined and stored in the cold, in the dark, as chlorophylls are unstable molecules once removed from the grana (chlorophyll solution bleaches in direct sunlight)

glass rod

aluminium foil

chlorophyll solution

Figure 13.2 Steps in the extraction of plant pigments

Chlorophyll as a mixture – introducing chromatography

Plant chlorophyll consists of a mixture of pigments, two of which are chlorophylls. **Chromatography** is the technique we use to separate components of mixtures, especially when working with small samples. It is an ideal technique for separating biologically active molecules since biochemists are often able to obtain only very small quantities.

The process of chromatography involves a support medium known as the stationary phase. Often, this consists of absorptive paper (**paper** chromatography), powdered solid (**column** chromatography) or a thin film of dried solid (**thin-layer** chromatography). The mixture to be separated is dissolved in a suitable solvent and then loaded onto the stationary phase at a spot near to but not at one edge. Each drop is allowed to dry. The loading and drying sequence is repeated several times.

After that, in both paper chromatography and thin-layer chromatography, the lower edge of the stationary phase is introduced into a suitable chromatography solvent, known as the moving phase. This occurs in an enclosed space – a boiling tube or small glass tank, for example. The solvent is allowed to travel through the stationary phase by capillarity, passing through the loaded spot.

Question

3 In Figure 13.3, how R_f values are calculated is explained. Suggest why it may be useful to know the R_f value of a component in a mixture?

As the solvent front moves up the chromatogram, components of a mixture move at different rates. This is because they have different solubilities in the solvent. Also, the components of the mixture being separated interact with the material of the stationary phase to differing degrees.

Before the solvent front reaches the opposite end of the chromatogram, the process is stopped. The level reached by the solvent front is marked. The chromatogram is then dried. The positions of *coloured* components of a mixture are easily located (as in the case of chlorophyll). The locations of any components that are not coloured have to be identified; various techniques are available.

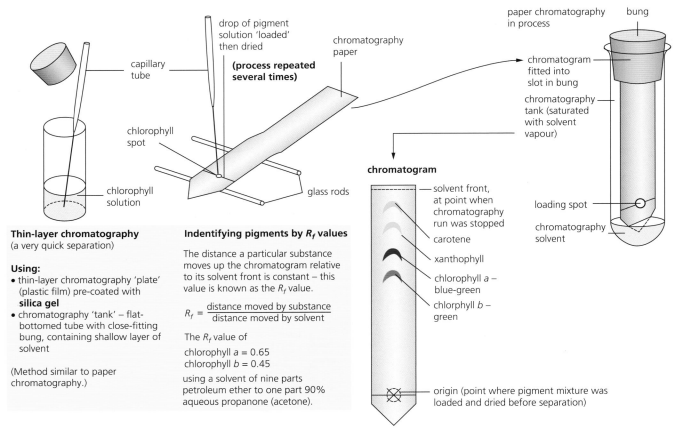

Figure 13.3 Preparing and running a chromatogram

Look at the chromatogram in Figure 13.3. Green leaves contain a mixture of photosynthetic pigments. These include two types of chlorophyll, known as **chlorophyll *a*** and **chlorophyll *b***. The others pigments belong to a group of compounds called **carotenoids**. Incidentally, larger samples of these individual pigments can be obtained by column chromatography, for example. Such samples may be large enough to investigate the properties of individual pigment molecules.

The chemistry of the chlorophylls is particularly important because they are directly involved in the energy transfer processes in the chloroplasts. The structure of chlorophyll is shown in Figure 13.4. *Look at the structure of chlorophyll now.*

The other pigments, the carotenoids, are described as **accessory pigments**. This is because they have the role of passing the light energy they absorb on to chlorophyll, before it can be involved in photosynthesis. (One of the caroteniods is a yellow pigment called carotene – an important source of vitamin A in human diets.

The chlorophyll pigments occur in the grana, held in the lipids and proteins of the thylakoid membranes. Chlorophyll is large – a 'head and tail' molecule, in fact. The 'heads' are hydrophilic and associate with proteins in the thylakoid membrane. The 'tails' are hydrophobic and associate with lipids in the thylakoid membranes. Also found here with these pigments are the enzymes and electron-carrier molecules involved in the steps of photosynthesis where light energy becomes chemical energy. We will meet these shortly.

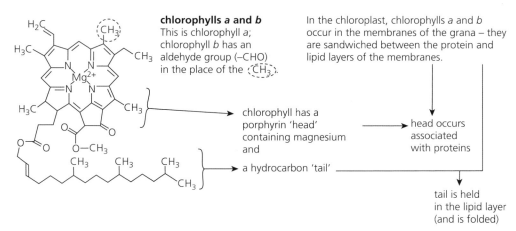

In the chloroplast, chlorophylls *a* and *b* occur in the membranes of the grana – they are sandwiched between the protein and lipid layers of the membranes.

chlorophylls *a* and *b*
This is chlorophyll *a*; chlorophyll *b* has an aldehyde group (–CHO) in the place of the CH₃.

chlorophyll has a porphyrin 'head' containing magnesium and

a hydrocarbon 'tail'

head occurs associated with proteins

tail is held in the lipid layer (and is folded)

Figure 13.4 The structure of chlorophyll

Question

4 What feature of the chlorophyll molecule prevents it being soluble in water – and why?

5 Why is chromatography a useful technique for the investigation of cell biochemistry.

The absorption and action spectra of chlorophyll

We have seen that light consists of a roughly equal mixture of all the visible wavelengths, namely violet, blue, green, yellow, orange and red (Figure 13.1). Now the issue is, how much of each wavelength does chlorophyll absorb? We call this information an absorption spectrum.

The **absorption spectra** of chlorophyll pigments are obtained by measuring their absorption of violet, blue, green, yellow, orange and red light, in turn. The results are plotted as a graph showing the amount of light absorbed over the wavelength range of visible light, as shown in Figure 13.5a. You can see that chlorophyll absorbs blue and red light most strongly. Other wavelengths are absorbed less so or not at all. It is the chemical structure of the chlorophyll molecule that causes absorption of the energy of blue and red light.

The **action spectrum** of chlorophyll is the wavelengths of light that bring about photosynthesis. This may be discovered by projecting different wavelengths, in turn and for a unit of time, on aquatic green pondweed. This is carried out in an experimental apparatus in which the rate of photosynthesis can be measured. The gas evolved by a green plant in the light is largely oxygen, and the volume given off in a unit of time is a measure of the rate of photosynthesis. Suitable apparatus is shown in Figure 13.14 (page 273), of which more later.

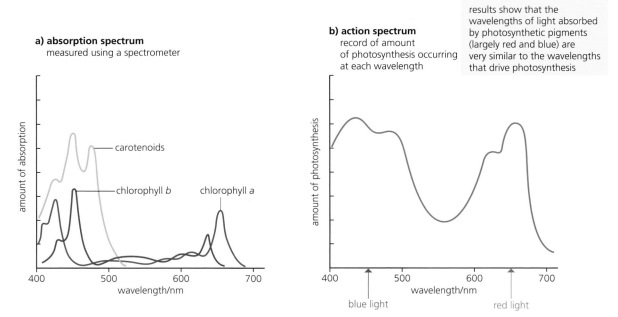

a) absorption spectrum
measured using a spectrometer

carotenoids
chlorophyll *b*
chlorophyll *a*

amount of absorption

wavelength/nm

b) action spectrum
record of amount of photosynthesis occurring at each wavelength

results show that the wavelengths of light absorbed by photosynthetic pigments (largely red and blue) are very similar to the wavelengths that drive photosynthesis

amount of photosynthesis

wavelength/nm

blue light   red light

Figure 13.5 Absorption and action spectra of chlorophyll pigments

The rate of photosynthesis at different wavelengths may then be plotted on a graph on the same scale as the absorption spectrum (Figure 13.5b). We have seen that blue and red light are most strongly absorbed by chlorophyll. From the action spectrum we see that it is these wave lengths that give the highest rates of photosynthesis.

Chlorophyll in solution versus chlorophyll in the chloroplast

The chlorophyll that has been extracted from leaves and dissolved in an organic solvent still absorbs light. However, chlorophyll in solution cannot use light energy to make sugar. This is because, in the extraction process, chlorophyll has been separated from the membrane systems and enzymes that surround it in chloroplasts. These are also essential for carrying out the biochemical steps of photosynthesis, as we shall now discover.

What happens in photosynthesis?

Fascinating 'detective work' by biochemists has shown us how photosynthesis works, step by step. Not surprisingly, photosynthesis consists of a complex set of many reactions taking place in chloroplasts in the light. However, these reactions divide naturally into **two stages**.

It was investigations into the effect of temperature on the rate of photosynthesis that first lead to the discovery that photosynthesis is a two-stage process. (We will return to the issue of the effects of external factors that limit the rate of photosynthesis later in this topic.) Meanwhile, the two inter-connected stages are:

- **a light-dependent stage**. This is a photochemical step which, like all photochemical reactions, is largely unaffected by temperature. It involves the splitting of water by light energy and the formation of reducing power (reduced hydrogen acceptor molecules). Also energy transfer to ATP occurs and oxygen is released as a waste product.
- **a light-independent stage**. This involves biochemical reactions which are catalysed by enzymes and are highly temperature sensitive. These reactions are dependent upon the supply of reduced hydrogen acceptor (reduced NADP) and ATP from the light-dependent stage. Carbon dioxide is 'fixed' to form 3-carbon sugars and these sugars are then used to synthesise other molecules.

These stages are summarised *in outline*, in the flow-diagram in Figure 13.6.

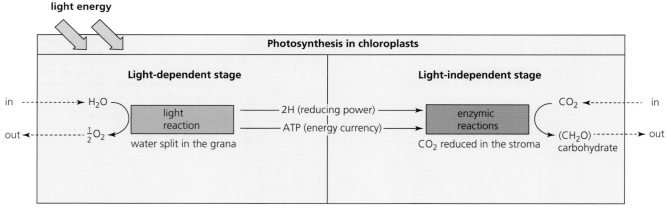

Figure 13.6 The two stages of photosynthesis: a summary

The light-dependent stage

In the light-dependent stage, light energy is used to split water (a step known as **photolysis**). Hydrogen is then removed by a hydrogen acceptor, known as **NADP (nicotinamide adenine dinucleotide phosphate)**.

NADP is a molecule found in the organelles of plant cells. You will notice that NADP is very similar to the coenzyme NAD of respiration, but it carries an additional phosphate group, hence the acronym NADP (Figure 13.7).

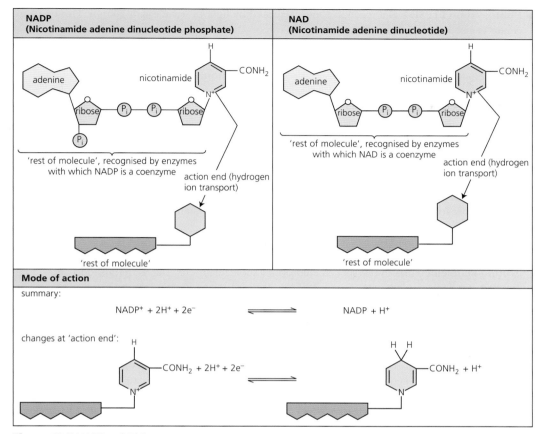

| NADP (Nicotinamide adenine dinucleotide phosphate) | NAD (Nicotinamide adenine dinucleotide) |

'rest of molecule', recognised by enzymes with which NADP is a coenzyme

action end (hydrogen ion transport)

'rest of molecule'

'rest of molecule', recognised by enzymes with which NAD is a coenzyme

action end (hydrogen ion transport)

'rest of molecule'

Mode of action

summary:

$$NADP^+ + 2H^+ + 2e^- \rightleftharpoons NADP + H^+$$

changes at 'action end':

$$-CONH_2 + 2H^+ + 2e^- \rightleftharpoons -CONH_2 + H^+$$

Figure 13.7 NADP and NAD

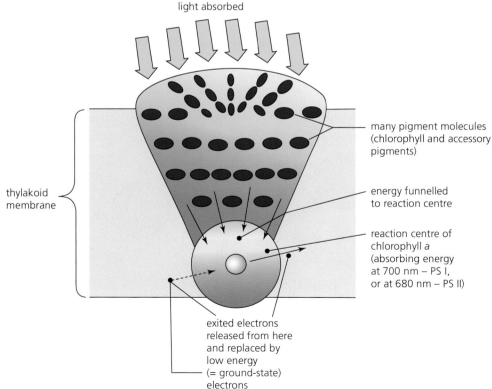

light absorbed

thylakoid membrane

many pigment molecules (chlorophyll and accessory pigments)

energy funnelled to reaction centre

reaction centre of chlorophyll *a* (absorbing energy at 700 nm – PS I, or at 680 nm – PS II)

exited electrons released from here and replaced by low energy (= ground-state) electrons

Figure 13.8 The structure of a photosystem

At the same time that reduced NADP is formed in the light-dependent stage, ATP is generated from **ADP** and phosphate. Light energy is needed for this, too, of course. This energy comes from photoactivated chlorophyll, and the process is known as **photophosphorylation**. Also, oxygen is given off as a waste product of the light-dependent stage.

This stage of photosynthesis occurs in the grana of the chloroplasts.

We will look at the structures involved (the photosystems) next, and then at the steps by which it occurs.

The photosystems

Chlorophyll molecules do not occur haphazardly in the grana. Rather, they are grouped together in structures called **photosystems**, held in the thylakoid membranes of the **grana** (Figure 13.8).

In each photosystem, several hundred chlorophyll molecules plus accessory pigments (carotenoids) are arranged, grouped together. All these pigment molecules harvest light energy, and they funnel the energy to a single chlorophyll molecule of the photosystem. This molecule is known as the **reaction centre**. The different pigments around the reaction centres absorb light energy of slightly different wavelengths, but all 'feed into' the reaction centre.

Now we also need to note that there are two types of photosystem present in the thylakoid membranes of the grana. These differ in the wavelength of light that the chlorophyll of the reaction centre absorbs:

- **Photosystem I (PSI)** has a reaction centre activated by light of wavelength 700 nm. This reaction centre is also referred to as P700.
- **Photosystem II (PSII)** has a reaction centre activated by light of wavelength 680 nm. This reaction centre is also referred to as P680.

Photosystems I and II have differing roles, as we shall see shortly. However, they occur grouped together in the thylakoid membranes of the grana, along with specific proteins, many of which function as **electron-carrier** molecules.

How light energy is transferred

In the light, pigment molecules of the photosystems harvest light energy and funnel it into a special chlorophyll molecule at the reaction centre of the photosystem. Photosystem II is activated by light energy of wavelength 680 nm, and photosystem I by wavelength 700 nm.

We will now follow what happens next, starting with photosystem II but remember, the two photosystems operate simultaneously.

1 The light energy reaching the reaction centre of photosystem II activates a particular pair of **ground-state electrons** in the special chlorophyll molecule there. These are raised to an **excited state** and immediately captured by an electron acceptor in the nearby chain of electron-carrier molecules (Figure 13.9). By the loss of electrons the chlorophyll molecule is oxidised, causing it to immediately acquire electrons from water. It is electrons in the ground state that fill the gap in the chlorophyll molecule vacated by the excited electrons. By this process, the water molecule is split. The remainder of the water molecule is released as **oxygen gas** (two oxygen atoms combine to form a molecule of oxygen) and **hydrogen ions**.

Meanwhile, the excited electrons are passed along a chain of electron carriers. As this happens, the energy level of the excited electrons progressively falls back to the ground state. However, the energy transferred from them as they flow along causes the pumping of hydrogen ions (protons) from the chloroplast's stroma into the thylakoid spaces. Here the hydrogen ions accumulate – incidentally, causing the pH to drop. The result is the creation of a proton gradient across the thylakoid membrane – of which more shortly.

Light energy reaches the reaction centre of photosystem I at the same time. Here, light energy again causes the ejection of particular ground-state electrons, raising them to an excited state. These, too, are captured by an electron acceptor in the nearby chain of electron-carrier molecules. Now there comes a difference.

The gap vacated by these electrons in photosystem I is filled by electrons that have been transferred from photosystem II. These have now fallen back to the ground state energy level.

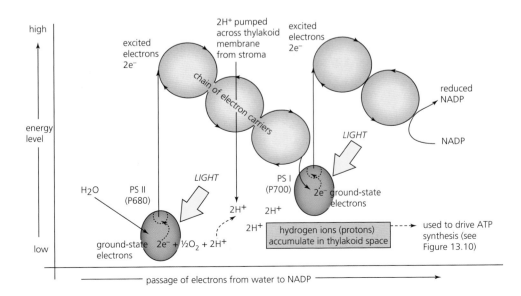

Figure 13.9 Z diagram showing the light-dependent stage of photosynthesis

Meanwhile, the excited electrons from photosystem I are passed along a short chain of electron carrier molecules and their energy is transferred to form a molecule of reduced NADP from NADP and hydrogen ions, in the stroma. In Figure 13.10 you can see that this occurs on the stromal side of the thylakoid membrane – an important point.

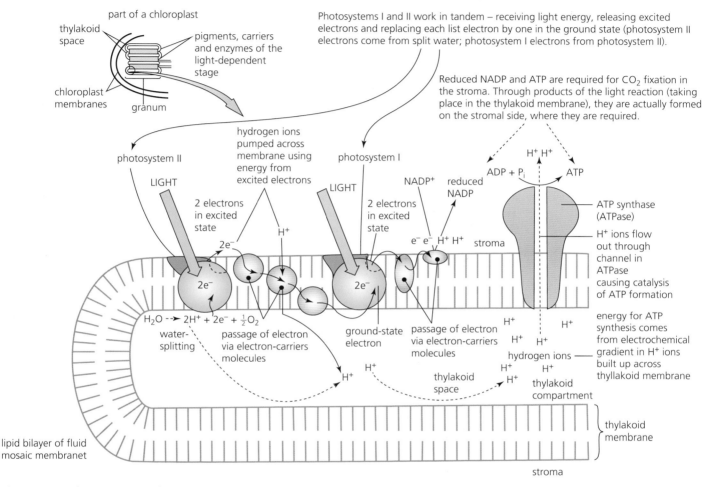

Figure 13.10 The organisation of the thylakoid membrane

6 Both reduced NADP and ATP, products of the light-dependent stage of photosynthesis, are formed on the side of thylakoid membranes that face the stroma. Suggest why this fact is significant.

7 How is the gradient in protons between the thylakoid space and the stroma generated?

2 Finally, ATP is generated, a process known as **photophosphorylation.** This occurs by **chemiosmosis.** (We previously met this same process in the mitochondria, page 248).

In the light-dependent stage, a gradient in hydrogen ions (protons) is built up between the thylakoid space and the stroma. These protons then diffuse back from the thylakoid space into the stroma through the pore in the **ATP synthase** enzyme molecules, causing the synthesis of ATP from ADP and a phosphate ion. This enzyme is also part of the photosystem proteins present in the thylakoid membrane (Figure 13.10). In this way, synthesis of ATP is coupled to this movement of protons. We call this **non-cyclic photophosphorylation** for reasons that are clear when we re-examine the Z diagram in Figure 13.9. Light energy is absorbed simultaneously by both photosystems and excited electrons are emitted by both reaction centres. But the overall pathway of each pair of electrons is a progression. They pass from water, first via photosystem II and associated electron acceptors and carriers to photosystem I and its associated acceptors and carriers, to reduced NADP. In the process, the energy of the excited electrons is used to build up the concentration of protons in the thylakoid space. Then, as the protons flow out via ATP synthase down their electrochemical gradient, ATP is synthesised.

Incidentally, we should notice that ATP synthesis occurs on the stromal side of the thylakoid membrane.

The products of the light-dependent reactions and their role

Now because the sequence of reactions of non-cyclic photophosphorylation are repeated again and again, at very great speed throughout every second of daylight, large amounts of the products of the light-dependent reactions (ATP and reduced NADP) are formed.

However, ATP and reduced NADP do not accumulate. They are essential components of steps in the **light-independent reaction** that occurs in the surrounding stroma. As a result, ADP, P_i and NADP are continuously released from the stroma in the light and reused in the grana. We shall see why this is so, next.

The light-independent reactions

In summary, in the light-independent stage, carbon dioxide is reduced to carbohydrate in the stroma of the chloroplasts. It occurs in three steps of a cyclic process, now known as the **Calvin cycle**, and requires a continuous supply of the products of the light reaction. The enzymes of the light-independent reaction are light activated.

In **Step One** carbon dioxide, which readily diffuses into the chloroplast, is combined with an acceptor molecule in the presence of a special enzyme, **ribulose bisphosphate carboxylase** (**rubisco** for short). The stroma is packed full of rubisco. This enzyme makes up the bulk of all the protein in a green plant – it is the most abundant enzyme present in the living world. The acceptor molecule is a five carbon sugar, **ribulose bisphosphate** (referred to as **RuBP**). Carbon dioxide is added in a process known as fixation. When one 'carbon' is added to five carbons should we not expect the product to have six carbons? In fact, two molecules of a three carbon compound, **glycerate-3-phosphate** (**GP**) are formed. (This molecule is sometimes known as PGA.) GP is essentially an organic acid – it is not at the energy level of a sugar.

In **Step Two**, GP is then **reduced** to **triose phosphate**, a three-carbon sugar. This is a high-energy level product and the reaction is brought about by reduced NADP and ATP. Reduced NADP and ATP are therefore essential for the formation of sugar in photosynthesis.

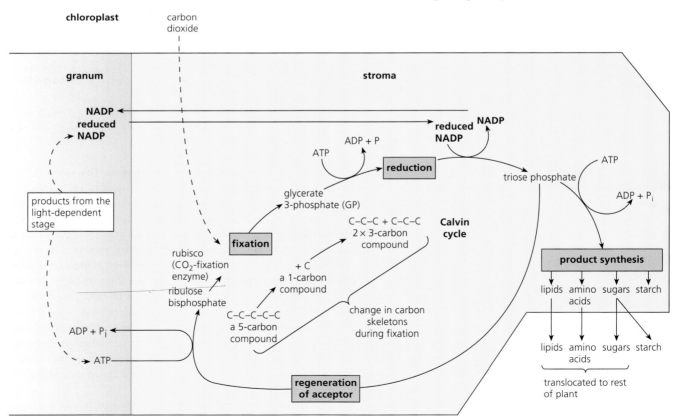

Figure 13.11 The reactions of the light-independent stage *in situ*

In **Step Three**, **regeneration of the acceptor molecule** occurs. For every six molecules of triose phosphate formed, five are used to form three molecules of RuBP. In this conversion, energy from ATP is also required. From the other molecule of triose phosphate other compounds are synthesized. These include **carbohydrates** (sugars and starch, and sucrose for translocation in the phloem), **lipids**, and **amino acids**. In these conversions energy from ATP is also required. How the products of photosynthesis sustain the whole **metabolism** of the plant is summarised in Figure 13.12.

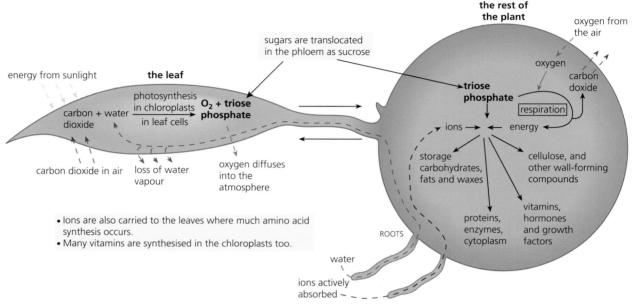

Figure 13.12 Photosynthesis: its place in plant nutrition

Photophosphorylation and the concentration of carbon dioxide

Sometimes, the concentration of carbon dioxide in a leaf in the light becomes very low. (These conditions **can** occur – we will return to the subject of limiting conditions later in this topic.) When this happens, fixation of carbon dioxide slows to a standstill. As a result, reduced NADP accumulates in the stroma and NADP disappears. Without a continuous supply of NADP, photosystems II and I are no longer able to operate together as they do in non-cyclic photophosphorylation.

In this situation, another type of ATP formation may occur – known as **cyclic photophosphorylation**. In cyclic photophosphorylation, excited electrons from the reaction centre in photosystem I are able to 'fall back' to where they came from. However, the route by which they do this is via the electron-carrier molecules that transfer energy from these excited electrons and pump protons into the thylakoid space. The modified Z diagram in Figure 13.13 illustrates this.

Consequently, some ATP synthesis by chemiosmosis is able to continue, despite the absence of carbon dioxide. This type of photophosphorylation is called 'cyclic' because the electrons have returned to the photosystem from which they originated. They are 'recycled' rather than used to form reduced NADP, but ATP is still generated.

Non-cyclic phosphorylation – a reminder
- The normal flow of electrons is from water, via PS II and electron carriers to PS I and electron carriers to NADP (see Figure 10.10).
- The excited electrons from PS II, as they lose the energy of the excited state, cause hydrogen ions (protons) to be pumped across into the thylakoid space.
- The gradient in protons between thylakoid space and the stroma causes protons to flow through ATP synthase and generate ATP.

Cyclic phosphorylation
- Occurs when reduced NADP accumulates in the stroma – e.g. when CO_2 concentration is low and the light-independent stage is blocked.
- Cyclic photophosphorylation results – excited electrons from PS I are captured by electron acceptors serving PS II, and then returned to a ground state (pumping protons as they go), to re-occupy their original space in PS I.
- The proton gradient between the thylakoid space and the stroma is maintained and ATP formation continues.

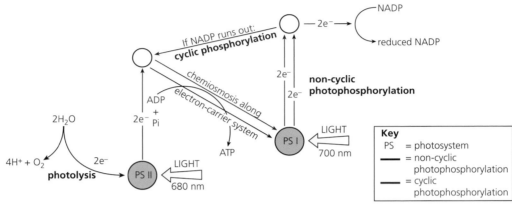

Figure 13.13 A comparison of cyclic and non-cyclic photophosphorylation

Questions

8 What are the ultimate fates of electrons displaced from the reaction centre of photosystem I
  a in non-cyclic photophosphorylation
  b in cyclic photophosphorylation?

9 In a school laboratory demonstration of the Hill reaction (page 277), what electron acceptor is used, and where do the electrons come from?

10 Many biologists are not surprised the enzyme rubisco is the most abundant enzyme present in the living world? Suggest why.

Extension

Finding the path of carbon in photosynthesis

Radioactive carbon (carbon-14) became available for biochemical investigations in 1945. One of the early applications was an investigation of photosynthetic carbon dioxide fixation. It was undertaken by Melvin Calvin, James Bassham and Andy Benson in California. Carbon dioxide labelled with carbon-14 ($^{14}CO_2$) is taken up by the cells and is fixed into the products of photosynthesis in just the same way as unlabelled carbon dioxide ($^{12}CO_2$) is. It was then possible to discover the sequence of metabolites which become labelled – in effect, the **path of carbon in photosynthesis**, from carbon dioxide to sugars and the other products of photosynthesis.

A culture of *Chlorella*, a unicellular alga, was used in these experiments, in place of mesophyll cells. This was because they have similar photosynthesis and they allow easy sampling. Samples of the photosynthesising cells, taken at frequent intervals after a pulse of $^{14}CO_2$ had been given were harvested and analysed. The intermediates that became progressively labelled with the carbon-14 were isolated by chromatography and then identified.

Extension

Chemosynthesis

Autotrophic organisms are 'self-feeding' organisms. They fall into two distinct groups, according to the source of energy used to build carbon dioxide into carbohydrates.

One group, the photosynthetic organisms, use light energy. We have focused on photosynthesis in this topic.

Another group of autotrophic organisms are unable to use light energy. Instead, a specific chemical reaction is catalysed in order to transfer the free energy required to synthesise organic compounds from carbon dioxide and water. Consequently, these organisms are known as **chemosynthetic**. Two important examples of chemosynthetic autotrophs that we meet elsewhere in this book are:
- the **nitrifying bacteria**, that occur in the soil. These microorganisms oxidise ammonium ions to nitrites or nitrites to nitrates. Both of these reactions provide the bacteria with energy needed to synthesis carbohydrates.
- the **iron bacteria**, which are sometimes used in industry in the extraction of metal from low-grade ores.

13.2 Investigation of limiting factors

Environmental factors influence the rate of photosynthesis. Investigating these shows how they can be managed in protected environments used in crop production.

By the end of this section you should be able to:

a) explain the term limiting factor in relation to photosynthesis
b) explain the effects of changes in light intensity, carbon dioxide concentration and temperature on the rate of photosynthesis
c) explain how an understanding of limiting factors is used to increase crop yields in protected environments, such as glasshouses
d) carry out an investigation to determine the effect of light intensity or light wavelength on the rate of photosynthesis using a redox indicator (e.g. DCPIP) and a suspension of chloroplasts (the Hill reaction)
e) carry out investigations on the effects of light intensity, carbon dioxide and temperature on the rate of photosynthesis using whole plants, e.g. aquatic plants such as *Elodea* and *Cabomba*

Question

11 A thermometer is not shown in the apparatus in Figure 13.14. Predict why one is required and state where it should be placed.

The environment and the rate of photosynthesis

The rate of photosynthesis can be measured in an aquatic plant, using a microburette – also called a photosynthometer (Figure 13.14). The experiment requires a freshly cut shoot of aquatic green pondweed which, when inverted, produces a vigorous stream of gas bubbles from the base. The bubbles tell us the pondweed is actively photosynthesising. The pondweed is placed in a very dilute solution of sodium hydrogencarbonate, which supplies the carbon dioxide (as hydrogencarbonate ions) required by the plant for photosynthesis. The quantity of gas evolved in a given time is measured by drawing the gas bubble that collects into the capillary tube and measuring its length. This length is then converted to a volume.

Using this apparatus the effects of external conditions such as light intensity, carbon dioxide concentration and temperature on the rate of photosynthesis have been investigated.

Note: this experiment can be simulated on a computer. Details are given on the disk that accompanies this book.

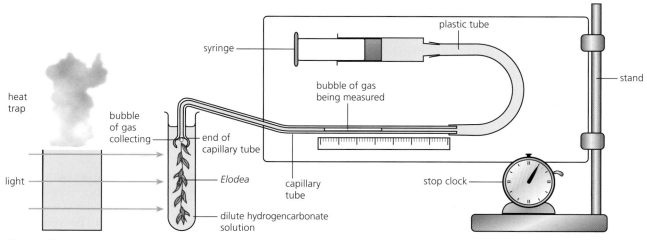

Figure 13.14 Measuring the rate of photosynthesis with a microburette

273

Light and carbon dioxide as limiting factors

Clearly, light is essential for photosynthesis. Green plant cells in the dark are unable to photosynthesise at all; under this condition the only gas exchange by the leaf tissues is the uptake of oxygen for aerobic respiration. Carbon dioxide is also lost.

As light starts to reach green cells, at dawn for example, photosynthesis starts, and some oxygen is produced. Eventually the light intensity increases to the point where oxygen production by photosynthesis is equal to oxygen consumption in respiration. Now the leaf is neither an oxygen importer nor exporter. This point is known as the **compensation point** (Figure 13.15).

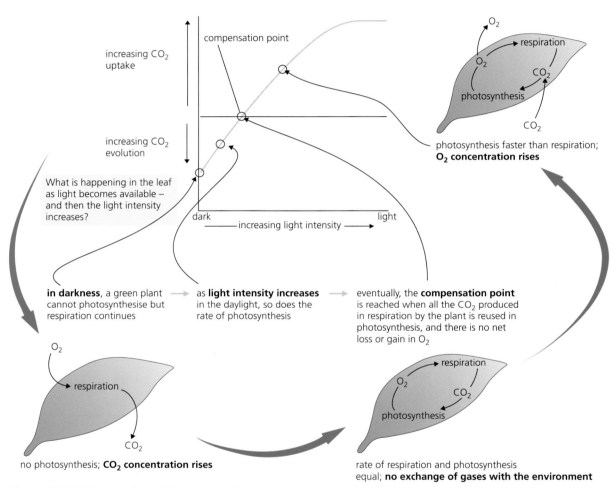

Figure 13.15 Light intensity and the compensation point

Then, as the light intensity increases further, the rate of photosynthesis also increases; the leaf becomes a net exporter of oxygen.

We can see the general effect of increasing light intensity on the rate of photosynthesis in Figure 13.15. At low light intensities the rate of photosynthesis increases in proportion to the increasing light. Here the low light intensity (i.e. the lack of sufficient light) is limiting the rate of photosynthesis. Light is a factor limiting the rate of photosynthesis under this condition.

However, as light intensity is raised further, a point is reached where increasing light intensity produces no further effect on the rate of photosynthesis. Now some factor other than light is limiting photosynthesis.

What may now be limiting the rate?

The limiting factor at that point is disclosed when the investigation is repeated at either higher light (Figure 13.16) or carbon dioxide concentrations, but not at the same time.

Look at this graph in Figure 13.16 now and follow the annotations.

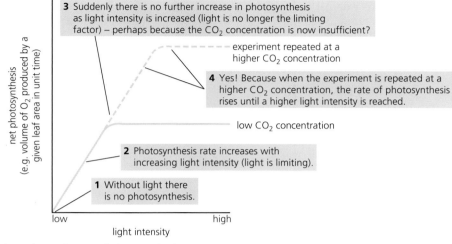

Understanding this graph
Read the axis labels ($x + y$ axes) and look at the shapes of the curves. Then follow the points numbered 1 to 4.

3 Suddenly there is no further increase in photosynthesis as light intensity is increased (light is no longer the limiting factor) – perhaps because the CO_2 concentration is now insufficient?

experiment repeated at a higher CO_2 concentration

4 Yes! Because when the experiment is repeated at a higher CO_2 concentration, the rate of photosynthesis rises until a higher light intensity is reached.

low CO_2 concentration

2 Photosynthesis rate increases with increasing light intensity (light is limiting).

1 Without light there is no photosynthesis.

net photosynthesis (e.g. volume of O_2 produced by a given leaf area in unit time)

light intensity

Figure 13.16 The effect of light intensity on the rate of photosynthesis

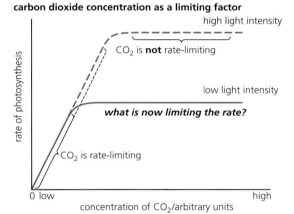

carbon dioxide concentration as a limiting factor

high light intensity

CO_2 is **not** rate-limiting

low light intensity

what is now limiting the rate?

CO_2 is rate-limiting

rate of photosynthesis

concentration of CO_2/arbitrary units

Figure 13.17 The effect of carbon dioxide concentration on photosynthesis

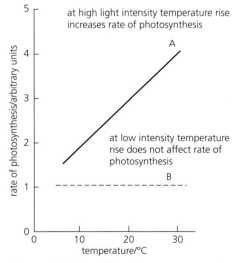

at high light intensity temperature rise increases rate of photosynthesis

A

at low intensity temperature rise does not affect rate of photosynthesis

B

rate of photosynthesis/arbitrary units

temperature/°C

Figure 13.18 The effect of temperature on the rate of photosynthesis

The results of these earlier, physiological investigations fit well with our current understanding. Photosynthesis is a biochemical process involving a series of interconnected reactions (Figure 13.19). All these reactions contribute to the overall rate of photosynthesis. However, at any one time, the rate of photosynthesis will be limited by the slowest of these reactions – the overall rate will be limited by the factor that is in shortest supply. This factor, whichever one it is, is known as the **limiting factor**.

Clearly, a limited supply of either light (Figure 13.16) or carbon dioxide (Figure 13.17) could determine the rate of photosynthesis, since both of these are essential.

Both **light intensity** and **carbon dioxide concentration** can be limiting factors in photosynthesis.

The effect of temperature on the rate of photosynthesis

The effect of temperature on the rate of photosynthesis, under controlled laboratory conditions, is especially informative. This is because, one effect of increasing temperature, say in the range from 10°C to 30°C, depends upon **light intensity**. Under low light intensities, a rise in temperature has little effect. Under higher intensities, the rise in temperature increases the rate of photosynthesis significantly. This relationship is shown in Figure 13.18.

How can this effect be explained?

Under low light intensity, light is obviously the limiting factor. Now a light-dependent reaction, like all photochemical events, is temperature indifferent. In the same way, a light-independent reaction (like all biochemical steps catalysed by enzymes) is highly temperature sensitive.

The graph in Figure 13.18 was interpreted as showing that photosynthesis is made up of two sequential stages, a light-dependent stage and a light-independent stage. Later, this idea was confirmed, once photosynthesis was investigated by entirely different, biochemical techniques, including the use of chloroplasts isolated from a leaf, for example.

12 a Suggest why we should expect that high external temperatures cause the rate of photosynthesis to fall off very rapidly?

b What may effectively prevent over-heating of green leaves?

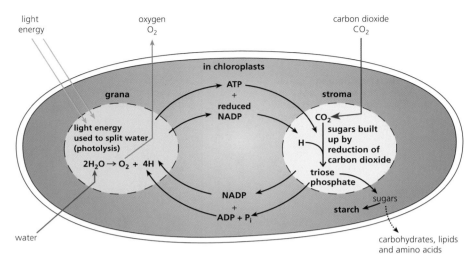

Figure 13.19 Photosynthesis: a summary

Limiting factors and 'crop yield' in intensive agriculture

Growers of commercial crops have used knowledge of factors limiting the rate of photosynthesis to increase crop yield. These applications are made in intensive cultivation of plants in glasshouses (frequently referred to as 'greenhouses') in particular. This is because external conditions – the immediate environment of crops, are more easily controlled there, compared to those in outdoor cultivation. The conditions most easily controlled are:

- **Light intensity**: banks of lamps providing photosynthetically active radiation (the bands of electromagnetic radiation that chloroplasts absorb and exploit) may be used.
- **Heat**: moderate increases in ambient temperatures enhance plant growth.
- **Additional carbon dioxide**: for the majority of greenhouse crops, net photosynthesis increases as CO_2 levels are increased from 340 to 1000 parts per million (Figure 13.20).

The highly significant benefits of carbon dioxide supplementation may be offset to some extent by the expense incurred in altering the growth environment in this way, and by the raised costs due to the need to monitor and control the environment. For example, infra-red gas analysers are needed to monitor this environmental factor, linked to a computer to maintain pre-set, minimum and maximum levels of atmospheric carbon dioxide. The crops concerned must be able to command a premium price in nearby markets for this approach to be sustainable.

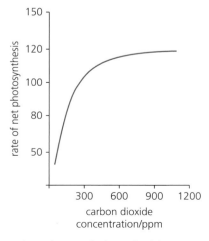

Figure 13.20 Ambient CO_2 levels and net photosynthetic productivity

Practical: Investigating the rate of photosynthesis using an isolated chloroplast suspension – the Hill reaction

Introducing the Hill reaction

Chloroplasts can be isolated from green plant leaves and suspended in buffer solution of the same concentration as the cytosol (an isotonic buffer), see Figure 13.21. If the suspended chloroplasts are undamaged, they function much as they do in the intact leaf. These isolated chloroplasts can be used to investigate the splitting of water, for example. However, this requires the natural electron-acceptor enzymes and electron-carrier molecules to be present, or the substitution of an alternative.

In the research laboratory, a sensitive piece of apparatus called an **oxygen electrode** is used to detect the oxygen given off by isolated chloroplasts.

In your laboratory, a **hydrogen-acceptor dye** that changes colour when it is reduced is a practical alternative. The dye, DCPIP (dichlorophenolindophenol) is an example. DCPIP does no harm when added to chloroplasts in a suitable buffer solution, but changes colour from blue to a colourless state when it is reduced. The splitting of water by light energy (photolysis) is the source of hydrogen that turns DCPIP colourless. The photolysis of water and the reduction of the dye are represented by the equation:

$$2DCPIP + 2H_2O \rightarrow 2DCPIPH_2 + O_2$$

This demonstration is referred to as the **Hill reaction** after the research scientist who first demonstrated this 'reducing power' of illuminated chloroplasts. It is a demonstration of the light-dependent stage outside of the leaf.

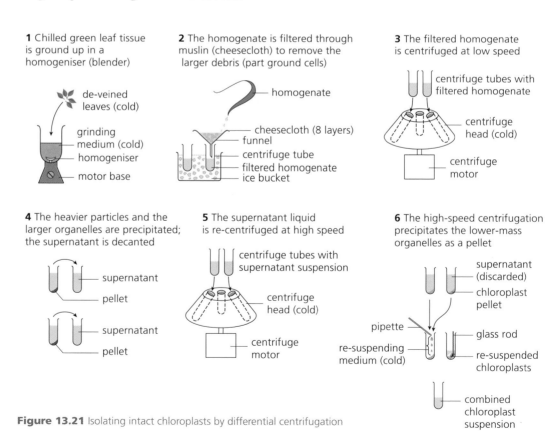

1 Chilled green leaf tissue is ground up in a homogeniser (blender)

de-veined leaves (cold)
grinding medium (cold)
homogeniser
motor base

2 The homogenate is filtered through muslin (cheesecloth) to remove the larger debris (part ground cells)

homogenate
cheesecloth (8 layers)
funnel
centrifuge tube
filtered homogenate
ice bucket

3 The filtered homogenate is centrifuged at low speed

centrifuge tubes with filtered homogenate
centrifuge head (cold)
centrifuge motor

4 The heavier particles and the larger organelles are precipitated; the supernatant is decanted

supernatant
pellet
supernatant
pellet

5 The supernatant liquid is re-centrifuged at high speed

centrifuge tubes with supernatant suspension
centrifuge head (cold)
centrifuge motor

6 The high-speed centrifugation precipitates the lower-mass organelles as a pellet

supernatant (discarded)
chloroplast pellet
pipette
glass rod
re-suspending medium (cold)
re-suspended chloroplasts
combined chloroplast suspension

Figure 13.21 Isolating intact chloroplasts by differential centrifugation

277

Setting up reaction mixtures and collecting reliable data

Isolated chloroplasts are re-suspended in 0.1% phosphate buffer solution at pH 6.5. With a chloroplast suspension to hand, the reaction mixtures listed in Table 13.1 can be assembled in individual colorimeter tubes, adding the chloroplast suspension last (all volumes in cm^3), when ready to carry out the measurements. The measurements are made in a colorimeter with green filter in place. Tubes 2, 3 and 4 are all control tubes.

Table 13.1

Tube	Buffer (0.1%)	DCPIP (0.1% sol)	Distilled water	Chloroplast suspension	Treatment
1	8.0	0.2	0.0	0.5	Placed in the colorimeter and the scale adjusted to read 0.7 on the absorbance scale. Then the tube removed, exposed to bright light, and returned to the colorimeter at 15 second intervals until no further change in the absorbance occurs. This time is recorded.
2	8.0	0.2	0.0	0.5	Contents kept in complete darkness (wrapped in aluminium foil) for the same period of time that Tube 1 takes to show no further colour change. Then the foil is removed and the absorbance read in the colorimeter, and also recorded.
3	8.0	0.2	0.5	0.0	These tubes are treated in the same way as Tube 1, maintaining the cycle of exposure and recording of absorbance for the same period of time it takes for Tube 1 to show no further colour change.
4	8.0	0.0	0.2	0.5	

Discuss the results you anticipate with peers and your teacher, and then answer Question 13.

Question

13 Given that a chloroplast suspension will decolourise a blue DCPIP solution in the light, explain:
  a why the steps to the chloroplast extraction were carried out in an ice-cold water bath, where possible
  b why, for the experiment itself (tubes in the light and the dark), the reaction mixtures were allowed to warm to room temperature
  c why an isotonic buffer solution is used to re-suspend the chloroplasts, rather than distilled water
  d what colour change you anticipate in Tubes 1 and 2, if any
  e what issues the control tubes (3 and 4) will resolve

13.3 Adaptations for photosynthesis

All the stages of photosynthesis occur in the chloroplast. Some tropical crops have C_4 metabolism and adaptations to maximise carbon dioxide fixation.

By the end of this section you should be able to:

a) describe the relationship between structure and function in the chloroplast using diagrams and electron micrographs

b) explain how the anatomy and physiology of the leaves of C_4 plants, such as maize or sorghum, are adapted for high rates of carbon fixation at high temperatures in terms of:
  • the spatial separation of initial carbon fixation from the light dependent stage (biochemical details of the C_4 pathway are required in outline only)
  • the high optimum temperatures of the enzymes involved

Chloroplasts – the site of photosynthesis

Mitochondria are the site of many of the reactions of respiration, but **chloroplasts** are the organelles where all the reactions of photosynthesis occur. Chloroplast are one of the larger organelles found in plant cells, yet they typically measure only 4–10 µm long and 2–3 µm wide. As a result, whilst a single chloroplast can be seen in outline by light microscopy, for detail of fine structure electron microscopy is required.

Look at the electron micrograph of a chloroplast in Figure 1.20, page 20, now.

There is a **double membrane** around the chloroplast. The inner membrane intucks at various points, forming a system of branching membranes, called **thylakoid membranes**. These thylakoid membranes are organised into many flat, compact, disc-shaped piles, called **grana** (*singular: granum*). Between the grana are loosely arranged tubular membranes suspended in a watery matrix, called the **stroma**. Also present are starch grains, lipid droplets and ribosomes. Chlorophyll pigments occur bound to the membranes of the grana (Figure 13.22).

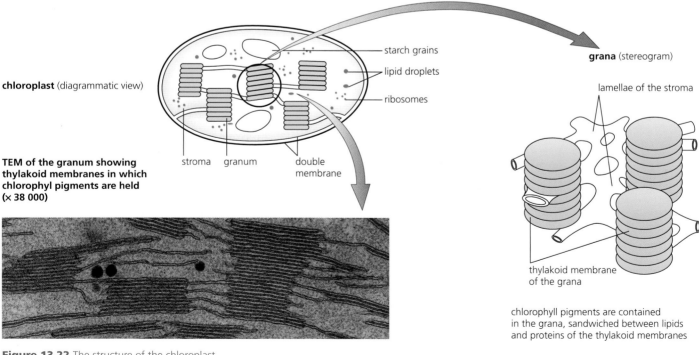

chloroplast (diagrammatic view)

starch grains
lipid droplets
ribosomes

stroma   granum   double membrane

grana (stereogram)

lamellae of the stroma

thylakoid membrane of the grana

chlorophyll pigments are contained in the grana, sandwiched between lipids and proteins of the thylakoid membranes

TEM of the granum showing thylakoid membranes in which chlorophyl pigments are held (× 38 000)

Figure 13.22 The structure of the chloroplast

Questions

14 Draw a diagram of part of a chloroplast to show its structure as revealed by electron microscopy. Annotate your diagram to indicate the adaptations of the chloroplast to its functions.

15 Distinguish between the following:
  a light-dependent reactions and light-independent reactions
  b photolysis and photophosphorylation.

16 Deduce the significant difference between the starting materials and the end-products of photosynthesis.

Chloroplast, structure in relation to function

We have seen that the grana are the site of the light-dependent reactions and the stroma the site of the light-independent reactions of photosynthesis. Table 13.2 identifies how the structure of the chloroplast facilitates these functions.

Examine Figure 13.22 and Table 13.2 and then answer Question 14.

Table 13.2 Structure and function in chloroplasts

Structure of chloroplast	Function or role
double membrane bounding the chloroplast	contains the grana and stroma, and is permeable to CO_2, O_2, ATP, sugars and other products of photosynthesis
photosystems with chlorophyll pigments arranged on thylakoid membranes of grana	provide huge surface area for maximum light absorption
thylakoid spaces within grana	restricted regions for accumulation of protons and establishment of the gradient
fluid stroma with loosely arranged thylakoid membranes	site of all the enzymes of fixation, reduction and regeneration of acceptor steps of light-independent reactions, and many enzymes of the product synthesis steps

Anatomy and physiology of C4 plants

The fixation of carbon dioxide in the light-independent step of photosynthesis was illustrated in Figure 13.11, page 270. We saw that carbon dioxide and the five-carbon acceptor molecule, ribulose bisphosphate, reacted together in the presence of an enzyme known as **rubisco**, in the stroma of the chloroplast. The first product is a three-carbon sugar, and because of this, that pathway of carbon fixation is known as the **C_3 pathway**. You can see an illustration of how rubisco functions in Figure 13.23 – and also the name of the three-carbon sugar product, should you have forgotten it.

Photosynthesis evolved when the Earth's atmosphere had an extremely low concentration of oxygen but a high concentration of carbon dioxide. This atmosphere was quite different from our atmosphere today. Since those early days, the concentration of oxygen in the atmosphere has increased (because of photosynthesis!). You will be aware that today's atmosphere contains about 21 per cent oxygen.

Despite these changes, rubisco mostly functions normally, particularly in green plants of temperate regions. However, in conditions experienced by tropical and sub-tropical plants, this is not always the case. Plants of tropical and sub-tropical climates experience much higher sunlight levels and higher temperatures. In such an environment, light is not a limiting factor of photosynthesis, but carbon dioxide often does become a limiting factor, in full sunlight.

In these conditions it has been discovered that **oxygen** from the air becomes a **competitive inhibitor** affecting the active site of rubisco. The unfortunate result is the breakdown of the five-carbon acceptor molecule, ribulose bisphosphate, to a two-carbon compound, of little value to the plant (and one molecule of the normal three-carbon sugar product). This phenomenon is known as **photorespiration**. It reduces the yield from photosynthesis. You can see this competitive inhibition event of rubisco and the products formed in Figure 13.23.

Photosynthesis in some tropical plants

In response, many tropical plants now have an additional mechanism of carbon dioxide fixation. This mechanism can deliver a higher concentration of carbon dioxide to the chloroplasts where photosynthetic fixation occurs. By this mechanism, photorespiration is reduced greatly. As a result, more sugar is produced by the chloroplasts.

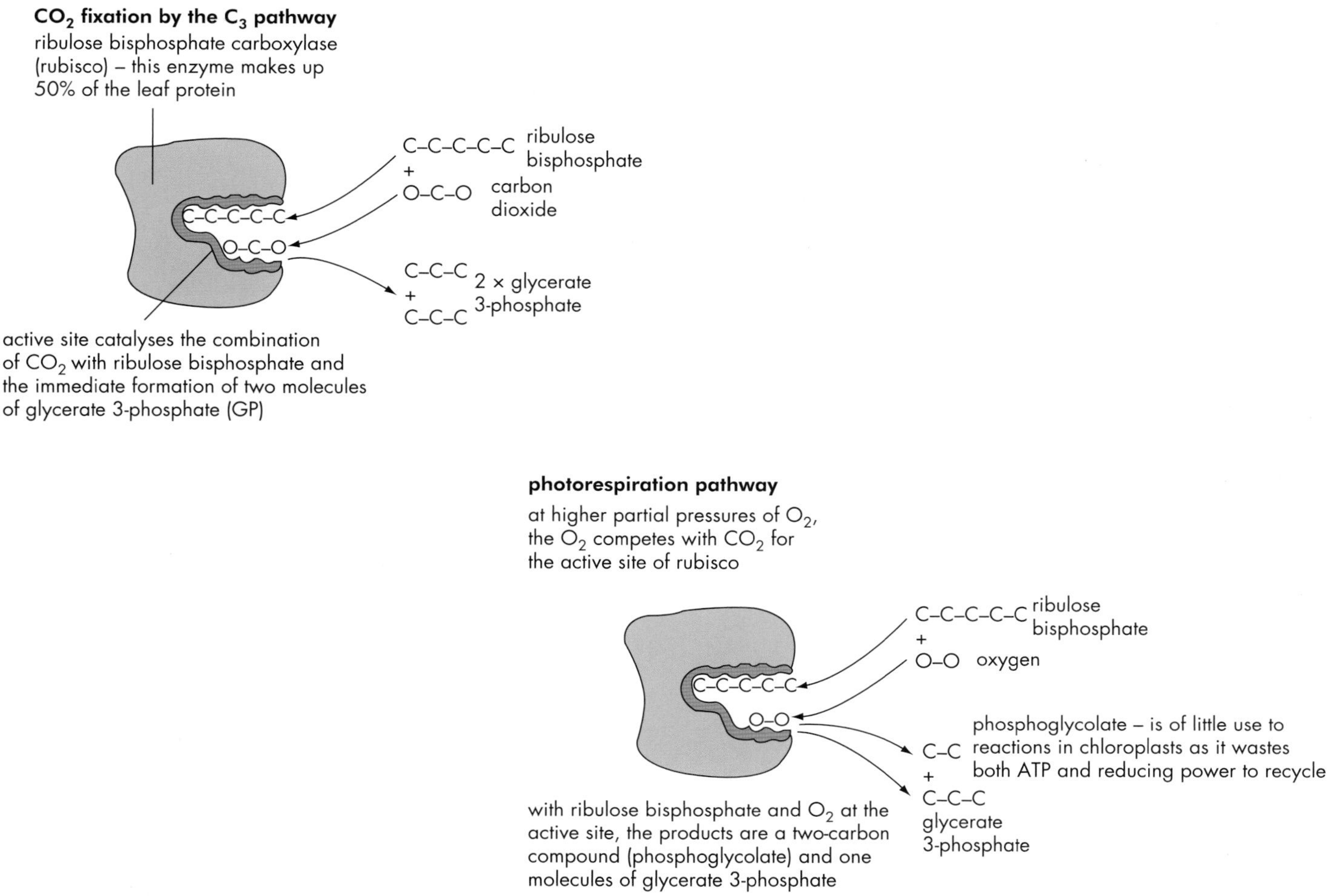

CO$_2$ fixation by the C$_3$ pathway
ribulose bisphosphate carboxylase
(rubisco) – this enzyme makes up
50% of the leaf protein

C–C–C–C–C ribulose
bisphosphate
+
O–C–O carbon
dioxide

C–C–C 2 × glycerate
+ 3-phosphate
C–C–C

active site catalyses the combination
of CO$_2$ with ribulose bisphosphate and
the immediate formation of two molecules
of glycerate 3-phosphate (GP)

photorespiration pathway
at higher partial pressures of O$_2$,
the O$_2$ competes with CO$_2$ for
the active site of rubisco

C–C–C–C–C ribulose
bisphosphate
+
O–O oxygen

phosphoglycolate – is of little use to
C–C reactions in chloroplasts as it wastes
both ATP and reducing power to recycle
+
C–C–C
glycerate
3-phosphate

with ribulose bisphosphate and O$_2$ at the
active site, the products are a two-carbon
compound (phosphoglycolate) and one
molecules of glycerate 3-phosphate

Figure 13.23 Rubisco in photosynthesis and photorespiration

The plants with this mechanism are called **C$_4$ plants**, to separate them from all other green plants that fix carbon dioxide only by the Calvin cycle and produce a three-carbon sugar (triose phosphate) as the first main products – the C$_3$ plants.

C$_4$ plants have a different arrangement of the photosynthetic cells in their leaves. Here, the main photosynthetic cells (called **bundle sheath cells**) are surrounded by **special mesophyll cells** where carbon dioxide is fixed into a four-carbon organic acid (called malate).

So, in C$_4$ plants in the light, the four-carbon organic acid formed in the mesophyll cells is moved into the bundle sheath cells as soon as it is formed. In the bundle sheath cells the organic acid breaks down, producing carbon dioxide and so ensuring its concentration is high (Figure 13.24). At this concentration, carbon dioxide out-competes oxygen at the active site of rubisco.

All the enzymes involved in the C$_4$ pathway have high optimum temperatures and so function efficiently under ambient temperatures in the tropics, in daylight. As a result of this mechanism, C$_4$ plants operate more efficiently at the high temperatures and high light intensities found in tropical regions of the globe. Many tropical grasses and plants of tropical and sub-tropical origins, such as the crop plants sugar cane, maize and sorghum are C$_4$ plants.

Sorghum is another member of the grass family. We have already examined this plant as a xerophyte (page 143). It is a plant that grows well in hot, semi-arid tropical environments of high light intensities, where annual rainfall is typically only 400–600 mm – conditions much too dry for maize. In fact, *Sorghum* has become the fifth most important world cereal, after wheat, rice, maize and barley. Like maize, it is a C$_4$ plant. Examine carefully the features of the *Sorghum* leaf that are the site of C$_4$ carbon dioxide fixation (Figure 13.25).

Question

17 In C$_4$ plants, carbon dioxide is fixed in the bundle sheath cells as well as the mesophyll cells. How does carbon dioxide fixation differ in these two (adjacent) locations?

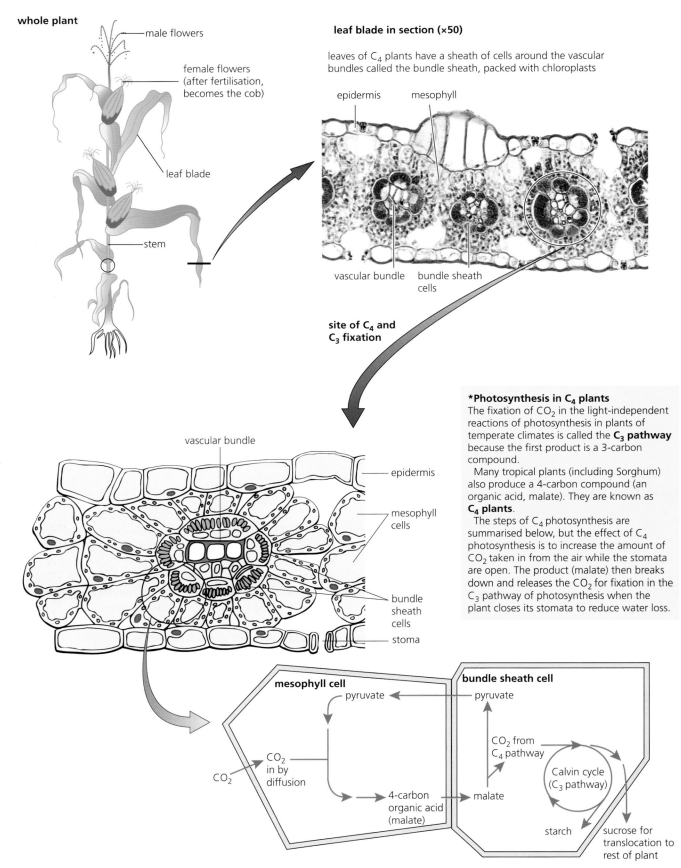

whole plant

male flowers

female flowers
(after fertilisation,
becomes the cob)

leaf blade

stem

leaf blade in section (×50)

leaves of C_4 plants have a sheath of cells around the vascular
bundles called the bundle sheath, packed with chloroplasts

epidermis   mesophyll

vascular bundle   bundle sheath
cells

**site of C_4 and
C_3 fixation**

vascular bundle

epidermis

mesophyll
cells

bundle
sheath
cells

stoma

***Photosynthesis in C_4 plants**
The fixation of CO_2 in the light-independent
reactions of photosynthesis in plants of
temperate climates is called the **C_3 pathway**
because the first product is a 3-carbon
compound.
   Many tropical plants (including Sorghum)
also produce a 4-carbon compound (an
organic acid, malate). They are known as
C_4 plants.
   The steps of C_4 photosynthesis are
summarised below, but the effect of C_4
photosynthesis is to increase the amount of
CO_2 taken in from the air while the stomata
are open. The product (malate) then breaks
down and releases the CO_2 for fixation in the
C_3 pathway of photosynthesis when the
plant closes its stomata to reduce water loss.

mesophyll cell

pyruvate

CO_2
in by
diffusion

CO_2

4-carbon
organic acid
(malate)

bundle sheath cell

pyruvate

CO_2 from
C_4 pathway

malate

Calvin cycle
(C_3 pathway)

starch

sucrose for
translocation to
rest of plant

Figure 13.24 The site and steps of C_4 photosynthesis

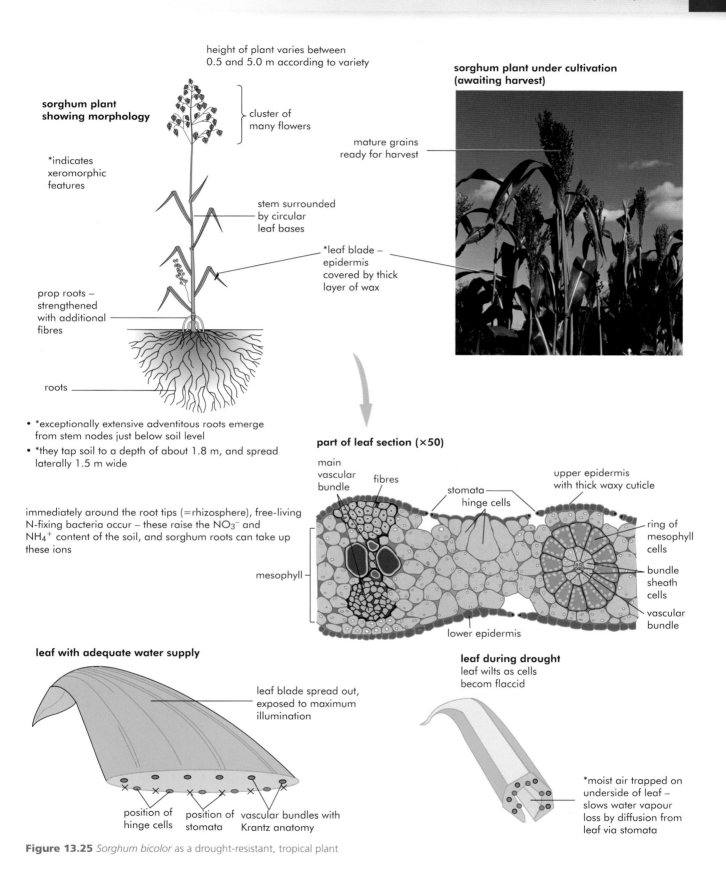

Figure 13.25 *Sorghum bicolor* as a drought-resistant, tropical plant

height of plant varies between 0.5 and 5.0 m according to variety

sorghum plant showing morphology

cluster of many flowers

*indicates xeromorphic features

stem surrounded by circular leaf bases

*leaf blade – epidermis covered by thick layer of wax

prop roots – strengthened with additional fibres

roots

- *exceptionally extensive adventitous roots emerge from stem nodes just below soil level
- *they tap soil to a depth of about 1.8 m, and spread laterally 1.5 m wide

immediately around the root tips (=rhizosphere), free-living N-fixing bacteria occur – these raise the NO_3^- and NH_4^+ content of the soil, and sorghum roots can take up these ions

sorghum plant under cultivation (awaiting harvest)

mature grains ready for harvest

part of leaf section (×50)

main vascular bundle

fibres

stomata

hinge cells

upper epidermis with thick waxy cuticle

ring of mesophyll cells

bundle sheath cells

vascular bundle

mesophyll

lower epidermis

leaf with adequate water supply

leaf blade spread out, exposed to maximum illumination

position of hinge cells

position of stomata

vascular bundles with Krantz anatomy

leaf during drought
leaf wilts as cells becom flaccid

*moist air trapped on underside of leaf – slows water vapour loss by diffusion from leaf via stomata

Summary

- **Photosynthesis** is an **energy transfer** process in which green plants manufacture **carbohydrates** from **carbon dioxide and water**, using energy from **sunlight**. **Oxygen** is the waste product. **Chloroplasts**, the organelles in which photosynthesis occurs, are chiefly found in the mesophyll cells of the leaves, particularly the **palisade mesophyll** cells. Plants use the products of photosynthesis to **manufacture** the other carbohydrates, lipids, proteins and **all other metabolites**, enzymes and pigments required. For this they require energy transferred by respiration and selected mineral ions, absorbed from the soil. Consequently, plant nutrition is described as **autotrophic** (self-feeding).

- The **leaf** is the 'factory' for photosynthesis, with palisade cells in positions to absorb maximum light. The cells are supplied with water from the **xylem vessels**, and with carbon dioxide from the air by diffusion through open **stomata**. The products of photosynthesis are exported from the leaf in the **sieve tubes** in the phloem.

- Chloroplasts are enclosed by a **double membrane**. The **inner membrane** intucks to form the membrane systems of the chloroplast (**thylakoid** membranes). These are arranged in compact, disc-shaped piles called **grana**, around which are loosely arranged thylakoid membranes of the **stroma**.

- Chlorophyll consists of a mixture of pigments, two of which are chlorophylls. They can be separated by **chromatography**. Measurements of the **absorption spectrum** of chlorophyll and the **action spectrum** of photosynthesis show that it is the **blue** and **red** components of white light that are selectively absorbed and are most effective in bringing about photosynthesis.

- Photosynthesis can be divided into two, linked stages.

  - A **light-dependent stage** occurs in the grana and results in the formation of reduced NADP and ATP. It involves the photolysis of water and the release of oxygen.

- A **light-independent stage** occurs in the stroma and involves the fixing of carbon dioxide to form trioses, using the products of the light-dependent stage. Trioses are used in a variety of biochemical reactions within the chloroplast to form monosaccharides (glucose and fructose) that are converted to sucrose for translocation or to starch for storage. Other reactions use minerals from the soil to form amino acids that are translocated and used for protein synthesis.

- In green plants the volume of oxygen-enriched gas given off in the light is a measure of the **rate of photosynthesis**. The volume of gas produced by aquatic plants under different conditions can be measured using a microburette (photosynthometer).

- As photosynthesis consists of a number of interconnected reactions, the slowest will determine the overall rate. The factor limiting its rate is then described as the **limiting factor**. A limited supply of either carbon dioxide or light will limit the overall rate of photosynthesis, since both are essential for photosynthesis.

- The autotrophic nutrition of the green plant also sustains other living things, since other organisms feed on plants or plant products, directly or indirectly.

- Plants, including crop plants show adaptations to different environments. These include **C₄ carbon dioxide metabolism** as part of photosynthesis in many successful tropical and sub-tropical plants. This mechanism is a response to the harmful effects of high oxygen concentrations and low carbon dioxide concentrations on the carbon dioxide-fixing enzyme, rubisco, at high ambient temperatures. The structural adaptations shown by mesophyll tissue in the leaves of sorghum and maize provide the venue for **C₄ carbon dioxide fixation**.

Examination style questions

1 An investigation into the effect of temperature on the rate of the light-dependent stage of photosynthesis was carried out using isolated chloroplasts.

Samples of chloroplasts suspended in buffer were mixed with a coloured electron acceptor and exposed to light. The colour changes from blue to colourless as electrons are taken up by the electron acceptor.

a) i) Sketch a graph to predict the results of the investigation. [2]

ii) Identify two key variables that must be controlled in this investigation. For each explain how it might be controlled. [4]

iii) Outline a procedure to find the rate of reaction for this investigation. [2]

b) In a further investigation, small quantities of ADP and inorganic phosphate were added to the isolated chloroplasts before testing.

Suggest an hypothesis being tested by this further investigation. [1]

[Total: 9]

(Cambridge International AS and A Level Biology 9700, Paper 51 Q1 November 2009)

2 a) By means of a large, fully labelled diagram, describe the structure of a chloroplast, as disclosed by electron microscopy. [6]

b) Annotate your diagram to identify the function or role of the structures you have labelled. [4]

c) In investigations of the physiology and biochemistry of photosynthesis, suspensions of single-celled algae (simple, photosynthetic organisms found growing in fresh water environments) are often substituted for whole, green terrestrial plants. Outline the advantages of this choice to the experimenter. [6]

[Total: 16]

3 a) Outline how the essential products of the light dependent stage of photosynthesis are formed by non-cyclic photophosphorylation from water, light energy, ADP and phosphat, and NADP. [10]

b) Describe the steps of the Calvin cycle, establishing how the products of the light-dependent stage are used, and identifying the products formed. [10]

[Total: 20]

4 a) The enzyme rubisco is the most abundant, naturally occurring protein. Where does rubisco occur, and what reaction does it catalyse? Suggest why its abundance is to be expected. [4]

b) By reference to rubusco, explain the principle of competitive inhibition of an enzyme. [4]

c) The reactions of photorespiration reduce the yield of photosynthesis. Why is this so? [4]

d) By reference to a named C_4 plant, explain, with use of a diagram, the arrangement of cells in a representative part of a leaf that is the site of C_4 metabolism. Annotate your diagram to identify the locations of the steps to C_4 metabolism. [8]

[Total: 20]

5 a) Describe the structure of photosystems and explain how a photosystem functions in **cyclic** photophosphorylation. [9]

b) Explain briefly how reduced NADP is formed in the light-dependent stage of photosynthesis and is used in the light-independent stage. [6]

[Total: 15]

(Cambridge International AS and A Level Biology 9700, Paper 04 Q10 November 2007)

6 a) Show how both the absorption spectrum and the action spectrum of a particular species of fresh water algae may be measured simultaneously.

b) Draw and fully label the absorption and action spectra you would expect to be produced, using a single graph.

c) Describe in outline the steps by which the chlorophyll present in a sample of green leaves can be obtained and the component pigments separated by means of paper chromatography.

14 Homeostasis

Cells function most efficiently if they are kept in near constant conditions. Cells in multicellular animals are surrounded by tissue fluid. The composition, pH and temperature of tissue fluid are kept constant by exchanges with the blood as discussed in the topic on Transport in mammals. In mammals, core temperature, blood glucose concentration and blood water potential are maintained within narrow limits to ensure the efficient operation of cells. Prior knowledge for this topic includes an understanding that waste products are excreted from the body – a role that is fulfilled by the kidneys – and an outline of the structure and function of the nervous and endocrine systems. In plants, guard cells respond to fluctuations in environmental conditions and open and close stomata as appropriate for photosynthesis and conserving water.

14.1 Homeostasis in mammals

Homeostasis in mammals requires complex systems to maintain internal conditions near constant.

By the end of this section you should be able to:

a) discuss the importance of homeostasis in mammals and explain the principles of homeostasis in terms of internal and external stimuli, receptors, central control, co-ordination systems, effectors (muscles and glands)

b) define the term negative feedback and explain how it is involved in homeostatic mechanisms

c) outline the roles of the nervous system and endocrine system in co-ordinating homeostatic mechanisms, including thermoregulation, osmoregulation and the control of blood glucose concentration

h) explain how the blood glucose concentration is regulated by negative feedback control mechanisms, with reference to insulin and glucagon

i) outline the role of cyclic AMP as a second messenger with reference to the stimulation of liver cells by adrenaline and glucagon

j) describe the three main stages of cell signalling in the control of blood glucose by adrenaline as follows:
   • hormone-receptor interaction at the cell surface
   • formation of cyclic AMP which binds to kinase proteins
   • an enzyme cascade involving activation of enzymes by phosphorylation to amplify the signal

k) explain the principles of the operation of dip sticks containing glucose oxidase and peroxidase enzymes, and biosensors that can be used for quantitative measurements of glucose in blood and urine

Maintaining a constant internal environment – homeostasis

Living things face changing and sometimes hostile environments. Some external conditions change slowly, others dramatically. For example, temperature changes quickly on land exposed to direct sunlight, but the temperature of water exposed to sunlight changes very slowly.

How do organisms respond to environmental changes?

Some animals are able to maintain their internal environment, keeping it more or less constant, allowing them to continue normal activities, at least over quite a wide range of external conditions. These are the '**regulators**'. For example, mammals and birds maintain a high and almost constant body temperature. Their bodies are kept at or about the optimum temperature for the majority of the enzymes that drive their metabolism. Their muscles contract efficiently and the nervous system co-ordinates responses precisely, even when external conditions are unfavourable.

Regulators are often able to avoid danger and they may also benefit from the vulnerability of prey organisms which happen to be '**non-regulators**'. So regulators may have greater freedom in choosing where to live. They can exploit more habitats with differing conditions than 'non-regulators' can, too. But whether an organism is a 'regulator' or not, some of control of the internal environment is essential.

Homeostasis is the name we give to the ability to maintain a constant internal environment. Homeostasis means 'staying the same'. Mammals are excellent examples of animals that keep their internal conditions remarkably constant. Their internal environment is the blood circulating in the body and the fluid circulating among cells (tissue fluid) that forms from the blood plasma, delivering nutrients and removing waste products whilst bathing the cells. Mammals successfully regulate and maintain the pH, the concentrations of oxygen, carbon dioxide and glucose, the temperature and the water content of their blood. All these are maintained at constant levels or within very narrow limits.

How is homeostasis brought about and maintained?

polar bear on ice

otter in fresh water

whale in the sea

camels in the desert

bat in the air

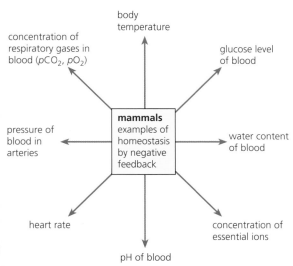

Mammals are a comparitively recent group in terms of their evolutionary history, yet they have successfully settled in significant numbers in virtually every type of habitat on Earth. This success is directly linked to their ability to control their internal environment by homeostasis.

Figure 14.1 Homeostasis in mammals

287

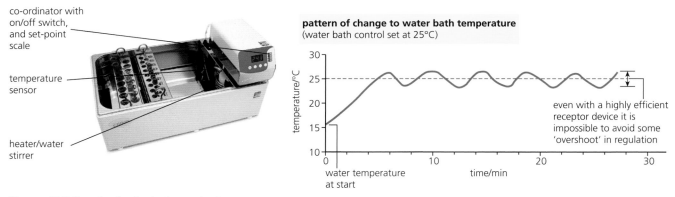

components of a negative feedback control system

input –
change to
the *system*

receptor
measures level
of the *variable*

co-ordinator
level of operation is set here, and
information from receptor *received* and
compared with set value, and *commands
to effector* despatched from here

effector
brings about a second
change to system
(in opposite direction
to the input)

output –
condition restored
to *set value*

feedback loop
establishes the change has been corrected,
and causes the *effector to be switched off*

the laboratory water bath unit, an example
of a self-regulating system

co-ordinator with
on/off switch,
and set-point
scale

temperature
sensor

heater/water
stirrer

pattern of change to water bath temperature
(water bath control set at 25°C)

water temperature
at start

time/min

even with a highly efficient
receptor device it is
impossible to avoid some
'overshoot' in regulation

Figure 14.2 Negative feedback: the mechanism

Negative feedback – the mechanism of homeostasis

Negative feedback is a type of control in which the conditions being regulated are brought back to a set value as soon as it is detected that they have deviated from it. We see this type of mechanism at work in a laboratory water bath. Analysis of this familiar example will show us the components of a negative feedback system.

A negative feedback system requires a **receptor** device that measures the value of the variable (in this case, the water temperature of the water bath) and transmits this information to a **co-ordinator** (in this case, the control unit). The co-ordinator compares data from the receptor with a **pre-set value** (the desired water temperature of the water bath). When the value is below the required value the co-ordinator switches on an **effector** device (a water heater in the water bath) so that the temperature starts to increase. Once the water reaches the required temperature, data from the receptor to this effect is received in the co-ordinator, which then switches off the response (the water heater). How precisely the variable is maintained depends on the sensitivity of the receptor, but negative feedback control typically involves some degree of 'overshoot'.

In mammals, regulation of body temperature, blood sugar level and the concentration of water and ions in blood and tissue fluid (osmoregulation) are all regulated by negative feedback. The receptors are specialised cells either in the brain or in other organs, such as the pancreas. The effectors are organs such as the skin, liver and kidneys. Information passes between them either by impulses in neurones within nerves of the **nervous system** or by **hormones** released into the blood by **endocrine organs**. The outcome is an incredibly precisely regulated internal environment. We shall return to examine these two systems in turn, shortly.

Question

1 Homeostasis is illustrated in the way our bodies regulate carbon dioxide level. (Remind yourself of the control of ventilation of the lungs by looking at Figure 9.8, page 177).

a Why is it essential that carbon dioxide in the blood does not exceed a certain level?

b In this example of homeostasis, identify where the receptor, co-ordinator and effectors occur in our bodies.

Homeostasis in action – control of body temperature

Heat may be transferred between an animal and the environment by **convection**, **radiation** and **conduction**, and the body loses heat in **evaporation** (Figure 14.3).

Figure 14.3 How heat is transferred between organism and environment

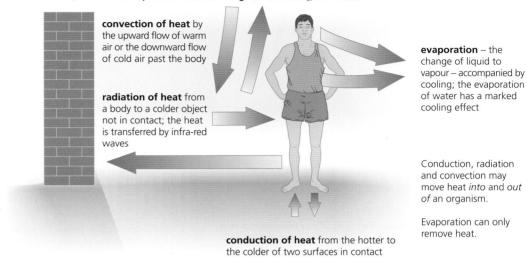

convection of heat by the upward flow of warm air or the downward flow of cold air past the body

radiation of heat from a body to a colder object not in contact; the heat is transferred by infra-red waves

evaporation – the change of liquid to vapour – accompanied by cooling; the evaporation of water has a marked cooling effect

Conduction, radiation and convection may move heat *into* and *out of* an organism.

Evaporation can only remove heat.

conduction of heat from the hotter to the colder of two surfaces in contact

The regulation of body temperature, known as **thermoregulation**, involves controlling the movement of heat across the body surface. Mammals maintain a high and relatively constant body temperature, using the heat energy generated by metabolism within their bodies (or by generating additional heat in the muscles when cold) and carefully controlling the loss of heat through the skin. An animal with this form of thermoregulation is called an **endotherm**, meaning 'inside heat'. Humans hold their inner body temperature (core temperature) just below 37 °C. In fact, human inner body temperature only varies between about 35.5 and 37.0 °C within a 24 hour period, when we are in good health.

Heat production in the human body

The major sources of heat are the biochemical reactions of metabolism that generate heat as a waste product. Heat is then distributed by the blood circulation. The organs of the body vary greatly in the amount of heat they yield (see Figure 12.13). When at rest, the bulk of our body heat (over 70 per cent) comes from other abdominal organs, mainly from the heart and kidneys, but also from the lungs and brain (which, like a computer central processing unit, needs to be kept cool). Of course, in times of intense physical activity, the skeletal muscles generate a great deal of heat as a waste product of respiration and contraction.

If the body experiences a sudden loss of heat, then heat production in increased. Heat output from the body muscles is raised by non-co-ordinated contraction of skeletal muscles, known as '**shivering**'. This raises muscle heat production about five times above basal rate.

The roles of the skin in thermoregulation

Heat exchange occurs largely at the surface of the body. Here in the skin, the following structures and mechanisms are combined in the regulation of heat loss (as detailed in Figure 14.4):

- at capillary networks, by the arteriole supplying them being dilated (vasodilation) when the body needs to loose heat, but constricted (vasoconstriction) when the body needs to retain heat
- by the hair erector muscles, which contract when heat must not be lost but relax when heat loss must be increased
- by the sweat glands, which produce sweat when the body needs to lose heat, but do not when the body needs to retain heat.

structure of the skn

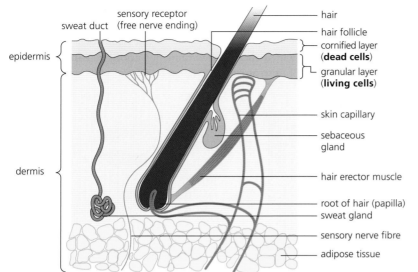

role of the sweat glands in regulating heat loss through the skin

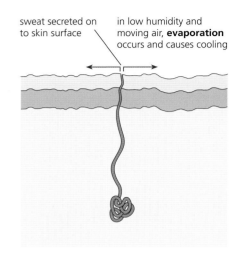

sweat secreted on to skin surface

in low humidity and moving air, **evaporation** occurs and causes cooling

role of capillaries in regulating heat loss through the skin

In skin that is especially exposed (e.g. outer ear, nose, extremities of the limbs) the capillary network is extensive, and the arterioles supplying it can be dilated or constricted.

warm conditions

cold conditions

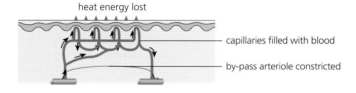
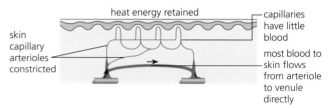

role of the hair in regulating heat loss through the skin

warm conditions

cold conditions

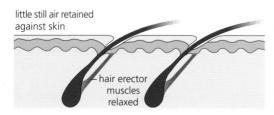
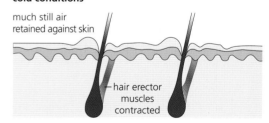

Figure 14.4 The skin and temperature regulation

The hypothalamus as the control centre

The 'control box' for temperature regulation in mammals is a region of the forebrain called the **hypothalamus**. It is called the thermoregulation centre, and it consists of a 'heat loss centre' and a 'heat gain centre' (Figure 14.5). Here, temperature-sensitive nerve cells (neurones) detect changes in the temperature of the blood flowing through the brain (*internal* stimuli). The thermoregulation centre of the hypothalamus also receives information via sensory nerves from temperature-sensitive receptors located in the skin (*external* stimuli), and in many internal organs.

The hypothalamus communicates with the rest of the body via the **nervous system**. For example, when the body temperature is lower than normal, the heat gain centre inhibits activity of the heat loss centre, and sends impulses to the skin, hair erector muscles, sweat glands and elsewhere that decrease heat loss (such as vasoconstriction of skin capillaries) and increase heat production (such as shivering, and enhanced 'brown fat' respiration). When body temperature is higher than normal the heat loss centre inhibits the heat gain centre activity, and sends impulses to the skin, hair erector muscles, sweat glands and elsewhere that increase heat loss (such as vasodilation of skin capillaries), and decrease heat production.

Another mechanism in thermoregulation

Thyxoxin is an iodine-containing **hormone** produced in the thyroid gland. On secretion it stimulates oxygen consumption and basal metabolic rate of the body organs. Variation in secretion of thyroxin helps to control body temperature (Figure 14.5).

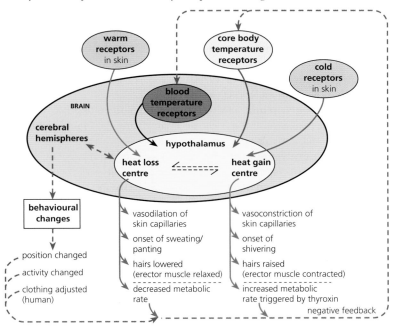

Figure 14.5 Temperature regulation of the body by the hypothalamus

The roles of the endocrine and nervous systems compared

In the processes of thermoregulation we have seen that both the nervous and endocrine systems are involved. The endocrine system and the nervous system work in distinctive and different ways in the control and co-ordination of body activities. However, many of the activities of the nervous and endocrine systems are **co-ordinated** by the pituitary gland, the master gland of the endocrine system, working in tandem with the hypothalamus of the brain. The hypothalamus secretes hormones that regulate the functioning of the pituitary. The hypothalamus also monitors the level of hormones in the blood and regulates secretion by negative feedback control.

Table 14.1 The endocrine and nervous systems compared

Endocrine system	Nervous system
Communication by chemical messengers transmitted in the bloodstream	Communication by electrochemical action potentials (impulses) transmitted via neurones
Hormones 'broadcast' all over the body but influence target cells and tissues only	Action potentials are targeted on specific cells
Causes changes in metabolic activity	Causes muscles to contract or glands to secrete
Have their effects over many minutes, several hours or longer	Produces effects within milliseconds
Effects tend to be long lasting	Effects tend to be short lived and reversible

Homeostasis in action – control of blood glucose concentration

Respiration is a continuous process in all living cells, and glucose is the principle respiratory substrate for most tissues. To maintain tissue respiration, cells need a regular supply of glucose. Transport of glucose to all cells is a key function of the blood circulation. Fortunately, blood glucose can be quickly absorbed across the cell membrane. In humans, the **normal level of blood glucose** is about 4 mM/L of blood, but it varies, typically between 3.6 and 5.8 mM/L. The lower values arise during an extended period without food, or after prolonged and heavy physical activity; the highest values occur after a meal rich in carbohydrate has been digested.

Most cells (including muscle cells) hold reserves in the form of glycogen. This polysaccharide is quickly converted to glucose during prolonged physical activity. However, glycogen reserves may be used up quickly. (In the brain, glucose is the only substrate the cells can use and there are no glycogen stores held in reserve there at all.)

The maintenance of a constant level of this monosaccharide in the blood plasma is the norm, but two extreme conditions can arise:

- **Hypoglycaemia**, in which our blood glucose falls below 2.0 mM/L. If this is not quickly reversed, we may faint. If the body, and particularly the brain, continue to be deprived of adequate glucose levels, convulsions and coma follow.
- **Hyperglycemia**, in which an abnormally high concentration of blood glucose occurs. Since a high concentration of any soluble metabolite lowers the water potential of the blood plasma, water is drawn immediately from the cells and tissue fluid by osmosis, back into the blood. As the volume of blood increases, water is excreted by the kidney to maintain the correct concentration of blood. As a result, the body tends to become dehydrated and the circulatory system is deprived of fluid. Ultimately, blood pressure cannot be maintained.

For these reasons, it is critically important that the blood glucose is held within set limits.

Question

2 Explain what liver cells receive from blood from the hepatic artery that is not present in blood from the hepatic portal vein.

Regulation of blood glucose

Regulation of blood glucose is the result of the actions of two hormones, released from groups of special cells, found in the pancreas.

After the digestion of carbohydrates in the gut, glucose is absorbed across the epithelium cells of the villi (finger-like extensions of the inner surface of the small intestine) into the hepatic portal vein (Figure 8.7, page 158). The blood carrying the glucose reaches the liver first. If the glucose level is too high, then glucose starts to be withdrawn from the blood and is stored as glycogen. However, not all of the glucose can be removed immediately. Blood circulating in the body immediately after a meal has a raised level of glucose (Figure 14.6).

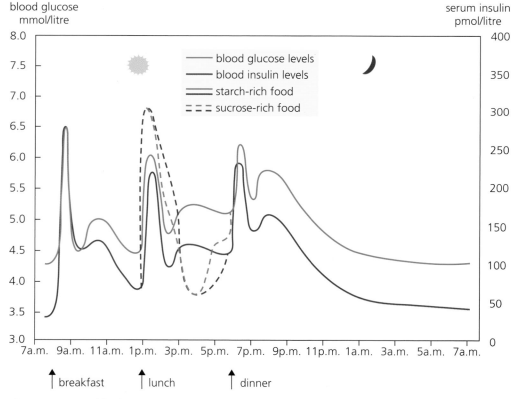

Figure 14.6 How blood glucose and insulin levels vary with time

At the pancreas, the presence of an excess of blood glucose is detected in patches of cells known as the **islets of Langerhans**. These islets are hormone-secreting glands (endocrine glands); they have a rich capillary network, but no ducts that would carry secretions away. Instead, they are transported all over the body by the blood. The islets of Langerhans contain two types of cell, **alpha (α) cells** and **beta (β) cells** (Figure 14.7).

TS of pancreatic gland showing an islet of Langerhans

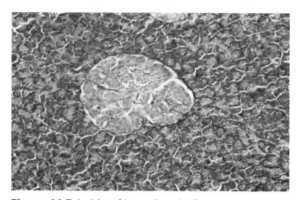

drawing of part of pancreatic gland

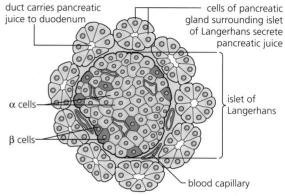

Figure 14.7 An islet of Langerhans in the pancreas

In the presence of a **raised blood glucose level**, the **beta cells** are stimulated. They secrete the hormone **insulin** into the capillary network. Insulin stimulates the uptake of glucose by cells all over the body, but especially by the liver and the skeletal muscle fibres. Another effect of insulin is to trigger the conversion of glucose to glycogen (**glycogenesis**) and of glucose to fatty acids and fats in liver cells. Insulin also promotes the deposition of fat around the body.

As the blood glucose level reverts to normal this is detected in the islets of Langerhans, and the beta cells respond by stopping insulin secretion. Meanwhile the hormone is excreted by the kidney tubules and the blood insulin level falls.

When the **blood glucose level falls below normal**, the **alpha cells** are stimulated. These secrete a hormone called **glucagon**. This hormone activates the enzymes that convert glycogen and amino acids to glucose (**gluconeogenesis**). Glucagon also reduces the rate of respiration.

As the blood glucose level reverts to normal, glucagon production ceases, and this hormone in turn is removed from the blood in the kidney tubules.

Question

3 Given the role of alpha cells and beta cells (the production and discharge of hormones), identify the organelles that are involved directly and list their specific roles.

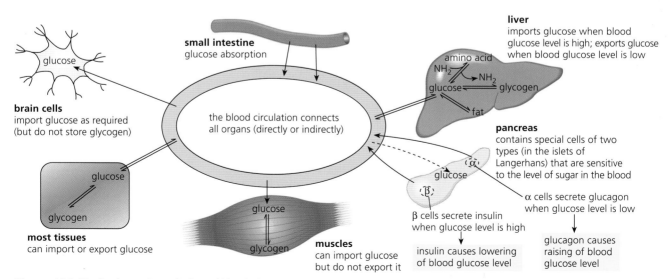

Figure 14.8 Distribution and metabolism of blood glucose

Figure 14.9 Glucose regulation
by negative feedback

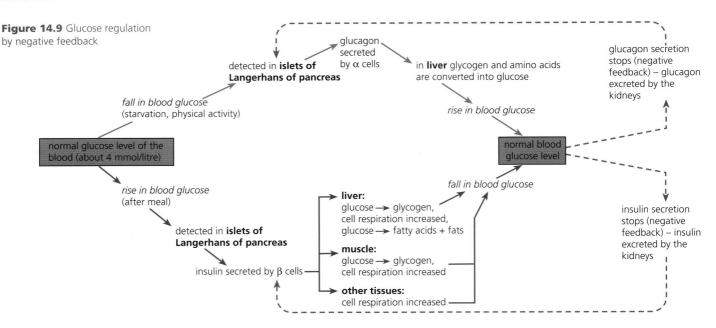

The role of cyclic AMP in the effects of adrenalin and glucagon on liver cells

Both adrenalin and glucagon are peptide hormones, which, whilst circulating in the bloodstream, stimulate liver cells to convert stored glycogen to glucose. The glucose formed then passes out from the liver cells and contributes to blood glucose levels.

The actions triggered by these two hormones are brought about within the liver cells without the hormones having passed through the cell's cell surface membrane. Instead, the hormones bind to specific receptors in the cell surface membrane, and this event triggers the following changes:

- First, another membrane-embedded protein, known as a G-protein, is activated.
- The activated G-protein in turn activates an enzyme known as adenylyl cyclase, also embedded in the cell surface membrane.

Figure 14.10 The role of cyclic AMP as a second messenger

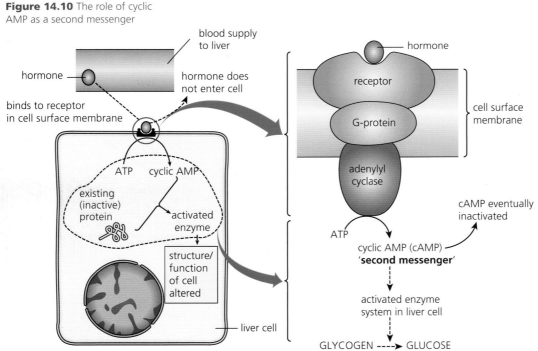

- Activated adenylyl cyclase catalyses the formation of cyclic AMP (cAMP) from ATP within the cell cytosol.

cAMP is known as a '**second messenger**'. It is a small, non-protein molecule that is water soluble, and so spreads quickly through the cytosol by diffusion. The presence of this second messenger triggers a cascade of reactions in liver cells in which specific enzymes are activated by reaction with ATP. The outcomes are:

- the hormone signal is amplified
- the glycogen stored in the liver cells is hydrolysed to glucose.

These events are summarised in Figure 14.10.

inactive kinase enzyme —↘ cAMP

active kinase enzyme

inactive phosphorylase enzyme ↘

active phosphorylase enzyme

glycogen ↘

glucose 1-phosphate

Figure 14.11 The three stages of cell-signalling that adrenalin triggers

The impact of adrenalin on liver cells as a three-stage process of cell-signalling

The impact of adrenalin on a liver cell is a **three-stage process** of cell signalling (Figure 14.11). After the first stage, when a hormone molecule interacts at the cell surface, subsequent events are activated by a succession of relay molecules, initiated by the formation of a second messenger, cAMP, within the cytosol.

The relay molecules are activated kinases and phosphorylases. These proteins catalyse specific types of reaction:

- **Kinases** are enzymes that transfer a phosphate group from ATP to an acceptor.
- **Phosphorylases** are enzymes that break down glucose-based polysaccharides (e.g. glycogen) to glucose 1-phosphate.

The critical role of the second and third stages of signal transduction is the **amplification** of the hormone signal. So, from one activated receptor molecule, 10 000 (10^4) molecules of cAMP are formed. As a result of the presence of cAMP in the cytosol, approximately 10^6 molecules of active phosphorylase enzyme will be formed, and these will then trigger the formation of perhaps 10^8 molecules of glucose 1-phosphate.

Table 14.2 Stages of signal transduction leading to amplification

Stage	Events	Consequences
1	hormone interaction at the cell surface	a single hormone initiates events
2	formation of cyclic AMP, which binds to kinase proteins ATP → cAMP	about 10^4 molecules of cAMP become active in the cytosol
3	an enzyme cascade involving activation of enzymes by phosphorylation to amplify the signal	approximately 10^8 molecules of glucose 1-phosphate are formed and become available to pass out into the bloodstream

Extension

The disease of diabetes

Diabetes is the name for a group of diseases in which the body fails to regulate blood glucose levels. **Type I diabetes** results from a failure of insulin production by the beta cells. It is also referred to as insulin-dependent diabetes, early onset diabetes or juvenile diabetes (because it typically appears early on in a person's life). **Type II diabetes** is a failure of the insulin receptor proteins on the cell surface membranes of target cells. It is also referred to as diabetes mellitus, late onset diabetes or adult onset diabetes.

As a consequence of either form of the disease, blood glucose regulation is more erratic and, generally, the level of glucose is permanently raised – to dangerous levels in untreated cases of Type I diabetes. Glucose is also regularly excreted in the urine. If this condition is not diagnosed and treated, it carries a risk of circulatory disorders, renal failure, blindness, strokes and heart attacks.

The production of **human insulin** for treatment of diabetes by the genetic engineering of bacteria is discussed in Topic 19.

Type I diabetes (insulin-dependent diabetes, early onset diabetes or juvenile diabetes)
affects young people, below the age of 20 years
due to the destruction of the β cells of the islets of Langerhans by the body's own immune system
symptoms:
- constant thirst
- undiminished hunger
- excessive urination

treatment:
- injection of insulin into the blood stream daily
- regular measurement of blood glucose level

Type II diabetes (diabetes mellitus, late onset diabetes or adult onset diabetes)
the common form (90% of all cases of diabetes are of this type)
common in people over 40 years especially if overweight, but this form of diabetes is having an increasing effect on human societies around the world, including on young people and even children in developed countries, seemingly because of poor diet
symptoms:
mild sufferers usually have sufficient blood insulin, but insulin receptors on cells have become defective
treatment:
largely by diet alone

Dip-sticks and biosensors for quantitative measurements of glucose

Dip sticks, such as Clinistix™ or, alternatively biosensors, are used to measure glucose levels in urine or blood.

Glucose is not present in the urine of a healthy person because, although it appears in the filtrate in the Bowman's capsule of the kidney, it is selectively reabsorbed in the proximal convoluted tubules (page 301). However, people with diabetes cannot control their blood sugar levels effectively. Glucose may appear in the urine when the blood sugar levels rise steeply, for example after a meal. Diabetics may need to inject insulin to reduce blood sugar levels. The urine may be tested to find out if an injection is needed using Clinistix™. Alternatively, a glucose biosensor may be used (see Figure 14.13).

The Clinistix™ strip contains two enzymes, glucose oxidase and peroxidase, together with a colourless hydrogen donor compound called chromogen. When the strip is dipped into the urine sample, if glucose is present it is oxidised to gluconic acid and hydrogen peroxide. The second enzyme catalyses the reduction of hydrogen peroxide and the oxidation of chromogen. The product is water and the oxidised dye, which is coloured. The more glucose present in the urine, the more coloured dye is formed. The colour of the test strip is then compared to the printed scale to indicate the amount of glucose in the urine (Figure 14.12).

An alternative approach for the diabetic is to measure the glucose level in the blood itself using a **glucose biosensor** (Figure 14.13). A biosensor is a device which makes use of a biological molecule (or sometimes a cell) to detect and measure a chemical compound.

The glucose biosensor has an immobilised enzyme, **glucose oxidase**, held between two membranes positioned at the tip of a platinum electrode. In use, the outer membrane is momentarily brought in contact with a tiny drop of blood, squeezed from a pinprick puncture of the skin at the tip of a finger. In contact with the immobilised enzyme, glucose in the blood plasma is immediately oxidised to gluconic acid and hydrogen peroxide. The electrode measures the drop in oxygen used to produce the hydrogen peroxide and an electrical signal is generated. The size of this signal is proportional to the concentration of glucose in the patient's blood. A digital read-out gives the concentration of blood glucose.

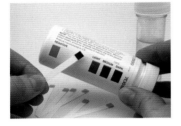

the principles

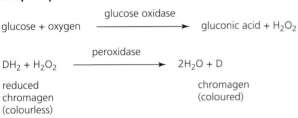

glucose + oxygen $\xrightarrow{\text{glucose oxidase}}$ gluconic acid + H_2O_2

$DH_2 + H_2O_2 \xrightarrow{\text{peroxidase}} 2H_2O + D$

reduced chromagen (colourless)

chromagen (coloured)

the process

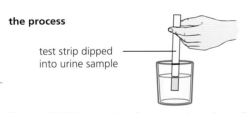

test strip dipped into urine sample

Figure 14.12 Measuring glucose in urine using a Clinistix™

(page 301)

Question

4 What are the advantages to a diabetic patient of measuring blood glucose by means of a biosensor, compared to the Clinistix™ method?

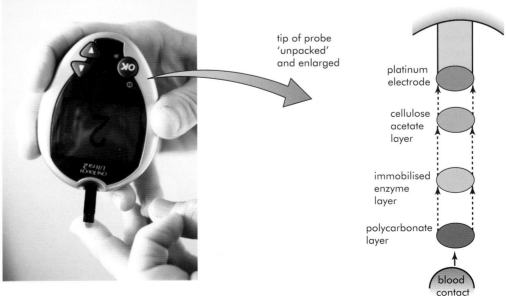

tip of probe 'unpacked' and enlarged

platinum electrode

cellulose acetate layer

immobilised enzyme layer

polycarbonate layer

blood contact

Figure 14.13 Biosensor for blood glucose testing

14.1 *continued* ... Kidneys – structure and function

The kidneys remove wastes from the blood and are the effectors for controlling the water potential of the blood.

By the end of this section you should be able to:

d) describe the deamination of amino acids and outline the formation of urea in the urea cycle
e) describe the gross structure of the kidney and the detailed structure of the nephron with its associated blood vessels using photomicrographs and electron micrographs
f) describe how the processes of ultrafiltration and selective reabsorption are involved with the formation of urine in the nephron
g) describe the roles of the hypothalamus, posterior pituitary, ADH and collecting ducts in osmoregulation
l) explain how urine analysis is used in diagnosis with reference to glucose, protein and ketones

Excretion, blood water balance and metabolic waste

Excretion: the elimination from the body of waste compounds produced during the metabolism of cells, including, for a human, carbon dioxide (excreted through the lungs) and urea (excreted through the kidneys in urine).

Excretion is a characteristic activity of all living things. It is essential because the chemical reactions of metabolism produce byproducts, some of which would be toxic if allowed to accumulate. Excretion is the removal from the body of the waste compounds produced during the metabolism of cells. Humans excrete carbon dioxide from the lungs and urea in urine produced by our kidneys.

In mammals, excretion is a part of the process of homeostasis. Excretion of nitrogenous compounds is important in animals because they do not store protein. Proteins in the diet are broken down to their constituent amino acids. Amino acids that are excess to requirements for protein synthesis are respired. The first step is called **deamination**. This occurs in the liver and is the removal from each amino acid of its amino group, which becomes **ammonia**. Ammonia is a very soluble and extremely toxic compound. However, the ammonia is promptly converted into a safer nitrogenous compound. In mammals that compound is **urea** – formed by reaction of ammonia with carbon dioxide (Figure 14.14). In dilute solution, urea is safely excreted from the body.

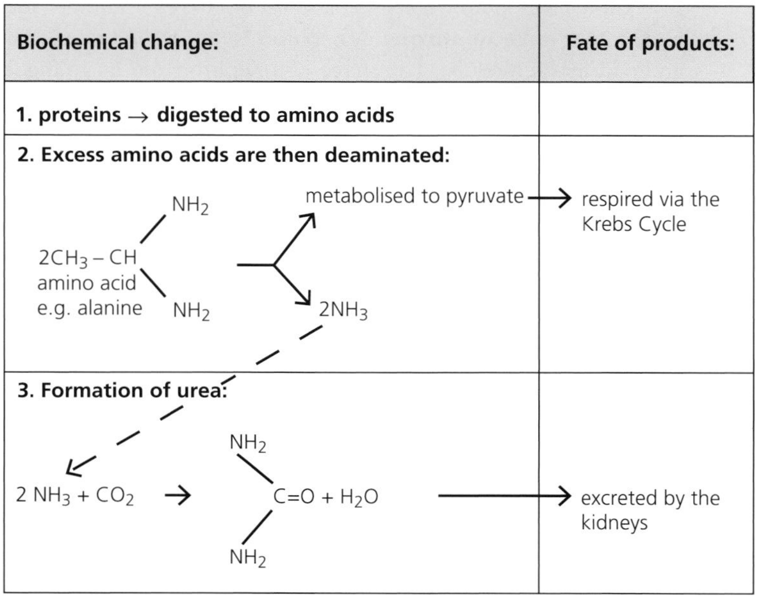

Biochemical change:	Fate of products:
1. proteins → digested to amino acids	
2. Excess amino acids are then deaminated:	metabolised to pyruvate → respired via the Krebs Cycle
3. Formation of urea:	excreted by the kidneys

Figure 14.14 Deamination

Associated with excretion, and very much part of it, is the process of osmoregulation. Osmoregulation is the maintenance of a proper balance in the water and dissolved substances in the blood. Excretion and osmoregulation together, are the work of our kidneys.

The kidney – an organ of excretion and osmoregulation

The role of our kidneys is to regulate the internal environment is by constantly adjusting the composition of the blood. Waste products of metabolism are transported from the metabolising cells by the blood circulation, removed from the blood in the kidneys and excreted in a solution called **urine**. At the same time, the concentrations of inorganic ions, such as sodium (Na^+) and chloride (Cl^-) ions, and water in the body are also regulated.

The position of the kidneys is shown in Figure 14.15. Each kidney is served by a **renal artery** and drained by a **renal vein**. Urine from the kidney is carried to the bladder by the **ureter** and from the **bladder** to the exterior by the **urethra**, when the bladder sphincter muscle is relaxed. Together these structures are known as the **urinary system**.

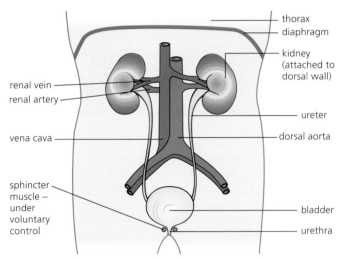

Figure 14.15 The human urinary system

In section, each kidney consists of an outer **cortex** and inner **medulla**, and these are made up of a million or more **nephrons**. A nephron is thin-walled tubule about 3 cm long, part in the cortex and part in the medulla. The shape of a nephron and its arrangement in the kidney are shown in Figure 14.16.

Blood vessels are closely associated with each of the distinctly-shaped regions of the nephrons. For example, the first part of the nephron is formed into a cup-shaped **renal capsule** and the capillary network here is known as the **glomerulus**. Collectively, these are known as the **Malpighian body**. They occur in the cortex. The **convoluted tubules** occur partly in the cortex and partly in the medulla, but notice that the extended **loops of Henle** and **collecting ducts** largely occur in the medulla.

Each region of the nephron has a specific role to play in the work of the kidney, and the capillary network serving the nephron plays a key part, too, as we shall now see.

The formation of urine

In humans, about 1–1.5 litres of urine are formed each day, typically containing about 40–50 g of solutes, of which **urea** (about 30 g) and **sodium chloride** (up to 15 g) make up the bulk. The nephron produces urine in a continuous process which we can conveniently divide into five steps, to show just how the blood composition is so precisely regulated.

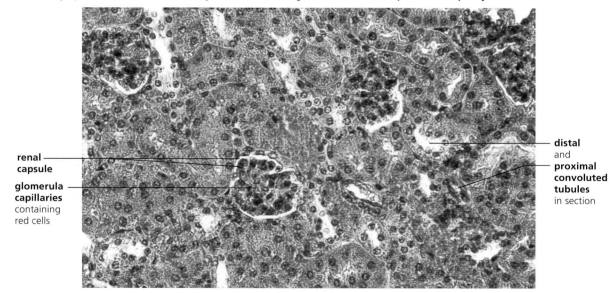

LS through kidney showing positions of nephrons in cortex and medulla

- fibrous capsule
- nephrons
- collecting duct
- renal artery
- pelvis – expanded origin of ureter
- medulla
- ureter
- cortex

nephron with blood capillaries

1 Malpighian body {
- renal capsule
- glomerulus

2 proximal convoluted tubule

4 distal convoluted tubule

- cortex
- medulla
- capillary network to convoluted tubules
- branch of renal artery
- branch of renal vein

3 loop of Henle
- descending limb
- ascending limb

- vasa recta (capillary to loop of Henle)

5 collecting duct

Roles of the parts of the nephron

1 Malpighian body = ultrafiltration
2 proximal convoluted tubule = selective reabsorption from filtrate
3 loop of Henle = water conservation
4 distal convoluted tubule = pH adjustment and ion reabsorption
5 collecting duct = water reabsorption

Photomicrograph of the cortex of the kidney in section, showing the tubules, renal capsules and capillary networks

- renal capsule
- glomerula capillaries containing red cells
- distal and proximal convoluted tubules in section

Figure 14.16 The kidney and its nephrons: structure and roles

Step 1: Ultrafiltration in the renal capsule

In the glomerulus, water and relatively small molecules of the blood plasma, including useful ions, glucose and amino acids pass out of the capillaries, along with urea, into the lumen of the capsule. This process is described as ultrafiltration because it is powered by the pressure of the blood.

The **blood pressure** is raised at this point by the input capillary (**afferent arteriole**) being wider than the output capillary (**efferent arteriole**). This increase in hydrostatic (blood) pressure exceeds the water potential of the plasma. This forces water and the soluble components of plasma that are able to pass out through the extremely fine sieve-like wall structure here, between the podocytes, into the capsule. This 'sieve' is made of two layers of cells (the endothelium of the capillaries of the glomerulus and the epithelium of the capsule), between which is a basement membrane. You can see this arrangement in Figure 14.17.

Notice that the cells of the capsule wall are called **podocytes** for they have feet-like extensions that form a network with tiny slits between them. Similarly, the endothelium of the capillaries has **pores**, too. This detail was discovered using the electron microscope, because these filtration gaps are very small indeed. So small, in fact, that not only are blood cells retained, but the majority of blood proteins and polypeptides dissolved in the plasma also remain in the circulating blood. It is the presence of the **basement membrane** that stops large proteins passing out. So, the fluid that has been filtered through into the renal capsule is very similar to blood plasma but with the significant difference that protein molecules are largely absent.

Question

5 a What is the source of the force that drives ultrafiltration in the glomerulus?
  b List the main components of the blood that are likely to be filtered in the nephron.

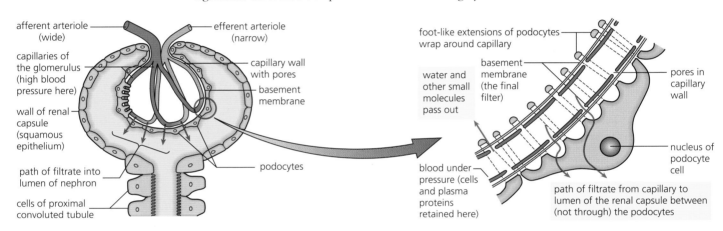

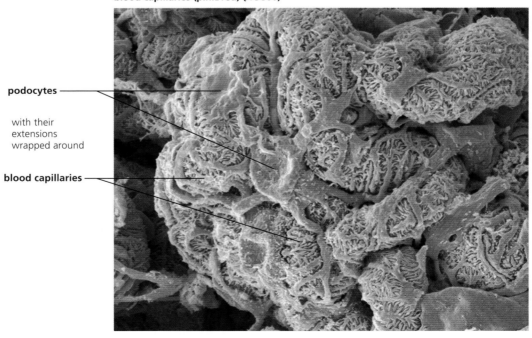

False-colour SEM of podocytes (pale purple) with their extensions wrapped around the blood capillaries (pink/red) (× 3500)

podocytes

with their extensions wrapped around

blood capillaries

Figure 14.17 The site of ultrafiltration

Step 2: Selective reabsorption in the proximal convoluted tubule

The proximal convoluted tubule is the longest section of the nephron and it is here that a large part of the filtrate is reabsorbed into the capillary network. The walls of the tubule are one cell thick and their cells are packed with mitochondria. (We would expect this, if active transport is part of the way reabsorption is brought about.) The cell surface membranes of the cells of the tubule wall in contact with the filtrate all have a 'brush border' of microvilli. These microvilli increase enormously the surface area where reabsorption occurs. The mechanisms of reabsorption are:

- **active transport** of **sugars** and **amino acids** across the cell surface membrane by the activity of special carrier proteins in a process known as co-transport. In this, the carrier protein uses the diffusion of hydrogen ions (protons) down their electrochemical gradient into the cell to drive the uptake of molecules of sugars such as glucose or sucrose, typically against their concentration gradient. Other essential metabolites are transported similarly.
- movement of mineral ions by a combination of **active transport**, **facilitated diffusion** and some **exchange of ions**
- **diffusion** of urea
- movement of proteins by **pinocytosis**
- some movement of water by **osmosis**.

The electrochemical gradient in H+ ions between the exterior and the interior of the cell drives the active transport of metabolites (e.g. sugars, amoni acids) across the membrane as the H+ flow down their electrochemical gradient via the co-transporter pump.

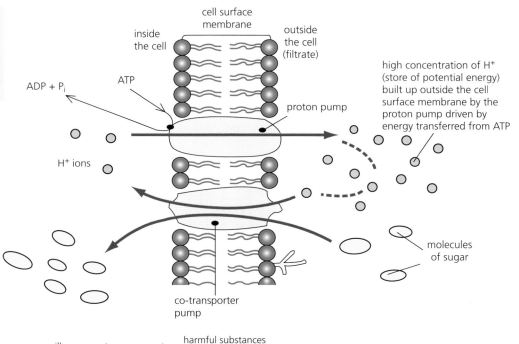

Figure 14.18 Co-transport: active transport driven by a concentration gradient

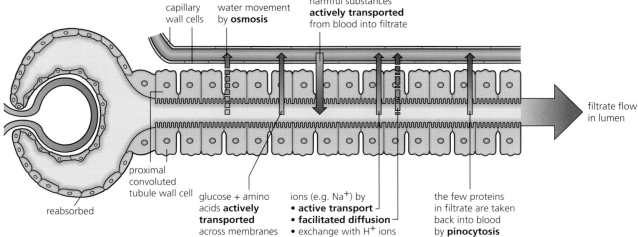

Figure 14.19 Reabsorption in the proximal convoluted tubule

Step 3: Water conservation in the loop of Henle

The function of the loop of Henle is to enable the kidneys to conserve water. Since urea is expelled from the body in solution, some water loss during excretion is inevitable. There is a potential problem here. Water is a major component of the body and it is often a scarce resource for terrestrial organisms. It is important, therefore, that mammals are able to form urine that is more concentrated than the blood (when necessary), thereby reducing the water loss to a minimum. Human urine can be up to five times as concentrated as the blood.

The structure of the loop of Henle with its **descending** and **ascending limbs**, together with a parallel blood supply, the **vasa recta**, is shown in Figure 14.20. The vasa recta is part of the same capillary network that surrounds a nephron. The role of the loops of Henle and their capillary loops is to create and maintain an osmotic gradient in the **medulla of the kidney**. The gradient across the medulla is from a less concentrated salt solution near the cortex to the most concentrated salt solution at the tips of the pyramid region of the medulla (Figure 11.5). The pyramid region of the medulla consists mostly of the collecting ducts. The osmotic gradient allows water to be withdrawn from the collecting ducts if circumstances require it. How this is occurs we shall discuss shortly. First, there is the question of how the gradient itself is created by the loops of Henle.

The gradient is brought about by a mechanism known as a **counter-current multiplier**. The principles of counter-current exchange here involve exchange between fluids flowing in opposite directions in two systems. The annotations in Figure 14.20 help show how the counter-current mechanism works. In following these annotations, remember that the descending and ascending limbs lie close together in the kidney.

Look first at the *second* half of the loop, the **ascending limb**. The energy to create the gradient is transferred from ATP to drive ion pumps in the wall cells of the ascending limbs. Here, sodium and chloride ions are pumped out of the filtrate into the fluid between the cells of the medulla, called the interstitial fluid. The walls of the ascending limbs are unusual in being impermeable to water. So water in the ascending limb is retained in the filtrate as salt is pumped out.

Opposite is the *first* half of the loop, the **descending limb**. This limb is fully permeable to water and also to most salts. Here, water passes out into the interstitial fluid by osmosis, due to the salt concentration in the medulla. At the same time and for the same reason, sodium, chloride and other ions tend to pass in.

Exchange in this counter-current multiplier is a dynamic process occurring down the whole length of the loop. At each level of the loops, the salt concentration in the descending limb is slightly higher than the salt concentration in the adjacent ascending limb. As the filtrate flows, the concentrating effect is multiplied and so the fluid in and around the hairpin bend of the loops of Henle is the saltiest.

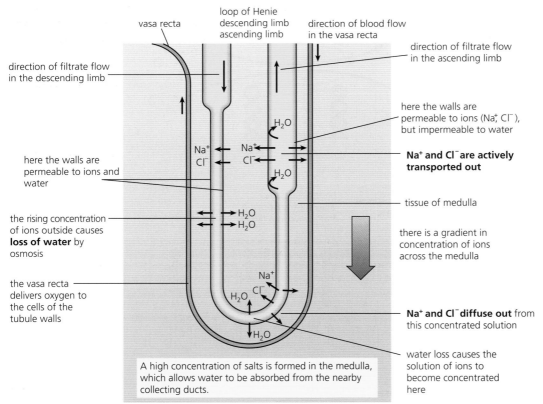

A high concentration of salts is formed in the medulla, which allows water to be absorbed from the nearby collecting ducts.

Figure 14.20 The functioning of the loop of Henle

Question

7 What is the essential feature of a counter-current flow system?

The role of the vasa recta is first to deliver oxygen to and remove carbon dioxide from the metabolically active cells of the loop of Henle. As it does this, the blood in the vasa recta also becomes saltier as it flows *down* beside the ascending limb and less salty as it flows back up and out of the medulla. In this way the cells of the loop are serviced without removing the accumulated salts from the medulla. The vasa recta does absorb water that has passed into the medulla at the collecting ducts. We discuss the working of these ducts in step 5, on the next page.

Step 4: Blood pH and ion concentration regulation in the distal convoluted tubule

The cells of the walls of the distal convoluted tube are of the same structure as those of the proximal convoluted tubule, but their roles differ somewhat.

Here the cells of the tubule walls adjust the composition of the blood, in particular, the **pH**. Any slight tendency for the pH of the blood to change is initially prevented by the blood proteins. Protein acts as a pH buffer. (You can read about how a pH buffer works in Appendix 1: Background chemistry for biologists, on the CD).

However, if the blood does begin to deviate from pH 7.4, then here in the distal convoluted tubule there is a controlled secretion of hydrogen ions (H^+) combined with reabsorption of hydrogencarbonate ions (HCO_3^-). Consequently, the pH of the blood remains in the range pH 7.35–7.45, but the pH of urine can vary widely, in the range pH 4.5–8.2.

Also in the distal convoluted tubule, the **concentration of useful ions is regulated**. In particular, the concentrations of potassium ions (K^+) is adjusted by secretion of any excess present in the plasma into the filtrate. Similarly, the concentration of sodium ions (Na^+) in the body is regulated by varying the quantity of sodium ions reabsorbed from the filtrate.

Step 5: Water reabsorption in the collecting ducts

When the intake of water exceeds the body's normal needs then the urine produced is copious and dilute. We notice this after we have been drinking a lot of water. On the other hand, when we have taken in very little water or when we have been sweating heavily (part of our temperature regulation mechanism) or if we have eaten very salty food perhaps, then a small volume of concentrated urine is formed.

Osmoregulation, the control of the water content of the blood (and therefore of the whole body) is a part of homeostasis – another example of regulation by negative feedback.

How exactly is this brought about?

The **hypothalamus**, part of the floor of the forebrain (Figure 14.21), controls many body functions. The composition of the blood is continuously monitored here, as it circulates through the capillary networks of the hypothalamus. Data is also received at the hypothalamus from sensory receptors located in certain organs in the body. All these inputs enable the hypothalamus to accurately control the activity of the pituitary gland.

The **pituitary gland** is situated below the hypothalamus, but is connected to it (Figure 14.21). The pituitary gland as a whole produces and releases hormones (it is part of our endocrine system, *see below*) – in fact it has been called the master hormone gland. In the process of osmoregulation, it is the posterior part of the **pituitary** that stores and releases **antidiuretic hormone** (**ADH**) – amongst others. (Other parts of the pituitary secrete hormones regulating a range of other body activities and functions.)

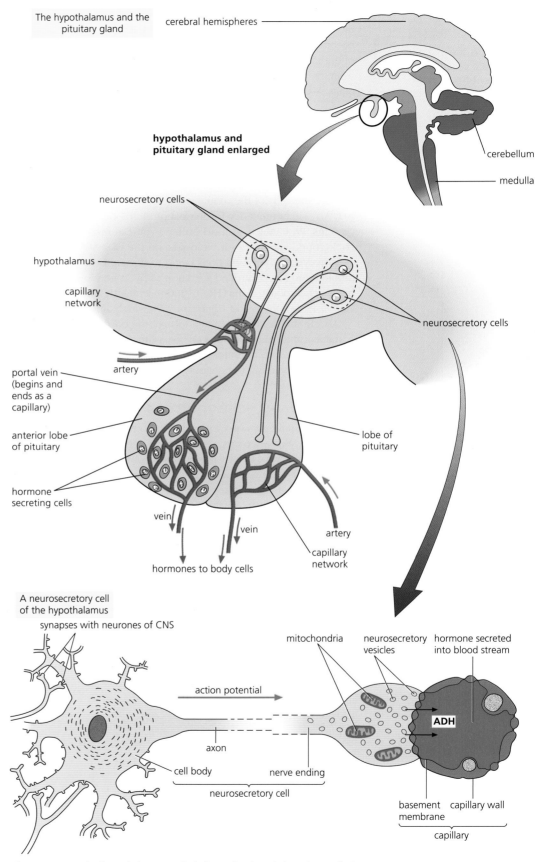

The hypothalamus and the pituitary gland

cerebral hemispheres

cerebellum

medulla

hypothalamus and pituitary gland enlarged

neurosecretory cells

hypothalamus

capillary network

neurosecretory cells

artery

portal vein (begins and ends as a capillary)

anterior lobe of pituitary

lobe of pituitary

hormone secreting cells

vein

vein

artery

capillary network

hormones to body cells

A neurosecretory cell of the hypothalamus synapses with neurones of CNS

mitochondria

neurosecretory vesicles

hormone secreted into blood stream

action potential

ADH

axon

cell body

nerve ending

neurosecretory cell

basement membrane

capillary wall

capillary

Figure 14.21 The hypothalamus and pituitary gland, and the release of ADH

Antidiuretic hormone (ADH) is actually produced in the hypothalamus and stored in vesicles at the ends of neurosecretory cells in the posterior pituitary gland. When nerve impulses from the hypothalamus trigger the release of ADH into the capillary networks in the posterior pituitary, ADH circulates in the blood stream. However, the targets of this hormone are the walls of the collecting ducts of the kidney tubules.

When the water content of the blood is low, antidiuretic hormone (ADH) is secreted from the posterior pituitary gland. When the water content of the blood is high, little or no ADH is secreted.
How does ADH change the permeability of the walls of the collecting ducts?

The cell surface membranes of the cells that form the walls of the collecting ducts contain a high proportion of channel proteins that are capable of forming an open pore running down their centre. You can see the structure of a fluid mosaic membrane and its channel proteins in Figure 4.3 (page 76) and Figure 4.9 (page 82).

When ADH is present in the blood circulating past the kidney tubules, this hormone causes the protein channels present in the collecting duct cell surface membranes to be open. As a result, much water diffuses out into the medulla and very little diffuses from the medulla into the collecting ducts.
Can you explain why? You may need to go back to step 3 above.

The water entering the medulla is taken up and redistributed in the body by the blood circulation. Only small amounts of very concentrated urine are formed. Meanwhile, as ADH circulates in the blood, the actions of the liver continually remove this hormone and inactivate it. This means that the presence of freshly released ADH has a regulatory effect.

When ADH is absent from the blood circulating past the kidney tubules, the protein channels in the collecting duct cell surface membranes are closed. The amount of water that is retained by the medulla tissue is now minimal. The urine become copious and dilute.

Urine analyses in medical diagnosis

An analysis of a urine sample for the presence of glucose, protein or ketone bodies is an aid to the early diagnosis of serious but treatable conditions, as summarised in Table 14.2.

Table 14.2 The value of urine analyses in medical diagnosis

Metabolite	Tested for by	Health significance
glucose	Clinistix™ or biosensor	Presence of glucose may be an indicator of **diabetes**.
protein	dipstick test for albumin	Presence of protein is an indicator of **proteinura** – a possible indicator of kidney disease arising from glomerula membrane damage, or severe hypertension.
ketones	dipstick test – Ketostix™	Ketones are a normal product of fatty acid metabolism, but are not normally present in the urine. However, if the body is short of glucose, then fats are used as an energy source and ketones may then appear in the urine. Typically this is the result of **type 1 diabetes**, or **advanced/untreated type 2 diabetes**, but **eating disorders** are another possibility.

Questions

8 a List the components of a negative feedback system and identify these components in the process of osmoregulation.

  b How is the effect of ADH on the collecting ducts fed back to the co-ordinator?

9 Why is too much water rarely a problem for plant cells, but potentially hazardous to animal cells?

Water reabsorption in the collecting ducts

When we have:
• drunk a lot of water
  the hypothalamus detects this and stops the posterior pituitary gland secreting ADH.

When we have:
• taken in little water
• sweated excessively
• eaten salty food
  the hypothalamus detects this and directs the posterior pituitary gland to secrete ADH.

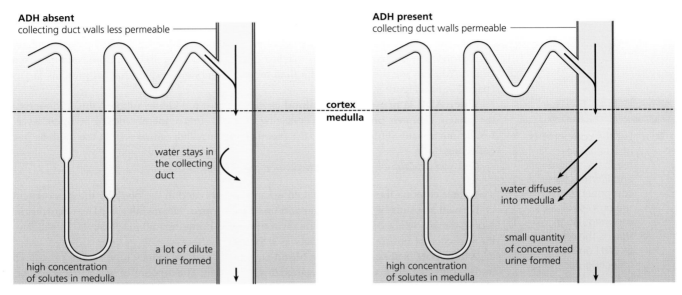

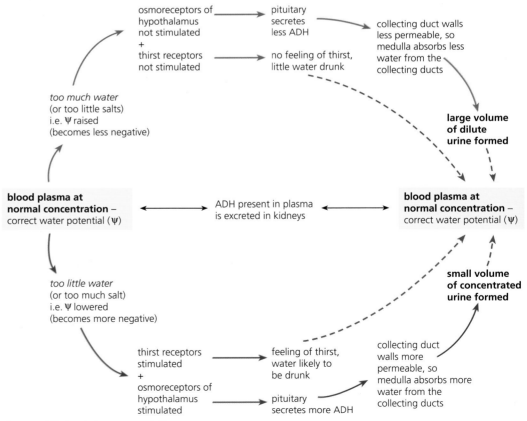

Figure 14.22 The collecting ducts and their role in osmoregulation

14.2 Homeostasis in plants

Stomatal aperture is regulated in response to the requirements for uptake of carbon dioxide for photosynthesis and conserving water.

By the end of this section you should be able to:

a) explain that stomata have daily rhythms of opening and closing and also respond to changes in environmental conditions to allow diffusion of carbon dioxide and regulate water loss by transpiration

b) describe the structure and function of guard cells and explain the mechanism by which they open and close stomata

c) describe the role of abscisic acid in the closure of stomata during times of water stress (including the role of calcium ions as a second messenger)

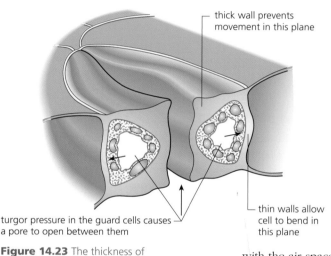

thick wall prevents movement in this plane

thin walls allow cell to bend in this plane

turgor pressure in the guard cells causes a pore to open between them

Figure 14.23 The thickness of guard cell walls and opening/closing movements

The stomata – structure and role

The tiny pores of the epidermis of leaves through which gas exchange can occur are known as stomata. Most stomata occur in the leaves, but some do occur in stems. In the broad-leaved plants, stomata are typically concentrated in the lower epidermis of the leaves. In the narrow, pointed leaves typical of many grasses for example, stomata may be equally distributed on both surfaces. The structure of stomata was introduced in Topic 7 (Figure 7.11 on page 139).

Remind yourself of their structure, now.

Each stoma consists of two elongated **guard cells**. These cells are attached to ordinary epidermal cells that surround them and are securely joined together at each end, but are detached and free to separate along their length, forming a pore between them. When open, stomatal pores connect the atmosphere external to the leaf with the air spaces between the living cells of leaf and stem. All these air spaces are interconnected. This is the pathway of diffusion of carbon dioxide into the leaf, and of water vapour out.

Stomata open and close due to changes in **turgor pressure** of the guard cells. They open when water is absorbed by the guard cells from the surrounding epidermal cells. The guard cells then become fully turgid and they push into the epidermal cell besides them. This is because of the variable thickness of cellulose in the walls, and because of the way cellulose fibres are laid down there (Figure 14.23). A pore then develops between the guard cells. When water is lost and the guard cells become flaccid, the pore closes again. This has been demonstrated experimentally (see Figure 14.25 on the next page).

Stomata tend to open in daylight and close in the dark (but there are exceptions to this diurnal pattern). This pattern of opening and closing of stomata, over a 24-hour period, is shown in Figure 14.24.

This daily pattern is over-ridden, however, if and when the plant becomes short of water and starts to wilt. For example, in very dry conditions when there is an inadequate water supply, stomata inevitably close relatively early in the day (because turgor cannot be maintained). This closure curtails water vapour loss by transpiration and halts further wilting. Adequate water reserves from the soil may be taken up subsequently, thereby allowing the opening of stomata again, for example, on the following day. The effect of this mechanism is that stomata regulate transpiration in that they prevent excessive water loss when this is threatened. Of course, when the stomata are forced to close, inward diffusion of carbon dioxide is interrupted.

the pattern of opening and closing of stomata over 24 hours (as observed in potato leaf)

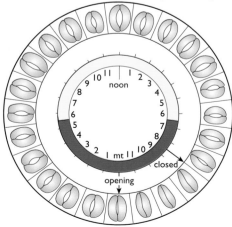

Figure 14.24 The diurnal pattern of stomatal opening and closing

In an experimental demonstration that turgor pressure of the guard cells causes the opening of the stomatal pore, a microdissection needle was inserted into a guard cell.

Observation: on release of the turgor pressure in one guard cell, the distinctive shape of the cell when the pore is open was lost. 'Half' of the pore disappeared.

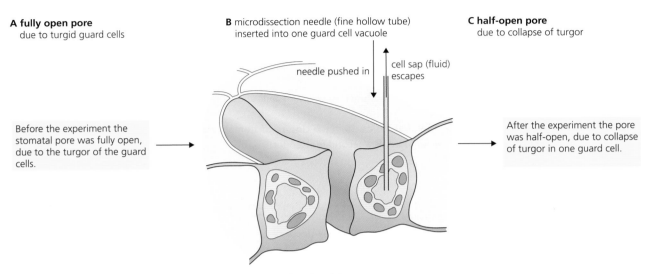

A fully open pore
due to turgid guard cells

B microdissection needle (fine hollow tube) inserted into one guard cell vacuole

needle pushed in

cell sap (fluid) escapes

C half-open pore
due to collapse of turgor

Before the experiment the stomatal pore was fully open, due to the turgor of the guard cells.

After the experiment the pore was half-open, due to collapse of turgor in one guard cell.

Figure 14.25 The opening of stomata – it's turgor pressure that does it!

The opening and closing of stomata

We have seen that, because of the structure of guard cell walls, stomata open when water is absorbed by the guard cells from the surrounding epidermal cells. Fully turgid guard cells push into the epidermal cell besides them, and the pore develops.

The guard cells contain chloroplasts (all the other epidermal cells do not), but opening is not due to the slow build-up of sugar by photosynthesis in these chloroplasts, leading to the turgor pressure change of opening. Opening is a much quicker process.

Stomatal opening depends on two biochemical changes:

1 **Potassium ions** (K^+ – a cation) are pumped into the guard cell vacuole, from surrounding cells, by proteins of the cell surface membranes, triggered by light (blue wavelengths). Calcium ions play a part in this process.
2 Starch, stored in the guard cells, is **converted to organic acids**, particularly malate. These anions accompany the K^+ cations in the guard cell vacuole.

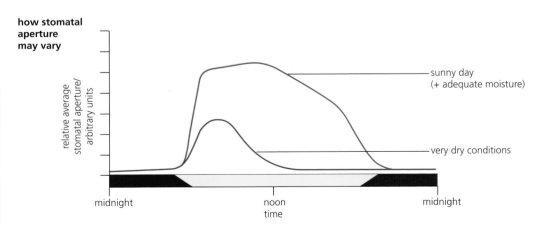

how stomatal aperture may vary

relative average stomatal aperture/ arbitrary units

sunny day (+ adequate moisture)

very dry conditions

midnight

noon
time

midnight

Figure 14.26 Stomatal opening and environmental conditions

The accumulation of these substances in the guard cell vacuole causes the water potential there to become more negative. So, net uptake of water from the surrounding ordinary epidermal cells occurs – making the guard cells extremely turgid (Figure 14.27).

Closing is brought about by the reversal of these steps in the dark. Alternatively closure occurs when triggered by the stress hormone abscisic acid (**ABA**), produced in leaf cells during wilting. We will discuss this situation next.

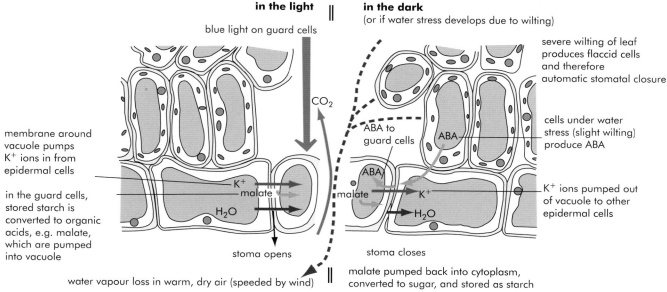

in the light

blue light on guard cells

in the dark
(or if water stress develops due to wilting)

severe wilting of leaf produces flaccid cells and therefore automatic stomatal closure

CO_2

ABA to guard cells

ABA

cells under water stress (slight wilting) produce ABA

membrane around vacuole pumps K^+ ions in from epidermal cells

ABA

in the guard cells, stored starch is converted to organic acids, e.g. malate, which are pumped into vacuole

K^+
malate

H_2O

malate

K^+

H_2O

K^+ ions pumped out of vacuole to other epidermal cells

stoma opens

stoma closes

water vapour loss in warm, dry air (speeded by wind)

malate pumped back into cytoplasm, converted to sugar, and stored as starch

Figure 14.27 How opening and closing of stomata occurs

Abscisic acid and water stress in the leaf

Abscisic acid (ABA) is a plant growth inhibitor substance. The discovery of growth inhibitors in plants came from investigations of dormancy in buds and seeds. ABA is formed in mature leaves, in ripe fruits and seeds, and is present in dormant seeds. It accumulates in leaves of deciduous trees in late summer, just before their winter buds become dormant, and also in deciduous tropical plants at the onset of the dry season.

Of particular interest is the fact that ABA is also involved in the **closure of stomata under conditions of drought**. Green plants that are experiencing drought produce ABA in their chloroplasts, including those in the guard cells, and this 'stress hormone' triggers stomatal closure. This response involves calcium ions as the 'second messenger', as follows.

ABA triggers release of calcium ions from the cell sap in the guard cell vacuoles into the guard cell cytosol, via activated calcium ion channels in the tonoplast. Consequently, the concentration of calcium ions in the cytosol is abruptly increased, and directly triggers:

1 a sudden efflux of anions such as chloride ions from the cytosol, across the cell surface membrane, by activated anion channels

2 a sudden efflux of potassium ions from the cytosol, across the cell surface membrane, by activated potassium ion pumps.

3 inhibition of efflux of potassium ions into the guard cells (a feature of the stomatal opening mechanism).

As a result, the water potential of the guard cells is abruptly raised. It is now higher than the water potential of the surrounding cells, and a net efflux of water occurs from the guard cells. The guard cells become flaccid and the pores close.

Summary

- **Homeostasis** is the maintenance of a constant internal environment despite fluctuating external conditions. The ability to do this efficiently has allowed mammals as a group to live in very diverse habitats. **Negative feedback control** mechanisms operate in homeostasis. When departure from the normal or set value of conditions within an organism is detected (such as an abnormal blood sugar level), responses are set in motion which restore normal conditions and switch off the 'disturbance' signal.

- The **nervous system** and the endocrine system have interconnected roles in co-ordinating homeostatic mechanisms. The nervous system consists of receptors (sense organs) linked to effectors (muscles or glands) by neurones. **Hormones** are produced in endocrine glands, transported all over the body in the blood, and effect specific target organs.

- **Thermoregulation**, the process of body temperature control, is brought about by regulating heat loss and heat gain by the body, and is an example of a homeostatic mechanism. As endotherms, our main heat source comes from the level of respiratory metabolism in the tissues. The body temperature is held at or close to the optimum for the action of metabolic enzymes.

- **Blood glucose** regulation is essential because cells need a more or less constant supply, especially brain cells. Too high a glucose level lowers the water potential of the blood to the point where the circulation takes water by osmosis from tissue fluid and dehydrates cells and tissues. Too low a glucose level may lead to loss of consciousness and coma.

- The **kidney** is an organ of **excretion and osmoregulation**. In excretion the waste products of metabolism are removed from the blood and made ready for discharge from the body. In osmoregulation, the balance of water and solutes in body fluids is maintained at a constant level despite variations in intake. Kidney tubules (nephrons) work by pressure filtration of some of the liquid and soluble components of blood, followed by selective reabsorption of useful substances from the filtrate, active secretion of unwanted substances and adjustments to water and ion content according to their status in the body.

- Adrenalin and glucagon, peptide hormones that stimulate liver cells to convert stored glycogen to glucose, work by binding to external receptors on liver cells cell surface membranes, and indirectly triggering cyclic AMP (cAMP) within the cell cytosol. This '**second messenger**' triggers a cascade of reactions in liver cells in which the hormone signal is amplified. The impact of adrenalin on a liver cell is a **three-stage process** of cell signalling resulting in the **amplification** of the hormone signal. So, from one activated receptor molecule, perhaps 10^8 molecules of glucose 1-phosphate are formed from glycogen.

- Plants carry out gaseous exchange between their living cells and the air spaces between them, which are continuous throughout the plant. Carbon dioxide gas enters and oxygen and water vapour leave the leaves through the **pores** in **stomata**. Stomata consist of guard cells surrounding a pore which permits this gaseous exchange between leaves and environment. There is a daily rhythm of opening and closing of guard cells that is over-ruled in the event of excess water loss (leading to water stress within the plant). It is the plant growth regulator, **abscisic acid** that then triggers premature closure of stomata, so conserving water.

Examination style questions

1. **a)** Describe how nitrogenous waste products are formed and explain why they need to be removed from the body. [6]
   b) Describe how the kidney removes metabolic wastes from the body. [9]

   [Total: 15]

   (Cambridge International AS and A Level Biology 9700, Paper 04 Q7 November 2005)

2. **a)** After the digestion of carbohydrates in the gut, glucose is absorbed into the blood stream and the concentration of blood glucose initially rises. Describe how a surge in blood glucose is normally regulated by the body and what happens to the bulk of this metabolite.
   b) Draw a flow diagram showing the steps by which the concentration of blood glucose is held constant by a process of negative feedback.

   [Total: 20]

3. **a)** Give an illustrate account of the structure of a stoma. [7]
   b) Outline the mechanism by which the stomatal pore opens. [6]
   c) How do stress conditions within leaves due to water shortage affect stomatal opening, and what part do plant growth regulators play in this response? [7]

   [Total: 20]

4. **a)** Identify the distinctive roles of the endocrine and nervous systems in homeostasis, illustrating your answer by reference to the mechanism of thermoregulation in mammals. [10]
   b) Outline the role of a 'second messenger' in the impact of a peptide hormone such as adrenalin on a liver cell. [10]

   [Total: 20]

15 Control and co-ordination

All the activities of multicellular organisms require co-ordinating, some very rapidly and some more slowly. The nervous system and the endocrine system provide co-ordination in mammals. Similar co-ordination systems exist in plants.

↻ 15.1 Control and co-ordination in mammals

The nervous system provides fast communication between receptors and effectors.

Transmission between neurones takes place at synapses.

By the end of this topic you should be able to:

a) compare the nervous and endocrine systems as communication systems that co-ordinate responses to changes in the internal and external environment

b) describe the structure of a sensory neurone and a motor neurone

c) outline the roles of sensory receptor cells in detecting stimuli and stimulating the transmission of nerve impulses in sensory neurones

d) describe the functions of sensory, relay and motor neurones in a reflex arc

e) describe and explain the transmission of an action potential in a myelinated neurone and its initiation from a resting potential

f) explain the importance of the myelin sheath (salutatory conduction) in determining the speed of nerve impulses and the refractory period in determining their frequency

g) describe the structure of a cholinergic synapse and explain how it functions, including the role of calcium ions

h) outline the roles of synapses in the nervous system in allowing transmission in one direction and in allowing connections between one neurone and many others

The need for communication systems

The ability to detect change and to make responses is essential for the survival of living things. It is as much a feature of single-celled organisms as it is of flowering plants and mammals. We see this when an *Amoeba* detects a suitable food organism and captures it by phagocytosis.

Large and complex organisms detect changes in the external environment as well as changes within the body. They need to communicate this information to parts of the body where appropriate responses will be made.

We call changes that bring about responses **stimuli**. The stimulus is detected by a **receptor** and an **effector** brings about a response. Since response often occurs in a different part of the body, efficient internal communication is also essential. In mammals, internal communication involves both the **nervous** and **endocrine** (hormone-producing) systems. We will examine these next.

Introducing the nervous system

The nervous system is built from specialised nerve cells called **neurones**. Neurones are grouped together to form the **central nervous system**, which consists of the **brain and spinal cord**. To and from the central nervous system run nerves of the **peripheral nervous system**. Communication between the central nervous system and all parts of the body occurs via neurones in these nerves.

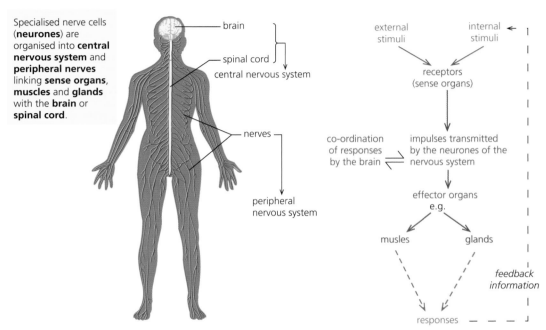

Specialised nerve cells (**neurones**) are organised into **central nervous system** and **peripheral nerves** linking **sense organs**, **muscles** and **glands** with the **brain** or **spinal cord**.

Figure 15.1 The organisation and co-ordination of the mammalian nervous system

Neurones – structure and function

Each neurone has a **cell body** containing the nucleus and the bulk of the cytoplasm. From the cell body, extremely fine **cytoplasmic processes** run. Many of these processes are very long indeed. They are specialised for the transmission of information in the form of **impulses**. Impulses are transmitted at speeds between 30 and 120 metres per second in mammals, so nervous co-ordination is extremely fast and responses are virtually immediate. Another feature of the nerve impulse is that it travels to particular points in the body. Consequently, the effects of impulses are localised rather than diffuse. (This makes communication by nerves different from that by hormones, as we shall see.)

The three types of neurones are shown in Figure 15.2.

Question

1 Compare motor, sensory and relay neurones by means of a concise table.

- **Motor neurones** have a cell body that lies within the brain or spinal cord. Many highly-branched cell processes extend from the cell body. These are called **dendrites**. Dendrites receive impulses from other neurones and conduct them *towards* the cell body. A single long **axon** transmits impulses *away* from the cell body. The function of a motor neurone is to transmit impulses from the central nervous system to effector organs, such as muscles or glands.
- **Relay neurones** (also known as inter-neurones) have numerous, short fibres. Each fibre is a thread-like extension of a nerve cell. Relay neurones occur in the central nervous system. They relay impulses to other neurones.
- **Sensory neurones** have a single long **dendron**, which brings impulses *towards* the cell body, and a single **axon** which carries impulses *away* from the cell body. Sensory neurones transmit impulses from receptors to the spinal cord or brain.

Surrounding the neurones there are different types of supporting cells called **glia cells**. These are also an important part of the nervous system. In the brain and spinal cord there are several types of glia cells. In the peripheral nervous system, the axons of many neurones are enclosed by glia cells called **Schwann cells**. These wrap themselves around the axon with many layers of cell surface membrane, forming a structure called a **myelin sheath** (Figure 15.2). The sheath consists largely of lipid with some protein. Between each pair of Schwann cells is a tiny, uncovered junction in the myelin sheath, called a **node of Ranvier**. The myelin sheath and its junctions help increase the speed at which impulses are conducted – a point we will return to shortly.

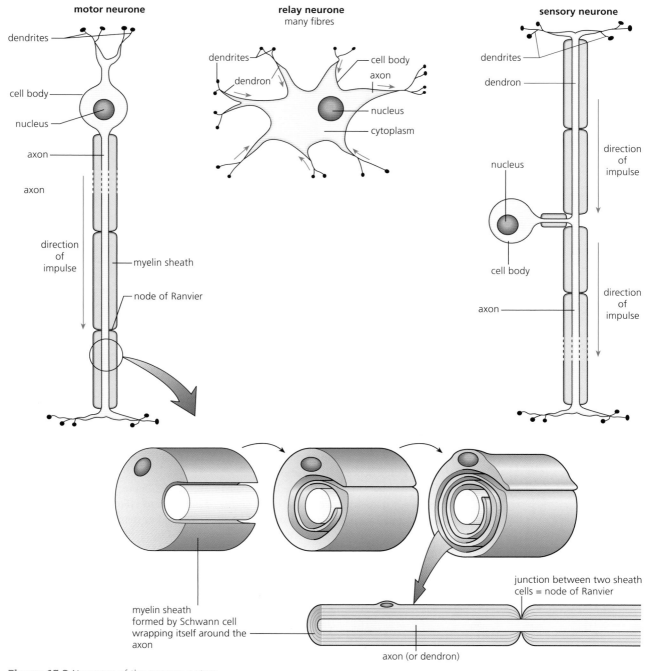

Figure 15.2 Neurones of the nervous system

Organisation among neurones – reflex arcs and reflex action

The transmission of nerve impulses is not a haphazard process. It involves organised pathways among the neurones. These pathways are called **reflex arcs**. The reflex arc consists of a **receptor**, which may be a sensory organ, connected by **neurones** to an **effector**, which may be a muscle or gland. A generalised reflex arc is shown in Figure 15.3.

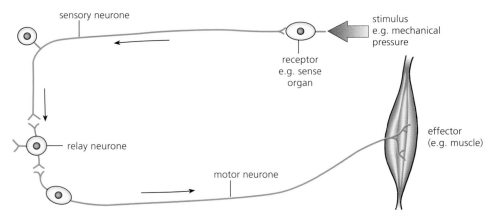

Figure 15.3 The layout of a reflex arc: the structural basis of reflex action

How does a reflex arc work?

The action begins when a sense organ detects a stimulus, which is a form of energy such as sound, light or mechanical pressure. This may become an impulse transmitted by a neurone that serves that sense cell. Once generated, the impulse is transmitted along dendrons and axons of a sequence of neurones of the reflex arc to an effector organ. When it arrives at the effector, the impulse causes a response – for example, it may cause a muscle to contract or a gland to secrete.

Extension

Reflex arcs and the nervous system

What has just been described is the simplest form of response in the nervous system. In mammals there is a complex nervous system. Many neurones connect reflex arcs with the **brain**. The brain contains a highly organised mass of relay neurones, connected with the rest of the nervous system by motor and sensory neurones.

With a nervous system of this type, complex patterns of behaviour are common, in addition to many reflex actions. This is because:
- impulses that originate in a reflex arc also travel to the brain
- impulses may originate in the brain and be conducted to effector organs.

Consequently much activity is **initiated** by the brain, rather than being simply a response to external stimuli. Also, reflex actions may be **overruled** by the brain and the response modified (as when we decide not to drop an extremely hot object because of its value).

So, the nervous system of mammals has roles in both **quick, precise communication** between the sense organs that detect stimuli and the muscles or glands that cause changes and in the **complex behaviour patterns** that mammals display.

Sensory receptors and the conversion of energy into impulses

All cells are sensitive to changes in their environment, but **sense cells** are specialised to detect stimuli and to respond by producing an impulse (an **action potential**). Specialised sense cells are called **receptors**. Different types of sensory receptors exist in the body.

Table 15.1 The sense organs of mammals

Sense data (form of energy)	Type of receptor	Location in the body
Mechanoreceptors: respond to mechanical stimulation		
light touch	touch receptors	mostly in dermis of skin
touch and pressure	touch and pressure receptors	dermis of skin
movement and position	stretch receptors, e.g. muscle spindles, proprioceptors	skeletal muscle
sound waves and gravity	sensory hair cells	cochlea and other parts of the inner ear
blood pressure	baroreceptors	aorta and carotid artery
Thermoreceptors: respond to thermal stimulation		
temperature change in the skin	nerve endings	dermis of skin
internally	cells of hypothalamus	brain
Chemoreceptors: responding to chemical stimulation		
chemicals in the air	sense cells of olfactory epithelium	nose
taste	taste buds	tongue
blood oxygen and carbon dioxide concentrations and pH	carotid body	carotid artery
osmotic concentration of the blood	osmoregulatory centre in hypothalamus	brain
Photoreceptors: respond to electromagnetic stimulation		
light	rod and cone cells of retina	eye

Question

2 Use Table 15.1 to list the various types of stimuli (sense data) that originate from conditions within the human body that are detected by particular receptors.

The Pacinian corpuscle

At rest, the nerve ending within the Pacinian corpuscle maintains a resting potential of $-70\,mV$, typical of the membrane of a neurone. When localised, strong pressure is applied to a Pacinian corpuscle it causes the layers of collagen of the capsule to be deformed. The outcome is a temporary change to the permeability of the membrane at the nerve ending. Sodium ions flow in, the membrane is depolarised and the interior starts to become less negative. The stronger the stimulus, the greater the depolarisation. If a threshold value is reached, it triggers an impulse in the sensory neurone that serves the sense cell.

The property of a sense cell is to transfer the energy of a particular type of stimulus into the electrochemical energy of an impulse, which is then conducted to other parts of the nervous system. The stimulus that the sense cell responds to is some form of energy, **mechanical**, **chemical**, **thermal** or **light** (**photic**). The sense organs of mammals are listed in Table 15.1 under headings of the stimuli they respond to.

Figure 15.4 The structure and function of a Pacinian corpuscle

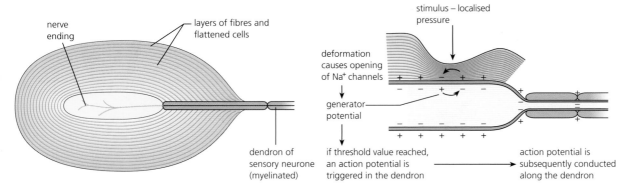

315

Transmission of an impulse

An impulse is transmitted along nerve fibres, but it is *not* an electrical current that flows along the 'wires' of the nerves. Rather, the impulse is a momentary reversal in electrical potential difference in the membrane, caused by the rapid movements of sodium and potassium ions into and out of the axon.

The resting potential

A neurone that is not transmitting an impulse is said to be at rest. Actually, this not the case; during the 'resting' interval between the transmission of impulses, the membrane of a neurone actively creates and maintains an electrical potential difference between the inside and the outside of the fibre. The result is that, in a 'resting' neurone there is a potential difference between the inside and the outside of the fibre of about −70 mv. (The discovery of the resting potential came from the use of an external and internal microelectrode on an isolated axon, the amplification of the signal and its display on a cathode ray oscilloscope).

Two processes together produce the resting potential difference across the neurone membrane.

- There is **active transport** of potassium ions (K^+) in across the membrane and of sodium ions (Na^+) out across the membrane. The ions are transported by a sodium–potassium pump, with transfer of energy from ATP. So potassium and sodium ions gradually concentrate on opposite sides of the membrane. Of course, this in itself makes *no change* to the potential difference across the membrane.
- There is also facilitated diffusion of potassium ions out and sodium ions back in. The important point here is that the membrane is *far more permeable* to potassium ions flowing out than to sodium ions returning. This causes the tissue fluid outside the neurone to contain many more positive ions than are present in the cytoplasm inside. As a result, the inside becomes more and more negatively charged compared with the outside; the resting neurone is said to be **polarised**. The difference in charge, or potential difference, is about −70 mV. This is known as the **resting potential**. Figure 15.5 is a summary of how it is set up.

Figure 15.5 The establishment of the resting potential

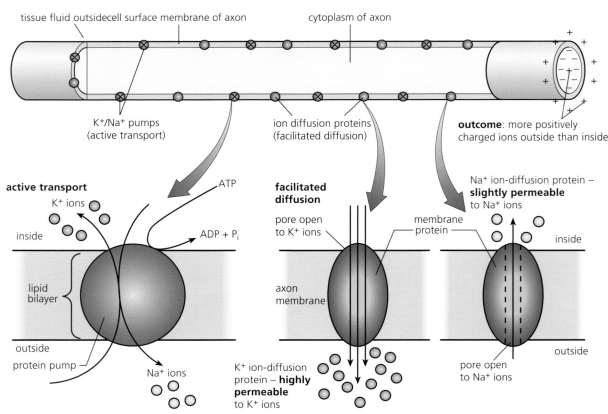

The action potential

The next event in a neurone, sooner or later, is the passage of an impulse. An impulse or **action potential** is triggered by a stimulus received at a receptor cell or sensitive nerve ending.

In the action potential, the energy transferred by this stimulus causes a temporary and local reversal of the resting potential. The result is that the membrane is briefly **depolarised** at this point. How does this happen?

The change in potential across the membrane occurs through pores in the membrane. They are called **ion channels** because they can allow ions to pass through. One type of channel is permeable to sodium ions and another to potassium ions. These channels are globular proteins that span the entire width of the membrane. They have a central pore with a gate which can open and close – they are **gated channels**. During a resting potential these channels are all closed.

The energy of the stimulus first opens the gates of the **sodium channels** in the cell surface membrane. This allows sodium ions to diffuse in, down their electrochemical gradient. This influx of sodium ions is very rapid indeed. So the cytoplasm inside the neurone fibre quickly becomes progressively more positive with respect to the outside. This charge reversal continues until the potential difference has altered from $-70\,mV$ to $+40\,mV$. At this point, an action potential has been created in the neurone fibre.

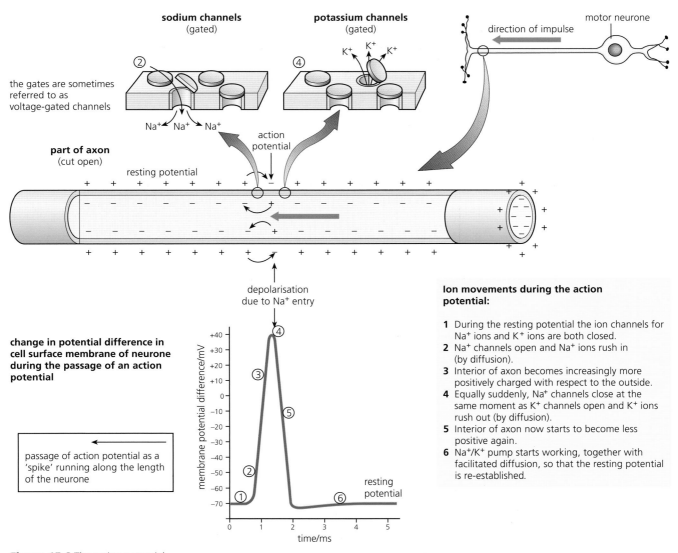

Ion movements during the action potential:

1 During the resting potential the ion channels for Na^+ ions and K^+ ions are both closed.
2 Na^+ channels open and Na^+ ions rush in (by diffusion).
3 Interior of axon becomes increasingly more positively charged with respect to the outside.
4 Equally suddenly, Na^+ channels close at the same moment as K^+ channels open and K^+ ions rush out (by diffusion).
5 Interior of axon now starts to become less positive again.
6 Na^+/K^+ pump starts working, together with facilitated diffusion, so that the resting potential is re-established.

Figure 15.6 The action potential

The action potential then travels along the whole length of the neurone fibre. At any one point it exists for only two thousandths of a second (2 milliseconds), before the membrane starts to re-establish the resting potential. So action potential transmission is exceedingly quick – an example of **positive feedback**, in fact.

Almost immediately an action potential has passed, the sodium channels close and the potassium channels open. So potassium ions can exit the cell, again down an electrochemical gradient, into the tissue fluid outside. This causes the interior of the neurone fibres to start to become less positive again. Then the potassium channels also close. Finally, the resting potential of −70 mV is re-established by the action of the sodium–potassium pump and the process of facilitated diffusion.

The refractory period

For a brief period, following the passage of an action potential, the neurone fibre is not excitable. This is the **refractory period**. It lasts only 5–10 milliseconds in total. During this time, firstly there is a large excess of sodium ions inside the neurone fibre and further influx is impossible. As the resting potential is progressively restored, however, it becomes increasingly possible for an action potential to be generated again. Because of this refractory period, the maximum frequency of impulses is between 500 and 1000 per second.

The 'all or nothing' principle

Obviously stimuli are of widely different natures, such as the difference between a light touch and the pain of a finger hit by a hammer. A stimulus must be at or above a minimum intensity, known as the **threshold of stimulation**, in order to initiate an action potential at all. Either a stimulus depolarises the membrane sufficiently to reverse the potential difference in the cytoplasm (−70 mV to +40 mV), or it does not. If not, no action potential is generated. With all sub-threshold stimuli, the influx of sodium ions is quickly reversed, and the resting potential is re-established.

However, as the intensity of the stimulus increases the frequency at which action potentials pass along the fibre increases. (Note that individual action potentials are all of standard amplitude.) For example, with a very persistent stimulus, action potentials pass along a fibre at an accelerated rate, up to the maximum possible permitted by the refractory period. This means the effector (or the brain) recognises the intensity of a stimulus from the **frequency** of action potentials (Figure 15.7).

Questions

3 Distinguish between positive and negative feedback.

4 What is the source of energy used to
  a establish the resting potential
  b power an action potential?

5 Why is an axon unable to conduct an impulse immediately after an action potential has been conducted?

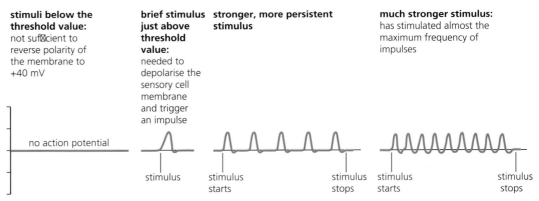

stimuli below the threshold value:
not sufficient to reverse polarity of the membrane to +40 mV

brief stimulus just above threshold value:
needed to depolarise the sensory cell membrane and trigger an impulse

stronger, more persistent stimulus

much stronger stimulus:
has stimulated almost the maximum frequency of impulses

no action potential

stimulus

stimulus starts | stimulus stops

stimulus starts | stimulus stops

Figure 15.7 Weak and strong stimuli and the threshold value

Speed of conduction of the action potential

We noted earlier that the presence of a myelin sheath affects the speed of transmission of the action potential. The junctions in the sheath, known as the nodes of Ranvier, occur at 1–2 mm intervals. Only at these nodes is the axon membrane exposed. Elsewhere along the fibre, the electrical resistance of the myelin sheath prevents depolarisations. Consequently, the action potentials are forced to jump from node to node. This is called **saltatory conduction** (from the Latin *saltare* meaning 'to leap'). This is an advantage, as it greatly speeds up the rate of transmission.

Not all neurones have myelinated fibres. In fact, non-myelinated dendrons and axons are common in non-vertebrate animals. Here, transmission is normally much slower, because the action potential flows steadily, all along the fibres. However, among non-myelinated fibres it is a fact that large diameter axons transmit action potentials much more speedily than do narrow ones. Certain non-vertebrates like the squid and the earthworm have giant fibres, which allow fast transmissions of action potentials (not as fast as in myelinated fibres, however). We saw that the original investigation of action potentials was carried out on such giant fibres.

Question

6 Why do myelinated fibres conduct impulses faster than non-myelinated fibres of the same size?

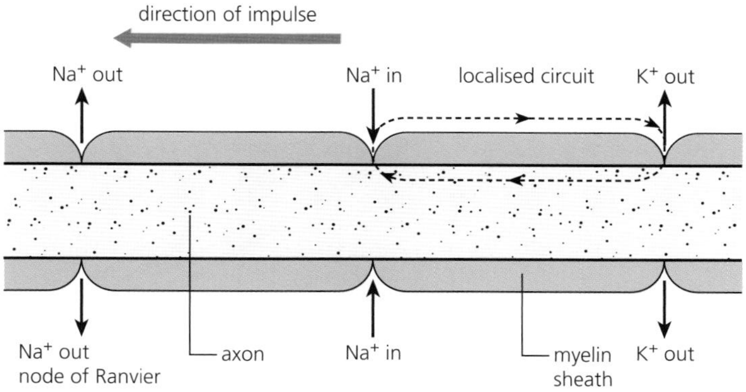

Figure 15.8 Saltatory conduction

Synapses – the junctions between neurones

Where two neurones meet they do not actually touch. A tiny gap, called a **synapse**, is the link point between neurones. Synapses consist of the swollen tip (**synaptic knob**) of the axon of one neurone (the **pre-synaptic neurone**) and the dendrite or cell body of another neurone (the **post-synaptic neurone**). Between these is the **synaptic cleft**, a gap of about 20 nm.

The practical effect of the synaptic cleft is that an action potential cannot cross it. Here, transmission occurs by particular chemicals, known as **transmitter substances**. These substances are all relatively small, diffusible molecules. They are produced in the Golgi apparatus in the synaptic knob and held in tiny vesicles before use.

Acetylcholine (ACh) is a commonly occurring transmitter substance. The neurones that release acetylcholine are known as **cholinergic neurones**. Another common transmitter substance is **noradrenalin** (released by **adrenergic neurones**). In the brain the commonly occurring transmitters are glutamic acid and dopamine.

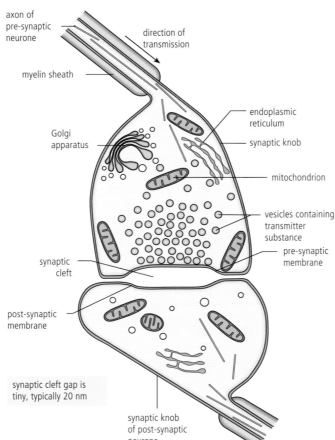

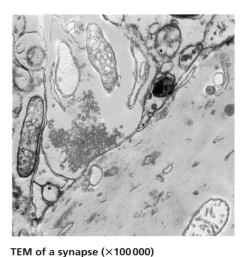

TEM of a synapse (×100 000)

Figure 15.9 A synapse in section

Steps to synapse transmission

It may help to follow the steps in Figure 15.10.

1 The arrival of an action potential at the synaptic knob opens **calcium ion (Ca²⁺) channels** in the **pre-synaptic membrane**. Calcium ions flow in from the synaptic cleft.

2 The calcium ions cause **vesicles of transmitter substance** to fuse with the pre-synaptic membrane and they release a transmitter substance into the synaptic cleft.

3 The transmitter substance diffuses across the synaptic cleft and binds with a **receptor protein**. In the **post-synaptic membrane** there are specific receptor sites for each transmitter substance. Each of these receptors also acts as a channel in the membrane which allows a specific ion (sodium, potassium or some other ion) to pass. The attachment of a transmitter molecule to its receptor instantly **opens the ion channel**.

   When a molecule of the transmitter substance attaches to its receptor site, sodium ion channels open. As the sodium ions rush into the cytoplasm of the post-synaptic neurone, **depolarisation** of the post-synaptic membrane occurs. As more and more molecules of the transmitter substance bind, it becomes increasingly likely that depolarisation will reach the **threshold level**. When it does, an **action potential** is generated in the post-synaptic neurone. This process of build-up to an action potential in post-synaptic membranes is called **facilitation**.

4 The transmitter substance on the receptors is quickly **inactivated**. For example, the enzyme cholinesterase hydrolyses acetylcholine to choline and ethanoic acid. These molecules are inactive as transmitters. This reaction causes the ion channel of the receptor protein to close and so allows the resting potential in the post-synaptic neurone to be re-established.

5 Meanwhile, the inactivated products of the transmitter re-enter the pre-synaptic knob, are **resynthesised** into transmitter substance and packaged for reuse.

Question

7 In the synaptic knob, what is the role of:
   a the Golgi apparatus
   b mitochondria?

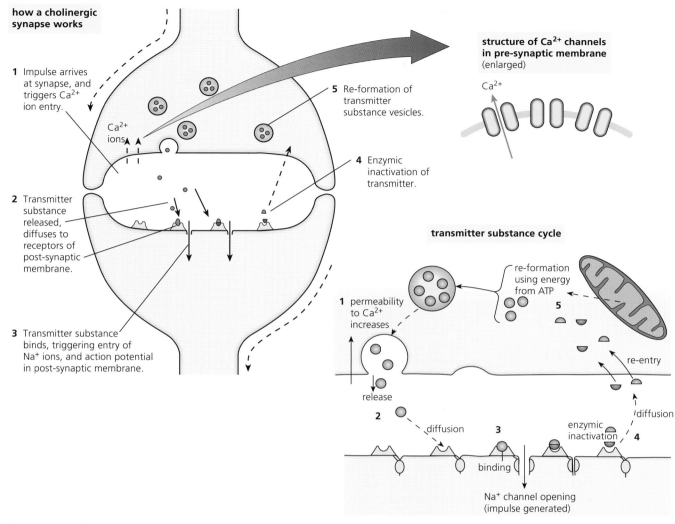

how a cholinergic synapse works

1 Impulse arrives at synapse, and triggers Ca²⁺ ion entry.

Ca²⁺ ions

2 Transmitter substance released, diffuses to receptors of post-synaptic membrane.

3 Transmitter substance binds, triggering entry of Na⁺ ions, and action potential in post-synaptic membrane.

5 Re-formation of transmitter substance vesicles.

4 Enzymic inactivation of transmitter.

structure of Ca²⁺ channels in pre-synaptic membrane (enlarged)

Ca²⁺

transmitter substance cycle

1 permeability to Ca²⁺ increases

release

2

diffusion

3

binding

Na⁺ channel opening (impulse generated)

re-formation using energy from ATP

5

re-entry

enzymic inactivation

diffusion

4

Figure 15.10 Chemical transmission at a synapse

An alternative type of synapse

In this section on synapses, it is an **excita-tory synapse** that has been described. That is, the incoming action potential *excited* the post-synaptic membrane and generated an action potential that was then transmitted along the post-synaptic neurone. Some synapses, however, have the opposite effect. These are known as **inhibitory synapses**.

Why have synapses between neurones?

Since synapses have the disadvantage of very slightly slowing down the transmission of action potentials, we may assume they also provide distinct advantages to the operation of nervous communication in organisms. In fact there are a number of advantages.

● They filter out low-level stimuli of limited importance.
● They protect the effectors (muscles and glands) from over-stimulation, since continuous transmission of action potentials eventually exhausts the supply of transmitter substances for a period of time. This is called synapse fatigue.
● They facilitate flexibility of response by the central nervous system, particularly by the brain.
● They allow integration of information since the post-synaptic neurone may receive action potentials from different types of neurone. Both excitatory pre-synaptic neurones and inhibitory pre-synaptic neurones exist. The post-synaptic neurone summates all the action potentials, thereby integrating action potentials from more than one source neurone or sense organ, for example.

Drugs may interfere with the activity of synapses

Some drugs **amplify** the processes of synaptic transmission, in effect they increase post-synaptic transmission. Nicotine and atropine have this effect.

Other drugs **inhibit** the processes of synaptic transmission, in effect decreasing synaptic transmission. Amphetamines and beta-blocker drugs have this effect.

15.1 *continued...* Neuro-muscular junctions and muscle contraction

Motor nerve endings make connections with muscle fibres and trigger muscle contraction.

By the end of this section you should be able to:

i) describe the roles of neuromuscular junctions, transverse system tubules and sarcoplasmic reticulum in stimulating contraction in striated muscle

j) describe the ultrastructure of striated muscle with particular reference to sarcomere structure

k) explain the sliding filament model of muscular contraction including the roles of troponin, tropomyosin, calcium ions and ATP

The structures of striated muscle

Skeletal muscle fibres appear striped under the light microscope, so skeletal muscle is also known as striated muscle. Striated muscle consists of bundles of muscle fibres (Figure 15.11). The remarkable feature of a muscle fibre is the ability to shorten to half or even a third of the relaxed or resting length. Actually, each fibre is itself composed of a mass of **myofibrils**, but we need the electron microscope to see this important detail.

The ultrastructure of skeletal muscle

Skeletal muscle fibres are multinucleate and contain specialised endoplasmic reticulum. By electron microscope we can see that each muscle fibre consists of very many parallel myofibrils within a plasma membrane known as the sarcolemma, together with cytoplasm. The cytoplasm contains **mitochondria** packed between the myofibrils. The sarcolemma infolds to form a system of transverse tubular endoplasmic reticulum (sometimes referred to as **T tubules**), and all parts of the **sarcoplasmic reticulum**. This is arranged as a network around individual myofibrils. The arrangements of myofibrils, sarcolemma and mitochondria, surrounded by the sarcoplasmic membrane, are shown in Figure 15.12.

skeletal muscle cut to show bundles of fibres

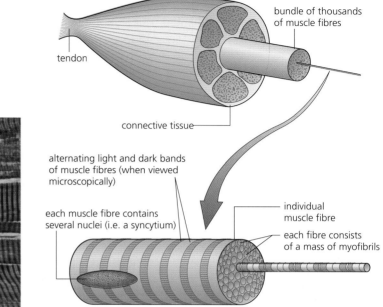

tendon

bundle of thousands of muscle fibres

connective tissue

alternating light and dark bands of muscle fibres (when viewed microscopically)

each muscle fibre contains several nuclei (i.e. a syncytium)

individual muscle fibre

each fibre consists of a mass of myofibrils

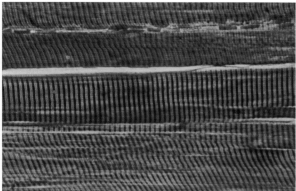

Figure 15.11 The structure of skeletal muscle

electron micrograph of TS through part of a muscle fibre, HP (x36 000)

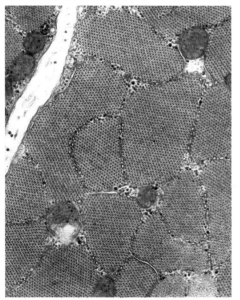

Figure 15.12 The ultrastructure of a muscle fibre

stereogram of part of a single muscle fibre

sarcoplasmic membrane

sarcoplasm (cytoplasm of muscle cell)

myofibril

mitochondrion

sarcoplasmic reticulum with transverse tubules

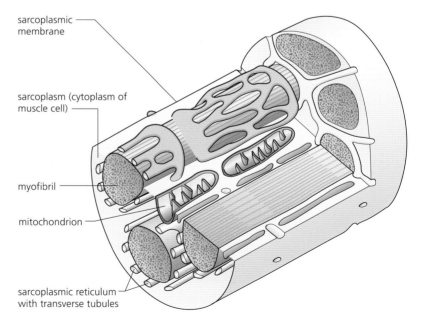

323

The striped appearance of skeletal muscle is due to an interlocking arrangement of two types of protein filaments, known respectively as **thick** and **thin filaments** – they make up the myofibrils. These protein filaments are aligned, giving the appearance of stripes (alternating **light and dark bands**). This is shown in the more highly magnified electron micrograph and interpretive drawing in Figure 15.13.

Electron micrograph of an individual sarcomere (×34 000)

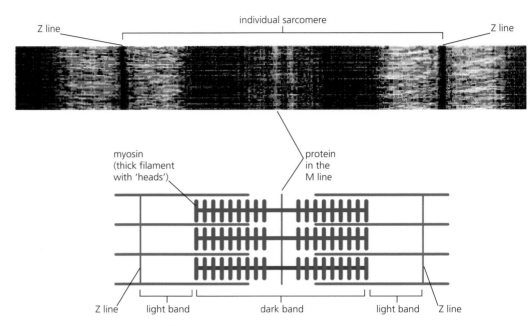

Figure 15.13 The ultrastructure of a myofibril

The thick filaments are made of a protein called **myosin**. They are about 15 nm in diameter. The longer, thin filaments are made of another protein, **actin**. Thin filaments are about 7 nm in diameter, and are held together by transverse bands, known as **Z lines**. Each repeating unit of the myofibril is, for convenience of description, referred to as a **sarcomere**. So we can think of a myofibril as consisting of a series of sarcomeres attached end to end.

How a motor nerve ending makes connection with a muscle fibre

Striated muscle fibres are innervated by a motor neurone nerve ending, at a structure known as a motor end plate or **neuromuscular junction**. This is a special type of synapse, but the transmitter substance is the familiar acetylcholine.

On arrival of an action potential at the neuromuscular junction, vesicles of acetylcholine are released and the transmitter molecules bind to receptors on the sarcoplasm (this is the plasma membrane of the muscle fibre). This triggers the release of calcium ions from the sarcoplasmic reticulum, into the cytoplasm around the myofibrils, via the T-tube system (Figure 15.SS). These calcium ions then remove the blocking molecules on binding sites of the actin filaments (see below). This starts the **contraction** process. When action potentials stop arriving at the muscle fibres, calcium ions return to the sarcoplasmic reticulum and the binding sites are again covered by blocking molecules.

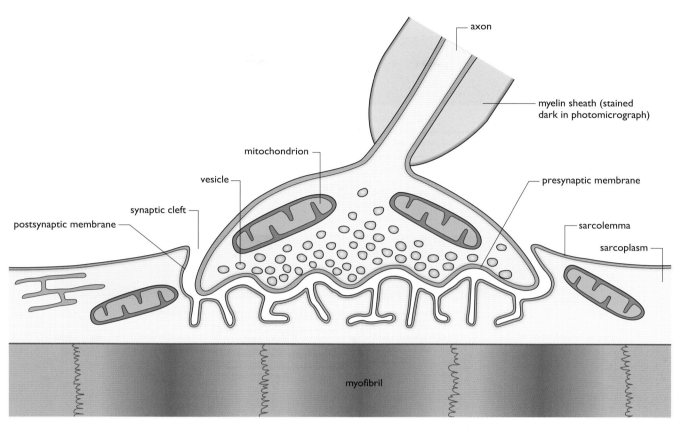

Figure 15.14 The structure of a neuromuscular junction

Question

8 Explain the relationship to a muscle of:
   a a muscle fibre
   b a myofibril
   c a myosin filament.

Skeletal muscle contracts by sliding of the filaments

When skeletal muscle contracts, the actin and myosin filaments **slide past each other**, in response to nervous stimulation, causing shortening of the sarcomeres (Figure 15.15). This occurs in a series of steps, sometimes described as a ratchet mechanism. A great deal of **ATP** is used in the contraction process.

Shortening is possible because the thick filaments are composed of many myosin molecules, each with a **bulbous head**, which protrudes from the length of the myosin filament. Along the actin filament are a complementary series of binding sites to which the bulbous heads fit. However, in muscle fibres at rest, the binding sites carry **blocking molecules** (a protein called **tropomyosin**), so binding and contraction are not possible.

Calcium ions play a critical part in the muscle fibre contract mechanism, together with the proteins tropomyosin and **troponin**. The contraction of a sarcomere is described in the following four steps.

stretched/relaxed:

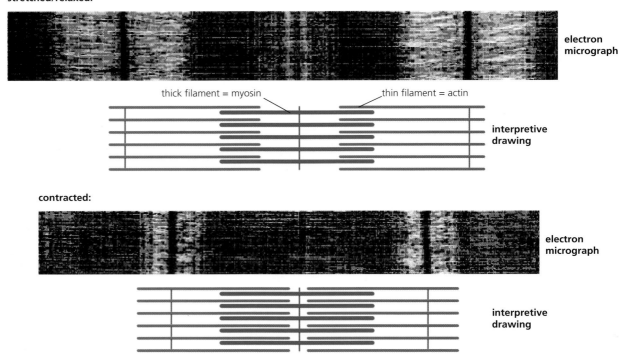

Figure 15.15 Muscle contraction of a single sarcomere

1 The myofibril is stimulated to contract by the arrival of an action potential. This triggers release of calcium ions from the sarcoplasmic reticulum, which surround the actin molecules. Calcium ions now react with an additional protein present (troponin) which, when so activated, triggers the removal of the blocking molecule, tropomyosin. The binding sites are now exposed.

2 Each bulbous head to which ADP and Pi are attached (called a **charged bulbous head**) reacts with a binding site on the actin molecule beside it. The phosphate group (P_i) is shed at this moment.

3 The ADP molecule is then released from the bulbous head, and this is the trigger for the **rowing movement** of the head, which tilts at an angle of about 45°, pushing the actin filament along. At this step, the **power stroke**, the myofibril has been shortened (contraction).

4 Finally, a fresh molecule of ATP binds to the bulbous head. The protein of the bulbous heads includes the enzyme ATPase, which catalyses the hydrolysis of ATP. When this reaction occurs, the ADP and inorganic phosphate (P_i) formed remain attached, and the bulbous head is now **'charged' again**. The charged head detaches from the binding site and straightens.

This cycle of movements is shown is Figure 15.16. The cycle is repeated many times per second, with thousands of bulbous heads working along each myofibril. ATP is rapidly used up, and the muscle may shorten by about 50% of its relaxed length. However, when the action potential stimulation stops, the muscle cell relaxes. Now, the filaments slide back to their original positions. Ion pumps in the sarcoplasmic reticulum pump calcium ions back inside, and so the calcium ion concentration surrounding the myosin filaments falls. Blockage of binding sites by tropomyosin is restored.

Arrival of nerve impulse releases Ca^{2+} from the sarcoplasmic reticulum. Ca^{2+} ions cause removal of blocking molecule from binding sites. Each myosin molecule has a bulbous head that reacts with $ATP \rightarrow ADP + P_i$ which remain bound.

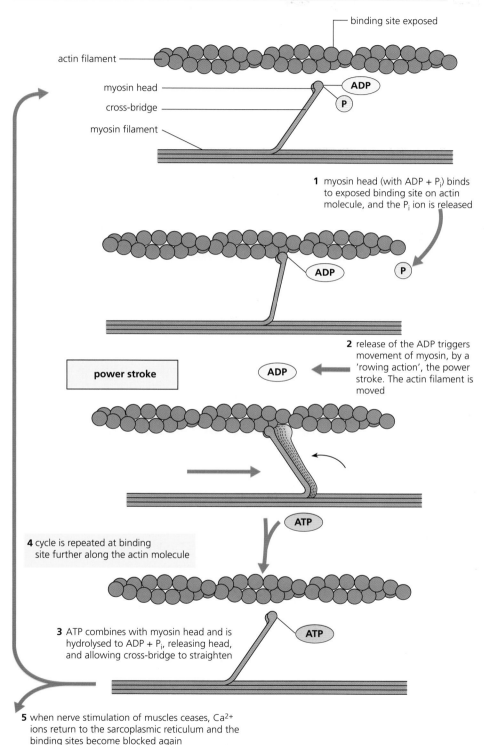

The mechanism of muscle contraction:

- The myofibril is stimulated to contract by nervous stimulation. This triggers release of calcium ions from the sarcoplasmic reticulum, around the actin molecules. Calcium ions react with the blocking molecules, removing them so **the binding sites are now exposed.**

- Each bulbous head to which ADP and Pi is attached (called a charged myosin head) reacts with a binding site on the actin molecule next to it. The phosphate ion (P_i) is shed at this moment.

- Then the ADP molecule is released from the myosin head. This is the trigger for the **'rowing movement'** of the head, which tilts by an angle of about 45°, pushing the actin filament along. At this step, **the power stroke**, the myofibril has been shortened (**contraction**).

- Finally, a fresh molecule of ATP binds to the myosin head. The protein of the myosin head includes the enzyme ATPase, which catalyses the hydrolysis of ATP. When this reaction occurs, the ADP and inorganic phosphate (P_i) formed remain attached. The myosin head is now 'charged' again. The charged head detaches from the binding site and straightens.

Figure 15.16 The sliding-filament hypothesis of muscle contraction

↻ 15.1 *continued*... The endocrine system

The endocrine system is a slower system that controls long-term changes. Fertility may be controlled by use of hormones.

By the end of this section you should be able to:

l) explain the roles of the hormones FSH, LH, oestrogen and progesterone in controlling changes in the ovary and uterus during the human menstrual cycle

m) outline the biological basis of contraceptive pills containing oestrogen and/or progesterone

Hormonal control – introducing the endocrine system

Endocrine gland: a gland containing specialised secretory cells that release a hormone into the bloodstream at a distance from the hormone's target organ.

Hormones are **chemical messengers**. They are produced and secreted from the cells in glands known as ductless or **endocrine glands**. These contain capillary networks and specialised secretory cells that make and release hormones. Hormones are released from the cells of the ductless gland, directly into the blood stream, when stimulated to do so. Hormones are then transported indiscriminately about the body via the blood circulation system, but they act only at specific sites, appropriately called **target organs**. Cells of the target organ possess specific receptor molecules on the external surface of their cell surface membrane to which the hormone molecules bind. A hormone typically works by triggering changes to specific metabolic reactions in their target organs. Although present in small quantities, hormones are extremely effective in the control and co-ordination of several body activities. The position of the main endocrine glands and the hormones produced are summarised in Figure 15.17.

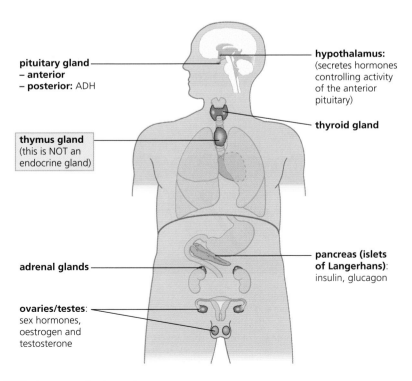

pituitary gland
– anterior
– posterior: ADH

hypothalamus:
(secretes hormones controlling activity of the anterior pituitary)

thyroid gland

thymus gland
(this is NOT an endocrine gland)

adrenal glands

pancreas (islets of Langerhans): insulin, glucagon

ovaries/testes: sex hormones, oestrogen and testosterone

Figure 15.17 The human endocrine system

Hormones circulate in the bloodstream only briefly because in the liver they are continually being broken down. Any breakdown products no longer of use to the body are excreted in the kidneys. So, long-acting hormones must be secreted into the bloodstream continuously to be effective – and they are.

You can see immediately that hormonal control of body function is quite different from nervous control; the latter is concerned with quick, precise communication, whereas hormones mostly work by causing specific changes in metabolism or development, often over an extended period of time. However, these contrasting systems are co-ordinated by the brain. Our nervous system and hormones work together.

The structure of endocrine glands can be contrasted with that of other glands in our body which deliver their secretions through tubular ducts, such as the salivary glands in the mouth and sweat glands in the skin. These ducted glands are called exocrine glands. Their secretions pass out of the gland via ducts and they have altogether different roles in the body.

Question

9 Hormones, on secretion, are transported all over the body. Explain how the effects of hormones are restricted to particular cells or tissues.

The role of hormones in the menstrual cycle

We have seen that **hormones** are chemical messenger substances produced in the cells of the ductless or **endocrine glands**. From these, hormones are released into the blood and travel to all parts of the body. However, hormones have their effects on specific **target cells**.

The onset of **puberty** is triggered by a hormone secreted by a part of the brain called the **hypothalamus**. It produces and secretes a releasing hormone, the target organ of which is the nearby **pituitary gland**. Here it causes production and release of two hormones. These are called **follicle stimulating hormone (FSH)** and **luteinising hormone (LH)**. They are so named because their roles in sexual development were discovered in the female (although they do operate in both sexes).

The first effects of FSH and LH are the enhanced secretion of the sex hormones **oestrogen** and **testosterone** by the ovaries and testes. Then, in the presence of FSH, LH and the respective sex hormone(s), there follows sexual development of the body and the preparation of the body for its role in sexual reproduction. This process is now illustrated in the female reproductive system.

In the female, the secretion of oestrogen and another hormone, **progesterone**, do not occur at a steady rate; instead their secretion is a recurring (cyclical) event. Together with FSH and LH, the continually changing concentration of all four hormones brings about a repeating cycle of changes that we call the **menstrual cycle**. The menstrual cycle consists of two cycles, one in the **ovaries** and one in the **lining of the uterus** (the **endometrium**). The ovarian cycle is concerned with the monthly preparation and shedding of an egg cell from an ovary. The uterine cycle is concerned with the build up of the lining of the uterus. 'Menstrual' means *monthly*; the combined cycles take on average 28 days.

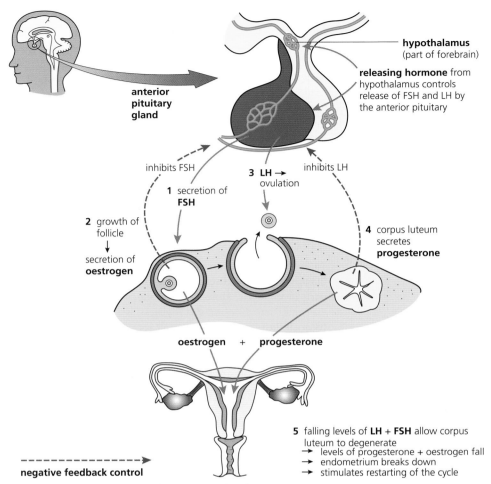

anterior pituitary gland

hypothalamus
(part of forebrain)

releasing hormone from hypothalamus controls release of FSH and LH by the anterior pituitary

inhibits FSH

3 LH → ovulation

inhibits LH

1 secretion of **FSH**

2 growth of follicle
↓
secretion of **oestrogen**

4 corpus luteum secretes **progesterone**

oestrogen + progesterone

5 falling levels of **LH + FSH** allow corpus luteum to degenerate
→ levels of progesterone + oestrogen fall
→ endometrium breaks down
→ stimulates restarting of the cycle

negative feedback control

Figure 15.18 Hormone regulation of the menstrual cycle

Question

10 Explain what is meant by *negative feedback control*. Can you think how positive feedback might differ from negative feedback?

By convention, the start of the menstrual cycle is taken as the **first day of menstruation** (bleeding), which is the shedding of the endometrial lining of the uterus. The steps, also summarised in Figure 15.18 above, are as follows:

1 FSH is secreted by the pituitary gland, and stimulates development of several immature egg cells (in secondary follicles) in the ovary. Only one will complete development into a mature secondary oocyte (now in the Graafian follicle).

2 The developing follicle then secretes oestrogen. Oestrogen is a hormone that has two targets:
- in the uterus oestrogen stimulates the build up of the **endometrium**, which prepares the uterus for a possible implantation of an embryo, if fertilisation takes place.
- in the pituitary gland, oestrogen inhibits the further secretion of FSH. This prevents the possibility of further follicles being stimulated to develop. It is an example of **negative feedback control**.

3 Meanwhile, the concentration of oestrogen continues to build up, increasing to a peak value just before the mid-point of the cycle. This high and rising level of oestrogen suddenly stimulates the secretion of LH and, to a slightly lesser extent, FSH, by the pituitary gland. LH stimulates ovulation (the shedding of the mature secondary oocyte from the Graafian follicle) and the secondary oocyte is released from the ovary.

4 As soon as the ovarian follicle has discharged its oocyte, LH also stimulates the conversion of the vacant follicle into an additional, temporary gland, called a corpus luteum. The corpus luteum secretes progesterone and, to a lesser extent, oestrogen. Progesterone has two targets:
- in the uterus progesterone continues the build up of the endometrium, further preparing it for a possible implantation of an embryo, if fertilisation takes place.
- in the pituitary gland progesterone inhibits further secretion of LH, and also of FSH. This is a second example of **negative feedback control**.

5 The levels of FSH and LH in the bloodstream now decrease rapidly. Low levels of FSH and LH allow the corpus luteum to degenerate. As a consequence, the levels of progesterone and oestrogen also fall. Soon the level of these hormones is so low that the extra lining of the uterus is no longer maintained. The **endometrium breaks down** and is lost through the vagina in the first five days or so of the new cycle.

The falling levels of progesterone again cause the secretion of FSH by the pituitary and a new cycle is underway. The changing levels of the hormones during the menstrual cycle and their effects on the ovaries and the lining of the uterus are show in Figure 15.19.

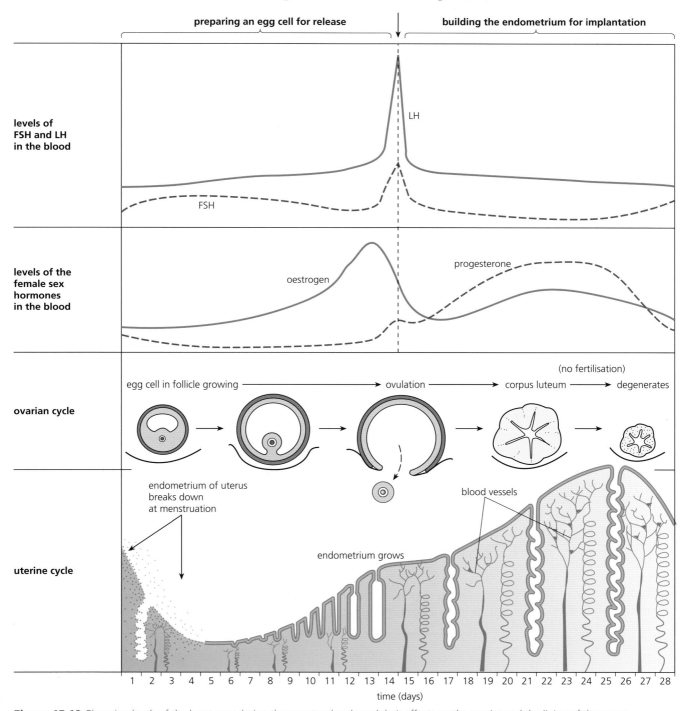

Figure 15.19 Changing levels of the hormones during the menstrual cycle and their effects on the ovaries and the lining of the uterus

Question

11 Identify the critical hormonal changes that:
  a trigger ovulation
  b cause degeneration of the corpus luteum.

Extension

If the egg is fertilised (the start of a pregnancy), then the developing embryo itself becomes an endocrine gland, secreting a hormone that circulates in the blood and maintains the corpus luteum as an endocrine gland for at least 16 weeks of pregnancy.

When eventually the corpus luteum does break down, the **placenta** takes over as an endocrine gland, secreting oestrogen and progesterone. These hormones continue to prevent ovulation and maintain the endometrium.

Controlling human reproduction

Conception is a process that starts with fertilisation and ends with implantation of an embryo resulting in **pregnancy**. **Contraception** allows control over human reproduction. There are many methods and they work in different ways. Some are more effective than others. One method is the contraceptive pill.

The biological basis of the contraceptive pill

The combined oral contraceptive pill, more often known as 'the pill', contains two hormones. These are chemically very similar to the two natural hormones women produce in their ovaries – oestrogen and progesterone. The effect of the pill is to:
- prevent ovulation, the pill stops the ovaries releasing an egg each month
- thicken the mucus in the cervix making it difficult for sperm to reach the egg
- make the lining of the uterus thinner so it is less likely to accept a fertilised egg.

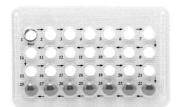

Figure 15.20 A commercial pack of 'the pill'

For the oral contraceptive pill to be effective, the woman must take one pill at the same time each day for 21 days. She then takes an inert pill (a placebo – containing an iron supplement, as a lot is lost in menstruation) for seven days, or she may take no pill at all. During this time menstruation occurs. This 'period' is usually lighter and shorter than is normally experienced.

In the body, the hormones from the pill prevent ovulation by decreasing the release of FSH and LH from the pituitary gland (without completely stopping their release). This restricts the growth of follicles in the ovaries and so a secondary oocyte does not grow and is not released. The endometrium still builds up in the uterus and then, later, breaks down again, but menstruation is shorter and lighter because the corpus luteum has not grown (in the absence of ovulation).

Nothing is without risk, but very many women who have access to it find the pill beneficial to their lifestyle, health and range of life choices. The likely benefits and possible risks are identified in Table 15.2.

Question

12 What do you understand by a 'placebo'?

Table 15.2 The risks and benefits of the pill

Risks	Benefits

• Some women develop nausea, headaches, tiredness and mood swings.
• It may trigger a rise in blood pressure, leading to an increased risk of thrombosis (blood clots are life-threatening).
• There is an increased risk of breast cancer, but this is only to a slight extent.
	• There is a reduced risk of:
	  • ovarian cysts
	  • cancer of the ovaries or uterus.
	• Menstruation is more regular for many, and pre-menstrual tension may be lessened.
	• The quality of the mucus plug in the cervix reduces the risk of bacterial entry to the uterus and therefore decreases the chance of infection.

The 'morning after' pill

This 'emergency' contraceptive medication contains synthetic progesterone that works in the body in the following ways:

- it delays ovulation or it inhibits ovulation if that event has not yet occurred
- it irritates the endometrium so that implantation does not occur.

Typically, the morning after pill is taken if a women has forgotten to take her combined oral contraceptive pill or if another contraceptive device has been omitted or used and failed (a condom may have broken). It may also be used if a woman has been raped.

In these 'morning after' situations, the sense of 'emergency' is a (quite natural) state of mind, rather than a biological fact. This is because fertilisation occurs in the oviduct, if at all. A fertilised egg then passes down to the uterus, a journey that takes 5–7 days. Only then can implantation in the wall of the uterus occur. So there is an interval of about 72 hours in which a pregnancy can be avoided. However, if fertilisation has taken place, then this particular contraceptive measure is a 'chemical abortion'.

The social and ethical implications of contraception

Overpopulation and the Earth's resources

The current human population explosion began at about the beginning of the Industrial Revolution, some 200 years ago and is ongoing. Birth rates are high in many communities and also for many there is an increasing life expectancy. This is most notable among people of the developed nations but it is a trend in many human populations (Figure 15.21).

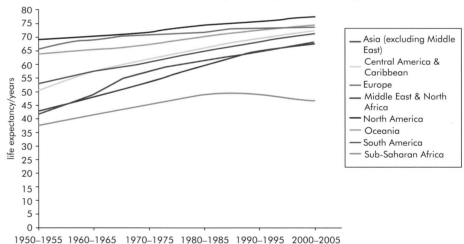

Figure 15.21 Life expectancy 1950–2005

People who make a study of human population growth (demographers) have predicted that by 2050 the Earth may have 10 billion human inhabitants – that is very many more than we currently have. The projections suggest that the world population will stabilise at this level.

Can such a total human population be sustained by the Earth's resources? Many people feel that it cannot. To them, the current destruction of habitats and devastation of biological diversity observed on all continents is directly or indirectly due to our attempts to provide for the lifestyles that existing humans seek (or seek to maintain). It is unacceptable to them that the environment is so abused. To these people, one immediate task is to reduce the birth rate. Contraception helps to make this possible.

15.2 Control and co-ordination in plants

Plant co-ordination systems involve rapid responses as in the case of the Venus fly trap, but also complex interactions between plant growth regulators, such as auxin and gibberellin.

Plants respond quite differently to different concentrations of plant growth regulators.

By the end of this section you should be able to:

a) describe the rapid response of the Venus fly trap to stimulation of hairs on the lobes of modified leaves and explain how the closure of the trap is achieved

b) explain the role of auxin in elongation growth by stimulating proton pumping to acidify cell walls

c) describe the role of gibberellin in the germination of wheat or barley

d) explain the role of gibberellin in stem elongation including the role of the dominant allele, *Le*, that codes for a functioning enzyme in the gibberellin synthesis pathway, and the recessive allele, *le*, that codes for a non-functional enzyme

The responding plant – communication and control in flowering plants

Sensitivity is a characteristic of all living things, but the responses of plants and animals differ. Most plants are anchored organisms, growing in one place and remaining there, even if the environment is not wholly favourable. As a result, plant responses are often less evident than those of animals, but they are no less important to the plant. Plants are highly sensitive to environmental clues and external and internal stimuli, and respond to changes. Yet plants have no nervous system and no muscle tissue. Consequently, plant sensitivity may lead to responses less dramatic than those of animals. Rapid movements by plants are extremely rare (but see Figure 15.22c); plant responses are mostly **growth movements**. An example of this is the growth of young stems towards light. Another is the growth of the main root down into the soil in response to gravity. Occasionally, movements are due to turgor changes (Figure 15.22d). However, there is no doubt that plant responses are also precise and carefully regulated responses.

a) The response of a plant stem to light

b) Leaf fall in a woody plant of temperate climate

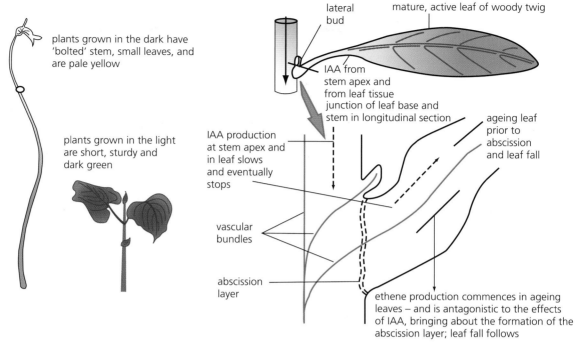

plants grown in the dark have 'bolted' stem, small leaves, and are pale yellow

plants grown in the light are short, sturdy and dark green

lateral bud

mature, active leaf of woody twig

IAA from stem apex and from leaf tissue
junction of leaf base and stem in longitudinal section

IAA production at stem apex and in leaf slows and eventually stops

ageing leaf prior to abscission and leaf fall

vascular bundles

abscission layer

ethene production commences in ageing leaves – and is antagonistic to the effects of IAA, bringing about the formation of the abscission layer; leaf fall follows

c) The response of the sensitive plant *Mimosa pudica* to touch

d) Closure of a leaf of the canivorous plant *Drosera rotundiflora*, to trap an insect

Figure 15.22 Examples of plant sensitivity and responses

Among the internal factors that play a part in plant sensitivity, the most important are the substances we call **plant growth regulators** and their effects. There are five major types of compound naturally occurring in plants that we call growth regulators. The effects of these chemicals are rather different from those of animal hormones (see Table 15.3).

Plant growth regulators:
- occur in very low concentrations in plant tissues, so sometimes it is hard to determine their precise role
- are produced in a variety of tissues, rather than in discrete endocrine glands
- often interact with each other in controlling growth and sensitivity – some interact to reinforce each other's effects (synergism), others oppose each other's effects (antagonism)
- if they move from their site of synthesis (and not all do), they may diffuse from cell to cell, be actively transported or be carried in the phloem or xylem
- may have profoundly differing effects, depending on whether they are present in low or high concentration
- may have different effects according which tissue they are in or at different stages of development of a single tissue.

Question

13 What is the difference between growth and development?

Table 15.3 The plant regulators: discovery, roles and synthesis. The structural formulae show how chemically diverse these substances are – *they do not require memorising.*

Auxins, principally Indoleacetic acid (IAA)	
Discovery Initially by Darwin, investigating coleoptiles and their curvature towards a unilateral light source. Later by others, including the devising of a biological assay to find the concentrations of auxins in plant organs (since these are at low concentrations).	indole acetic acid (the principal auxin)
Roles • Promotion of extension growth of stems and roots (at different concentrations) • Dominance of terminal buds • Promotion of fruit growth • Inhibition of leaf fall	
Synthesis At stem and root tips and in young leaves (from the amino acid tryptophan)	

Gibberellins (GAs)	
Discovery Initially, in dwarf rice plants, which, when infected with a fungus (*Gibberella* sp.), responded by growing to full height.	gibberellins e.g. gibberellic acid
Roles • Promotion of extension growth of stems • Delay of leaf senescence and leaf fall • Inhibition of lateral root initiation • Switching on of genes to promote germination	
Synthesis In the embryos of seeds and in young leaves (except in genetically dwarf varieties) There are several types of gibberellin	

Abscisic acid (ABA)	
Discovery During investigation of bud and seed dormancy and in abscission of fruits	abscisic acid
Roles • A stress hormone • Triggering of stomatal closure when leaf cells are short of water • Induction of bud and seed dormancy	
Synthesis In most organs of mature plants, in tiny amounts.	

Interactions of plant growth regulators

Sometimes plant growth regulators have their effects acting on their own, but in other situations, they interact with other growth regulators. When one substance enhances the effect of the other, this is known as **synergism**. For example, gibberellins (GAs) and indoleacetic acid (IAA) work synergistically in stem extension growth.

Alternatively, a growth regulator may reduce the effect of another. This situation is known as **antagonism**. Some important interactions involve the other growth regulators (cytokinins – their main role is in cell differentiation, and ethylene – its main role is in fruit ripening).

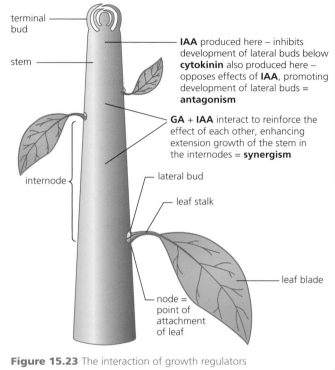

terminal bud

stem

IAA produced here – inhibits development of lateral buds below
cytokinin also produced here – opposes effects of **IAA**, promoting development of lateral buds = **antagonism**

GA + IAA interact to reinforce the effect of each other, enhancing extension growth of the stem in the internodes = **synergism**

internode

lateral bud

leaf stalk

leaf blade

node = point of attachment of leaf

Figure 15.23 The interaction of growth regulators

The Venus fly trap

The Venus fly trap (*Dionaea muscipula*) is a carnivorous plant, native to the subtropical wetlands of the East Coast of the USA.

Remember, plants typically obtain the ions they need for their metabolism by active uptake from the soil solution. However, in some habitats, the decay of dead organisms (by which the ions are released into the soil solution) may be very slow, and when they are released, the ions tend to be washed away by the high rainfall. In these environments carnivorous plants that can trap minibeasts such as flying insects prosper. They take their essential ions from the decaying or digested bodies of these victims, thus enabling them to survive in these unfavourable growing conditions where other plants often fail. The Venus fly trap is an example of this.

Look at Figure 15.24 now.

The margins of the leaves of this plant have short, stiff hairs, which when they are bent by an insect large enough to do so, trigger the two halves of the leaf to snap shut, trapping the insect. Then, short, glandular hairs on the inner leaf surface secrete hydrolytic protease enzymes that digest the soft tissue of the insect. Active uptake of the ions released then takes place into the leaf cells. Later, the trap reopens and the undigested matter is blown away.

The rapid closure of the Venus flytrap is one of the fastest movements observed in the plant kingdom. It occurs in the complete absence of a nervous system. It is due, at least in part, to an extremely sudden change in turgidity of cells in the hinge region of the leaf blade. However, how this rapid loss in turgor is brought about is not clear. Another hypothesis is a sudden acid-induced wall loosening in 'motor cells' of the leaf – implying that proton pumps are involved.

Figure 15.24 Venus fly trap, with close-up view of leaves, one 'closed' with a 'prisoner'!

How auxin stimulates elongation growth

We have seen that auxin has a major role in the growth of the shoot apex, where it promotes the elongation of cells. Auxin has this effect on growth and development by direct action on the components of growing cells, including the walls. For example auxin-triggered cell elongation occurs due to the stimulating effect of auxin on proton pumps in the cell surface membranes of cells that lie beside the cellulose walls (Figure 15.25). The result is that the walls become acidic, and the lowered pH initiates the breakage of cross-links between cellulose microfibrils and binding polysaccharides that had helped to make the cellulose inflexible. The wall's resistance to stretching due to turgor pressure is decreased and elongation occurs.

Note that auxin transport across cells is polar, with its entry into the cell being passive (by diffusion), but its efflux is active (ATP-driven). Auxin efflux pumps set up concentration gradients in plant tissues, in fact.

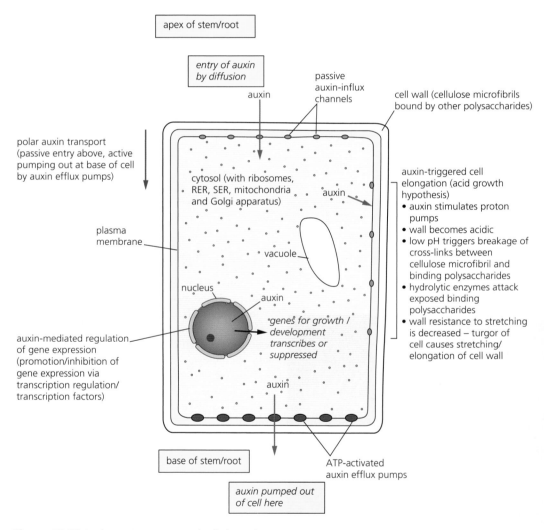

Figure 15.25 Auxin, proton pumps and cell elongation

The role of gibberellin in the germination of barley

Many higher plants have been found to contain **gibberellic acid (GA)**. In fact, several different forms of the molecule have been discovered and these are often referred to collectively as **gibberellins**. Gibberellins are formed in chloroplasts, in the embryo of seeds and in young leaves. Gibberellins are absent from dwarf varieties of pea plants, but when artificially supplied to dwarf varieties, they become tall.

The highest concentrations of gibberellins occur in fruits and seeds as they are formed and in seeds as they germinate. In the seed, gibberellins switch on genes that promote germination and go on to promote additional mitosis, thereby increasing the numbers of cells formed. Gibberellins also have a role in the mobilisation of stored food in germinating barley fruits, as illustrated in Figure 15.26, on the next page. During germination, the uptake of water activates the gene for gibberellic acid synthesis. This gibberellic acid then switches on the genes that control the synthesis of hydrolytic enzymes. Hydrolytic enzymes catalyse the mobilisation of the food reserves, causing the release of sugars, amino acids and fatty acids for germination and growth.

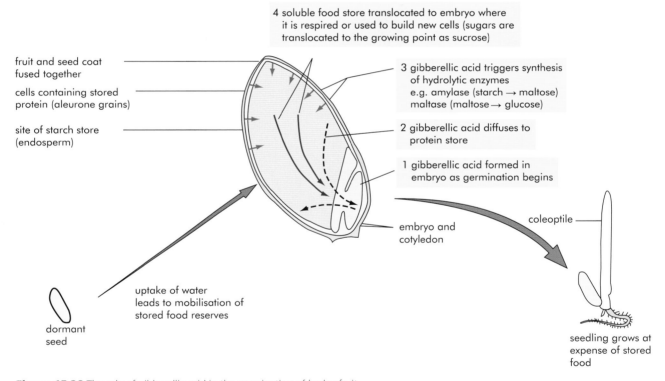

4 soluble food store translocated to embryo where it is respired or used to build new cells (sugars are translocated to the growing point as sucrose)

fruit and seed coat fused together

cells containing stored protein (aleurone grains)

site of starch store (endosperm)

3 gibberellic acid triggers synthesis of hydrolytic enzymes
e.g. amylase (starch → maltose)
maltase (maltose → glucose)

2 gibberellic acid diffuses to protein store

1 gibberellic acid formed in embryo as germination begins

embryo and cotyledon

coleoptile

uptake of water leads to mobilisation of stored food reserves

dormant seed

seedling grows at expense of stored food

Figure 15.26 The role of gibberellic acid in the germination of barley fruit

The role of gibberellin in stem elongation – the gene involved

Gregor Mendel (1822–84) conducted experiments into the inheritance of a contrasting characteristics of the garden pea plant, *Pisum sativum*. For example, the height of stem in the garden pea plant, which may be either 'tall' (say about 48 cm), or 'dwarf' (about 12 cm). Ultimately, it was established that this characteristic is controlled by a single gene with two alleles.

We now know that this height difference between tall and dwarf pea plants is due to the ability (or otherwise) to convert a precursor molecule of gibberellin into an active form in the cells of the pea plant. In fact, tall pea plants carry at least one allele (***Le***), which codes for a protein that functions normally in the gibberellin-synthesis pathway, catalysing the formation of gibberellin. That allele is 'dominant', so it may be present as a single allele or with two, but in either case, gibberellin will be present and the plant will be tall.

Dwarf pea plants carry two alleles (*le*) that code for a non-functional protein. This enzyme fails in the gibberellin-synthesis pathway. Plants with two recessive alleles remain dwarf, lacking the supply of gibberellin that enables extension growth in stem length to occur.

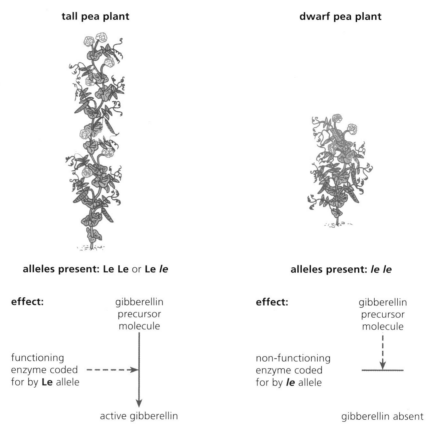

Figure 15.27 The role of specific alleles in the control of height in the garden pea

Summary

- The **nervous system** contains **receptors** (e.g. sense organs) linked to **effectors** (muscles or glands) by **neurones**, known collectively as **reflex arcs**. In mammals, reflex arcs are co-ordinated in the spinal cord and may be connected to the brain. Neurones are the basic units of the nervous system. The nerve fibres of neurones, the **dendron** and **axon**, are typically protected by a **myelin sheath**, which also speeds conduction of the impulse.

- An impulse or **action potential** is a temporary reversal of the electrical potential difference that is maintained across the membrane of the nerve fibres. Conduction is extremely fast and, after an action potential, there is a brief period when the fibre is not excitable (**refractory period**). Action potentials are transmitted between neurones across tiny gaps at **synapses**. Transmission here is chemical, involving diffusion of a specific transmitter substance.

- Receptors contain sensory cells or dendrites of sensory neurones that respond to stimuli with the production of action potentials. The **brain** receives and integrates incoming information from sensory receptors, sends impulses to effectors, stores information as a memory bank, and initiates activities in its own right.

- **Hormones** work with the animal's nervous system in the control and co-ordination of the body. Hormones are produced in endocrine glands, transported all over the body in the blood and affect specific target organs.

(continued)

Summary *(continued)*

- The **menstrual cycle** consists of two cycles, one in the **ovaries** and one in the **lining of the uterus**. The hormones controlling these cycle are **follicle stimulating hormone (FSH)** and **luteinising hormone (LH)** from the pituitary and **oestrogen** and **progesterone** from the ovaries. At the start, **FSH** stimulates **follicle development** to the point at which one eventually becomes an **ovarian follicle**. **Oestrogen** from this follicle triggers thickening of the **endometrium** and reduces the secretion of FSH. The rising levels of oestrogen cause **ovulation** and the empty follicle becomes a **corpus luteum**. The corpus luteum secretes progesterone. Build up of the endometrium continues until **fertilisation fails** to occur. The corpus luteum then breaks down, oestrogen and progesterone levels fall and **menstruation** follows.

- The **combined oral contraceptive pill** contains two hormones, chemically similar to oestrogen and progesterone. Provided the pill is taken daily for 21 days it:

  - prevents ovulation – the pill stops the ovaries releasing an egg each month
  - thickens the mucus in the cervix making it difficult for sperm to reach the egg
  - makes the lining of the uterus thinner so it is less likely to accept a fertilised egg.

- **Plant sensitivity** and co-ordination is mediated by **growth regulators** (including auxins, gibberellins and abscisic acid) which are different from animal hormones.

- Plant responses are mainly **growth responses**. For example, the tips of plant stems grow towards the light, and plant stems and root tips respond to gravity so the aerial shoot grows up and the roots down. Auxins (principally **indoleacetic acid**) maintain apical dominance of stem; gibberellins are active in stem elongation and in germination and abscisic acid is involved in the closure of stomata.

Examination style questions

1 Fig. 1.1 shows the changes in potential difference (p.d.) across the membrane of a neurone over a period of time. The membrane was stimulated at time **A** and time **B** with stimuli of different intensities.

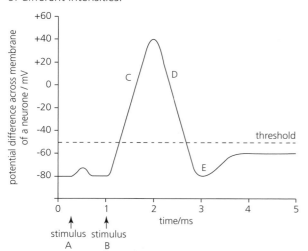

Fig. 1.1

a) Stimulus **B** resulted in an action potential. Describe what is occurring at **C**, **D** and **E**. [6]

b) Suggest why stimulus **A** did not result in an action potential being produced whereas stimulus **B** did. [2]

[Total: 8]

(Cambridge International AS and A Level Biology 9700, Paper 04 Q8 November 2007)

2 a) Outline the ways in which the endocrine **and** nervous systems carry out their roles in control and co-ordination in animals. [8]

b) Describe the part played by auxins in apical dominance in a plant shoot. [7]

[Total: 15]

(Cambridge International AS and A Level Biology 9700, Paper 41 Q9 June 2011)

3 a) The steroid hormones oestrogen and progesterone are secreted by the ovary. State precisely the sites of secretion of each. [2]

b) The most effective oral contraceptives for general use are the so-called combined oral contraceptives (COCs), which contain both oestrogen and progesterone. Explain how COCs produce their effects. [4]

c) Describe two **social** implications of the use of contraceptives. [2]

[Total: 8]

(Cambridge International AS and A Level Biology 9700, Paper 04 Q2 June 2009)

4 a) By means of a labelled diagram show the structure of a synapse. [5]

b) Describe the sequence of steps by which an action potential crosses a cholinergic synapse. [10]

c) Outline the roles synapses play in the functioning nervous system. [5]

[Total: 20]

16 Inherited change

Genetic information is transmitted from generation to generation to maintain the continuity of life. In sexual reproduction, meiosis introduces genetic variation so that offspring resemble their parents but are not identical to them. Genetic crosses reveal how some features are inherited. The phenotype of organisms is determined partly by the genes they have inherited and partly by the effect of the environment. Genes determine how organisms develop and gene control in bacteria gives us a glimpse of this process in action.

16.1 Passage of information from parent to offspring

Diploid organisms contain pairs of homologous chromosomes.

The behaviour of maternal and paternal chromosomes during meiosis generates much variation amongst individuals of the next generation.

By the end of this section you should be able to:

a) explain what is meant by homologous pairs of chromosomes
b) explain the meanings of the terms haploid and diploid and the need for a reduction division (meiosis) prior to fertilisation in sexual reproduction
c) outline the role of meiosis in gametogenesis in humans and in the formation of pollen grains and embryo sacs in flowering plants
d) describe, with the aid of photomicrographs and diagrams, the behaviour of chromosomes in plant and animal cells during meiosis, and the associated behaviour of the nuclear envelope, cell surface membrane and the spindle
e) explain how crossing over and random assortment of homologous chromosomes during meiosis and random fusion of gametes at fertilisation lead to genetic variation including the expression of rare, recessive alleles

Diploid organisms contain homologous pairs of chromosomes

We introduced the structure and features of chromosomes in Topic 5:

- The **number of chromosomes** per **species** is fixed.
- The shape of a chromosome is characteristic; they are of **fixed length** and with a narrow region, called the **centromere**, somewhere along their length.
- The centromere is always in **the same position** on any given chromosome, and this position and the length of the chromosome is how it can be identified in photomicrographs.
- Chromosomes hold the hereditary factors, **genes**. A particular gene always occurs on the same chromosome in the same position.
- The position of a gene is called a **locus**.
- Each gene has two or more forms, called **alleles**.

The **loci** are the positions along the chromosomes where genes occur, so alleles of the same gene occupy the same locus.

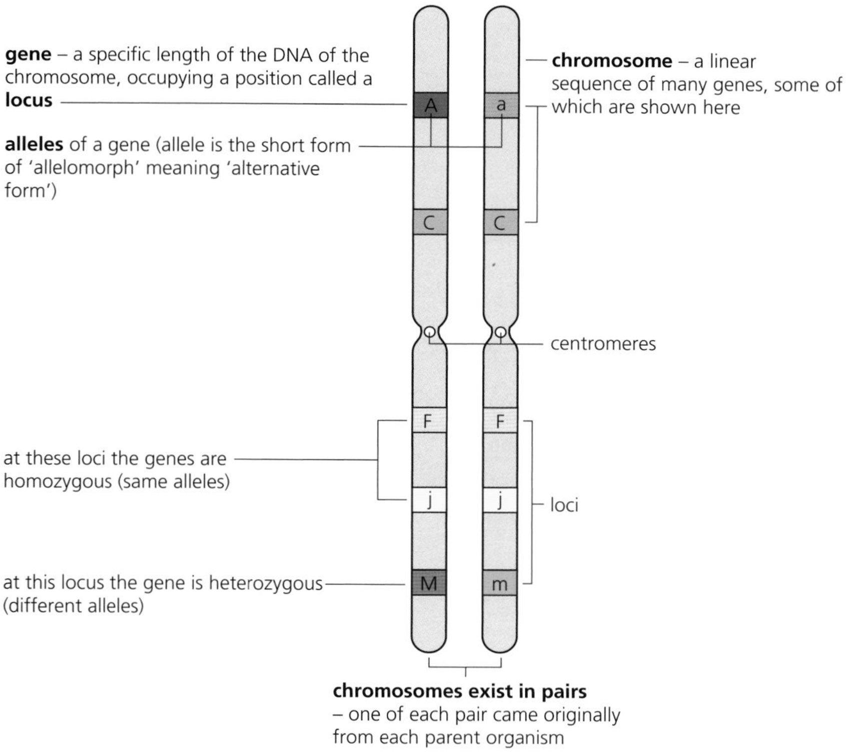

gene – a specific length of the DNA of the chromosome, occupying a position called a **locus**

alleles of a gene (allele is the short form of 'allelomorph' meaning 'alternative form')

chromosome – a linear sequence of many genes, some of which are shown here

centromeres

at these loci the genes are homozygous (same alleles)

loci

at this locus the gene is heterozygous (different alleles)

chromosomes exist in pairs
– one of each pair came originally from each parent organism

Figure 16.1 The genes and alleles of a homologous pair of chromosomes

We have also seen that the **nuclear division** that occurs during growth, known as **mitosis**, results in the exact duplication of these chromosomes in the daughter nuclei formed.

Examination of the chromosomes at certain stages in mitosis discloses another important feature. Chromosomes occur in pairs. These pairs of chromosomes are known as **homologous pairs** (Figure 16.1). These homologous pairs of chromosomes arise in the first place because one member of the pair has come from the male parent and the other from the female parent. Homologous pairs are then maintained by the exact replication that takes place prior to each mitotic division. A cell with a full set of chromosomes (all homologous pairs) in its nucleus is said to be in the **diploid** condition.

Diploid: a eukaryotic cell or organism containing two complete sets of chromosomes (two copies of each homologous chromosome), shown as $2n$, such as a human body (somatic) cell.

Haploid: a eukaryotic cell or organism containing only one complete set of chromosomes (only one of each homologous pair of chromosomes), shown as n, such as a human sperm or secondary oocyte.

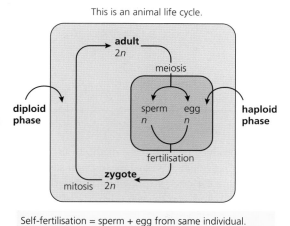

This is an animal life cycle.

Self-fertilisation = sperm + egg from same individual.
Cross-fertilisation = sperm + egg from different individuals.

Figure 16.2 Meiosis and the diploid life cycle

Meiosis – the reductive division

There is a different type of nuclear division that results in daughter nuclei each containing half the number of chromosomes of the parent cell.

In **sexual reproduction**, two sex cells (called **gametes**) fuse; this is **fertilisation**. The resulting cell is called the zygote, and it will grow and develop into a new individual.

Because two nuclei fuse when sexual reproduction occurs, the chromosome number is doubled at that time. For example, human gametes (each egg cell and sperm) have just 23 chromosomes each. When the male and female gametes fuse, the full component of 46 chromosomes is restored.

The doubling of the number chromosome each time sexual reproduction occurs (every generation) is avoided by a nuclear division that halves the chromosome number. This reductive division is called **meiosis**. The sex cells formed following meiosis contain a single set of chromosomes, a condition known as **haploid**. When haploid gametes fuse the resulting cell has the diploid condition again.

Question

1 a Explain why chromosomes occur in homologous pairs in diploid cells.
  b How do the products of mitosis and meiosis differ?

The event of meiosis in the formation of gametes (a process known as **gametogenesis**) in humans is shown in Figure 16.3. In the formation of gametes in flowering plants (the pollen grains and embryo sacs) meiosis features similarly. The differences between mitosis and meiosis are summarised in Figure 5.4, page 102.

Note that meiosis features in the maturation stage of garnete formation.

Note that meiosis features in the maturation stage of garnete formation.

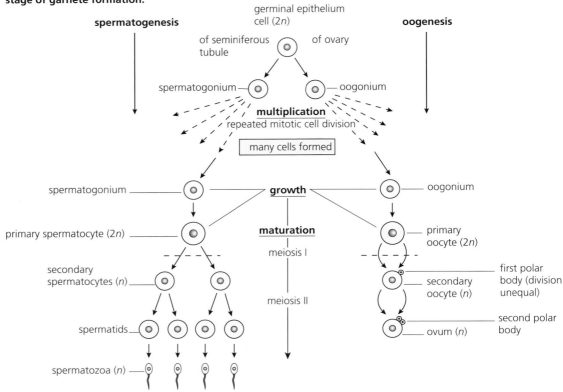

Figure 16.3 Meiosis is gametogenesis in humans

Extension

Asexual reproduction

Organisms reproduce asexually or sexually and many reproduce by both these methods. In **asexual reproduction** a single organism produces new individuals. Asexual means 'non-sexual'; no gametes are formed in asexual reproduction. The cells of the new offspring are produced by **mitosis** (page 102), so the progeny are identical to the parent and to each other. Identical offspring produced by asexual reproduction are known as **clones**.

Mammals reproduce by sexual reproduction only, but many flowering plants reproduce by both asexual and sexual reproduction.

The process of meiosis

Once started, meiosis proceeds steadily as a continuous process of nuclear division. However, the steps of meiosis are explained in four distinct phases (**prophase**, **metaphase**, **anaphase** and **telophase**). This is just for convenience of analysis and description, there are no breaks between the phases in nuclear divisions.

The behaviour of the chromosomes in the phases of meiosis is shown in Figure 16.4. For clarity, the drawings show a cell with a single pair of homologous chromosomes.

Meiosis involves **two divisions of the nucleus**, known as **meiosis I** and **meiosis II**. Both of these divisions *superficially* resemble mitosis. For example, as happens in mitosis, chromosomes are copied, appearing as sister **chromatids** during interphase, before meiosis occurs. However, early in meiosis I, **homologous chromosomes pair up**. By the end of meiosis I, homologous chromosomes have separated again, but the chromatids they consist of do not separate until meiosis II. So, meiosis consists of two nuclear divisions but only one replication of the chromosomes.

Meiosis I

Prophase I

What happens to chromosomes during prophase I is quite complex and difficult to see when you are looking at dividing cells under the microscope. However, you can follow the changes in Figure 16.5.

The chromosomes appear under the light microscope as single threads with many tiny bead-like thickenings along their lengths. These thickenings represent an early stage in the process of shortening and thickening by coiling that continues throughout prophase. In fact, each chromosome has already replicated and consists of two chromatids but the individual chromatids are not visible as yet.

The formation of pairs of homologous chromosomes

As the chromosomes continue to thicken, homologous chromosomes are seen to come together in specific pairs, point by point, all along their length. Remember, in a diploid cell each chromosome has a partner that is the same length and shape and with the same linear sequence of genes.

The homologous chromosomes continue to shorten and thicken. Later in prophase the individual chromosomes can be seen to be double-stranded, as the sister chromatids become visible.

Crossing over

Within the homologous pair, during the coiling and shortening process, breakages of the chromatids occur frequently. Breakages occur in non-sister chromatids at the same points along the length of each. Broken ends rejoin more or less immediately, but where the rejoining is between non-sister chromatids, swapping of pieces of the chromatids has occurred, hence the name 'crossing over'. This results in new combinations of genes on the chromosomes.

The point of the join between different chromatids is called a **chiasma** (*plural*: **chiasmata**). Virtually every pair of homologous chromosomes forms at least one chiasma at this time, and two or more chiasmata in the same homologous pair is very common.

In the later stage of prophase I the attraction and tight pairing of the homologous chromosomes ends, but the attractions between sister chromatids remains for the moment. This attraction of sister chromatids keeps the homologous pairs together. The chromatids are now at their shortest and thickest.

In late prophase I the two centrioles of the centrosome present in animal cells (page 19) duplicate. The two centromeres then start to move apart as a prelude to spindle formation. Plant cells do not have centrioles.

Finally, the disappearance of the nucleoli and **nuclear envelope** marks the end of prophase I.

Figure 16.4 Photomicrograph of homologous chromosomes held together by chiasmata

prophase I (early)
During interphase the chromosomes replicate into chromatids held together by a centromere (the chromatids are not visible). Now the chromosomes condense (shorten and thicken) and become visible.

MEIOSIS I

prophase I (mid)
Homologous chromosomes pair up as they continue to shorten and thicken. Centrioles duplicate.

prophase I (late)
Homologous chromosomes repel each other. Chromosomes can now be seen to consist of chromatids. Sites where chromatids have broken and rejoined, causing crossing over, are visible as chiasmata.

metaphase I
Nuclear envelope breaks down. Spindle forms. Homologous pairs line up at the equator, attached by centromeres.

anaphase I
Homologous chromosomes separate and whole chromosomes are pulled towards opposite poles of the spindle, centromere first.

telophase I
Nuclear envelope re-forms around the daughter nuclei. The chromosome number has been halved. The chromosomes start to decondense.

there is no interphase between **MEIOSIS I** and **MEIOSIS II**

prophase II
The chromosomes condense and the centrioles duplicate.

MEIOSIS II

metaphase II
The nuclear envelope breaks down and the spindle forms. The chromosomes attach by their centromere to spindle fibres at the equator of the spindle.

anaphase II
The chromatids separate at their centromeres and move to opposite poles of the spindle.

telophase II
The chromatids (now called chromosomes) decondense. The nuclear envelope re-forms. The cytoplasm divides.

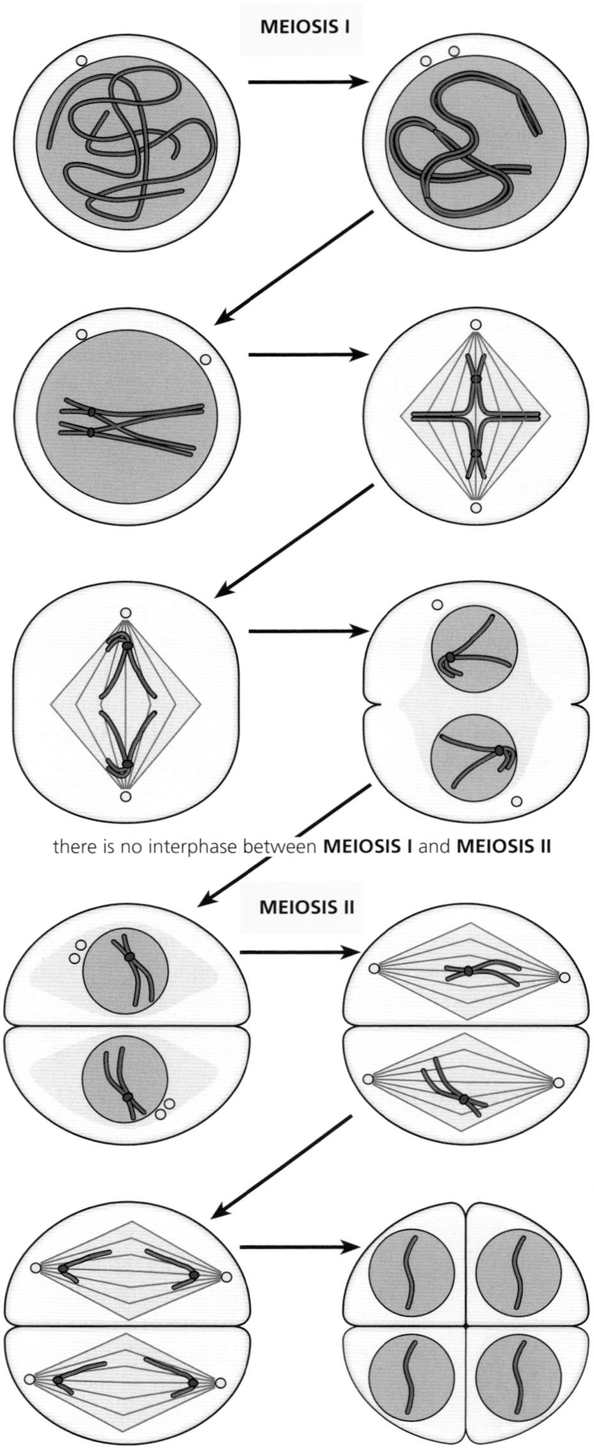

Figure 16.5 The process of meiosis

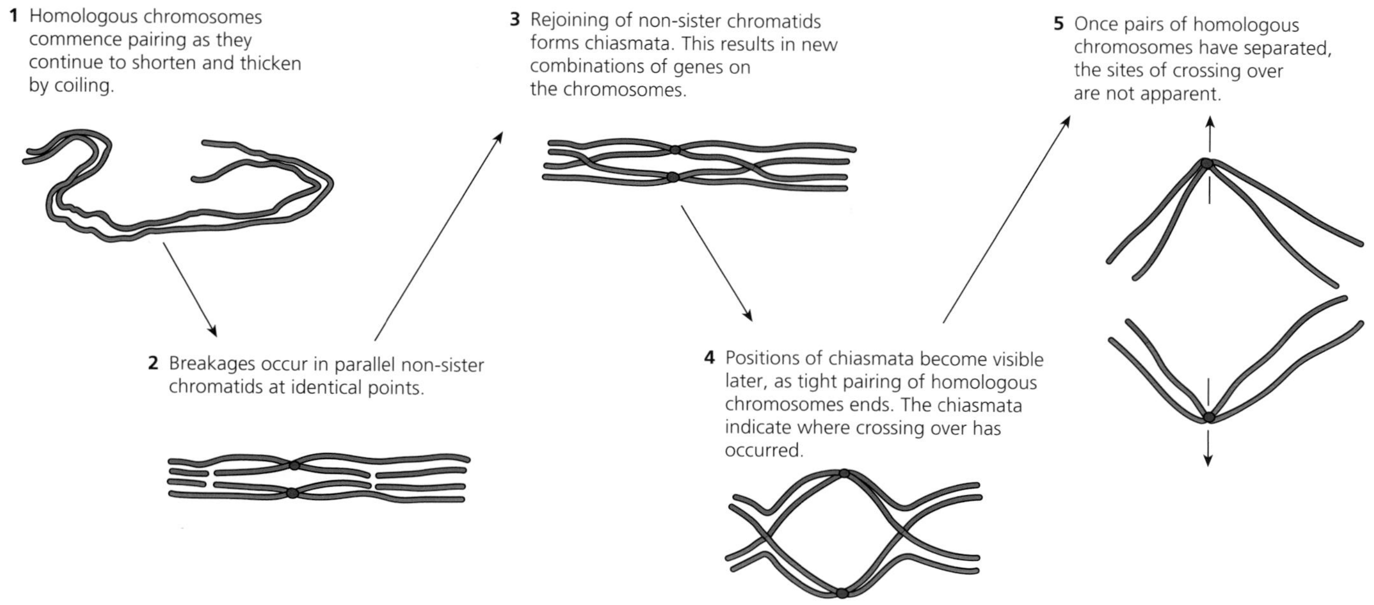

1 Homologous chromosomes commence pairing as they continue to shorten and thicken by coiling.

2 Breakages occur in parallel non-sister chromatids at identical points.

3 Rejoining of non-sister chromatids forms chiasmata. This results in new combinations of genes on the chromosomes.

4 Positions of chiasmata become visible later, as tight pairing of homologous chromosomes ends. The chiasmata indicate where crossing over has occurred.

5 Once pairs of homologous chromosomes have separated, the sites of crossing over are not apparent.

Figure 16.6 The formation of chiasmata

Metaphase I

Next, the spindle forms, and the homologous pairs become attached to individual microtubules of the spindle by their centromeres. The homologous pairs are now arranged at the equatorial plate of the spindle framework – we say that they line up at the centre of the cell. By the end of metaphase I, the members of the homologous pairs start to repel each other and separate. However, at this point they are held together by one or more chiasmata and this gives a temporary but unusual shape to the homologous pair.

Anaphase I

The chromosomes of each homologous pair now move to opposite poles of the spindle, but with the individual chromatids remaining attached by their centromeres. The attraction of sister chromatids has lapsed, and they separate slightly – both are clearly visible. However, they do not separate yet; they go to the same pole. Consequently, meiosis I separates homologous pairs of chromosomes but not the sister chromatids of which each is composed.

Telophase I

The arrival of homologous chromosomes at opposite poles signals the end of meiosis I. The chromosomes tend to uncoil to some extent, and a nuclear envelope re-forms around both nuclei. The spindle breaks down, but these two cells do not go into interphase, rather they continue into meiosis II, which takes place at right angles to meiosis I. There may, however, be a delay between meiosis I and meiosis II, as in oogenesis (page 344).

Question

2 What is the major consequence of there being no interphase between meiosis I and meiosis II?

3 At precisely which points in the process of meiosis (Figure 16.5) do the following events occur?

  a the nuclear envelope breaks down and the spindle forms

  b homologous chromosomes pair up

  c the presence of chiasmata becomes visible

  d chromatids separate at their centromeres

  e homologous chromosomes separate

  f haploid nuclei form

Meiosis II

Meiosis II is remarkably similar to mitosis:

- **Prophase II**
  The nuclear envelopes break down again, and the chromosomes shorten and re-thicken by coiling. Centrioles, if present, duplicate and move to opposite poles of the cell. By the end of prophase II the spindle apparatus has re-formed, but at right angles to the original spindle.
- **Metaphase II**
  The chromosomes line up at the equator of the spindle, attached by their centromeres.
- **Anaphase II**
  The centromeres divide and the chromatids move to opposite poles of the spindle, centromeres first.
- **Telophase II**
  Nuclear envelopes form around the four groups of chromatids, so that four nuclei are formed. Now there are four cells, each with half the chromosome number of the original parent cell. Finally, the chromatids – now recognised as chromosomes, uncoil and become apparently dispersed as **chromatin**. Nucleoli re-form.

The process of meiosis is now complete, and is followed by division of the cytoplasm (cytokinesis, page 101).

Meiosis and genetic variation

There are differences in the genetic information carried by different gametes. This variation arises in meiosis and is highly significant for the organism, as we shall see. The reason why the four haploid cells produced by meiosis differ genetically from each other is because of two important features of meiosis. These are:

- **the independent assortment of maternal and paternal homologous chromosomes**. The way the chromosomes of each homologous pair line up at the equator of the spindle in meiosis I is entirely random. Which chromosome of a given pair goes to which pole is unaffected by (independent of) the behaviour of the chromosomes in other pairs. Figure 16.7 shows a parent cell with only four chromosomes, for clarity. Of course, the more homologous pairs there are in the nucleus, the greater the variation that is possible. In humans there are 23 pairs of chromosomes, so the number of possible combinations of chromosomes that can be formed as a result of independent assortment is 2^{23}. This is over 8 million.
- **the crossing over of segments of individual maternal and paternal homologous chromosomes**. This results in new combinations of genes on the chromosomes of the haploid cells produced, as illustrated in Figure 16.8. Crossing over generates the possibility of an almost unimaginable degree of variation. For example, if we were to assume for sake of discussion that there are 30 000 individual genes on the human chromosome complement, all with at least two alternative alleles, and that crossing over is equally likely between all of these genes, then there are $2^{30\,000}$ different combinations! Of course, all these assumptions will be inaccurate to varying extents, but the point that virtually unlimited recombinations are possible is established.

Finally, further genetic variation in the progeny of organisms that reproduce sexually is assured by the **random fusion** of male and female gametes in sexual reproduction.

A note on 'recombinants'

Offspring with combinations of characteristics different from those of their parents are called **recombinants**. Recombination in genetics is the re-assortment of alleles or characters into different combinations from those of the parents. We have seen that recombination occurs for genes located on separate chromosomes (unlinked genes) by chromosome assortment in meiosis (Figure 16.7), and for genes on the same chromosomes (linked genes) by crossing over during meiosis (Figure 16.8).

We shall see in the next section, how crossing over, random assortment of homologous chromosomes during meiosis, and random fusion of gametes at fertilisation may also lead to the expression of rare, recessive alleles, such as those that are responsible for albinism, sickle cell anaemia, haemophilia and Huntington's disease.

Allele: one of two or more alternative nucleotide sequences at a single gene locus, so alleles are variant forms of a gene. For example, the alleles of the ABO blood group gene are found at a locus on chromosome 9, with the alleles including I^A, I^B and I^O. Diploid body cells contain two copies of each homologous chromosome, so have two copies of chromosome 9, and so have two copies of the gene. These may be the same allele (homozygous), for example $I^A I^A$, or $I^B I^B$ or $I^O I^O$, or they may be different alleles (heterozygous), for example $I^A I^B$, or $I^A I^O$ or $I^B I^O$. The gene for producing the haemoglobin β-polypeptide has a number of alleles. Two of these are the normal allele, Hb^A, and the sickle cell allele, Hb^S, giving $Hb^A Hb^A$ and $Hb^S Hb^S$ as homozygous genotypes and $Hb^A Hb^S$ as a heterozygous genotype.

Question

4 What are the essential differences between mitosis and meiosis?

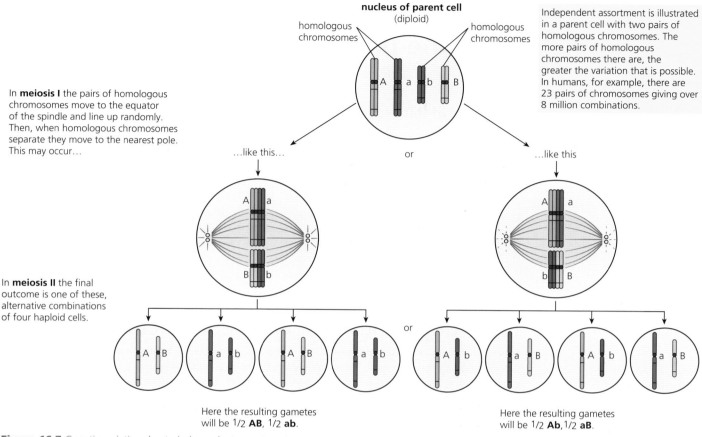

In **meiosis I** the pairs of homologous chromosomes move to the equator of the spindle and line up randomly. Then, when homologous chromosomes separate they move to the nearest pole. This may occur…

Independent assortment is illustrated in a parent cell with two pairs of homologous chromosomes. The more pairs of homologous chromosomes there are, the greater the variation that is possible. In humans, for example, there are 23 pairs of chromosomes giving over 8 million combinations.

In **meiosis II** the final outcome is one of these, alternative combinations of four haploid cells.

Here the resulting gametes will be 1/2 **AB**, 1/2 **ab**.

Here the resulting gametes will be 1/2 **Ab**, 1/2 **aB**.

Figure 16.7 Genetic variation due to independent assortment

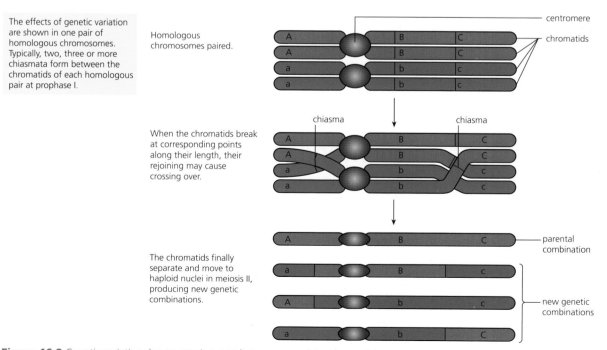

The effects of genetic variation are shown in one pair of homologous chromosomes. Typically, two, three or more chiasmata form between the chromatids of each homologous pair at prophase I.

Homologous chromosomes paired.

When the chromatids break at corresponding points along their length, their rejoining may cause crossing over.

The chromatids finally separate and move to haploid nuclei in meiosis II, producing new genetic combinations.

Figure 16.8 Genetic variation due to crossing over between non-sister chromatids

16.2 The roles of genes in determining the phenotype

Patterns of inheritance are explained by using genetic diagrams. The results of genetic crosses are analysed statistically using the chi-squared test.

By the end of this section you should be able to:

a) explain the terms gene, locus, allele, dominant, recessive, codominant, linkage, test cross, F1 and F2, phenotype, genotype, homozygous and heterozygous

b) use genetic diagrams to solve problems involving monohybrid and dihybrid crosses, including those involving autosomal linkage, sex linkage, codominance, multiple alleles and gene interactions

c) use genetic diagrams to solve problems involving test crosses

d) use the chi-squared test to test the significance of differences between observed and expected results

Inheriting genes in sexual reproduction

In sexual reproduction haploid gametes are formed by meiosis. At fertilisation, male and female gametes fuse to form a zygote. As a consequence, the zygote, a diploid cell, has two sets of chromosomes (homologous pairs), one from each parent. Thus there are two alleles of each gene present in the new individual.

Genotype and phenotype

The alleles that an organism carries (present in every cell) make up the **genotype** of that organism. A genotype in which the two alleles of a gene are the same is said to 'breed true' or to be **homozygous** for that gene. In Figure 16.12, the parent pea plants (P generation) were either homozygous tall or homozygous dwarf. Alternatively, if the alleles are different the organism is **heterozygous** for that gene. In Figure 16.12, the F_1 generation were all heterozygous tall.

So the genotype is the genetic constitution of an organism. Alleles interact in various ways (and with environmental factors, as we shall see later). The outcome is the phenotype. The **phenotype** is the way in which the genotype of the organism is expressed. It is the appearance of the organism. Here the dwarfness and tallness of the plants were the phenotypes observed.

The monohybrid cross

The mechanism of inheritance was successfully investigated *before* chromosomes had been observed or genes were known about. It was **Gregor Mendel** who made the first important discoveries about heredity (Figure 16.9).

An investigation of the inheritance of a single contrasting characteristic is known as a **monohybrid cross**. It had been noticed that the garden pea plant was either tall or dwarf. How this contrasting characteristic is controlled is made clear by analysing the monohybrid cross.

In Figure 16.11 the steps of this investigation into the inheritance of height in the pea plants is summarised. Note that the experiment began with plants that always '**bred true**', that is the tall plants produced progeny that were all tall and the dwarf plants produced progeny that were all dwarf, when each was allowed to self-fertilise. Self-fertilisation is the normal condition in the garden pea plant.

Extension

Gregor Mendel – the founder of modern genetics

Gregor Mendel was born in eastern Europe in 1822, the son of a peasant farmer. As a young boy he worked to support himself through schooling, but at the age of 21 he was offered a place in the monastery at Bruno (now in the Czech Republic). The monastery was a centre of research in natural sciences and agriculture, as well as in the humanities. Mendel was successful there. Later, he became Abbot.

Mendel discovered the principles of heredity by studying **the inheritance of seven contrasting characteristics of the garden pea plant**. These did not 'blend' on crossing, but retained their identities, and were inherited in fixed mathematical ratios.

He concluded that hereditary factors (we now call them genes) determine these characteristics, that these factors occur in duplicate in parents, and that the two copies of the factors segregate from each other in the formation of gametes.

Figure 16.9 Gregor Mendel

Today we often refer to Mendel's laws of heredity, but Mendel's results were **not presented as laws** – which may help to explain the difficulty others had in seeing the significance of his work *at the time*.

Mendel was successful because:

- his experiments were carefully planned, and used large samples
- he carefully recorded the numbers of plants of each type but expressed his results as ratios
- in the pea, contrasting characteristics are easily recognised
- by chance, each of these characteristics was controlled by a single factor (gene)* rather than by many genes, as most human characteristics are
- pairs of contrasting characters that he worked on were controlled by factors (genes) on separate chromosomes*
- in interpreting results, Mendel made use of the mathermatics he had learnt.

*Genes and chromosomes were not known then.

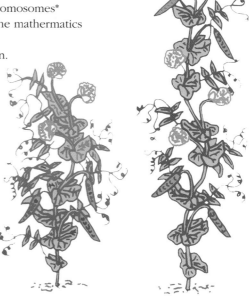

  round v. wrinkled seeds

green v. yellow cotyledons (seed leaves)

dwarf v. tall plants

Figure 16.10 Features of the garden pea

Question

5 At the time of Mendel's experiments it was thought that the characteristics of parents 'blended' in their offspring. What features of Mendel's methods enabled him to avoid this error?

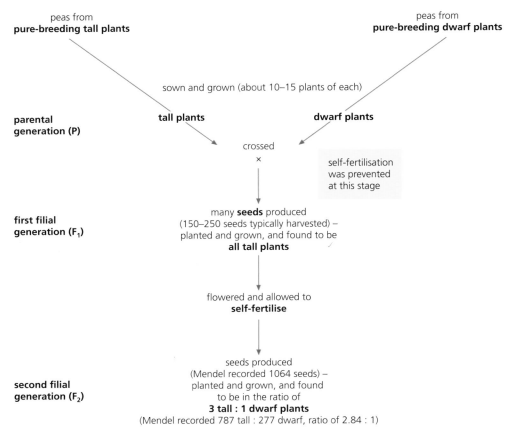

Figure 16.11 The steps of the monohybrid cross

In the monohybrid cross it appears that the 'dwarf' characteristic disappeared in the first generation (F_1 generation) but then reappeared in the F_2 generation. We can explain what is going on if there is a factor controlling 'dwarfness'. This 'factor' remained intact from one generation to another. However, it does not have any effect (we say it does not 'express itself') in the presence of a similar factor for 'tall'. In other words, 'tall' is a **dominant** characteristic and 'dwarf' is a **recessive** characteristic. So there must be two independent factors for height, one from one parent and the other factor from the other parent. A sex cell (gamete) must contain only one of these factors.

Today we see that these 'factors' are **alleles** of the gene for height in the pea plant. When the tall plants of the F_1 generation were allowed to self-fertilise, the resulting members of the F_2 generation were in the ratio of 3 tall plants to 1 dwarf plant. The genetic diagram in Figure 16.12 is the way we explain what is going on.

From the monohybrid cross we see that:
- within an organism are genes (called 'factors' when first discovered) controlling characteristics such as 'tall' and 'dwarf'
- there are two alleles of the gene in each cell
- one allele has come from each parent
- each allele has an equal chance of being passed on to an offspring
- the allele for tall is an alternative form of the allele for dwarf
- the allele for tall is dominant over the allele for dwarf.

We call this the **Law of Segregation**.

> During meiosis, pairs of alleles separate so that each cell has one allele of a pair.
> Allele pairs separate independently during the formation of gametes.

Phenotype: the physical, detectable expression of the particular alleles of a gene or genes present in an individual. It may be possible to see the phenotype (e.g. human eye colour) or tests may be required (e.g. ABO blood group). When the phenotype is controlled by a small number of alleles of a particular gene, it may be genetically determined (e.g. human eye colour), giving rise to **discontinuous variation**. When the phenotype is controlled by the additive effects of many genes (polygenic), it may be affected by the environment as well as genes (e.g. human height), giving rise to **continuous variation**.

Genotype: the particular alleles of a gene at the appropriate locus on both copies of the homologous chromosomes of its cells (for example, $I^A I^B$). It is sometimes described as the genetic constitution of an organism with respect to a gene or genes.

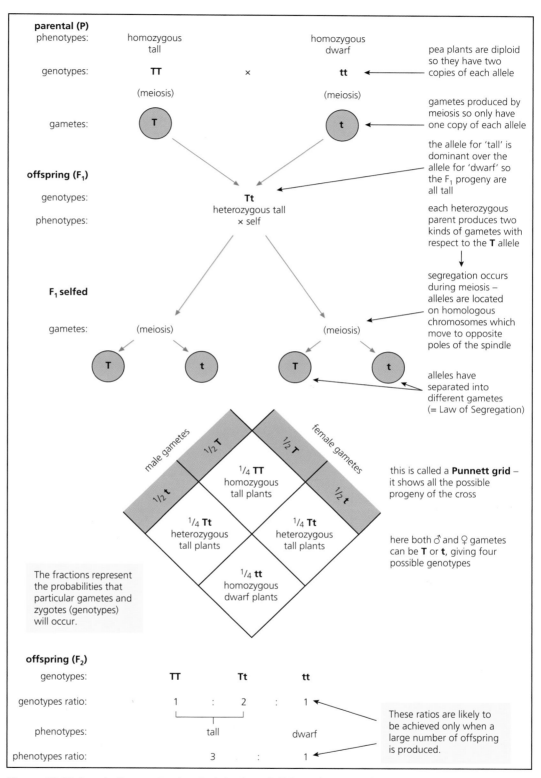

Question

6 From the Punnett grid in Figure 16.12, a ratio of 3 tall to 1 dwarf pea plants was predicted. In fact, a ratio of 2.84 : 1 occurred (Figure 16.11). Suggest what chance events may influence the actual ratios of offspring obtained in breeding experiments like these.

Figure 16.12 Genetic diagram showing the behaviour of alleles in the monohybrid cross

The test cross

If an organism shows a recessive characteristic in its phenotype (for example, it is a dwarf pea plant) it must have a homozygous genotype (**tt**). However, if it shows the dominant characteristic (for example, it is a tall pea plant) then it may be either homozygous for the dominant allele (**TT**) or heterozygous for the dominant allele (**Tt**). In other words, **TT** and **Tt** look alike. They have the same phenotype but different genotypes.

You can tell plants like these apart only by the offspring they produce in a particular cross. When tall heterozygous plants (**Tt**) are crossed with homozygous recessive plants (**tt**), the cross yields 50 per cent tall and 50 per cent dwarf plants (Figure 16.13). This type of cross has become known as a **test cross**. If the offspring produced were all tall this would show that the tall parent plants were homozygous plants (**TT**). Of course, sufficient plants have to be used to obtain these distinctive ratios.

Homozygous: a term describing a diploid organism that has the same allele of a gene at the gene's locus on both copies of the homologous chromosomes in its cells (e.g. Hb^A Hb^A) and therefore produces gametes with identical genotypes (all Hb^A). A homozygote is an organism that is homozygous.

Heterozygous: a term describing a diploid organism that has different alleles of a gene at the gene's locus on each of the homologous chromosomes in its cells (e.g. Hb^A Hb^S) and therefore produces gametes with two different genotypes (½ Hb^A and ½ Hb^S). A heterozygote is an organism that is heterozygous.

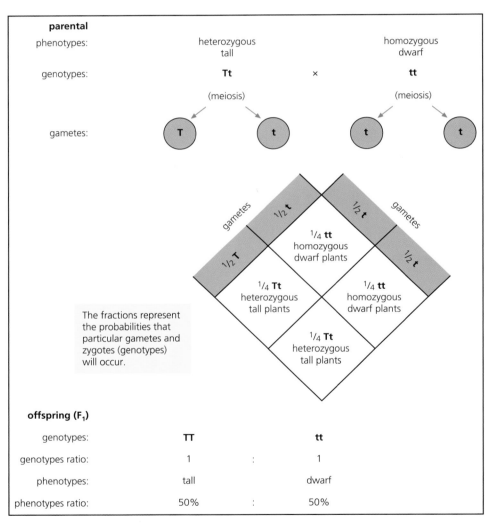

Figure 16.13 Genetic diagram showing the behaviour of alleles in the monohybrid test cross

Modification of the 3:1 ratio

In certain types of monohybrid cross the 3:1 ratio is not obtained. Two of these situations are now illustrated.

Codominance – when both alleles are expressed

In the case of some genes, both alleles may be expressed, rather than one being dominant and the other recessive in the phenotype, as has been the case up to now. When both alleles are expressed codominantly we know that both are transcribed into messenger RNA.

An example of **codominance** is the case of the haemoglobin β-polypeptide gene, where the phenotype of a person who has the genotype $Hb^A Hb^A$ is unaffected by sickle cell disorder, the phenotype of a person who has the genotype $Hb^A Hb^S$ is the less severe sickle cell trait and the phenotype of a person who has the genotype $Hb^S Hb^S$ is the more severe sickle cell anaemia. You can see that, in a cross between parents both of whom have the sickle cell trait, the probability that any offspring will have sickle cell anaemia is one in four (Figure 16.14).

Codominant: alleles that are both expressed if they are present together in a heterozygous organism. For example, alleles I^A and I^B of the ABO blood group gene are codominant. Therefore, in a heterozygous person, $I^A I^B$, both alleles are expressed and the blood group is AB. In the case of the haemoglobin α-polypeptide gene, codominance means that the phenotype of a person who has $Hb^A Hb^A$ is unaffected by sickle cell disorder, the phenotype of a person who has $Hb^A Hb^S$ is the less severe sickle cell trait and the phenotype of a person who has $Hb^S Hb^S$ is the more severe sickle cell anaemia.

Question

7 Construct for yourself (using pencil and paper) a monohybrid cross between cattle of a variety with a gene for coat colour with codominant alleles, 'red' and 'white' coat. Homozygous parents produce 'roan' offspring (red and white hairs together). Predict what offspring you expect and in what proportions when a sibling cross (equivalent to 'selfing' in plants) occurs between roan offspring.

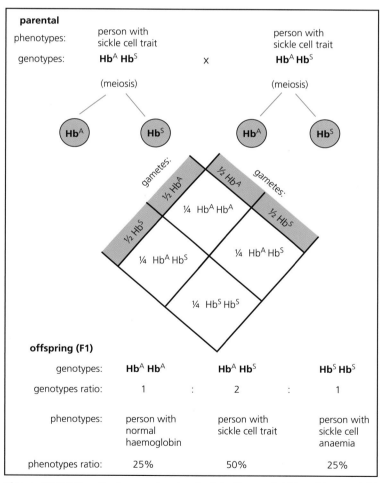

Figure 16.14 Genetic diagram showing the inheritance of sickle cell haemoglobin – an example of codominance

More than two alleles exist for a particular locus

To begin this discussion we considered the inheritance of a gene for which there are just two forms (two alleles), like the 'height' gene of the garden pea, which has tall and dwarf alleles. We represented this situation in a genetic diagram using a single letter (**T** or **t**) according to whether we were representing the dominant or the recessive allele. However, we now know that not all genes are like this. In fact, most genes have more than two alleles, and these are cases of **multiple alleles**.

With multiple alleles, we choose a single capital letter to represent the locus at which the alleles may occur and represent the individual alleles by a single capital letter in a superscript position (Table 16.1) – as with codominant alleles.

An excellent example of multiple alleles is those controlling the ABO blood group system of humans. Our blood belongs to one of either A, B, AB or O group. Table 16.1 lists the possible phenotypes and the genotypes that may be responsible for each.

So, the ABO blood group system is determined by various combinations of alternative alleles. In each individual only two of these three alleles exist, but they are inherited as if they were alternative alleles of a pair. Alleles I^A and I^B are dominant to I^O which is recessive. Alleles I^A and I^B are **codominant alleles**. Figure 16.15 shows one possible cross between parents with different blood groups.

Locus: the position of a gene or other specific piece of DNA (such as a marker) on a chromosome. The same gene is always found at the same locus of the same chromosome (unless there has been a mutation). The locus is designated by the chromosome number, its arm, and its place. For example, the gene associated with ABO blood groups is at locus 9q34, meaning the gene is found on chromosome 9, on the long arm (q) at region 34. The gene associated with sickle cell anaemia is at locus 11p15.5, meaning chromosome 11, short arm (p), region 15.5.

Table 16.1 The ABO blood groups – phenotypes and genotypes

Phenotype	Genotypes
A	$I^A I^A$ or $I^A I^O$
B	$I^B I^B$ or $I^B I^O$
AB	$I^A I^B$
O	$I^O I^O$

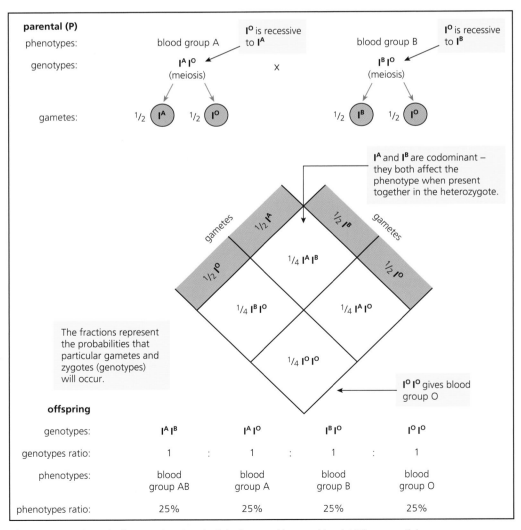

Figure 16.15 Genetic diagram showing the inheritance of human blood ABO group alleles

Reviewing the genetic terms learnt so far

Table 16.2 lists the essential genetic terms and defines each, succinctly.

Cover the right hand column with a piece of paper, temporarily, and see if you can define each correctly, in your own words.

Table 16.2 Essential genetic terms

Haploid	a eukaryotic cell or organism containing only one complete set of chromosomes (only one of each homologous chromosome)
Diploid	a eukaryotic cell or organism containing two complete sets of chromosomes (two copies of each homologous chromosome)
Gene	a length of DNA that codes for a particular polypeptide or protein
Allele	one of two or more alternative nucleotide sequences at a single gene locus; alleles are variant forms of genes
Locus	the position on the chromosome where a gene occurs; alleles of the same gene occupy the same locus
Genotype	the genetic constitution of an organism with respect to a gene or genes
Phenotype	the physical, detectable expression of the particular alleles of a gene or genes present in an individual
Homozygous	a term describing a diploid organism that has the same allele of a gene at the gene's locus on both copies of the homologous chromosomes in its cell
Heterozygous	a term describing a diploid organism that has different alleles of a gene at the gene's locus on both copies of the homologous chromosomes in its cell
Dominant allele	an allele that affects the phenotype of the organism whether present in the heterozygous or homozygous condition
Recessive allele	an allele that affects the phenotype of the organism only when the dominant allele is absent (i.e. the individual is homozygous recessive)
Codominant alleles	alleles that are both expressed if they are present together in a heterozygous person
Test cross	testing a suspected heterozygote by crossing it with a known homozygous recessive

Question

8 One busy night in an understaffed maternity unit, four children were born about the same time. Then the babies were muddled up by mistake; it was not certain which child belonged to which family. Fortunately the children had different blood groups: A, B, AB and O.

The parents' blood groups were also known:

- Mr and Mrs Jones A and B
- Mr and Mrs Lee B and O
- Mr and Mrs Gerber O and O
- Mr and Mrs Santiago AB and O

Staff were able to decide which child belonged to which family. How was this done?

Sex chromosomes

In humans, inheritance of sex is controlled by specific chromosomes known as the sex chromosomes. Each of us has one pair of **sex chromosomes** (either XX or XY chromosomes) along with the 22 other pairs. (These others are known as **autosomal chromosomes**.)

Female gametes produced by meiosis all carry an X chromosome, but half of male gametes carry an X chromosome and half carry a Y chromosome. At fertilisation, a female gamete may fuse with a male gamete carrying an X chromosome, which results in a female offspring. Alternatively, the female gamete may fuse with a male gamete carrying a Y chromosome, which results in a male offspring.

So, the sex of offspring in humans (and all mammals) is decided by which type of male gamete fuses with the female gamete – that is, by the male.

Note also that we would expect equal numbers of male and female offspring to be produced by a breeding population over time.

How human X and Y chromosomes control sex

Both male and female embryos develop identically in the uterus, initially. At the seventh week of pregnancy, however, if a Y chromosome is present in the embryonic cells a cascade of developmental events is triggered, leading to the growth of male sex organs.

On the Y chromosome is the prime male-determining gene. This gene codes for a protein – the **testis determining factor** (**TDF**). TDF functions as a molecular switch; on reaching the embryonic gonad tissues, TDF initiates the production of a relatively low level of testosterone. The effect of

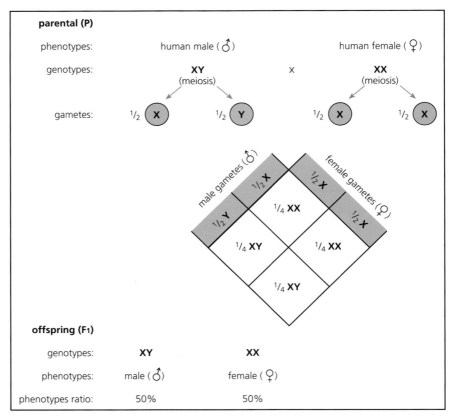

Figure 16.16 X and Y chromosomes and the determination of sex

this hormone at this stage is to inhibit the development of female sex organs and to cause the formation of the testes, scrotum and penis. In the absence of a Y chromosome an ovary develops. Then, partly under the influence of hormones from the ovary, the female reproductive structures develop.

Sex linkage

It is obvious that any gene present on the sex chromosomes is likely to be inherited with the sex of the individual. These genes are said to determine **sex-linked characteristics**. However, the inheritance of these sex-linked genes is different from the inheritance of genes on the other chromosomes present. This is because the X chromosome is much longer than the Y chromosome, so many of the genes on the X chromosome are absent from the Y chromosome. In a male (XY), an allele present on the X chromosome is most likely to be on its own and will be apparent in the phenotype **even if it is a recessive allele**.

Meanwhile, in a female, a single recessive gene is often masked by a dominant allele on the other X chromosome, and in these cases the recessive allele is not expressed. A human female can be homozygous or heterozygous with respect to sex-linked characteristics, whereas males will have only one allele.

'Carriers'

In a heterozygous individual with one dominant allele and one recessive allele of a gene, the recessive allele will not have an effect on their phenotype. The individual is known as a **carrier**. They 'carry' the allele but it is not expressed. It is masked by the presence of the dominant allele.

Consequently, female carriers are heterozygous for sex-linked recessive characteristics. In the case of a male (XY), the unpaired alleles of the X chromosome are all expressed. The alleles on the short Y chromosome are mostly concerned with male structures and male functions.

However, there are some recessive genetically-inherited conditions caused by recessive alleles on the X chromosome. Examples of these are Duchene muscular dystrophy, red–green colour blindness and haemophilia. Notice that these examples are **genetically-inherited conditions**.

The consequence of the presence of one of these alleles may be quite different for males and females. This is because, if a single recessive allele is present in a male human, the allele will be expressed. Meanwhile, a female must be homozygous recessive for a sex-linked characteristic for the allele to be expressed. A female who is heterozygous for one of these will be a 'carrier'. We can illustrate this point by looking at the inheritance of the conditions of red–green colour blindness and of haemophilia.

Red–green colour blindness

A person who is red–green colour blind sees the colours green, yellow, orange and red all as the same colour. The condition affects about 8 per cent of males, but only 0.4 per cent of females in the human population. This is because a female may be homozygous, with both alleles for normal development of cones in the retina ($X^B X^B$), or she may be heterozygous, with an allele for normal colour vision and a mutant allele for abnormal development of these cones ($X^B X^b$). For a female to be red–green colour blind, she must be homozygous recessive for this allele ($X^b X^b$) and this occurs extremely rarely. On the other hand a male with a single recessive allele for red–green colour blindness ($X^b Y$) will be affected.

The inheritance of red–green colour blindness is illustrated in Figure 16.17. It is helpful for those who are red–green colour blind to recognise their inherited condition. Red–green colour blindness is detected by the use of multicoloured test cards.

Question

9 **a** How is the genetic constitution of a female who is red–green colour blind represented?
   b Explain why it is impossible to be a 'carrier' male.

Colour blindness is detected by multicoloured test cards. A mosaic of dots is arranged on the cards so that those with normal colour vision see a pattern that is not visible to those with colour blindness.

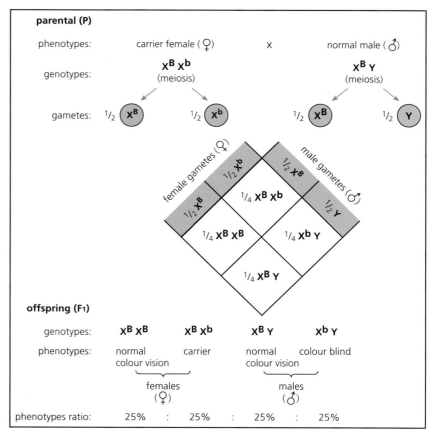

Figure 16.17 The detection and inheritance of red-green colour blindness

The dihybrid cross

Another early investigation by Mendel into the inheritance of variation involved **two pairs of contrasting characters**, again using the garden pea plant. This is now referred to as a **dihybrid cross**.

For example, pure-breeding pea plants from **round seeds** with **yellow cotyledons** (seed leaves) were crossed with pure-breeding plants from **wrinkled seeds** with **green cotyledons** (P generation). All the progeny (F_1 generation) had **round seeds with yellow cotyledons**.

When plants grown from these seeds were allowed to self-fertilise the following season, the resulting seeds (F_2 generation) – of which there were more than 500 to be classified and counted – were of four phenotypes and were present in the ratio shown in Table 16.3 and in Figure 16.18.

Table 16.3 Expected ratio of offspring from a dihybrid cross

Phenotype	round seed / yellow cotyledons	round seed / green cotyledons	wrinkled seed / yellow cotyledons	wrinkled seed / green cotyledons
Ratio	9	3	3	1

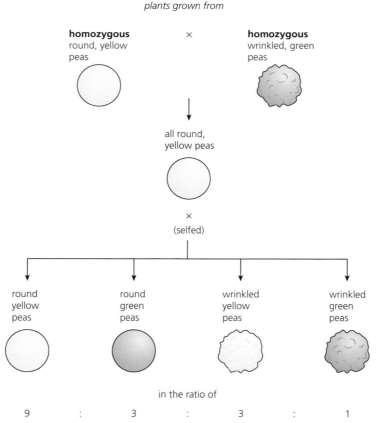

Figure 16.18 A dihybrid cross

It was noticed that two new combinations (**recombinations**), not represented in the parents, appeared in the progeny: either 'round' or 'wrinkled' seeds can occur with either 'green' or 'yellow' cotyledons.

Thus the two pairs of factors were inherited independently. That means either one of a pair of contrasting characters could be passed to the next generation. This tells us that a heterozygous plant must produce four types of gametes in equal numbers. This is explained in Figure 16.19.

From the dihybrid cross we can conclude a **Law of Independent Assortment**:

> During meiosis the separation of one pair of alleles is independent of the separation of another pair of alleles.

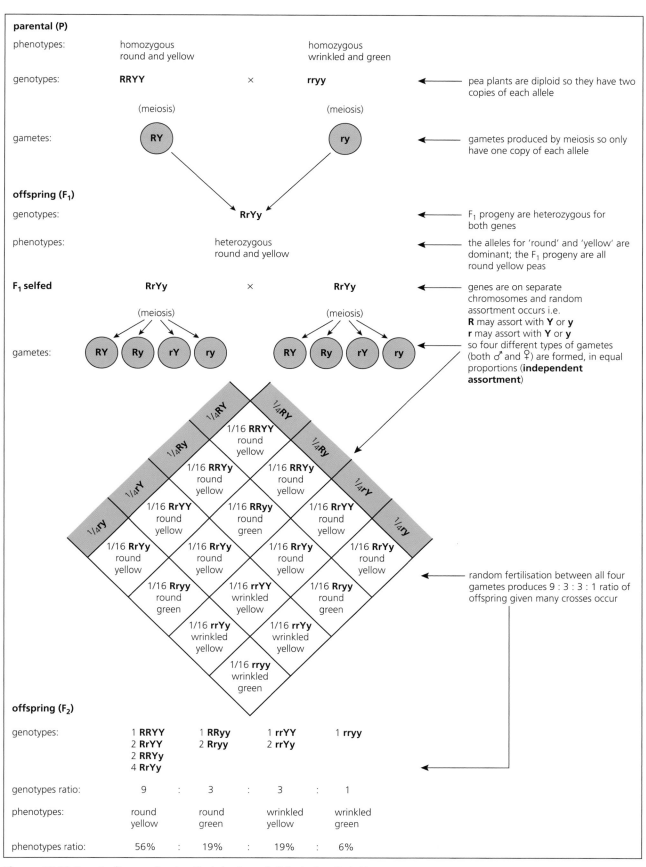

parental (P)

phenotypes: homozygous round and yellow | homozygous wrinkled and green

genotypes: **RRYY** × **rryy** ← pea plants are diploid so they have two copies of each allele

(meiosis) | (meiosis)

gametes: RY | ry ← gametes produced by meiosis so only have one copy of each allele

offspring (F₁)

genotypes: **RrYy** ← F₁ progeny are heterozygous for both genes

phenotypes: heterozygous round and yellow ← the alleles for 'round' and 'yellow' are dominant; the F₁ progeny are all round yellow peas

F₁ selfed **RrYy** × **RrYy** ← genes are on separate chromosomes and random assortment occurs i.e. **R** may assort with **Y** or **y**, **r** may assort with **Y** or **y**

(meiosis) | (meiosis)

gametes: RY Ry rY ry | RY Ry rY ry ← so four different types of gametes (both ♂ and ♀) are formed, in equal proportions (**independent assortment**)

¼RY ¼Ry ¼rY ¼ry | ¼RY ¼Ry ¼rY ¼ry

1/16 **RRYY** round yellow
1/16 **RRYy** round yellow
1/16 **RRYy** round yellow
1/16 **RrYY** round yellow
1/16 **RRyy** round green
1/16 **RrYY** round yellow
1/16 **RrYy** round yellow
1/16 **RrYy** round yellow
1/16 **RrYy** round yellow
1/16 **RrYy** round yellow
1/16 **Rryy** round green
1/16 **rrYY** wrinkled yellow
1/16 **Rryy** round green
1/16 **rrYy** wrinkled yellow
1/16 **rrYy** wrinkled yellow
1/16 **rryy** wrinkled green

← random fertilisation between all four gametes produces 9 : 3 : 3 : 1 ratio of offspring given many crosses occur

offspring (F₂)

genotypes:
1 **RRYY** | 1 **RRyy** | 1 **rrYY** | 1 **rryy**
2 **RrYY** | 2 **Rryy** | 2 **rrYy**
2 **RRYy**
4 **RrYy**

genotypes ratio: 9 : 3 : 3 : 1

phenotypes: round yellow | round green | wrinkled yellow | wrinkled green

phenotypes ratio: 56% : 19% : 19% : 6%

Figure 16.19 Genetic diagram showing the behaviour of alleles in the dihybrid cross

Questions

10 In Figure 16.19, identify the progeny that are recombinants.

11 When a tall, cut-leaved tomato plant and a dwarf, potato-leaved tomato plant were crossed all their progeny were tall and cut-leaved. These were subsequently crossed and the progeny of this second generation were in the ratio:

  • tall and cut-leaved   9

  • tall and potato-leaved   3

  • dwarf and cut-leaved   3

  • dwarf and potato-leaved   1

  Draw a fully annotated genetic diagram to show genotypes of the parents and progeny.

12 In a breed of cocker spaniel black coat (**B** allele) is dominant to red coat (**b** allele) and solid pattern (**S** allele) is dominant to spotted pattern (**s** allele).

  a What are the possible genotypes of a spaniel with a solid-patterned black coat?

  b How could you find out the genotype of this animal?

The dihybrid test cross

Look back to page 350 to remind yourself of the issue the monohybrid test cross sorts out.

Why is a test cross sometimes necessary?

In the case of dihybrid inheritance, too, whilst homozygous recessive genotypes such as wrinkled green peas (**rryy**) can be recognised in the phenotype, homozygous round yellow peas (**RRYY**) and heterozygous round yellow peas (**RrYy**) look exactly the same (Figure 16.18). They can only be distinguished by the progeny they produce, as illustrated in a dihybrid test cross (Figure 16.20).

Unlike the monohybrid test cross where the outcome is a ratio of 1 : 1, in the dihybrid test cross the outcome is a ratio of 1 : 1 : 1 : 1.

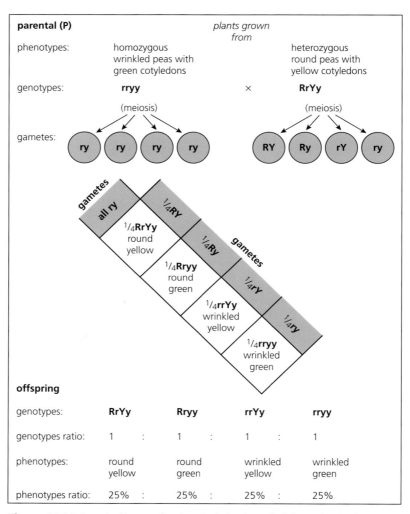

Figure 16.20 Genetic diagram showing the behaviour of alleles in the dihybrid test cross

Introducing *Drosophila*

The fruit fly (*Drosophila*) commonly occurs around rotting vegetable material. The most common form is often called the 'wild type', to differentiate it from the various naturally occurring mutant forms.

Drosophila has become a useful experimental animal in the study of genetics because:

- *Drosophila* has four pairs of chromosomes (Figure 16.21)
- from mating to emergence of adult flies (generation time) only takes about 14 weeks
- one female produces hundreds of offspring
- fruit flies are relatively easily handled, cultured on sterilised artificial medium in glass bottles and they can be temporarily anaesthetised for setting up cultures and sorting progeny.

wild type *Drosophila*

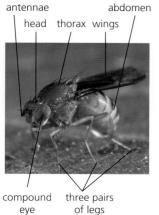

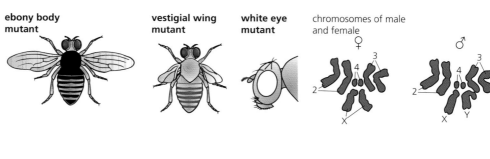

Figure 16.21 Wild type *Drosophila* and some common mutants

A dihybrid cross can be shown in *Drosophila,* for example by crossing normal flies (wild type) with flies homozygous for vestigial wings and ebony body (Figure 16.22). These characteristics are controlled by genes that are not on the sex chromosomes. The genes concerned in this *Drosophila* cross are on different autosomal chromosomes.

Questions

13 Define what is meant by the term mutant organism (or cell).

14 a Construct a genetic diagram for the dihybrid cross shown in Figure 16.22, using the layout given in Figure 16.19.

 b Determine the genotypes of the offspring of the F$_2$ generation.

 c Identify which of these are recombinants.

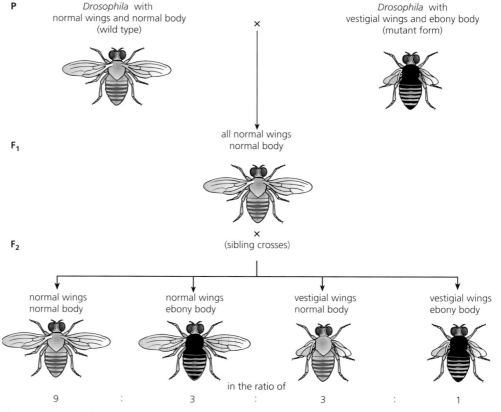

Figure 16.22 A dihybrid cross in *Drosophila*

Drosophila and the dihybrid cross

Mendel had died in 1884 – in relative obscurity, at least as far as the scientific community was concerned. Few seemed to know of his scientific work and those who were aware did not appear to understand his results. But his papers were rediscovered in 1900, as a result of careful literature searches by people keen to advance our understanding of inheritance. Mendel's results came to be confirmed and then extended by many others, using a range of species.

Drosophila melanogaster (the fruit fly) was first selected by an American geneticist called **Thomas Morgan** in 1908 as an experimental organism for the investigation of genetics in an animal. Morgan was awarded a Nobel Prize in 1933 because by his experiments he had:

- shown that Mendel's 'factors' are linear sequences of genes on chromosomes (what is now called the **Chromosome Theory of Inheritance**)
- discovered sex chromosomes and sex linkage
- demonstrated crossing over and the exchange of alleles between chromosomes resulting from chiasmata formed during meiosis.

Probability and chance in genetic crosses

We can see that there is an expected ratio of offspring of $9:3:3:1$ when the dihybrid cross is carried out. Actually, the offspring produced in many dihybrid cross experiment do not *exactly* agree with the expected ratio. This is illustrated by the results of an experiment with mutant forms of *Drosophila* in Table 16.4.

Table 16.4 Observed and expected numbers of offspring from a dihybrid cross

Phenotypes in F₂ generation	normal wings, normal body	normal wings, ebony body	vestigial wings, normal body	vestigial wings, ebony body	Total numbers
Predicted ratio	9	3	3	1	
Expected numbers of offspring	313	104	104	35	556
Actual numbers of offspring	315	108	101	32	556

Clearly, these results are fairly close to the ratio of $9:3:3:1$, but they are not exactly in the predicted ratio, precisely. *What, if anything, went 'wrong'?*

Well, we can expect this ratio among the progeny only if three conditions are met:
- fertilisation is entirely random
- there are equal opportunities for survival among the offspring
- very large numbers of offspring are produced.

In this experiment with *Drosophila*, the exact ratio may not have been obtained because, for example:
- more male flies of one type may have succeed in fertilising females than of the other type
- more females of one type may have died before reaching egg-laying condition than of the other type
- fewer eggs of one type may have completed their development than of the other types.

Similarly, in breeding experiments with plants such as the pea plant, exact ratios may not be obtained because of parasite damage or by the action of browsing predators on the anthers or ovaries in some flowers or because some pollen types fail to be transported by pollinating insects as successfully as others, perhaps.

The chi-squared test

Experimental geneticists are often in the situation of asking 'Do the observed values differ significantly from the expected outcome?' – for example, for the sort of reasons discussed above.

This question is resolved by a simple statistical test, known as the **chi-squared (χ^2) test**. This is used to estimate the probability that any difference between the observed results and the expected results is due to chance. If it is not due to chance, then there may be an entirely different explanation and the phenomenon would need further investigation.

The chi-squared statistic is calculated using this formula:

$$\chi^2 = \frac{\Sigma(O - E)^2}{E} \quad \text{where:} \quad \begin{array}{l} O = \text{observed result} \\ E = \text{expected result} \\ (\Sigma \text{ means 'the sum of').} \end{array}$$

The chi-squared test applied

We can test whether the observed values obtained from the dihybrid cross between wild type (normal) *Drosophila* flies and flies homozygous for vestigial wings and ebony body (Table 16.5) differ significantly from the expected outcome.

First we calculate χ^2 (In this example, χ^2 is 0.47).

Table 16.5 χ^2 calculation

Category	Predicted ratio	Observed number, O	Expected number, E	$O - E$	$(O - E)^2$	$\chi^2 = \dfrac{\Sigma(O - E)^2}{E}$
normal wings, normal body	9	315	312.75	2.25	5.0625	0.016
normal wings, ebony body	3	108	104.25	3.75	14.0625	0.135
vestigial wings, normal body	3	101	104.25	−3.25	10.5625	0.101
vestigial wings, ebony body	1	32	34.75	−2.75	7.5625	0.218
Total		**556**	**556**			**0.470**

We now consult a table of the **distribution of χ^2** to find the **probability** of obtaining a deviation by chance alone as large as (or larger than) the one we have.

The table takes account of the number of independent comparisons involved in our test. Here there were four categories and therefore there were three comparisons made – we call this 'three **degrees of freedom (d.f.)**'.

Table 16.6 Table of χ^2 distribution

Degrees of freedom	Probability greater than:						
	0.99	0.95	0.90	0.10	0.05	0.01	0.001
d.f. = 1	0.0001	0.0039	0.0158	2.706	3.841	6.635	10.83
d.f. = 2	0.0201	0.103	0.211	4.605	5.991	9.210	13.82
d.f. = 3	0.115	0.352	0.584	6.251	7.815	11.345	16.27

Question

15 In the dihybrid test cross between homozygous dwarf pea plants with terminal flowers (ttaa) and heterozygous tall pea plants with axial flowers (TtAa), the progeny were:

- tall with axial flowers: 55
- tall with terminal flowers: 51
- dwarf with axial flowers: 49
- dwarf with terminal flowers: 53

Use the χ^2 test to determine whether or not the difference between these observed results and the expected results is significant.

The probability given by the table suggests that the differences between the results we expected and the outcome we observed is due to chance.

- If the value of χ^2 is **bigger** than the critical value highlighted (a probability of 0.05) then we can be at least 95 per cent confident that the difference between the observed and expected results is significant.
- If the value of χ^2 is **smaller** than the critical value highlighted (a probability of 0.05) then we can be confident that the differences between the observed and expected results are due to chance.

In biological experiments we take a **probability of 0.05 or larger** to indicate that the differences between observed (O) and expected (E) results are **not significant**. We can say they are due to chance.

In this example, the value (0.47) lies between a probability of 0.95 and 0.90. This means that a deviation of this size can be expected 90–95 per cent of the times the experiment is carried out. There is clearly no significant deviation between the observed (O) and the expected (E) results.

The chi-squared test is similarly applicable to the results of test crosses and monohybrid crosses. However, in any chi-squared test that does not confirm that the results conform to the anticipated result (i.e. if it gives us a value for χ^2 that is **bigger** than the critical value meaning that the **probability** of getting that result is **smaller** than 0.05), then we must reconsider our explanation of what is occurring. In this outcome, the statistical test gives no clue as to the actual position or behavior of the alleles concerned. Further genetic investigations are required.

How environment may affect the phenotype

If plants of a tall variety of pea are deprived of nutrients (nitrates and phosphates, for example) in the growing phase of development, full size may not be reached. A 'tall' plant may *appear* dwarf. The same occurs in humans who have been seriously and continuously underfed and malnourished as growing children.

Many characteristics of organisms are affected by both the environment and their genotype. In fact the **phenotype** is the product of **genotype and influences of the environment**.

A striking example of this occurs in the honey bee (*Apis mellifera*). In a colony of honey bees there are three phenotypes (**workers**, **drones** and **queen** – Figure 16.23), but only two genotypes. The drones are the community's males and they develop from unfertilised eggs (their genotype is haploid). The queen and the workers develop from fertilised eggs and have identical genotypes. The queen, who is a much larger organism, differs from her workers only by the diet she is fed in the larval stage (an environmental factor). Her protein-rich food is not given to the larvae that will be workers. This makes a significant difference in their phenotypes.

Queen

Worker

Drone

- the queen lays about 1500 eggs per day, each in a wax cell prepared for her by the workers

- most eggs are fertilised

- occasionally eggs that are unfertilised are laid by the queen. These still develop into larvae (with a single set of chromosomes, i.e. haploid). These larvae are fed on 'royal jelly' (briefly), followed by pollen and nectar; they **develop into drones**

- a very few larvae are fed on 'royal jelly' (protein-rich food prepared by the nurse workers) and **develop into new queens**

- a queen normally survives for 2–5 years but mates only once, storing sperms sufficient for her lifetime in her abdomen

- larvae fed on pollen and nectar **develop into workers** (females with non-functioning reproductive organs)

- workers survive for about 6 weeks in the summer period

- workers undertake a sequence of duties, typically:

  1. as 'nurse' workers:
     - tending growing larvae

  2. as 'outside' workers:
     - surveying for feeding sites
     - guard duty at hive entrance
     - foraging for food and water
     - communicating about new food sites to fellow workers

- drones are fertile males that survive for about 5 weeks

- drones do no work in the hive

- their sole role is to compete to fertilise a new queen on her nuptial flight, after which they are killed or die

A honey bee community typically consists of 20 000–80 000 individuals, of which the vast majority are workers.

Figure 16.23 Honey bees – two genotypes but three phenotypes

16.2 *continued*... How mutations may affect the phenotype – studies in human genetics

Studies of human genetic conditions have revealed the links between genes, enzymes and the phenotype.

By the end of this section you should be able to:

e) explain that gene mutation occurs by substitution, deletion and insertion of base pairs in DNA and outline how such mutations may affect the phenotype

f) outline the effects of mutant alleles on the phenotype in the following human conditions: albinism, sickle cell anaemia, haemophilia and Huntington's disease

g) explain the relationship between genes, enzymes and phenotype with respect to the gene for tyrosinase that is involved with the production of melanin

How mutations may affect the phenotype

DNA can change. An abrupt change in the structure, arrangement or amount of DNA of chromosomes may result in a change in the characteristics of an organism (its phenotype) or of an individual cell. This is because the change, called a **mutation**, may result in the alteration or non-production of a cell protein (because of a change in the **messenger RNA** which codes for it, pages 124–5). This issue was introduced in Topic 6.

Gene mutations can occur spontaneously as a result of errors in normal cell processes such as DNA replication *but this is rare*. Alternatively, gene mutations can be caused by environmental agents we call **mutagens**. These can include ionising radiation in the form of X-rays, cosmic rays, and radiation from radioactive isotopes (alpha particles, beta particles and gamma rays). Any of these can cause breakage of the DNA molecule. Non-ionising mutagens include UV light and various chemicals, including carcinogens in tobacco smoke (tar compounds). These act by modifying the chemistry of the base pairs of DNA. The different ways this may happen are shown in Table 16.7.

Table 16.7 Different ways the base sequence in a gene may be altered

Change within the gene	Likely impact or effect
Base addition: one or more bases are **added** to a sequence e.g. GAG CCT GAG → GAG **T**CC TGA G	These bring about '**frame shifts**' in the code. (Look up which amino acids this sequence coded for originally in Figure 6.10, page 119. What does it code for now?)
Base deletion: one or more bases are **lost** from a sequence e.g. GAG **C**CT GAG → GAG CTG AG	Often, the effect is to change the properties of the protein, particularly its tertiary structure (page 47) – or to cause no protein to be coded for at all.
Base substitution: one or more bases are **substituted**: e.g. CCT G**A**G CCT → CCT G**T**G CCT	Many of the amino acids have two or more triplet codes, so a substitution may have no effect. Alternatively, a substitution may cause an alternative amino acid to be built into a protein, changing its tertiary structure and properties, as is the case in the sickle cell mutation in haemoglobin – see below.

Base substitution mutation and sickle cell anaemia

The smallest form of gene mutation occurs when one base is replaced by another – a **base substitution**. An example of this type of gene mutation is the cause of the human condition known as **sickle cell anaemia**. This was illustrated in Figure 6.17 (page 125). The gene that codes for the amino acid sequence of the beta chain of haemoglobin occurs on chromosome 11 and is prone to a substitution of the base adenine (A) by thymine (T) in a codon for the amino acid glutamic acid. As a consequence, the amino acid valine appears at that point, instead.

Turn back to pages 125–6 now and note the consequence of this substitution at the transcription and translation stages.

The effects on the properties of the haemoglobin molecule of this simple change of one amino acid residue for another within the whole haemoglobin molecule are dramatic. The molecules with this unusual haemoglobin (known as **haemoglobin Hb^s**) tend to clump together and form long fibres that distort the red blood cells into sickle shapes. In this condition they transport little oxygen and the sickle cells may even block smaller vessels. People who are heterozygous for the mutated allele (**Hb Hb^s**) have less than 50 per cent of the sickle-shaped red blood cells. The person is said to have **sickle cell trait** and they are only mildly anaemic. However, people who are homozygous for sickle-cell haemoglobin (**Hb^s Hb^s**) have a serious problem and are described as having **sickle cell anaemia**. Problems with the heart and kidneys are common in people affected with sickle cell anaemia.

We have already noted the potential advantage in having the sickle cell trait. Malaria is the most important of all insect-borne diseases (see Figure 10.7, page 197). The malarial pathogen *Plasmodium* completes its lifecycle in red blood cells, but it cannot do so in red blood cells containing the abnormal form of haemoglobin (**Hb^s**). People with sickle cell trait are protected to a significant extent. Where malaria is endemic in Africa, possession of one mutant allele (i.e. having sickle cell trait but *not sickle cell anaemia*) is advantageous.

Turn back to Figure 10.9 (page 199) now and note the correlations between the distributions of the sickle cell allele and of the distribution of malaria in Africa.

The nucleotide sequence in DNA, the amino acid sequence in proteins and phenotype

We recognise that a person with sickle cell trait has a different phenotype from that of a person with normal haemoglobin. For example, the former has an improved chance of surviving malaria in regions where this disease is endemic.

We have just seen that sickle cell anaemia is due to the substitution of a single nucleotide in one codon of the gene for the protein in the beta chain of haemoglobin. So sickle cell anaemia is a case of a chance change in the nucleotide sequence in DNA affecting the amino acid sequence in a protein and the phenotype of the organism.

Investigating human genetics

Studying human inheritance by experimental crosses (by selecting parents, sibling crosses, and the production of large numbers of progeny) is out of the question. Instead we may investigate the pattern of inheritance of a particular characteristic by researching a family pedigree, where appropriate records of the ancestors exist. A human pedigree chart uses a set of rules. These are identified in Figure 16.24.

Question

16 Explain how the cause of sickle cell trait and sickle cell anaemia supports the concept of a gene as a linear sequence of bases.

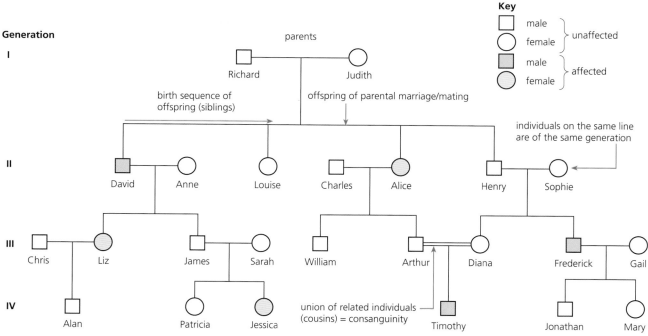

Figure 16.24 An example of a human pedigree chart

We can use a pedigree chart to detect conditions likely to be due to dominant and recessive alleles. In the case of a characteristic due to a **dominant** allele, the characteristic tends to occur in one or more members of the family in every generation. On the other hand, a **recessive** characteristic is seen infrequently, often skipping many generations.

Albinism

Albinism is a rare inherited condition of humans (and other mammals) in which the individual has a block in the biochemical pathway by which the pigment melanin is formed. Albinos have white hair, very light coloured skin and pink eyes. Albinism shows a pattern of recessive monohybrid inheritance in humans (Figure 16.25). In the chart shown of a family with albino members, albinism occurs infrequently, skipping two generations altogether.

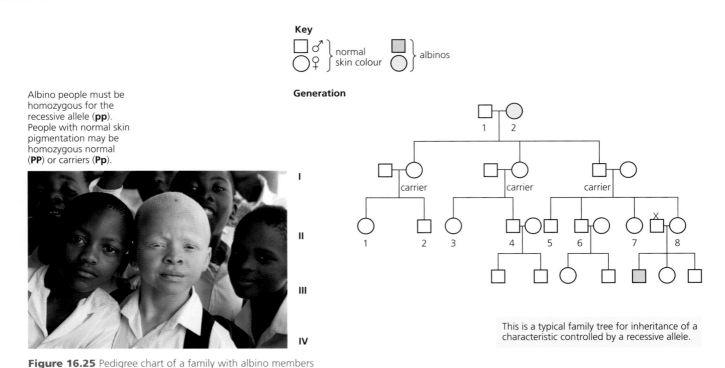

Albino people must be homozygous for the recessive allele (**pp**). People with normal skin pigmentation may be homozygous normal (**PP**) or carriers (**Pp**).

Key

normal skin colour

albinos

Generation

This is a typical family tree for inheritance of a characteristic controlled by a recessive allele.

Figure 16.25 Pedigree chart of a family with albino members

Biochemical basis of albinism – the relationship between gene, enzyme and phenotype

A mutation in the gene for the enzyme tyrosinase, in which glutamine is substituted for arginine at a particular locus, results in a defective gene that fails to code for the tyrosinase protein. As a consequence, active tyrosinase and the pigment melanin are absent from pigment-forming cells in the body. This has evident impacts on the phenotype. The mutated allele is recessive, so the albino people are homozygous for the mutant allele.

Huntington's disease

Huntington's disease (HD) is due to an autosomal **dominant** allele on chromosome 4. The mutation takes the form of repeats of three nucleotides (CAG) between 36 and 120 times at a particular locus in the gene. The function of the gene, or rather of the protein it codes for, is unclear, but it appears to be active in the brain.

The disease is extremely rare (1 case per 20 000 live births). Affected individuals are almost certainly heterozygous for the mutant (defective) gene. Appearance of the symptoms is usually delayed until the age of 40–50 years, by which time the affected person – unaware of the presence of the disease – may have passed a copy of the dominant allele to one or more of his or her children. Any child of an affected person has a 50 per cent chance of inheriting the condition.

The disease takes the form of progressive mental deterioration, which is accompanied by involuntary muscle movements (twisting, grimacing and staring in 'fear'). A person with HD then loses all control of his or her mental and physical abilities, and death occurs within 10 years. The disease is named after the American doctor who first investigated the condition. There is no known treatment.

Haemophilia

In the circulatory system of a mammal, if an injury occurs there is a risk of uncontrolled bleeding. This is normally overcome by the **blood clotting mechanism** that causes any gap in a blood vessel to be plugged. Haemophilia is a rare genetically-determined condition in which the blood does not clot normally. The result is frequent, excessive bleeding.

There are two forms of haemophilia, known as haemophilia A and haemophilia B. They are due to a failure to produce adequate amounts of particular blood proteins essential to the complex blood clotting mechanism. Today, haemophilia is effectively treated by the administration of the clotting factor the patient lacks.

Haemophilia is a sex-linked condition because the genes controlling the production of the blood proteins concerned are located on the X chromosome. Haemophilia is caused by a recessive allele. As a result, haemophilia is largely a condition of the male, since in him a single X chromosome carrying the defective allele ($X^h Y$) will result in disease. For a female to have haemophilia, she must be homozygous for the recessive gene ($X^h X^h$), but this condition is usually fatal *in utero*. It may result in a natural abortion.

A female with one X chromosome with the recessive allele and one with the dominant allele ($X^H X^h$) is a carrier. She has normal blood clotting but for any daughters she has with a normal male, there is a 50 per cent chance of them being carriers; for any sons there is a 50 per cent chance of them being haemophiliac (Figure 16.26).

Question

19 Haemophilia is sex-linked condition that results from the inheritance of an allele carried on the X chromosome. Explain why haemophilia is more common in males.

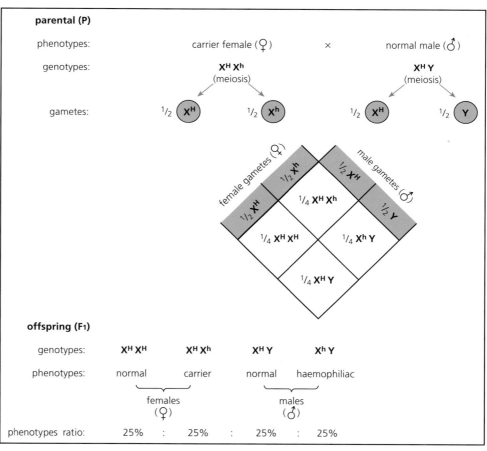

Figure 16.26 Genetic diagram showing the inheritance of haemophilia

16.3 Gene control

Some genes are transcribed all the time to produce constitutive proteins; others are only 'switched on' when their protein products are required.

By the end of this section you should be able to:

a) distinguish between structural and regulatory genes and between repressible and inducible enzymes
b) explain genetic control of protein production in a prokaryote using the lac operon
c) explain the function of transcription factors in gene expression in eukaryotes
d) explain how gibberellin activates genes by causing the breakdown of DELLA protein repressors, which normally inhibit factors that promote transcription

Regulation of gene expression

We have seen that chromosomes are effectively a linear series of genes. We define a gene as a specific region of a chromosome that is capable of determining or influencing the development of a specific characteristic of an organism. It is a specific length of the DNA double helix, hundreds or (more typically) thousands of base pairs long. Remember, every living cell carries a full complement of genes; this genetic complement constitutes the **genome** of the cell or organism.

Consequently, cells and organisms must regulate the particular genes that they express at any given time or stage of development. Unicellular organisms and multicellular organisms may continually switch on and switch off genes in response to signals from their external and internal environments. Whilst some genes are actively transcribed throughout the life of the cell, others are activated (we say they are '**expressed**') only at a particular stage in the life of the cell, or when the substance they act on (their substrate molecule) is present, for example. Very many of our genes have to be deliberately activated, as required. In a multicellular organism, every nucleus contains the coded information relating to the development and maintenance of all mature tissues and organs. For example, the genes concerned with development of the organism from the zygote are some of the first to be expressed, but most probably for a limited period only. So, both during development and later in the life of the organism, this genetic information is used selectively. Typically, less than 25 per cent of the protein-coding genes in human cells are expressed at any time. The expression of genes is related to when and where the proteins they code for are needed (Table 16.8).

Table 16.8 Gene regulation – some examples

When and why genes are expressed	
Expressed all the time	genes responsible for routine, continuous metabolic functions (for example, respiration), common to all cells, continuous throughout life
Expressed at a selected stage in cell or tissue development	as cells derived from stem cells are developing into muscle fibres or neurones, for example
Expressed only in the mature cell	genes responsible for antibody production in a mature plasma cell, after these have been cloned
Expressed on receipt of an internal or external signal	when a particular hormone signal, metabolic signal, or nerve impulse is received by the cell, such as the gene for insulin production in β cells in the islets of Langerhans

How is this regulation brought about?

Regulation of gene action is partly regulated by the genes themselves. Whilst the role of many genes is to code for specific enzymes or proteins required by working cells at some stage, others are different – their role is to regulate other genes. Consequently, we can recognise two categories of gene:

- **Structural genes** code for a structural protein, or an enzyme, or an RNA molecule not involved in regulation. They are required by cells to create or maintain the structure, or enable the functioning, of the cell or organism. They are transcribed into proteins.
- **Regulatory genes** are involved in controlling the expression of one or more genes. They may code for a protein or for an RNA molecule.

Genes are also regulated by 'signals' from the environment – typically from within the cell. These act on the enzymes or on the genes that code for them. Consequently, we can recognise two categories of enzymes:

- **Inducible enzymes** are produced under specific conditions, such as the presence of a particular substrate. This substance controls the expression of one or more genes, structural genes, involved in the metabolism of that substance. In the absence of the substrate, gene action is switched off.
- **Repressible enzymes** are generally produced continuously, but their production can be halted. They are formed unless a signal, such as an excess of product, turns their production off.

Case studies in gene regulation

1 Genetic control of protein production in a prokaryote, using the *lac* operon

The *lac* operon (lactose operon) is a mechanism that ensures that the enzymes required for the metabolism of lactose are produced only in the presence of lactose. It has been observed in bacteria, such as **E. coli**.

This mechanism involves a **repressor molecule** (coded for by a regulator gene), an **operator gene** situated close to the gene that is being regulated (a **structural gene**, in this case coding for the lactose metabolising enzyme), and a **promoter gene**. In the absence of lactose, the repressor binds to the operator and prevents transcription of the structural gene. However, if lactose is present, for example, when it becomes available in the medium in which the bacteria is growing, then the lactose molecule reacts with the regulator protein, preventing its binding with the operator gene. As a result, the lactose-metabolising gene is transcribed, and lactose is metabolised. Once all the lactose has been used up, the repressor molecule blocks transcription again. The mechanism is illustrated in Figure 16.27.

This type of mechanism occurs in prokaryotes (bacteria and cyanobacteria) only. In eukaryotes the regulator-gene mechanisms are more complicated

Figure 16.27 The lac operon and gene regulation

In the presence of the substrate lactose:

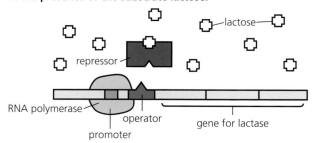

1 Lactose has combined with the repressor and inhibited its action.

2 RNA polymerase has bound to the promoter and is about to express the structural gene for lactase. mRNA for lactase synthesis will be transcribed, and pass straight to ribosomes in the surrounding cytoplasm, where lactase is synthesised.

3 Eventually, all the lactose will have been metabolised, and the repressor will be free to bind to the operator again.

The gene is turned on.

In the absence of the substrate lactose:

1 There is no lactose to combine with the repressor.

2 The repressor binds with the operator.

3 The RNA polymerase is obstructed from binding to the promoter.

The gene is turned off.

2 Regulation in eukaryotes by enhancers and transcription factors

In eukaryotes, before mRNA can be transcribed by the enzyme RNA polymerase it first binds together with a small group of proteins called **general transcription factors** at a sequence of bases known as the **promoter**. Promoter regions occur on DNA strands just before the start of a gene's sequence of bases. (The promoter is an example of a length of non-coding DNA with a special function.) Only when this transcription complex of proteins (enzyme + factors) has been assembled can transcription of the template strand of the gene begin. Once transcription has been initiated, the RNA polymerase moves along the DNA, untwisting the helix as it goes, and exposing the DNA nucleotides so that RNA nucleotides can pair, and the messenger RNA strand be formed and peel way.

The rate of transcription may be increased (or decreased) by the binding of specific transition factors on the enhancer site for the gene. The position of an enhancer site of a gene is shown in Figure 16.28. This is at some distance 'upstream' of the promoter and gene sequence. However, when **activator** proteins bind to this enhancer site a new complex is formed and makes contact with the polymerase–transcription factor complex. Then the rate of gene expression is increased.

Regulator transcription factors, activators and RNA polymerase are all proteins, and are coded for by other genes. It is clear that protein-protein interactions play a key part in the initiation of transcription in eukaryotes.

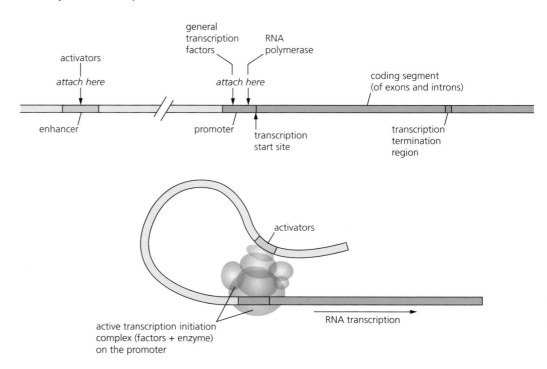

The lower part of the figure shows a portion of the upper DNA molecule with the activator (orange) region, promotor (green) region and coding (blue) region. The DNA is looped so that it comes into contact with different parts of the transcription initiation complex. The complex consists of several proteins and other factors hence the multiple overlapping spheres (blue). The activator (pink) is also part of the complex. These spheres have been drawn transparent so that contact with the DNA molecule can be shown.

Figure 16.28 Control sites and the initiation of transcription

How gibberellin activates genes

Gibberellin (GA) regulates and promotes many aspects of plant growth and development (page 336). *Remind yourself of some of the roles of GA now.*

It is now known that a significant cause of restraint on many aspects of plant growth and development is the presence of a group of nucleus-based proteins called DELLA proteins. These proteins are coded for by specific regulatory genes, and they function in the nucleus itself as powerful transcription factors. How these proteins regulate transcription is poorly understood, but it has been shown that, when GA levels rise in a plant, DELLA proteins are deactivated and the plant grows. For example, GA has this effect as it triggers expression of genes coding for hydrolytic enzymes that convert starch stores to glucose in germinating seeds such as maize and barley (Figure 15.26, page 339).

The way that GA controls the levels of growth-repressing DELLA proteins is apparently by labelling them with a molecule known as ubiquitin. With this marker attached, a DELLA protein is identified and promptly destroyed by a nucleus-based complex, which has the role of degrading unneeded or damaged proteins present in the nucleus.

Summary

- Genetics is the study of **inheritance**. Many characteristics of organisms are controlled by the **genes** which occur as a linear series along **chromosomes**. **Diploid organisms** contain two copies of each gene in each of their cells. Alternative forms of a gene are called **alleles**. In sexual reproduction, **gametes** are formed containing one copy of each gene. Offspring formed from the **zygote** resulting from **fertilisation** receive two copies of each gene (two alleles), one from each parent.

- Meiosis is the reductive nuclear division in which, after cell division, the nucleus present in each of the four daughter cells formed has half the chromosome number of the parent cell, typically one of each type. Meiosis is associated with sexual reproduction and the production of gametes.

- Alleles may be **dominant** or **recessive** or show **codominance**. An organism with two identical alleles of a gene is **homozygous** for that gene; an organism with different alleles of a gene is **heterozygous**. The genetic constitution of an organism is its genotype. The resulting appearance of the organism, its visible and measurable features, is called its **phenotype**.

- In a **monohybrid cross** the inheritance of a contrasting characteristic controlled by a single gene (such as tall and dwarf height in garden pea plants) is investigated. When parents homozygous for a contrasting characteristic are crossed, the first generation (F$_1$) will be heterozygous The characteristic they show, such as 'tall' in the pea, will establish that the allele for 'tall' is **dominant** over the allele for 'dwarf', which is **recessive**. When the F$_1$ generation self-fertilise (or sibling crosses are made in animals), a recessive characteristic re-emerges in a minority of the offspring, in the ration of 3 : 1 (provided many offspring are produced). This shows that only one allele of a gene is carried in a single gamete.

- Since the genotype of an organism showing dominant characteristics may be homozygous or heterozygous for the gene concerned, the genotype of this organism must be determined by a **test cross** with an organism that is homozygous recessive for the characteristic. In the monohybrid test cross the outcome is a ratio of 1 : 1 and in the dihybrid test cross the outcome is a ratio of 1 : 1 : 1 : 1.

- Sex is determined in humans by the **sex chromosomes**: XX in the female and XY in the male. The X chromosome is longer than the Y and has genes not present on the Y chromosome. In the male, recessive alleles on the single X chromosome cannot be masked as they often are by dominant alleles in the female. Consequently, a number of rare **recessive conditions** are sex-linked and occur more frequently in males. These include red–green colour blindness, haemophilia and Duchenne muscular dystrophy.

- Exceptions to the monohybrid ratio arise from alleles which are **codominant** (rather than there being one dominant and one recessive allele). In addition, the existence of more than two alleles for a gene, known as **multiple alleles**, complicates inheritance. The human ABO blood group is an example of the latter. It is determined by two of three possible alleles which gives four possible phenotypes, blood groups A, B, AB and O.

- A **dihybrid cross** concerns the inheritance of two pairs of contrasting characteristics located on different chromosomes. The results confirm that each allele of a gene is equally likely to be inherited with each allele of another gene, a reflection of the **independent assortment** of chromosomes that occurs in meiosis.

- The **environmental conditions** experienced by an organism may also influence the expression of alleles. A genetically tall organism deprived of sufficient nutrients and so develops a phenotype that is short is an obvious example. Characteristics influenced by the environment may show a degree of continuous variation.

- **Mutations** may also affect the phenotype. When the nucleotide sequence in the DNA of a chromosome is changed it may lead to alteration in the amino acid sequence of proteins formed in the cell. The consequence may be a different phenotype.

- Some genes are transcribed all the time, whilst others are only switched on when their products are required. **Regulatory genes** control the timing of the expression of other genes whose roles are to create and maintain the structure or functions of the cell or organism.

Examination style questions

1 Colour blindness is a condition characterised by the inability of the brain to perceive certain colours accurately.
- The most common form is termed red–green colour blindness (RGC).
- RGC results from a recessive allele.
- 0.6% of females worldwide have RGC.
- 8.0% of males worldwide have RGC.

Fig. 1.1 shows the occurrence of RGC in one family.

☐ = male   ◯ = female   ◼ = male with RGC

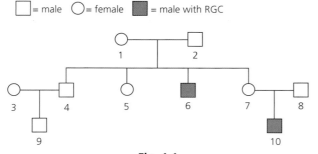

Fig. 1.1

a) Explain the meaning of the terms *allele* and *recessive*. [2]

b) Explain why females are less likely than males to have RGC. [2]

c) With reference to Fig. 6.1, and using the symbols **R** for the dominant allele and **r** for the recessive allele, state the genotypes of the individuals **1**, **4**, **6** and **7**. [4]

[Total: 8]

(Cambridge International AS and A Level Biology 9700, Paper 04 Q6 June 2008)

2 In mice there are several alleles of the gene that controls the intensity of pigmentation of the fur.

The alleles are listed below in order of dominance with **C** as the most dominant.

$$C = \text{full colour}$$

$$C^{ch} = \text{chinchilla}$$

$$Ch = \text{himalayan}$$

$$C^p = \text{platinum}$$

$$C^a = \text{albino}$$

The gene for eye colour has two alleles. The allele for black eyes, **B**, is dominant, while the allele for red eyes, **b**, is recessive.

A mouse with full colour and black eyes was crossed with a himalayan mouse with black eyes. One of the offspring was albino with red eyes.

Using the symbols above, draw a genetic diagram to show the genotypes and phenotypes of the offspring of this cross.

[Total: 6]

(Cambridge International AS and A Level Biology 9700, Paper 04 Q8 November 2008)

3 a) Explain what is meant by a gene mutation due to base substitution. [4]

b) In the inheritance of sickle cell anaemia, the normal allele is represented by **Hb** and the sickle cell allele by **Hb^S**. Draw and complete a genetic diagram to explain the genotypes and phenotypes and the proportions of each resulting from a cross between two individuals of genotype **Hb Hb^S**. [8]

c) A study of a local population in a central African country found that in a sample of 12 5000 people:

9449 were homozygous for normal haemoglobin (**Hb Hb**)

3020 had sickle cell trait (**Hb Hb^S**)

31 had sickle cell anaemia (**Hb^S Hb^S**)

Comment on the distribution of the sickle cell allele within this population, and on any environmental factors that may influence it. [8]

[Total: 20]

4 a) List four differences between the processes of mitosis and meiosis, and state briefly the significance of each. [8]

b) Where would you expect (i) mitosis and (ii) meiosis to occur in a fully grown flowering plant? [2]

c) For a cell with two chromosomes draw diagrams to show:

i) metaphase of mitosis

ii) metaphase I of meiosis

iii) prophase II of meiosis

iv) metaphase II of meiosis [4]

d) Define the term 'phenotype' and explain with reference to an example, how the environment may affect phenotype. [4]

[Total: 18]

17 Selection and evolution

Charles Darwin and Alfred Russel Wallace proposed a theory of natural selection to account for the evolution of species in 1858. A year later, Darwin published *On the Origin of Species* providing evidence for the way in which aspects of the environment act as agents of selection and determine which variants survive and which do not. The individuals best adapted to the prevailing conditions succeed in the 'struggle for existence'.

17.1 Variation

The variation that exists within a species is categorised as continuous and discontinuous. The environment has considerable influence on the expression of features that show continuous (or quantitative) variation.

By the end of this section you should be able to:

a) describe the differences between continuous and discontinuous variation and explain the genetic basis of continuous (many, additive genes control a characteristic) and discontinuous variation (one or few genes control a characteristic)

b) explain, with examples, how the environment may affect the phenotype of plants and animals

c) use the t-test to compare the variation of two different populations

d) explain why genetic variation is important in selection

Introducing variation

Individuals of a species are strikingly similar, which is how we may identify them, whether humans, buttercups or houseflies, for example. But individuals also show many differences, although we may have to look carefully in members of species other than our own. Within families there are remarkable similarities between parents and their offspring, but no two members of a family are identical, apart from identical twins. About these differences we can say:

- some differences may be controlled by genes – such as human blood groups
- other differences between individuals may be due to the effect of our environment, such as the fur colour of the Arctic hare or the Siamese cat
- other differences between individuals may be due to both genetics and environment, such as our body height and weight.

Another important point about variation is that it is of two types (Figure 17.1).

- **Discontinuous variation** arises when the characteristic concerned is one of two or more discrete types with no intermediate forms. Examples include the garden pea plant (tall or dwarf) and human ABO blood grouping (group A, B, AB or O). These are genetically determined.
- **Continuous variation** results in a continuous distribution of values. Height in humans is a good example. Continuous variation may be genetically determined, or it may be due to environmental and genetic factors working together.

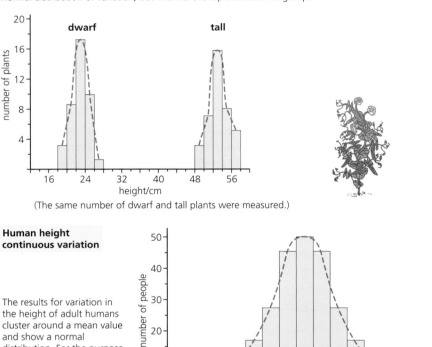

Dwarf and tall peas discontinuous variation

The heights of these plants fall into two discrete groups, in both of which there is a normal distribution of variation, but with no overlap between the groups.

(The same number of dwarf and tall plants were measured.)

Human height continuous variation

The results for variation in the height of adult humans cluster around a mean value and show a normal distribution. For the purpose of the graph, the heights are collected into arbitrary groups, each of a height range of 2 cm.

Figure 17.1 Discontinuous and continuous variation

Genetic basis of discontinuous variation

We began the story of genetics in Topic 16 with an investigation of the inheritance of height in the garden pea (page 351), where one gene with two alleles gave tall or dwarf plants. This clear-cut difference in an inherited characteristic is an example of **discontinuous variation** in that there is no intermediate form, and no overlap between the two phenotypes. There are other examples of discontinuous variation in the introduction to monohybrid and dihybrid inheritance in that topic.

Genetic basis of continuous variation

In fact, very few characteristics of organisms are controlled by a single gene. Mostly, characteristics are controlled by a number of genes. Groups of genes that together determine a characteristic are called polygenes. Polygenic inheritance is the inheritance of phenotypes that are determined by the collective effect of several genes. The genes that make up a polygene are often (but not necessarily always) located on different **chromosomes**. The effects of any one of these genes make a very small or insignificant effect on the phenotype, but the combined effect of all the genes of the polygene is to produce *infinite* variety among the offspring.

Many features of humans are controlled by polygenes, including body weight and height (Figure 17.1, above), but also, human skin colour.

The colour of human skin is due to the amount of the pigment called **melanin** that is produced in the skin. Melanin synthesis is genetically controlled. It seems that three, four *or more* separately inherited genes control melanin production. The outcome is a continuous distribution of skin colour from very pale (presence of no alleles coding for melanin production) to very dark brown (all 'skin colour' alleles coding for melanin production). In our illustration of polygenic inheritance of human skin colour, we have used only two independent genes. This is because dealing with, say, four genes becomes impossibly unwieldy, on the printed page. Also, the principle can be demonstrated clearly enough using just two genes (Figure 17.2).

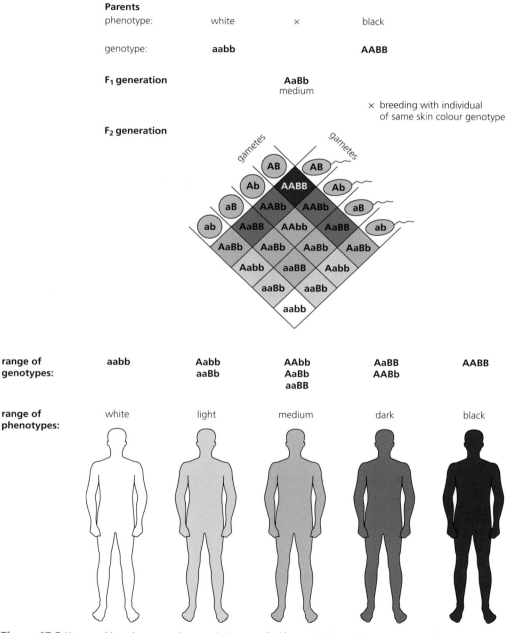

Figure 17.2 Human skin colour as a characteristic controlled by two independent genes – an illustration of polygenic inheritance

How the environment may affect the phenotype

Many characteristics of organisms are affected by both the environment and their genotype. In fact:

$$\text{phenotype} = \text{genotype} + \text{influences of the environment}$$

For example, if plants of a tall variety of a pea are deprived of nutrients (nitrates and phosphates, for example) in the growing phase of development, full size may not be reached. A 'tall' plant may appear dwarf.

When the wind-dispersed seeds of the dandelion plant (*Taraxacum officinale*) land and germinate in contrasting habitats such as in fully sunlit wasteland, in shaded woodland soil, or in a crevice on a wall, the forms of the resulting plants when fully grown reflect the different environmental influences (Figure 17.3).

in soil, in full sunlight   **in shaded woodland soil**   **in an epilithic site**

Figure 17.3 Dandelion plant (*Taraxacum officinale*) plants in contrasting habitats

A further, striking example of the influence of environment on adult form occurs in the honey bee (*Apis mellifera*). In a colony of honey bees there are three phenotypes (**workers**, **drones** and **queen**), but only two genotypes. This was shown in Figure 16.23, on page 366. The drones are the community's males and they develop from unfertilised eggs (their genotype is haploid). The queen and the workers develop from fertilised eggs and have identical genotypes. The queen, who is a much larger organism, differs from her workers only by the diet she is fed in the larval stage (an environmental factor). Her protein-rich food, prepared for her by the nurse worker bees in the colony, is not available to the larvae that will be workers. This makes a hugely significant difference in their phenotypes.

Using the *t*-test to compare the variation of two different populations

Statistical tests typically compare large, randomly selected representative samples of normally distributed data. In practice it is often the case that data can only be obtained from quite small samples.

The *t*-test may be applied to sample sizes of more than five and less than 30 of normally distributed data. It provides a way of measuring the overlap between two sets of data – a large value of '*t*' indicates little overlap and makes it highly likely there is a significant difference between the two data sets. An example will illustrate the method. However, you should note that you are not expected to calculate values of *t*.

Applying the *t*-test

An ecologist was investigating woodland microhabitats, contrasting the communities in a shaded position with those in full sunlight. One of the plants was ivy (*Hedra helix*), but relatively few occurred at the locations under investigation. The issue arose: were the leaves in the shade actually larger than those in the sunlight?

Leaf widths were measured, but because the size of the leaves varied with the position on the plant, only the fourth leaf from each stem tip was measured. The results from the plants available are shown in Table 17.1.

Table 17.1 Sizes of sun and shade leaves of *Hedera helix*

Size-class/mm	Sun leaves (A)	Shade leaves (B)
20–24	24	
25–29	26, 26	26
30–34	30, 31, 31, 32, 32, 33	33, 34
35–39	37, 38	35, 35, 36, 36, 37
40–44	43	41, 42
45–49		45

Steps to the *t*-test:

1 The **null hypothesis** (negative hypothesis) assumes the difference under investigation has arisen by chance; in this example the null hypothesis is:
'There is no difference in size between sun and shade leaves.'

The role of the statistical test is to determine whether to accept or reject the null hypothesis. If it is rejected here, we can have confidence that the difference in the leaf sizes of the two samples is statistically significant.

Next, check that the data are normally distributed. This is done by arranging the data for sun leaves and shade leaves as in Table 17.1 (and plotting a histogram, if necessary).

2 *You are not necessarily expected to calculate values of t.* This is a statistic, which, when required can be found by using a scientific or statistics calculator, or by means of a spreadsheet incorporating formulae. Actually, a formula for the *t*-test for unmatched samples (data sets **a** versus **b**) is:

$$t = \frac{\bar{x}a - \bar{x}b}{\sqrt{\dfrac{sa^2}{na} + \dfrac{sb^2}{nb}}}$$

where:
$\bar{x}_a$ = mean of data set **a**
$\bar{x}_b$ = mean of data set **b**
sa^2 = standard deviation for data set **a**, squared
sb^2 = standard deviation for data set **b**, squared
na = number of data in set **a**
nb = number of data in set **b**
$\sqrt{}$ = square root of

3 Once a value of *t* has been calculated (the value of *t* here is 2.10) we determine the degrees of freedom (**df**) for the two samples, using the formula:

$$df = (\text{total number of values in both samples}) - 2$$
$$= n_a + n_b - 2$$

In this case, df = 11 + 11 = 22.

Degrees of freedom (df)	decreasing value of *p* ⟶			
	p values			
	0.10	**0.05**	**0.01**	**0.001**
1	6.31	12.71	63.66	636.60
2	2.92	4.30	9.92	31.60
3	2.35	3.18	5.84	12.92
4	2.13	2.78	4.60	8.61
5	2.02	2.57	4.03	6.87
6	1.94	2.45	3.71	5.96
7	1.89	2.36	3.50	5.41
8	1.86	2.31	3.36	5.04
9	1.83	2.26	3.25	4.78
10	1.81	2.23	3.17	4.59
12	1.78	2.18	3.05	4.32
14	1.76	2.15	2.98	4.14
16	1.75	2.12	2.92	4.02
18	1.73	2.10	2.88	3.92
20	1.72	2.09	2.85	3.85
22	1.72	2.08	2.82	3.79
24	1.71	2.06	2.80	3.74
26	1.71	2.06	2.78	3.71
28	1.70	2.05	2.76	3.67
30	1.70	2.04	2.75	3.65
40	1.68	2.02	2.70	3.55
60	1.67	2.00	2.66	3.46
120	1.66	1.98	2.62	3.37
∞	1.64	1.96	2.58	3.29

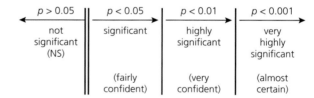

$p > 0.05$	$p < 0.05$	$p < 0.01$	$p < 0.001$
not significant (NS)	significant	highly significant	very highly significant
	(fairly confident)	(very confident)	(almost certain)

Figure 17.4 Critical values for the *t*-test

4 A table of **Critical values** for the *t*-test is given in Figure 17.4. We look down the column of significance levels (*p*) at the 0.05 level until you reach the line corresponding to df = 22. You will see that here, *p* = 2.08.

5 Since the calculated value of *t* (2.10) exceeds this critical value (2.08) at the 0.05 level of significance, it indicates that there is a lower than 0.05 probability (5%) that the difference between the two means is solely due to chance. Therefore, we can reject the null hypothesis, and conclude the difference between the two samples is significant.

6 For the experimenter, the significance of this statistic is there is a reason for the difference in the means, which can be further investigated and fresh hypotheses proposed.

Why variation is important in selection

Differences between individuals of a species very often reflect varying abilities and potentials to tackle problems, to be able to overcome difficulties, and to benefit from opportunities the environment may provide. As the environment changes, some of these individuals may be better placed to survive and prosper than are others. If none of the individuals can rise to and overcome new challenges in a rapidly changing environment, then the whole species is in danger of dying out. However, where some have the potential to adapt and survive, these individuals are likely to do so. We will see that this type of variation is important because it provides the genetic material for natural selection. Ultimately, genetic variation lies in the order of the bases of nucleotides in the genes.

 # 17.2 Natural and artificial selection

Populations of organisms have the potential to produce large numbers of offspring, yet their numbers remain fairly constant year after year.

By the end of this section you should be able to:

a) explain that natural selection occurs as populations have the capacity to produce many offspring that compete for resources; in the 'struggle for existence' only the individuals that are best adapted survive to breed and pass on their alleles to the next generation

b) explain, with examples, how environmental factors can act as stabilising, disruptive and directional forces of natural selection

c) explain how selection, the founder effect and genetic drift may affect allele frequencies in populations

d) use the Hardy–Weinberg principle to calculate allele, genotype and phenotype frequencies in populations and explain situations when this principle does not apply

Competition for resources

Resources of every sort are limited and organisms compete for them. Plants compete for space, light and mineral nutrients. Animals compete for food, shelter and a mate. To lose out in the competition for resources means the individual grows and reproduces more slowly. In extreme cases, they die.

It is also the case that organisms are capable of producing many more offspring than survive to be mature individuals (Table 17.2). The phrase '**struggle for existence**' neatly sums up the fact that there is an overproduction of offspring in the wild, but only limited resources to support them.

Table 17.2 Numbers of offspring produced

Organism	Number of offspring
Rabbit	8–12
Great tit	10
Cod	2–20 million
Honey bee (queen)	120 000
Poppy	6000
How many of these offspring survive to breed themselves?	

Consequently, in a stable population parent generations give rise to a single breeding pair of offspring, on average. All their other offspring may become casualties of the 'struggle'. So, populations do not show rapidly increasing numbers in most habitats, or at least, not for long.

Classifying 'competition' – some terms

Organisms interact with their non-living surroundings and in some situations it may be these conditions that limit the growth of a population. These are known as the **abiotic environment**. They include both the chemical and physical components of the environment – factors such as light, temperature, water and soil. All of these can be affected by the climate and seasonal changes.

On the other hand, the interactions that occur between organisms are known as **biotic factors**. These include competition for space and for resources and may involve predation or grazing and parasitism.

Case studies in competition

1 Plankton populations fluctuate naturally

In practice, populations typical show large fluctuations in their numbers over a period of time. We see a good example of this in the phytoplankton (primary producers) and zooplankton (primary consumers) observed in a fresh water lake in a part of the world with a temperate, seasonal climate. One such community was analysed for a 12-month period. *Look carefully at the curves on the graph in Figure 17.5.*

Question

1 Examine the results of the ecological study shown in Figure 17.5, and then suggest:

 a two mineral nutrients in the lake water that would be taken up by phytoplankton and facilitate their rapid growth rate in March and April

 b why the numbers of phytoplankton decrease rapidly by May and June, despite the favourable conditions of light and temperature

 c the most likely source of the increase in nutrients that begins in October.

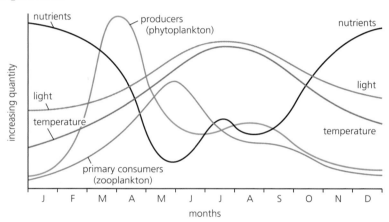

Figure 17.5 Plankton of a fresh water lake, with data on abiotic factors

We can see there is a significant 'bloom' in the phytoplankton population in the third and fourth months of the year and that a corresponding surge in the zooplankton population occurs shortly afterwards. Note that the abiotic factors of nutrients, light and temperature were also measured and recorded.

2 The introduction of the rabbit to Australian grasslands – an ecological disaster

A very few wild rabbits were imported from Britain and released in Australia in 1859. Before that time there were no rabbits in Australia, yet rabbits very rapidly became the dominant herbivorous mammals of the grasslands of Victoria and New South Wales. They had reached the Northern Territory of Australia by 1900.

Why was this so?

Apparently the environment in Australia provided the rabbits with all they required. At the same time, natural **predators** and parasites of the rabbit had not arrived in large numbers and become established. The resulting population explosion caused the destruction of the natural vegetation, the extinction of many plant species and erosion of top soil. The loss of many native marsupial species followed. Domestic livestock was threatened, too. This was because ten rabbits or fewer typically consume the equivalent that one adult sheep requires.

Rabbits are territorial and within their territory, a dominant male mates with most of the females. The gestation period of a rabbit is 28–30 days. A mature female may have eight or more young per litter and five or six litters are possible in a good season. Juvenile rabbits will migrate from their

Question

2 a How is it that a non-native (introduced) species may rapidly become the most common species in a new environment?

 b What factors may later result in the non-native species decreasing in numbers?

parental territory to establish new territories of their own depending on seasonal conditions and the availability of suitable conditions elsewhere.

In 1950, a biological control agent, the myxomatosis virus, was introduced. Initially it wiped out between 95 and 100 per cent of the rabbits in some areas. However, rabbit populations eventually recovered as resistance to the myxomatosis virus spread among the rabbits. Later, the introduction of other specific viral diseases helped control populations, but again, the rapid appearance of resistance has left rabbits as a formidable pest in Australia.

Figure 17.6 a) Overpopulation of rabbits in Australia, b) A terminal case of myxomatosis

3 Prey–predator population oscillation

Evidence for naturally-occurring changes in the populations of **prey** and predator came from the record of fur skins (pelts) received by the Hudson Bay Trading Company of Canada from trappers over a 100-year period. The data are for lynx (predator) and snowshoe hare (prey). At that time, hunters of these wild animals survived and made a living by trading in the skins they had collected from the wild mammals they had been able to trap in and around the forests of northern Canada each year. This type of evidence assumes that the size of the annual catch is directly related to the numbers of these mammals in the wild population in the years concerned.

Question

3 What annotations could you add to the graph in Figure 17.7:

  a to the rising side to a peak of a predator curve

  b to the declining side of a predator curve

  c to interpret and explain the results shown?

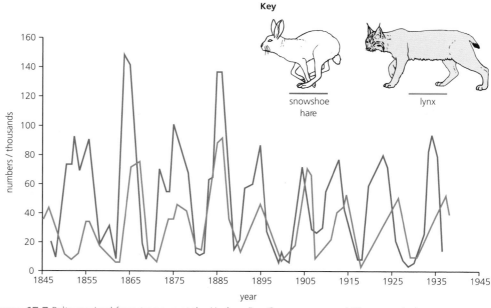

Figure 17.7 Pelts received from trappers at the Hudson Bay Company over a 100-year period

We can conclude from these three contrasting situations that natural populations have the potential to increase rapidly in numbers provided they have access to the essential resources they require. However, the size of a population is eventually limited by restraints that we can call **environmental factors**. These factors include space, light and the availability of food. The interactions of predators or parasites with a population may be equally influential in determining the size of a population. It is this never-ending competition that results in the majority of organisms failing to survive and reproduce. In effect, the environment appears to be able to support a certain number of organisms and the number of individuals in a species remains more or less constant over a period of time. Competition is a major fact of life for the living world.

Natural selection and speciation

Natural selection operates on individuals, or rather on their phenotypes. Phenotypes are the product of a particular combination of alleles, interacting with the effects of the environment of the organism. Consequently, natural selection causes changes to gene pools. For example, individuals possessing a particular allele or combination of alleles, may be more likely to survive, breed and pass on their alleles. Individuals that are less well adapted may not survive or be successful enough to pass on their genes. This process is referred to as differential mortality. Actually, whether the individual lives or dies is not important. What is relevant is whether or not their alleles are passed on to the next generation.

So, natural selection operates to change the composition of gene pools, but the effect of this varies. We can recognise different types of selection.

Stabilising selection

Stabilising selection occurs when environmental conditions are largely unchanging. Stabilising selection does not lead to evolution. It is a mechanism which maintains a favourable characteristic and the alleles responsible for them, and eliminates variants and abnormalities that are useless or harmful. Probably most populations undergo stabilising selection. The example shown in Figure 17.8 comes from birth records of human babies born between 1935 and 1946 in a London hospital. It shows there was an optimum birth weight for babies and those with birth weights heavier or lighter were at a selective disadvantage.

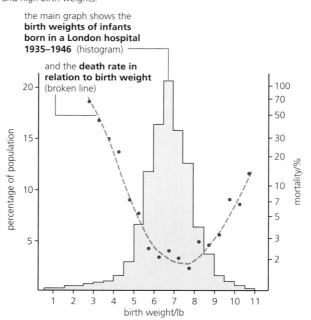
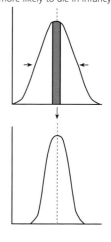

Figure 17.8 Birth weight and infant mortality: a case of stabilising selection

Directional selection

Directional selection may result from changing environmental conditions. In these situations the majority of an existing form of an organism may be no longer best suited to the environment. Some other, alternative phenotypes may have a selective advantage.

An example of directional selection is the development of resistance to an antibiotic by bacteria. Certain bacteria cause disease and patients with bacterial infections are frequently treated with an antibiotic to help them overcome the infection. Antibiotics are very widely used. However, in a large population of a species of bacteria, some may carry a gene for resistance to the antibiotic in use. Sometimes such a gene arises by spontaneous mutation. Sometimes the gene is acquired in a form of sexual reproduction between bacteria of different populations.

A 'resistant' bacterium has no selective advantage in the absence of the antibiotic and must compete for resources with non-resistant bacteria. But when the antibiotic is present most bacteria of the population are killed off. Resistant bacteria remain and create the future population, all of which now carry the gene for resistance to the antibiotic. The genome has been changed abruptly.

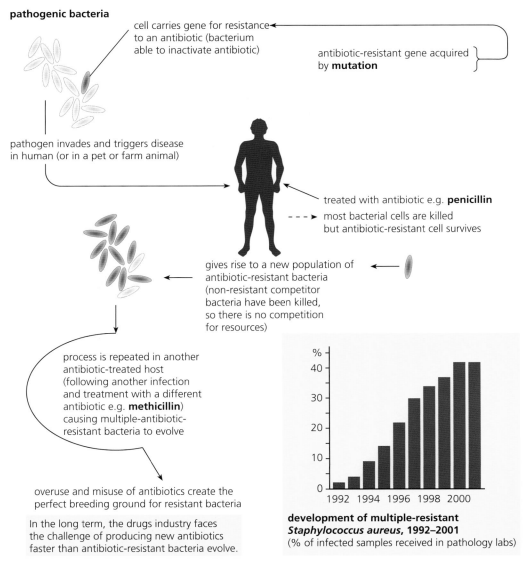

pathogenic bacteria

cell carries gene for resistance to an antibiotic (bacterium able to inactivate antibiotic)

antibiotic-resistant gene acquired by **mutation**

pathogen invades and triggers disease in human (or in a pet or farm animal)

treated with antibiotic e.g. **penicillin**

most bacterial cells are killed but antibiotic-resistant cell survives

gives rise to a new population of antibiotic-resistant bacteria (non-resistant competitor bacteria have been killed, so there is no competition for resources)

process is repeated in another antibiotic-treated host (following another infection and treatment with a different antibiotic e.g. **methicillin**) causing multiple-antibiotic-resistant bacteria to evolve

overuse and misuse of antibiotics create the perfect breeding ground for resistant bacteria

In the long term, the drugs industry faces the challenge of producing new antibiotics faster than antibiotic-resistant bacteria evolve.

development of multiple-resistant *Staphylococcus aureus*, 1992–2001 (% of infected samples received in pathology labs)

Figure 17.9 Multiple antibiotic resistance in bacteria

Disruptive selection

Disruptive selection occurs when particular environment conditions favour the extremes of a phenotypic range over intermediate phenotypes. As a result, it is likely that the gene pool will become split into two distinct gene pools. New species may be formed. This phenomenon has been illustrated by colonisation by plants of mine waste tips (Figure 17.10). These habitats often contain high concentrations of toxic metals such as copper and lead. Most plants are unable to grow on them. However, you can see from the photograph that some hardy grasses have colonised the contaminated soil. We can assume that these species have developed resistance to toxic metal ions. At the same time, their ability to grow and compete on uncontaminated soils has decreased. In the uncontaminated soil around the tip the grasses that have no resistance to toxic metal ions are able to flourish.

Grasses are wind-pollinated plants, so breeding between resistant grass plants and non-resistant grass plants goes on. When their seeds fall to the ground and attempt to germinate, disruptive selection may occur. Both the non-resistant plants germinating on contaminated soil and the resistant plants growing on uncontaminated soil fail to survive to reproduce. The result is increasing divergence of the populations, initially into two distinctly different varieties of grass plant. In time, new species may be formed.

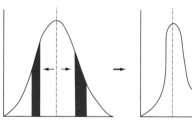

disruptive selection
favours two extremes of the 'chosen' characters at the expense of intermediate forms

Figure 17.10 Mining waste tip habitat and disruptive selection

Balancing selection

In directional selection and stabilising selection there is a tendency to reduce variation. Balancing selection has completely the opposite effect.

Balancing selection is a process which actively maintains multiple alleles in the gene pool of a population. It does this at frequencies *well above* that due to gene mutation. An example of this is **sickle cell trait**. This is a hereditary condition that causes a proportion of the red blood cells circulating in the body to be sickle-shaped. You can see a sample of red blood cells from the circulation of someone with sickle cell trait in Figure 6.17 (page 125). This illustration also explains the mutation by which the gene that codes for the amino acid sequence of the beta chain of haemoglobin is prone to a substitution of the base adenine (A) by thiamine (T) in a codon coding for the amino acid glutamic acid. As a consequence, the amino acid valine appears at that point instead. When the genetic code of this mutated gene has been transcribed into messenger RNA and then translated at ribosomes, molecules of an unusual form of haemoglobin, known as haemoglobin Hb^s results. The molecules with this unusual haemoglobin tend to clump together and form long fibres that distort the red blood cells into sickle shapes. In this condition the red blood cells transport little oxygen and they may even block smaller vessels. In people who are heterozygous for this allele, Hb Hb^s, less than 50 per cent of their red blood cells are sickle-shaped. Sickle cell trait is a non-lethal condition. However, people with it are mildly anaemic and we might expect them to be at some disadvantage in life. However, there are areas of the world where this is definitely not so.

The world distribution of **malaria** is shown in Figure 10.7 (page 197). This disease kills a large number of people each year. Notice that in regions of Africa and India malaria is endemic. Here, sickle cell trait confers an advantage. This is because the malarial parasite *Plasmodium* completes its lifecycle in red blood cells, but it cannot do so in red blood cells containing Hb^s. So a person who inherits the sickle cell allele from one parent and the normal haemoglobin allele from the other (a heterozygote) is resistant to the malarial parasite. However, people who are homozygous for the sickle cell allele (Hb^s Hb^s) have **sickle cell anaemia** and this affects their survival rate.

So we see balancing selection at work here. *Turn back to Figure 10.9 (page 199) now and note the correlation between the distribution of the sickle cell allele and of the distribution of malaria in Africa.*

In any area there are strong selection forces against the homozygous sickle cell condition; in areas where malaria is endemic there is also a strong selection force against people homozygous for normal haemoglobin. However, heterozygotes have a permanent advantage where malaria exists. Heterozygotes are better adapted than either of the homozygotes. Because of this selective advantage, the sickle-cell condition is an example of **balanced polymorphism** – the stable co-existence of two (or more) distinct types of allele in a species (or population).

Population genetics

Population genetics is the study of genes in populations. Populations are important to our argument now, because they are where evolution may occur.

A **population** is a group of individuals of a species, living close together and able to interbreed. So a population of garden snails might occupy a small part of a garden, say around a compost heap. A population of thrushes might occupy some gardens and surrounding fields. In other words, the area occupied by a population depends on the size of the organism and on how mobile it is, for example, as well as on environmental factors (eg. food supply, predation, etc.). The boundaries of a population may be hard to define, too. Some populations are fully 'open', with individuals moving in or out from nearby populations. Alternatively, some populations are more or less 'closed', that is, isolated communities, almost completely cut off from neighbours of the same species.

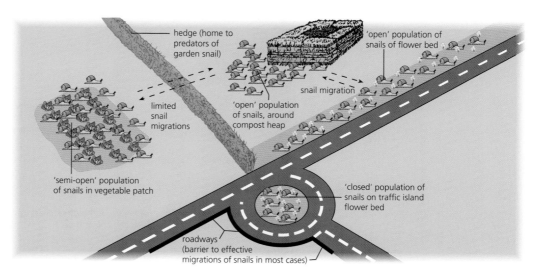

Figure 17.11 The concept of 'population'

Gene pool: all the genes and their different alleles, present in an interbreeding population.

Populations and gene pools

In any population, the total of the alleles of the genes located in the reproductive cells of the individuals make up a **gene pool**.

When breeding between members of a population occurs, a sample of the alleles of the gene pool will contribute to form the genomes (gene sets of individuals) of the next generation, and so

> **Allele frequency**: the commonness of the occurrence of any particular allele in a population.

on, from generation to generation. Remember, an allele is one of a number of alternative forms of a gene that can occupy a given locus on a chromosome. The frequency with which any particular allele occurs in a given population will vary.

When **allele frequencies** of a particular population are investigated they may turn out to be static and unchanging. Alternatively, we may find allele frequencies changing. They might do so quite rapidly with succeeding generations, for example.

When the allele frequencies of a gene pool remain more or less unchanged, then we know that population is static as regards its inherited characteristics. We can say that the population is not evolving.

However, if the allele frequencies of gene pool of a population are changing (i.e. the proportions of particular alleles are altered, we say 'disturbed' in some way), then we may assume that evolution is going on. For example, some alleles may be increasing in frequency because of an advantage they confer to the individuals carrying them. With possession of those alleles the organism is more successful. It may produce more offspring, for example. If we can detect change in a gene pool we may detect evolution happening, possibly even well before a new species is observed.

How can we detect change or constancy in gene pools?

The answer is, by a mathematical formula called the **Hardy–Weinberg formula**. Independently this principle was discovered by two people, in the process of explaining why dominant characteristics don't take over in populations, driving out the recessive form of that characteristic. (For example, at that time people thought [wrongly] that human eye colour was controlled by a single gene, and that an allele for blue eyes was dominant to the allele for brown eyes. 'Why doesn't the population become blue-eyed?', was the issue.)

Let the frequency of the dominant allele (**G**) be p, and the frequency of the recessive allele (**g**) be q.

The frequency of alleles must add up to 1, so $p + q = 1$.

This means in a cross, a proportion (p) of the gametes carries the **G** allele, and a proportion (q) of the gametes carries the **g** allele.

The offspring of each generation are given by the Punnett grid.

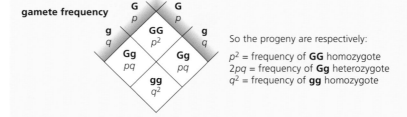

gamete frequency

So the progeny are respectively:

p^2 = frequency of **GG** homozygote
$2pq$ = frequency of **Gg** heterozygote
q^2 = frequency of **gg** homozygote

Hardy–Weinberg formula

If the frequency of one allele (**G**) is p, and the frequency of the other allele (**g**) is q then the frequencies of the three possible genotypes **GG**, **Gg** and **gg** are respectively p^2, $2pq$ and q^2.

In this way, Hardy and Weinberg developed the following equation to describe stable gene pools:

p^2	+	$2pq$	+	q^2	=	1
frequency of homozygous dominant individuals		frequency of heterozygous individuals		frequency of homozygous recessive individuals		total

Figure 17.12 Deriving the Hardy–Weinberg formula

The main problem of finding gene frequencies is that is not possible to distinguish between homozygous dominants and heterozygotes based on their appearance or phenotype. However, using the above equation, it is possible to calculate gene frequency from the number of homozygous recessive individuals in the population. This is q^2. We can find q by finding the square root. The result tells us the frequency of the recessive allele, and this can then be substituted into the initial equation $p + q = 1$ and we can find the frequency of the dominant allele.

Using the Hardy–Weinberg formula

We can use the Hardy–Weinberg formula to find the frequency of a gene in cases of dominance where we are unable to distinguish between the homozygous dominants and the heteroygotes on the basis of phenotype.

● **Example 1**:

In humans, the ability to taste the chemical phenylthiocarbamide (PTC) is conferred by the dominant allele **T**. Both the dominant homozygotes (**TT**) and the heterozygotes (**Tt**) are 'tasters'. The non-tasters are the homozygotes (**tt**).

In a sample of a local population of 200 people in Western Europe,130 (65 per cent) were tasters and 70 (35 per cent) were non-tasters.

Phenotypes	tasters	non-tasters
Genotypes	TT + Tt	tt
Frequency	0.65	0.35

Apply this data to the Hardy–Weinberg formula; we know the value of q^2 to be 0.35.

Taking the square root the value of $q = 0.59$.

So the frequency of the non-tasting alleles (**t**) in this European population was 0.59.

● **Example 2**:

The absence of the skin pigment, melanin, is a condition called albinism (Figure 16.25, page 370), a genetically controlled characteristic. An albino has the genotype **pp** (homozygous recessive), whereas people with normal pigmentation are homozygous (**PP**) or heterozygous (**Pp**). In a large population only one person in 10000 was albino.

From the equation above, homozygous recessives (**pp**) = q^2.

Thus $q^2 = 0.0001$, so $q = \sqrt{0.0001} = 0.01$.

So substituting into the initial equation $p + q = 1$,

$p + 0.01 = 1$, therefore $p = 0.99$.

Thus the Hardy–Weinberg formula has allowed us find the frequencies of alleles **P** and **p** in a population.

Incidentally, it has also shown that the frequency of 'carriers' of an allele for albinism in the population (**Pp**) is quite high (about 1 in 50 of the population) despite the fact that albinos make up only 1 in 10000. In other words, very many more people carry around an allele for 'albinism' than those who know they may do so.

Hardy–Weinberg and 'disturbing factors'

The Hardy–Weinberg principle predicts that the gene pool in a population does not change in succeeding generations. That is, genes and genotype frequencies normally remain constant in a breeding population, provided that:

● the breeding population under investigation is a large one
● there is random mating, with individuals of any genotype all equally likely to mate with individuals of any other genotype (e.g. no one genotype is being selectively predated)
● there is no introduction of new alleles into the population, either by mutations or by immigration of new breeding individuals.

But gene pools do change!

In some populations, the composition of the gene pool changes. This may be due to a range of factors, known as 'disturbing factors' in that they operate to alter the proportions of some alleles.

1 Selection.

We have already noted various ways in which natural selection operates to change the composition of gene pools. A further example will suffice here.

Selective predation of members of the population with certain characteristics that are genetically controlled, will lead to changing frequencies of certain alleles. For example, selective predation of snails with a particular shell coloration that makes them visible in (say) a grassland habitat, but effectively camouflaged in a woodland habitat (Figure 17.13).

The banded, coloured shells of the snail *Cepea nemoralis* are common sights in woods, hedges and grasslands. The shells may be brown, pink or yellow, and possess up to five dark bands. In woodland leaf litter the shells that are camouflaged are darker and more banded than those camouflaged among grasses.

woodland leaf litter

the **thrush** (*Turdus ericetorum*) selectively predates local populations of snails, using a stone as an 'anvil' to break the shell so that it can eat the snail

grass

Figure 17.13 Selective predation of snails – natural selection in operation

2 Genetic drift.

In a very large breeding population there is every likelihood that the sample of genes coming together in zygotes will be fully representative of the gene pool, and so of previous generations, too. On the other hand, in a small population the genes selected may not be representative of the gene pool as a whole.

In the event of sudden hostile physical conditions, such as extreme cold, devastating flooding or prolonged drought, a natural population may be reduced to a very few survivors. On the return of a favourable environment, the numbers of the affected species may quickly return to normal (typically, because of reduced competition for food or other resources). However, the new population has been built from a very small sample of the original, 'pre-disaster' population, with numerous 'first cousin' and backcross matings (causing fewer heterozygotes and more homozygotes) and with some alleles lost altogether.

3 Founder effect.

A small group of a large breeding population may become isolated, possibly due to a barrier that arises within part of the territory, or when a small group wander away, or are carried away by chance – such as by a tsunami. If these small samples of the original population are unrepresentative of the original population, a sudden change in gene frequency will have taken place.

Question

4 What factors may cause the composition of a gene pool to change?

4 Other 'disturbing factors are:

Emigration/immigration. This may introduce new genes into populations, for example.

Mutation. These are random, rare, spontaneous changes in the genes that occur in gonads, leading to the possibility of new characteristics in the offspring.

17.2 *continued...* Artificial selection

Humans use selective breeding (artificial selection) to improve features in ornamental plants, crop plants, domesticated animals and livestock.

By the end of this section you should be able to:

e) describe how selective breeding (artificial selection) has been used to improve the milk yield of dairy cattle

f) outline the following examples of crop improvement by selective breeding:
  • the introduction of disease resistance to varieties of wheat and rice
  • the incorporation of mutant alleles for gibberellin synthesis into dwarf varieties so increasing yield by having a greater proportion of energy put into grain
  • inbreeding and hybridisation to produce vigorous, uniform varieties of maize

Artificial selection is selection caused by humans. This process of selective breeding is usually a deliberate and planned process. Artificial selection involves identifying the largest, the best or the most useful of the progeny for the intended purpose and using them as the next generation of parents. Then, continuous removal of progeny showing less desired features, generation by generation, leads to deliberate genetic change. The genetic constitution of the population changes rapidly. Artificial selection is an on-going process to obtain higher yields, superior nutrient content and resistance to disease in very many of today's domestic animals and crop plants.

Charles Darwin started breeding pigeons as a result of his interest in variation in organisms (Figure 17.14). In *The Origin of Species* he noted there were more than a dozen varieties of pigeon which, had they been presented as wild birds to an expert on birds, would have been recognised as separate species. All these pigeons were descendents of the rock dove, a common wild bird.

Darwin argued that if so much change can be induced in so few generations, then species must be able to evolve into other species by the gradual accumulation of minute changes, as environmental conditions change and select some progeny and not others.

Question

5 Charles Darwin argued that the great wealth of varieties we have produced in domestication supports the concept of evolution. Outline how this is so.

rock dove (*Columbia livia*) from which all varieties of pigeon have been derived

jacobin

pouter

fantail

Figure 17.14 Pigeon varieties produced by selective breeding

By artificial selection, the plants and animals used by humans (such as in agriculture, transport, companionship and leisure) have been derived from wild organisms. The origins of artificial selection go back to the earliest developments of agriculture, although at this stage, successful practice was not based on an understanding of the theory of evolution (Figure 17.15).

The steps to domestication of wild animals

1 A wild population of a 'species' useful as a source of hides, meat, etc. is identified and people learn to distinguish it from related species. Herd animals (such as sheep and cattle) are naturally sociable and lend themselves to this.

2 Once a wild herd has been trapped inside a compound or field, there is a selective killing (culling) of the least suitable members of this herd in order to meet immediate needs for food and materials for living.

3 Breeding is encouraged among members of the herd showing the desired feature and they are protected from attack by wild predators.

4 From the offspring the individual with the most useful features is selected and used as the future breeding stock.

5 The breeding stock is maintained during unfavourable conditions, such as drought and floods.

6 After a long period of artificial selection, a domesticated herd, dependent on the herds people rather than living wild, is established. This leads to the possibility of trading individuals of a breeding stock with other groups of herds people living nearby who have similar stocks. This introduces a wider gene pool into the herd – and may introduce fresh characteristics.

Question

6 Explain the key difference between natural and artificial selection.

Wild sheep or mouflon (*Ovis musimon*) occur today on Sardinia and Corsica

Soay sheep of the outer Hebrides suggest to us what the earliest domesticated sheep looked like

Modern selective breeding has produced shorter animals with a woolly fleece in place of coarse hair and with muscle of higher fat content. Many breeds have lost their horns

Figure 17.15 From wild to domesticated species and the origins of selective breeding skills

Selective breeding of dairy cattle for improved milk yield

In highly developed agricultural systems, principally the USA and Western Europe, milk yield per cow has more than doubled in the past 50 years. Holstein Friesian cattle are today's most productive. This breed originated in Europe, in Holland and Germany, and proved to be the animals that could best exploit the abundant grass resources available (Figure 17.16).

Cows as ruminants:
Cellulose, the most abundant organic compound, is a major component of the diet of all herbivores. Mammals are unable to produce the enzyme cellulase, but many bacteria can. Herbivores exploit this facility of microorganisms in order to digest the cellulose in their diet. Cows have a four-chambered 'stomach'. The first compartment, the **rumen**, is a large fermentation tank containing microorganisms able to digest cellulose. Digestion of cellulose begins with grinding by jaws and teeth. Later on, in the cycle of fermentation, grass is also regurgitated to the mouth from the rumen for further grinding. Meanwhile, much of the sugar released by cellulase during fermentation in the rumen is turned into organic acids. Sugars and organic acids are the major source of energy for the cow, and are absorbed as soon as they are formed. The microorganisms present also synthesise their own amino acids and proteins from inorganic nitrogen ($-NO_3$ ions, NH_3). When, later, the contents of the rumen pass on to the true stomach, the cow digests the microorganisms as an additional source of protein.

Figure 17.16 A herd of Holstein cows

Many cows now produce more than 20 000 kg of milk per lactation. This has been achieved by selecting bulls from high-yield herds, and breeding them with cows that have the best milk production. The use of artificial insemination techniques and the maintenance of stocks of frozen semen have permitted the use of semen from the most promising of animals to have the widest impact. However, at the same time there has been a substantial reduction in fertility in milking herds. Each lactation starts following the birth of a calf, but first the cow has to be inseminated as the previous lactation comes to an end. This cycle has started to become problematical in some of the highest yield cows. Research now indicates there may be a genetic connection, and the problem is under intensive investigation.

Examples of crop improvement by selective breeding

Cereal grains are highly significant components of human diet. Plant matter forms the bulk of human food intake in both developed and less-developed countries, but it is plants of one family, the grasses, which we and most of our livestock depend on (Figure 17.17). This family includes cereals, which are the fruits (grain) of cultivated grass species. They are relatively easy to grow and the mature grains they yield are comparatively easy to store. Grains contain significant quantities of protein, as well as starch, as we shall shortly see. Fortunately for some of the huge and growing human population the Earth is increasingly required to support at this time, sheep and goats feed mostly on grass leaves, rather than requiring precious grain stocks.

On the other hand, cattle and poultry in the developed world are largely fed on grain such as maize. Because much arable land is used to grow grain for feeding livestock, others go hungry. In the US, 157 million tonnes of cereals, legumes and vegetable protein (all suitable for human consumption) are fed to livestock to produce just 28 million tonnes of animal protein (meat).

Table 17.3 The world's top ten food crops and the estimated edible dry matter (in millions of tonnes) they provide

Food crop		Edible dry matter provided
1	wheat	468
2	maize	429
3	rice	330
4	barley	160
5	soybean	88
6	cane sugar	67
7	sorghum	60
8	potato	54
9	oats	43
10	cassava	41

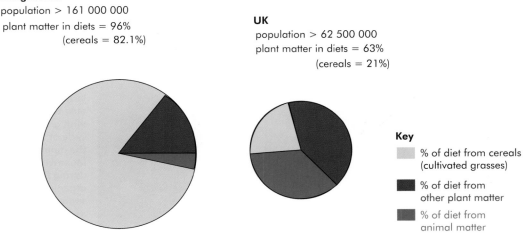

Bangladesh
population > 161 000 000
plant matter in diets = 96%
(cereals = 82.1%)

UK
population > 62 500 000
plant matter in diets = 63%
(cereals = 21%)

Key
% of diet from cereals (cultivated grasses)

% of diet from other plant matter

% of diet from animal matter

Figure 17.17 Plant matter as a proportion of the human diet

Actually, the number of plant species that humans depend upon for their food is quite limited – about 300 species out of a total number of flowering plant species that approaches 200 000. Of these 300 common food plants, just 17 species provide 90 per cent of our food. Furthermore, from Table 17.3 you can see that four grain plants contribute by far the most.

So, it is largely the grains of certain cultivated grasses which stand between humans and starvation – to state a fact rather dramatically!

Disease resistance in varieties of wheat and rice

Plants used in agriculture and horticulture have been obtained by genetic manipulations of wild plants. Today, new varieties have been developed. There are two ways in which this is done.

1 Artificial selection.
In traditional breeding projects, the most useful offspring for a particular purpose, such as resistance to specific pest and diseases, are selected and used as the next generation of parents. Offspring regarded as less useful are excluded. This technique has been applied with increasing sophistication, and is still in common use by researchers.

2 Genetic engineering (recombinant DNA technology).
Today, genetic modification is also achieved by the direct transfer of genes from one organism to the set of genes (the **genome**) of another, often unrelated organism. In green plants, introducing genes into cells is complicated by the presence of the cell wall. However, walls may be temporarily removed by the gentle action of enzymes, leaving a cell protoplast. Genes may be introduced into protoplasts, or differing protoplasts induced to fuse together, rather like gametes do. Alternatively, genetic modification can be brought about by the tumour-forming bacterium *Agrobacterium*. The gene for tumour formation occurs in a plasmid. Useful genes may be added to the plasmid, too. Then, the gall tissue the bacterium induces in a plant it attacks (and the new plants grown from it) may contain and express the new gene.

Today, new varieties of wheat and rice have been developed by genetic engineering, including varieties resistant to diseases. Most plant diseases are caused by fungal pathogens, and there are naturally synthesised peptides with antimicrobial properties, the genes for which have been engineered into high-yielding varieties. This technique is proving a valuable tool in the control of a wide range of fungal pathogens. The peptides have proved harmless to humans and other animals.

Breeding of dwarf varieties to increase the yield of grain

The cereals, wheat, barley and oats, are the most important arable crops grown by agricultural communities in the Northern Hemisphere. They are examples of cultivated grass plants adapted to temperate climates with normal amounts of rainfall. For example, wheat grows best in cool springs with moderate moisture for early growth, followed by sunny summer months that turn dry, for harvesting.

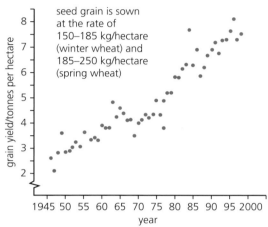

seed grain is sown at the rate of 150–185 kg/hectare (winter wheat) and 185–250 kg/hectare (spring wheat)

Figure 17.18 Productivity of wheat

Bread wheat (*Triticum aestivum*) is a hexaploid (2n = 42), and arose by natural crossings of various wild wheats (2n = 14) that occurred in the Fertile Crescent of the Middle East, about 8000 years ago. Today. **'winter' wheat** varieties are planted in the autumn, and root and tiller (produce side shoots at ground level) before the low temperatures of winter suspend growth. The crop matures in early summer. **'Spring' wheats** are sown after winter, and are adapted to a shorter growing season (but give lighter yields).

There are two classes of wheat varieties. **Hard wheats** are higher in proteins, are grown in areas of lower rainfall, and are used to produce bread that keeps. **Soft wheat** varieties are more starchy, are used to make 'French bread' and pastas, and are grown in more humid conditions.

Productivity in these crops has been enormously increased (Figure 17.18). This has been deliberately brought about by improvements in cultivation, by the widespread use of pesticides, the careful targeting of artificial fertilisers, and by genetic modifications. For example, the incorporation of mutant alleles for gibberellin synthesis (these prevent the action of gibberellin) has led to new dwarf varieties that put a greater proportion of energy into grain, at the expense of stem length.

Breeding of vigorous, uniformly growing varieties of maize

Maize is the second most important food crop (Table 17.3). It was probably originally domesticated in an area we now know of as Mexico. Today maize is grown all over the world, but most intensively in the USA. Modern varieties are hybrids – the original form of wild maize has long been extinct. Maize has characteristic wind-pollinated flowers with both male and female flowers on the same plant (Figure 17.19). The male flowers occur at the top (terminally) with pendulous stamens and versatile anthers. Below are the female flowers (the cobs), in the axils of the leaves. They have huge, feathery stigma (the tassels). The plant is capable of self-pollination, leading to self-fertilisation, but naturally tends to cross-pollinate. Self-fertilisation in maize largely prevented because male and female flowers mature at different times. Maize lends itself to crossings during controlled plant breeding experiments when adjacent rows of varieties (possibly previously inbred) are prevented from self-pollinating by removal of the tassels ('detasselling').

Today's growers and consumers seek in maize the following characteristics:

- plants that grow vigorously in the local environment (climate, soil), but that are resistant to diseases and that result in high yielding cobs
- crops of fairly standard size (height) and with cobs that are ready to harvest at the same time, to facilitate harvesting.

Plants originally grown in the wild would have had some of these characteristics, but not all. Selective breeding has been undertaken over the years to achieve the quality and consistency required. Remember, consistent inbreeding (producing generations by self-pollination) leads eventually to plants that lack vigour and fertility and have reduced size and yield. This is referred to as **inbreeding depression**. However, inbreeding does generate plants which show little variation (a standardised crop), for they will tend to have the same alleles for most genes.

male flowers

female flowers

Figure 17.19 A modern maize plant in flower

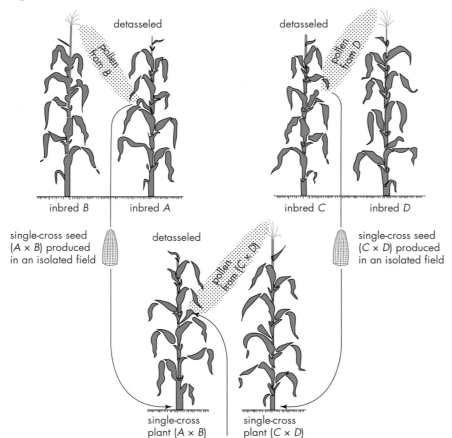

detasseled

pollen from B

inbred B   inbred A

detasseled

pollen from D

inbred C   inbred D

single-cross seed
(A × B) produced
in an isolated field

single-cross seed
(C × D) produced
in an isolated field

detasseled

pollen from (C × D)

single-cross
plant (A × B)

single-cross
plant (C × D)

double-cross seed (A × B) × (C × D)
is produced in an isolated field, sold to
the farmer, and planted to produced
high-yielding double-cross corn

Figure 17.20 The production of double-cross hybrid corn

When inbred lines are crossed the results are hybrids which exhibit **hybrid vigour** – the hybrids show greater yields and more vigorous growth. This is because, in them, fewer recessive alleles are expressed. However, if cross-fertilisation continues at random, the results are crops with much variation between individual plants.

The most successful breeding experiments have started with 'parent' generations which are selectively inbred to deliver high yield and other valuable characteristics such as disease resistance (achieved by exposure to the pest and use in further breeding programmes only of those plants which remain pest free). These parent generations are then involved in double-cross hybridisation (Figure 17.20). The result has been maize fruit that when planted and cultivated appropriately, produces crops with spectacularly increased yields, but without the loss of other favourable characteristics.

17.3 Evolution

Isolating mechanisms can lead to the accumulation of different genetic information in populations, potentially leading to new species.

Over prolonged periods of time, some species have remained virtually unchanged, others have changed significantly and many have become extinct.

By the end of this section you should be able to:

a) state the general theory of evolution that organisms have changed over time
b) discuss the molecular evidence that reveals similarities between closely related organisms with reference to mitochondrial DNA and protein sequence data
c) explain how speciation may occur as a result of geographical separation (allopatric speciation), and ecological and behavioural separation (sympatric speciation)
d) explain the role of pre-zygotic and post-zygotic isolating mechanisms in the evolution of new species
e) explain why organisms become extinct, with reference to climate change, competition, habitat loss and killing by humans

Evolution by natural selection – the ideas and arguments

By evolution we mean the gradual development of life in geological time. The word evolution is used widely, but in biology it specifically means the processes by which life has been changed from its earliest beginnings to the diversity of organisms we know about today, living and extinct. It is the development of new types of living organisms from pre-existing types by the accumulation of genetic differences over long periods of time.

Charles Darwin (1809–1882) was a careful observer and naturalist who made many discoveries in biology. After attempting to become a doctor (at Edinburgh University) and then a clergyman (at Cambridge University), he became, in 1831, the unpaid naturalist on an expedition to the southern hemisphere on a ship called HMS Beagle. On this five-year expedition around the world and in his later investigations he developed the idea of **organic evolution by natural selection**.

Darwin was very anxious about how the idea of evolution might be received and he made no moves to publish it until the same idea was presented to him in a letter by another biologist and traveller, **Alfred Russel Wallace**. Only then, in 1859, was *On the Origin of Species by Natural Selection* completed and published.

The arguments and ideas of 'The Origin of Species' are summarised in Table 17.4.

Table 17.4 Charles Darwin's ideas about the origin of species, summarised in four statements (S) and three deductions (D) from these statements

		Statements / deductions
S1		Organisms produce a far greater number of progeny than ever give rise to mature individuals.
S2		The number of individuals in species remains more or less constant.
	D1	**Therefore, there must be a high mortality rate.**
S3		The individuals in a species are not all identical, but show variations in their characteristics.
	D2	**Therefore, some variants will have more success than others in the competition for survival. So the parents for the next generation will be selected from those members of the species better adapted to the conditions of the environment.**
S4		Hereditary resemblance between parents and offspring is a fact.
	D3	**Therefore, subsequent generations will maintain and improve on the degree of adaptation of their parents by gradual change.**

Neo-Darwinism

Charles Darwin (and nearly everyone else in the scientific community of his time) knew nothing about Mendel's work on genetics. Chromosomes had not been reported, and the existence of genes, alleles and DNA were unknown.

Instead, biologists generally subscribed to the concept of 'blending inheritance' when mating occurred (which would reduce the genetic variation available for natural selection).

Neo-Darwinism is an essential restatement of the concepts of evolution by natural selection in terms of Mendelian and post-Mendelian genetics.

Today, modern genetics has shown us that blending generally does not occur and that there are several ways by which genetic variation arises in gamete formation and fertilisation. Neo-Darwinism is a restatement of the ideas of evolution by natural selection in terms of modern genetics. The ideas of Neo-Darwinism are summarised below.

Genetic variation arises by:
- **mutations**, including **chromosome mutations** and gene mutations
- **random assortment** of paternal and maternal chromosomes during meiosis, which occurs in the process of gamete formation
- **recombination of segments** of maternal and paternal **homologous chromosomes** during crossing over that occurs during meiosis in gamete formation
- the **random fusion of male and female gametes** in sexual reproduction (which was understood in Darwin's time).

Then, when genetic variation has arisen in organisms:
- it is expressed in their phenotypes
- some phenotypes are better able to survive and reproduce in a particular environment, whilst other fail to – a point known as 'differential survival'
- natural selection operates, determining the survivors and the genes that are perpetuated.

In time, this process may lead to new varieties and new species.

Questions

7 Put in your own words the ideas from modern genetics that provide a basis for the theory of the origin of species by natural selection.

8 Suggest the significance for the theory of evolution by natural selection of the realisation by geologists that the Earth was more than a few thousand years old?

Question

9 Deduce the importance of modern genetics to the theory of the origin of species by natural selection.

Survival of the fittest?

The operation of natural selection is sometimes summarised in the phrase '**survival of the fittest**', although these were not words that Darwin used, at least not initially.

To avoid the criticism that 'survival of the fittest' is a circular phrase (how can fitness be judged except in terms of survival?), the term 'fittest' is understood in a particular context. For example, the fittest of the wildebeest of the African savannah (hunted herbivores) may be those with the acutest senses, quickest reflexes and strongest leg muscles for efficient escape from predators. By natural selection of these characteristics, the health and survival of wildebeests is assured.

Molecular evidence for evolutionary relationships

Evidence from protein sequence data

All living things have DNA as their genetic material, with a genetic code that is virtually universal. The processes of 'reading' the code and protein synthesis, using RNA and ribosomes, are very similar in prokaryotes and eukaryotes, too. Processes such as respiration involve the same types of steps and similar or identical intermediates and biochemical reactions, similarly catalysed. ATP is the universal energy currency. Also, among the autotrophic organisms the biochemistry of photosynthesis is virtually identical.

This biochemical commonality suggests a common origin for life, as the biochemical differences between the living things of today are limited. Some of the earliest events in the evolution of life must have been biochemical, and the results have been inherited widely. However, large molecules like nucleic acids and the proteins they may code for are subjected to changes with time, but this change may be an aid to the study of evolution and relatedness. It is possible to measure the relatedness of different groups of organisms by the amount of difference between specific molecules such as DNA, proteins and enzyme systems – which is a function of time since particular organisms shared a common ancestor.

Variations in haemoglobin molecules that indicate relatedness

Haemoglobin, the β chain of which is built from 146 amino acid residues, shows variations in the sequence of amino acids in different species in which it occurs. Haemoglobin structure is determined by inherited genes, so the more closely related species are, the more likely their amino acids sequence match (Table 17.5). Variations are thought to arise by mutations of an 'ancestral' gene for haemoglobin. If so, the earlier that species diverged from a common ancestor, the more likely it is that differences may arise.

Similar studies have been made of the differences in the polypeptide chains of other protein molecules, including ones common to all eukaryotes and prokaryotes. One such is the universally occurring electron transport carrier, **cytochrome c**.

Table 17.5 Number of amino acid differences in β chain of haemoglobin compared to human haemoglobin

Species	Differences	Species	Differences
human	0	kangaroo	38
gorilla	1	chicken	45
gibbon	2	frog	67
Rhesus monkey	8	lamprey	125
mouse	27	sea slug (mollusc)	127

Biochemical variation used as an evolutionary clock

Biochemical changes like these may occur at a constant rate, and if so, may be used as a 'molecular clock'. If the rate of change can be reliably estimated, then they do record the time that has passed between the separations of evolutionary lines. In the case of the haemoglobin of vertebrate animals the haemoglobin 'clock' does appear to 'tick' regularly.

Immunological studies are another means of detecting differences in specific proteins of species, and therefore (indirectly) their relatedness. **Serum** is a liquid produced from blood samples from which blood cells and fibrinogen have been removed. Protein molecules present in the serum act as antigens if the serum is injected into animals with an immune system that lacks these proteins.

Typically, a rabbit is used when investigating relatedness to humans. The injected serum causes the production of antibodies against the injected proteins. Then, serum produced from the treated rabbit's blood (it now contains antibodies against human proteins) can be tested against serum from a range of animals. The more closely related the animal is to humans, the greater the precipitation observed (Figure 17.21).

The precipitation produced by reaction with human serum is taken as 100%. For each species in Table 17.6, the greater the precipitation, the more recently the species shared a common ancestor with humans. This technique, called comparative serology, has been used by taxonomists to establish phylogenetic links in a number of cases, in both mammals and non-vertebrates.

Immunological studies are a means of detecting differences in specific proteins of species, and therefore (indirectly) their **relatedness**.

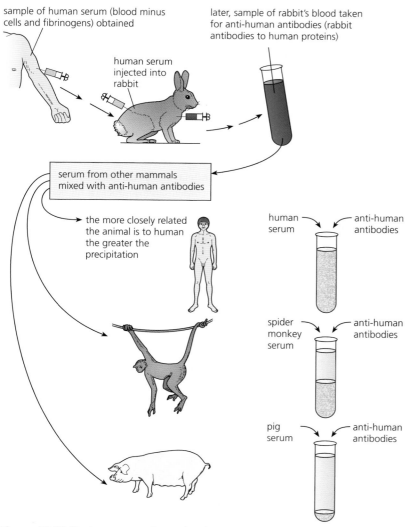

sample of human serum (blood minus cells and fibrinogens) obtained

human serum injected into rabbit

later, sample of rabbit's blood taken for anti-human antibodies (rabbit antibodies to human proteins)

serum from other mammals mixed with anti-human antibodies

the more closely related the animal is to human the greater the precipitation

human serum — anti-human antibodies

spider monkey serum — anti-human antibodies

pig serum — anti-human antibodies

Figure 17.21 The immune reaction and evolutionary relationships

Table 17.6 Relatedness investigated via the immune reaction

Species	Precipitation (%)	Difference from human (%)	Difference to common ancestor (half difference from human)	Postulated time since common ancestor (my) (*see below*)
human	100	–	–	–
chimpanzee	95	5	2.5	4
gorilla	95	5	2.5	4
orang-utan	85	15	7.5	13
gibbon	82	18	9	15
baboon	73	27	13.5	23
spider monkey	60	40	20	34
lemur	35	65	32.5	55
dog	25	75	37.5	64
kangaroo	8	92	46	79

We do not know of the common ancestor to these animals and the blood of that ancestor is not available to test anyway. But if the 584 amino acids that make up blood albumin change at a constant rate, then the percentage immunological 'distance' between humans and any of these animals will be a product of the distances back to the common ancestor plus the difference 'forward' again to any one of the listed animals. Hence the differences between a listed animal and humans can be halved to gauge the difference between a modern form and the common ancestor.

Since the radiation of the primates is known from geological and fossil evidence, the forward rate of change since the lemur gives the rate of the molecular clock – namely 35 per cent in 60 million years (my), or 0.6 per cent every million years. This calculation can now be applied to all the data (Table 17.6, column 5).

Mitochondrial DNA as a molecular clock

DNA has potential as a molecular clock, too. DNA in eukaryotic cells occurs in chromosomes in the nucleus (99 per cent) and in the mitochondria. **Mitochondrial DNA (mtDNA)** is a circular molecule, very short in comparison with nuclear DNA. Cells contain any number of mitochondria, typically between one hundred and a thousand.

Mitochondrial DNA has approximately 16 500 base pairs. Mutations occur at a very slow, steady rate in all DNA, but chromosomal DNA has with it enzymes that may repair the changes in some cases. These enzymes are absent from mtDNA.

Thus mtDNA changes 5–10 times faster than chromosomal DNA – involving about 1–2 base changes in every 100 nucleotides per million years. Consequently, the length of time since organisms belonging to different but related species have diverged can be estimated by extracting and comparing samples of their mtDNA.

Furthermore, at fertilisation, the sperm contributes a nucleus only (no cytoplasm). All the mitochondria of the zygote come from the egg cell. There is no mixing of mtDNA genes at fertilisation, and so the evidence about relationships from studying differences between samples of mtDNA is easier to interpret in the search for early evidence of evolution.

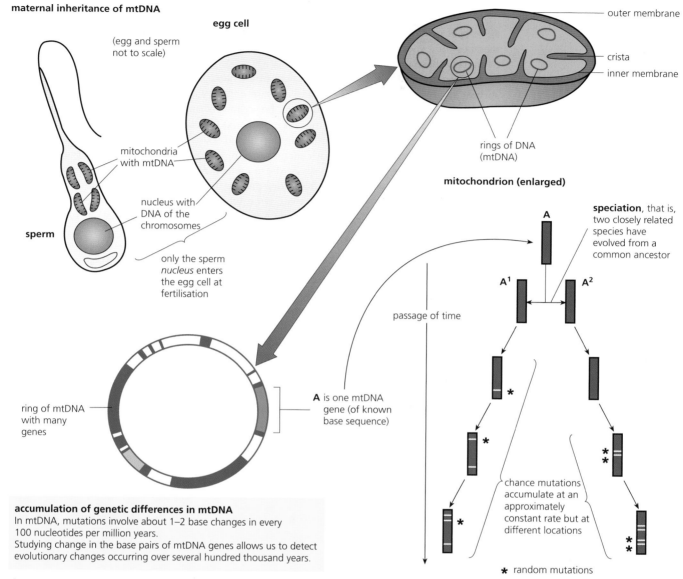

Figure 17.22 The use of mitochondrial DNA in measuring evolutionary divergence

Labels within the figure:

maternal inheritance of mtDNA

egg cell

(egg and sperm not to scale)

outer membrane

crista

inner membrane

mitochondria with mtDNA

nucleus with DNA of the chromosomes

rings of DNA (mtDNA)

mitochondrion (enlarged)

sperm

only the sperm *nucleus* enters the egg cell at fertilisation

speciation, that is, two closely related species have evolved from a common ancestor

A

A^1 A^2

passage of time

ring of mtDNA with many genes

A is one mtDNA gene (of known base sequence)

chance mutations accumulate at an approximately constant rate but at different locations

✱ random mutations

accumulation of genetic differences in mtDNA
In mtDNA, mutations involve about 1–2 base changes in every 100 nucleotides per million years.
Studying change in the base pairs of mtDNA genes allows us to detect evolutionary changes occurring over several hundred thousand years.

Speciation

Present-day plants and animals have arisen by change from pre-existing forms of life. This process has been called 'descent with modification' and 'organic evolution', but perhaps '**speciation**' is better because it emphasises that species change. So, what is a species?

When Linnaeus devised the binomial system of nomenclature in the 18th century there was no problem in defining species. It was believed that each species was derived from the original pair of animals created by God. Since species had been created in this way they were fixed and unchanging.

In fact the fossil record provides evidence that changes do occur in living things – human fossils alone illustrate this point. Today, as many different characteristics as possible are used in order to define and identify a species. The three main characteristics used are:

- **morphology** and **anatomy** (external and internal structure)
- **cell structure** (whether cells are eukaryotic or prokaryotic)
- **physiology** (blood composition, renal function) and **chemical composition** (comparisons of nucleic acids and proteins, and the similarities in proteins between organisms, for example).

Species: a group of organisms that are reproductively isolated, interbreeding to produce fertile offspring. Organisms belonging to a species have morphological (structural) similarities, which are often used to identify to which species they belong.

'Species' or 'variety'?

Since species may change (mostly a slow process), there is a time when the differences between members of a species become great enough to identify separate **varieties** or **sub-species**. Eventually these may become new species.

An illustration of this is the development of tolerance to heavy metal ions in plants able to survive and even flourish on the otherwise bare mining waste tips commonly found at sites where ores and minerals have been mined. Here, heavy metals such as zinc, copper, lead and nickel are often present as ions dissolved in the soil moisture at concentrations that generate toxic conditions for the plants present on the surrounding unpolluted soils.

We have seen that several heavy metal ions are essential for normal plant growth, but only in trace amounts. In mining waste these concentrations of ions are frequently exceeded and heaps left from 18th- and 19th-century mining activities in several countries around the world remain largely bare of plant cover, even when surrounding, unpolluted soils have dense vegetation cover. Seeds from these plants regularly fall on mining waste, but plants fail to establish.

However, careful observations of mining waste at many locations have disclosed the presence of local populations of plants that have evolved tolerance. One example is the grass *Agrostis tenuis* (Bent grass), populations of which are tolerant of otherwise toxic concentrations of copper. Biochemical and physiological mechanisms have evolved in tolerant species, including:

- the ability to selectively avoid uptake of heavy metal ions
- the accumulation of ions that enter in insoluble compounds in cell walls by the formation of stable complexes with cell wall polysaccharides
- the transport of toxic ions into the vacuoles of cells, the membrane of which is unable to pump them out again, so interaction with cell enzymes is avoided.

At what point will these copper-resistant forms of Bent grass be recognised as a separate species?

The evolution of this form of tolerance has been demonstrated in several species of terrestrial plants and also in species of seaweeds tolerant to the copper in the anti-fouling paints frequently applied to the hulls of ships.

Question

10 Explain the differences between a variety and a species. Find out about an example of both from an organism you are familiar with.

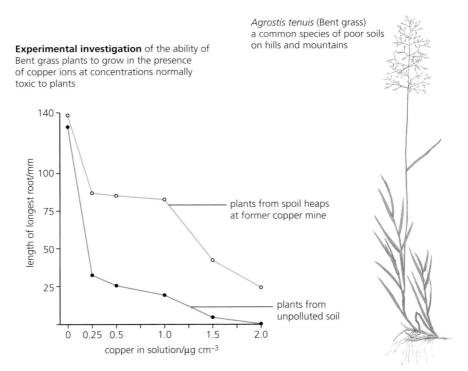

Experimental investigation of the ability of Bent grass plants to grow in the presence of copper ions at concentrations normally toxic to plants

Agrostis tenuis (Bent grass) a common species of poor soils on hills and mountains

plants from spoil heaps at former copper mine

plants from unpolluted soil

Figure 17.23 Copper ion tolerance in populations of Bent grass

Changing gene pools and speciation

A species can have localised populations, although the boundaries of this local population can be hard to define. Individuals in the local populations tend to resemble each other. They may become quite different from members of other populations. Local populations are very important because they are potentially a starting point for speciation. **Speciation** is the name we give to the process by which one species evolves into another.

Population genetics is the study of genes in populations. In any population, the total of the alleles of the genes located in the reproductive cells of the individuals make up a gene pool. A **gene pool** consists of all the genes and their different alleles present in an interbreeding population. When breeding between members of a population occurs, a sample of the alleles of the gene pool will contribute to form the **genomes** (gene sets of individuals) of the next generation, and so on, from generation to generation.

The frequency with which any particular allele occurs in a given population will vary. By **allele frequency** we mean how commonly any particular allele occurs in a population. When allele frequencies of a population are investigated they may turn out to be unchanging. When the allele frequencies of a gene pool remain more or less unchanged, then we know that population is static as regards its inherited characteristics. We can say that the population is not evolving.

Alternatively, allele frequencies may change quite rapidly from generation to generation. If the allele frequencies of genes in a population are changing, then we assume that evolution is going on. For example, some alleles may be increasing in frequency because of the advantage they give to the individuals carrying them. Because of these alleles the organism is more successful – it may produce more offspring, for example. If we can detect change in a gene pool we may be seeing evolution happening well before a new species is observed.

Speciation by geographic isolation

A first step to speciation may be when a local population (particularly a small, local population) becomes completely cut off in some way. Even then, many generations may elapse before the composition of the gene pool has changed sufficiently to allow us to call the new individuals a different species. Such changes in local gene pools are an early indication of speciation.

A population is occasionally suddenly divided by the appearance of a barrier, into two populations that are isolated from each other. Before separation, individuals shared a common gene pool, but after isolation, 'disturbing processes' like natural selection, mutation and random genetic drift may occur independently in both populations, causing them to diverge in their features and characteristics.

Geographic isolation between populations occurs when natural (or human-imposed) barriers arise and sharply restrict movement of individuals (and their spores and gametes, in the case of plants) between the divided populations (see figure 17.24 on the next page).

Geographic isolation also arises when motile or mobile species are dispersed to isolated habitats, as for example when organisms are accidentally rafted from mainland territories to distant islands. The 2004 tsunami generated examples of this in South East Asia. Violent events of this type have frequently occurred in the world's geological history.

Question

11 What factors may cause the composition of a gene pool to change? (Think about the changes that may go on in a population and between its members.)

1 isolation by a new, natural physical barrier
A natural habitat became divided when a river broke its banks and took a new route

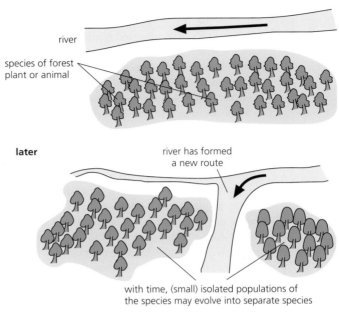

river

species of forest plant or animal

later

river has formed a new route

with time, (small) isolated populations of the species may evolve into separate species

2 isolation by a human-imposed barrier
A new road cuts through established habitats, separating local populations

Figure 17.24 Geographic barriers

Speciation by ecological and behavioural separation

Ecological and **behavioural separation** occur when members of a local population come to occupy different parts of a habitat leading to effective separation and therefore reproductive isolation. An example is seen in the terrestrial and marine giant iguana lizards of the Galapagos islands (Figure 17.25). These islands are about 500–600 miles from the South American mainland. The origin of these islands was volcanic; they appeared out of the sea about 16 million years ago, so we know they were uninhabited, initially. Today they have flora and fauna that relate to mainland species, including the iguana lizard. This species had no mammal competition when it arrived on the Galapagos, and it became the dominant form of vertebrate life. Today, two species of iguana lizard are present, one terrestrial and the other fully adapted to marine life. The latter is assumed to have evolved locally as a result of competition for space and food on the islands (both species are vegetarian), which drove some members of the population out of the terrestrial habitat.

Many organisms (e.g. insects and birds) may have flown or been carried on wind currents to the Galapagos from the mainland. Mammals are most unlikely to have survived drifting there on a natural raft over this distance, but many large reptiles can survive long periods without food or water.

immigrant travel to the Galapagos

The Galapagos Islands

Today the tortoise population of each island is distinctive and identifiable.

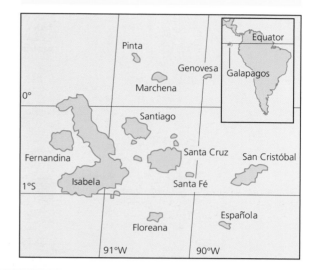

The **giant iguana lizards** on the Galapagos Islands became dominant vertebrates, and today are two distinct species, one still terrestrial, the other marine, with webbed feet and a laterally flattened tail (like the tail fin of a fish).

Figure 17.25 The Galapagos Islands and species divergence there.

Another example of behavioural and ecological separation leading to speciation has been demonstrated in the fauna of the Galapagos islands. Today there are 14–15 species of finch, and they have all been derived from a common ancestor, and have evolved, living in the same environment. These birds were noted by Charles Darwin (but they failed to attract his detailed attention) whilst on his visits to the islands. At a later date, detailed study of these birds was begun by the ornithologist, David Lack.

Lack studied the variation in the beaks of finches – a genetically controlled characteristic. Beak morphology reflects differences in feeding habits as members of the local population progressively focused on different diets, and evolved alternative feeding strategies. The evidence they provided for Darwin's theory of evolution by natural selection so impressed Lack that he coined the name 'Darwin's finches'. It has stayed with them, misleading though it is.

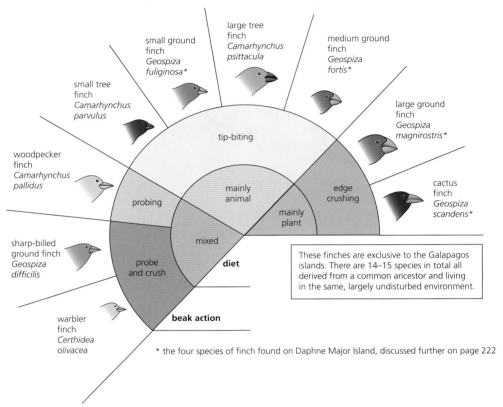

Figure 17.26 Adaptive radiation in Galapagos finches

Speciation — a summary

Apart from cases of instant speciation by polyploidy, species do not evolve in a rapid way. The process is usually gradual, taking place over a long period of time. In fact, speciation may occur over several thousand years, and in all cases requires 'isolation'.

A **deme** is the name we give to a small, isolated population. The individuals of a deme are not exactly alike, but they resemble one another more closely than they resemble members of other demes. This similarity is to be expected, partly because the members are closely related genetically (similar genotypes), and partly because they experience the same environmental conditions (which affect their phenotype).

We have noted above some examples of the ways demes become isolated. Reviewing these, we see they fall into two groups, depending on the ways isolation is brought about.

- Isolating mechanisms that involve spatial separation are known as **allopatric speciation** (meaning literally 'different country')
- Isolating mechanisms involving demes in the same location are known as **sympatric speciation** (meaning literally 'same country').

Table 17.7 Comparison of allopatric speciation and sympatric speciation

Allopatric speciation:	Sympatric speciation:
due to physical separation of the gene pool by geographic isolation, when motile or mobile species are dispersed to isolated habitats, preventing organisms of related demes or their gametes from meeting.	due to an isolating mechanism within a gene pool, preventing production of viable offspring between members of related demes in the same locality due to behavioural/ecological isolation.

Reproductive isolation

By definition, different species cannot interbreed and have fertile offspring – **gene flow** is prevented between them. When members of related demes have evolved to this point and have become fully reproductively isolated, we recognise them as members of different species. The barriers that prevent interbreeding between closely related species occur either before fertilisation can be attempted (pre-zygotic isolation) or after fertilisation has occurred (post-zygotic isolation). In Table 17.8 the underlying isolating mechanisms are summarised.

Table 17.8 Reproductive isolating mechanism

Pre-zygotic reproductive isolation	Post-zygotic reproductive isolation
Prevention of mating due to: • habitat differences that prevent meeting • behavioural differences, such as different mating rituals • temporal differences, such as being fertile at different times or seasons • mechanical differences, such as sex organs that are incompatible • gametic differences, such as failure of gametes to recognise each other, so preventing fertilisation.	• Hybrids formed are not viable and die prematurely. • Hybrids formed are infertile, for example, because the chromosomes cannot pair up in meiosis and produce haploid gametes. • Hybrids formed have low fertility. With each succeeding generation, fewer survive, leading to them all dying out.

Why organisms become extinct

The obliteration of the dinosaurs about 65 million years ago has been widely discussed. It is less well known that at that time almost half the genera of marine non-vertebrates also became extinct. Moreover, this was only one of several extinction events. The extinctions at the end of the Permian period, 250 million years ago, eliminated over 80 per cent of the marine, non-vertebrate genera. The sequence of mass extinctions, obtained by plotting the rate that genera have died out in a geological period is shown in Figure 17.27.

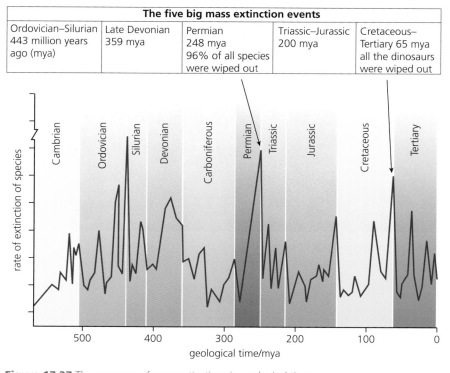

The five big mass extinction events				
Ordovician–Silurian 443 million years ago (mya)	Late Devonian 359 mya	Permian 248 mya 96% of all species were wiped out	Triassic–Jurassic 200 mya	Cretaceous–Tertiary 65 mya all the dinosaurs were wiped out

Figure 17.27 The sequence of mass extinctions in geological time

The significant forces driving extinctions are climate change, competition between species, whether interspecific or intraspecific, habitat loss, and by human destruction.

Climate change

Coral reefs are one of the 'theatres of environmental change' threatened by global warming. Corals are colonies of small animals embedded in a calcium carbonate shell that they secrete around themselves. They form their underwater structures in warm, shallow water where sunlight penetrates. Microscopic (photosynthetic) algae live sheltered and protected in the cells of corals. The relationship is one of mutual advantage (a form of symbiosis called mutualism), for the coral gets up to 90 per cent of its organic nutrients from these organisms. Coral reefs are the 'rainforests of the oceans' – the most diverse of ecosystems known. Although they cover less than 0.1 per cent of the surface of the oceans, these reefs are home to about 25 per cent of all marine species.

When under environmental stress (for example, high water temperature), the algae are expelled (causing loss of colour). In addition, much of the carbon dioxide that enters the atmosphere dissolves in the oceans. With the resulting ocean acidification, coral cannot absorb the ions they need to build or maintain their calcium carbonate skeletons. The coral starts to die, and the surrounding marine species, likewise. Mass bleaching events occurred in the Great Barrier Reef in 1999 and 2002. Today, coral reefs are dying all around the world. The effects from thermal stress are likely to be exacerbated under future climate scenarios.

Figure 17.28 Coral reefs – the crisis situation

Competition

Competition between organisms leading to the extinction of a species is difficult to establish. We might argue that any species that has become extinct and is known only as a fossil may be an example, but establishing a link between cause and effect may be impossible. In the laboratory, we can demonstrate how competition between two organisms that have the same diet or nutritional requirements and live in the same habitat (that is, they occupy the same niche) leads to the exclusion of one (Figure 17.29). This experiment is a demonstration of the competitive exclusion principle.

An experiment carried out by G. C. Gause in 1934 using species of *Paramecium*, a large protozoan common in fresh water. It feeds on **plankton**, the food source used in these experiments.

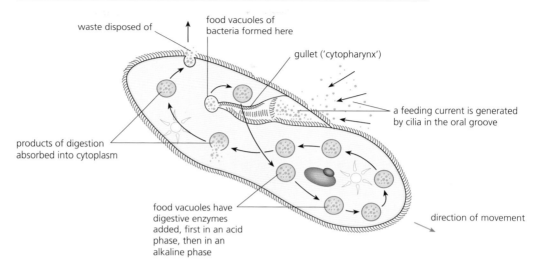

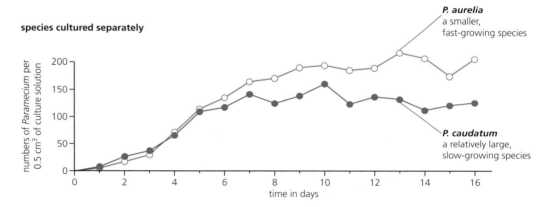

species cultured separately

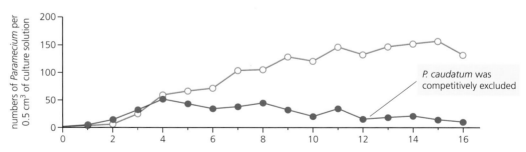

species cultured together

Figure 17.29 A demonstration of competitive exclusion

The history of the evolution of the genus *Homo* has many examples of species that became extinct (Figure 17.30). Can we know the actual causes of their extinctions? Surely we can only guess.

Figure 17.30 Some evolutionary stages in hominid evolution

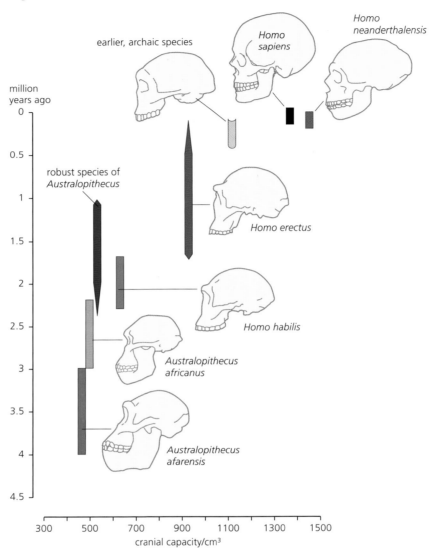

Habitat loss

Rainforests cover almost 2 per cent of the Earth's land surface, but they provide the habitats for almost 50 per cent of all living species. It has been predicted that if all non-vertebrates occurring in a single cubic metre of tropical rainforest soil were collected for identification, there would be present at least one completely previously unknown species. It is the case that tropical rainforests contain the greatest diversity of life of any of the world's biomes.

Now, tropical rainforests are being rapidly destroyed. Satellite imaging of the Earth's surface provides the evidence for this – if and where no other reliable sources of information are available. The world's three remaining tropical forests of real size are in **South America** (around the Amazon Basin), in **West Africa** (around the Congo Basin) and in the **Far East** (particularly but not exclusively on the islands of Indonesia).

The current rate of destruction is estimated to be about one hectare (100 m × 100 m – a little larger than a football pitch) every second. This means that each year an area larger than the British Isles (31 million hectares) is cleared. Whilst extinction is a natural process, this current rate is on a scale equivalent to that at the time of the extinction of the dinosaurs (an event 65 mya, at the Cretaceous–Tertiary boundary).

Destruction by humans

The **African elephant** (*Loxodonta africana*) once roamed most of the continent of Africa. In 1930, it was estimated there were 5–10 million African elephants, but by 1979 their numbers were reduced to 1.3 million. In 1989, when they were added to the international list of the most endangered species, there were about 600 000 remaining, less than 1 per cent of their original number. Although still relatively widely distributed south of the Sahara, populations are now fragmented. Many are restricted to National Parks and Reserves. One reason why African governments take measures to protect elephants at these venues is the importance of the tourist trade to their economies. Their National Parks bring in much-needed income, and 'ecotourism' does not deplete wildlife populations.

Today, elephants are threatened by loss of their habitats, by conflict with humans and by ivory poaching.

The **demand for ivory** threatens the largest adults with the biggest tusks. Old matriarchs (the oldest adult females who provide the 'social glue' for the herds) are particularly vulnerable. Their group existence makes them easier than solitary adult males for the poachers to locate. Other people continue to slaughter these magnificent animals too, for the 'bush meat' trade, for example.

Encroachment on human settlements is another problem. The hungry elephant that destroys crops is often hunted down and killed. Moreover, the increasing pressures of land use, as the wild spaces are deforested or cleared and converted to crop production, enhances this problem. In East Africa in particular, a high percentage of wildlife lives outside reserves, and so is not protected. Here, few elephants are now predicted to survive outside high-security areas. A similar trend is seen in the rest of Africa.

The outcome of these threats to the African elephant is predicted to be extinction in the wild.

Figure 17.31 An African elephant herd, observed in a National Nature Reserve

A well-documented example of extinction driven by human action is that of the **dodo**. The dodo was an inhabitant of the island of Mauritius in the Indian Ocean. It was a bird related to modern pigeons. Over geological time this distinctive organism had evolved to master a terrestrial habit. In the process it became a large sized bird (it was about a metre long and had a mass of approximately 20 kg.). The dodo nested on the ground and reared its young there. The diet was one of seeds and fruits that had fallen from the forest trees. It was one of many forest dwelling birds on the island – one of the 45 species for which there are early records, of which only 21 species have survived to this day.

The factors that contributed to the extinction of the dodos:

- Mauritius, an island far from any mainland, became a port-of-call for European explorers in the sixteenth century. Ships' crews restocked with fresh meat, and the dodo was one source. The animal was an easy victim; it had previously lacked significant natural enemies.
- Later, settlers brought cats, dogs and pigs, and inadvertently, rats, all alien species that fed on the young in the dodo's nest.
- Finally, natural habitats of the island were deliberately destroyed as land was cleared by human settlers for agriculture production.

Figure 17.32 The dodo – a reconstruction

Summary

- **Evolution** is the progressive change in living things in geological time, so that they become better able to survive in their environment. Evolution occurs by **natural selection** of chance differences. **Variation** arises by mutations of genes and chromosomes, by the production of different combinations of alleles that occurs during meiosis as a result of independent assortment and crossing over, and by the random nature of fertilisation.

- A **population** consists of all the organisms of the same species in a habitat that have the chance to interbreed. A population has the potential to increase in size with little constraint, initially, but is prevented from doing so indefinitely by **environmental factors**, which take many different forms.

- The **abiotic environment** with which organisms interact and which may limit the growth of the population includes climatic and soil factors. Interactions between organisms of the ecosystem, known as **biotic factors**, include competition for resources and may involve predation, grazing and parasitism.

- The total of all the alleles in a breeding population is known as a **gene pool**. The frequency of an allele in a population is known as the **allele frequency**. In the absence of 'disturbing factors' allele frequency does not change in a breeding population in succeeding generations. **Disturbing factors** that may alter the proportions of alleles are selective predation, migration, mutation and random genetic drift following a dramatic reduction in the size of a population.

- **New species** may form when a small part of a population becomes genetically isolated from others by a geographic or reproductive barrier. Alternatively, an abrupt change in the structure or number of chromosomes (a chromosome mutation) may cause an almost instant appearance of a new species.

- **Natural selection** may work to keep the characteristics of a species constant (stabilising selection), but if the environment changes then new forms may emerge (directional and disruptive selection). Balancing selection is a process which actively maintains multiple alleles in the gene pool of a population.

- Humans have obtained the animals and plants used in today's agriculture, transport and leisure pursuits by a process of domestication of wild organisms by means of **artificial selection. Selective breeding** is carried out by careful selection of the parents in breeding crosses and the selection of progeny with the required features.

Examination style questions

1 a) Explain what is meant by *artificial selection*. [4]
　b) In a plant breeding programme, corn, *Zea mays*, was bred in an attempt to produce a high yield of protein in the grain.

　　The results of this programme are shown in Fig. 1.1.

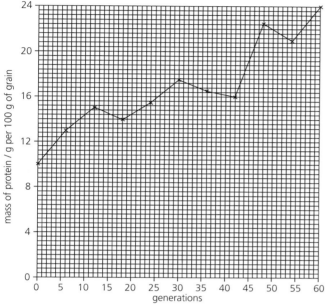

Fig. 1.1

i) With reference to Fig. 8.1, calculate the percentage increase in grain protein by the end of the experiment. Show your working. [2]

ii) Suggest why the protein yield does not increase steadily in each generation. [2]

[Total: 8]

(Cambridge International AS and A Level Biology 9700, Paper 04 Q8 June 2007)

2 a) Explain how changes in the nucleotide sequence of DNA may affect the amino acid sequence in a protein. [7]
　b) Explain how natural selection may bring about evolution. [8]

[Total: 15]

(Cambridge International AS and A Level Biology 9700, Paper 04 Q9 June 2009)

3 *Spartina* species (cord grass) are common plants of estuaries and salt marshes in many parts of the world. Two species, *S. maritima* (60 chromosomes – AA genome) and *S. alterniflora* (62 chromosomes – BB genome), once grew apart in different waters of the northern hemisphere. Now they occur together in many habitats. Today they have been joined by a new species of cord grass, *S. angelica* (122 chromosomes – AABB genome). This latter cord grass is a larger plant, and grows vigorously.

　a) It is assumed that *S. angelica* has evolved by a particular mechanism, involving the other two species. Describe this type of change, how it may have come about, and the steps that would have been involved. [12]

　b) Identify another plant species that has evolved by this mechanism. [2]

[Total: 14]

4 a) Define the terms 'gene pool' and 'differential mortality'. [2]
　b) Illustrate what you understand by stabilising selection by means of an example, and explain the suggestion that stabilising selection does not lead to evolution. [6]
　c) What is directional selection? By means of an example, show how directional selection may lead to new varieties of organism. [6]
　d) Disruptive or diversifying selection is said to result in balanced polymorphism. Elaborate this idea by means of a named example and by specifying the selection forces operating. [6]

[Total: 20]

18 Biodiversity, classification and conservation

The biodiversity of the Earth is threatened by human activities and climate change. Classification systems attempt to put order on the chaos of all the organisms that exist on Earth. Field work is an important part of a biological education to appreciate this diversity and find out how to analyse it. There are opportunities in this topic for students to observe different species in their locality and assess species distribution and abundance. Conserving biodiversity is a difficult task but is achieved by individuals, local groups, national and international organisations. Students should appreciate the threats to biodiversity and consider the steps taken in conservation, both locally and globally.

18.1 Biodiversity

> Biodiversity is much more than a list of all the species in a particular area.

By the end of this section you should be able to:

a) define the terms species, ecosystem and niche

b) explain that biodiversity is considered at three different levels:
  • variation in ecosystems or habitats
  • the number of species and their relative abundance
  • genetic variation within each species

c) explain the importance of random sampling in determining the biodiversity of an area

d) use suitable methods, such as frame quadrats, line transects, belt transects and mark-release-recapture, to assess the distribution and abundance of organisms in a local area

e) use Spearman's rank correlation and Pearson's linear correlation to analyse the relationships between the distribution and abundance of species and abiotic or biotic factors

f) use Simpson's Index of Diversity (D) to calculate the biodiversity of a habitat, using the formula

$$D = 1 - \left(\Sigma \left(\frac{n}{N} \right)^2 \right)$$ and state the significance of different values of D

> **Species**: A group of organisms of common ancestry that closely resemble each other structurally and biochemically, and which are members of natural populations that are actually or potentially capable of breeding with each other to produce fertile offspring, and which do not interbreed with members of other species.

Species, ecosystem and niche – core concepts in environmental study

Species

It was the Swedish botanist Karl Linnaeus (1707–78) who devised the binomial system of nomenclature in which every organism has a double name consisting of a Latinised generic name (**genus**) and a specific adjective (**species**). There was no problem in Linnaeus' day in defining species because it was believed that each species was derived from the original pair of animals created by God. Since species had been created in this way they were fixed and unchanging.

In fact the fossil record provides evidence that changes do occur in living things. 'Humanoid' and human fossils alone illustrate this point, as we shall see. Taxonomists now use as many different characteristics as possible in order to define and identify a species. The three main characteristics used are:

- external and internal structure (**morphology** and **anatomy**)
- **cell structure** (whether cells are eukaryotic or prokaryotic)
- **chemical composition** (comparisons of nucleic acids and proteins and the immunological reactions of organisms).

The last part of this definition cannot be applied to self-fertilising populations or to organisms that reproduce only asexually. Such groups are species because they look very similar (morphologically similar), and because they behave and respond in similar ways, with bodies that function similarly (they are physiologically similar).

But however we define the term, since species may change with time (mostly a slow process), there is a time when the differences between members of a species become significant enough to identify separate **varieties** or **sub-species**. Eventually these may become new species. All these points are a matter of judgment.

Ecosystem

One of the ideas that ecologists have introduced is that of the **ecosystem**. This is defined as a community of organisms and their surroundings, the environment in which they live and with which they live. An ecosystem is a basic functional unit of ecology since the organisms that make up a community cannot realistically be considered separate from their physical environment.

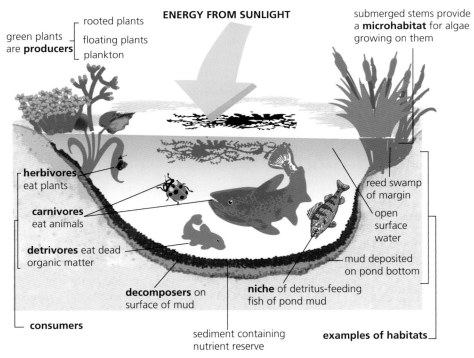

Figure 18.1 An example of an ecosystem – a lake

Niche

The niche is an ecological term that defines just how an organism **feeds**, where it **lives** *and* how it **behaves** in relation to other organisms in its habitat. The niche concept is useful because it identifies precise conditions which a species needs.

The principle of distinct niches can be illustrated by the two common and rather similar sea birds, the shag and the cormorant (Figure 18.2). These birds live and feed along the coastline and they rear their young on cliffs and rock systems. We can see they share the same habitat but their diets and behaviour differ. The cormorant feeds close to the shore on seabed fish, such as

---

Ecosystem: a unit made up of biotic and abiotic components interacting and functioning together, including all the living organisms of all types in a given area and all the abiotic physical and chemical factors in their environment, linked together by energy flow and cycling of nutrients. Ecosystems may vary in size but always form a functional entity: for example, a decomposing log, a pond, a meadow, a reef, a forest or the entire biosphere.

Question

1 Name one ecosystem in your locality that you might study and then list three abiotic factors affecting living things in that ecosystem.

Niche: the functional role or place of a species of organism within an ecosystem, including interactions with other organisms (such as feeding interactions), habitat, lifecycle and location, adding up to a description of the specific environmental features to which the species is well adapted.

flatfish. The shag builds its nest on much narrower cliff ledges. It also feeds further out to sea and captures fish and eels from the upper layers of the water. Since these birds feed differently and have different behaviour patterns, although they occur in close proximity, they avoid competition. **Competition** refers to the interaction between two organisms striving for the same resource. For competing organisms, the more their niches overlap, the more that both will strive to secure a finite resource. Hence many potential competitors have evolved different niches.

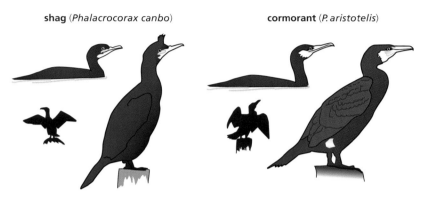

shag (*Phalacrocorax canbo*)   cormorant (*P. aristotelis*)

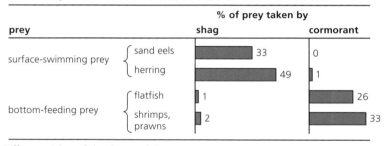

diet is a key difference in the niches of these otherwise similar birds

prey		% of prey taken by	
		shag	cormorant
surface-swimming prey	sand eels	33	0
	herring	49	1
bottom-feeding prey	flatfish	1	26
	shrimps, prawns	2	33

Figure 18.2 Different niches of the shag and the cormorant

Habitat

The habitat is the locality in which an organism occurs. It is where the organism is normally found. It is easy to give the habitat of some organisms, for instance the red squirrel is restricted to large blocks of coniferous forest in northern Asia and Europe. The seaweed bladder wrack is found only in the mid zone of the tidal seashore. On the other hand, the dandelion occurs widely – on a range of soils but also in crevices of walls and between paving stones.

If the area of the habitat is extremely small we call it a **microhabitat**. The insects that inhabit the crevices in the bark of a tree are in their own microhabitat. Conditions in a microhabitat are likely to be very different from conditions in the surrounding habitat.

Figure 18.3 The habitats of bladder wrack and the dandelion

Population: all of the organisms of one particular species within a specified area at a particular time, sharing the same gene pool and more or less isolated from other populations of the same species.

Population

By population we mean all the living things of the same species in a habitat at any one time. The members of a population have the chance of interbreeding, assuming the species concerned reproduces sexually. All the genes and their different alleles present in an interbreeding population is called the **gene pool**. The boundaries of populations of aquatic organisms occurring in a small pond are clearly limited by the boundary of the pond. The populations of the garden snail may be found in contrasting habitats. Some snails occur in extensive and ill-defined areas, around a compost heap, for example, whilst others occur in small restricted areas. Meanwhile, a songbird predator of the snail, the thrush, might occupy the whole garden and its surrounding fields and hedgerows. In other words, the area occupied by a population depends on the size of the organism and on how mobile it is, for example, as well as on environmental factors, such as food supply and predation.

So the boundaries of a population may be hard to define. Some populations are fully 'open', with individuals moving in or out from nearby populations. Alternatively, some populations are more or less 'closed', that is, isolated communities almost completely cut off from neighbours of the same species.

Figure 17.11, on page 389, describes the concept of 'population'. *Look back at it now.*

Community: all of the populations of all of the different species within a specified area at a particular time.

Community

A community consists of all the living things in a habitat, the total of all the populations, in fact. So, for example, the community of a well stocked pond would include the populations of rooted, floating and submerged plants, the populations of bottom-living animals, the populations of fish and non-vertebrates of the open water and the populations of surface-living organisms – typically a very large number of organisms.

Within any community there exists complex interactions between individuals – necessarily, organisms affect each other. This is because the resources required by plants and animals are in limited supply. Competitive interactions occur between members of the same species (**intraspecific competition**) and between members of different species (**interspecific competition**). We will re-visit aspects of these interactions when we discuss **food chains** and **food webs,** energy transfer between trophic levels and the cycling of nutrients within the environment.

Introducing biodiversity

There are vast numbers of living things in the world. The word '**biodiversity**', which refers to this, is a contraction of 'biological diversity'. It is a term we may use in three different ways.

1 Biodiversity – the total numbers of species known, and their relative abundance

Up to now, about 1.7 million species have been described and named. However, until very recently there has been no attempt to produce an international 'library of living things' where new discoveries are automatically checked (see below). Consequently, some known organisms may have been 'discovered' more than once.

Meanwhile, previously unknown species are being discovered all the time. In the UK alone, several hundred new species have been described in the past decade. We might have expected all the wildlife in these islands to be known, since Britain was one of the countries to pioneer the systematic study of plants and animals. Apparently, this is not the case; previously unknown organisms are still frequently found here.

Worldwide, the number of unknown species is estimated at between 3 and 5 million at the very least, and possibly as high as 100 million. So scientists are not certain just how many different types of organisms exist. Figure 18.4 is a representation of the proportions of known and unknown species estimated to exist in many of the major divisions of living things.

Question

2 Use the terms below to describe one or more of the features of a freshwater lake listed (a)–(g).
population, ecosystem, habitat, abiotic factor, community, biomass

a The whole lake

b All the frogs of the lake

c The flow of water through the lake

d All the plants and animals present

e The mud of the lake

f The temperature variations in the lake

g The total mass of vegetation growing in the lake

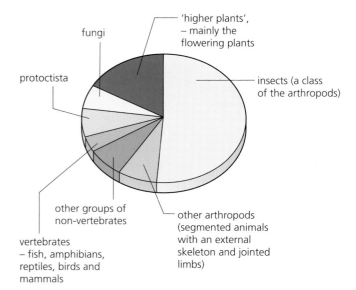

Of the 1.7 million described species, more than 50% are insects, and the higher plants, mostly flowering plants, are the next largest group. By contrast, only 4000 species of mammals are known, about 0.25% of all known species.

Figure 18.4 The relative number of animal and plant groups

Extension

The Species 2000 Programme

The **Species 2000** programme is part of a federation of databases and the organisations that are creating them, and was set up in 1994. It has the goal of creating a valid checklist of all of the world's species of plants, animals, fungi and microorganisms. This huge task is to be achieved by bringing together existing species databases covering each of the major groups of organisms. The UK arm of this initiative is based at the University of Reading. There, Species 2000 is in partnership with the Integrated Taxonomic Information System (ITIS) of North America, currently producing the **Catalogue of Life**, which is linked to the Global Biodiversity Information Facility (GBIF). You can keep up to date with developments at www.sp2000.org.

2 Biodiversity – variation in ecosystem or habitat

Ecosystems or habitats are units or components of the living world that we might choose to study. The biotic element present is likely to contain a huge community of organisms – the flora, fauna, and microorganisms that occur there. We may also refer to this specific, localised range of different organisms as 'biodiversity'.

3 Biodiversity – genetic variation within species

The gene pool that the members of each population represent is also referred to as an aspect of 'biodiversity'. We will return to this point later, when discussing the reasons why conservation of biodiversity is important.

> **Biodiversity:** the total number of different species living in a defined area, ecosystem or biome. It is also possible to consider the biodiversity of the Earth.

Investigating aspects of biodiversity

In the study of a habitat it may be important to collect accurate information on the size of populations present. A total count of all the members of a population is called a **census**. However, such an approach is usually totally impractical because of the size of the habitat, the movement of animals, and the huge numbers of organisms usually involved.

For practical reasons it is a **random sample** that is selected for this, involving either a part of the area of the habitat, or a restricted number of the whole population. By means of random sampling, every individual of the population has an equal chance of being selected, and so a representative sample is assured (Figure 18.5).

1 A map of the habitat (e.g. meadowland) is marked out with gridlines along two edges of the area to be analysed.

2 Co-ordinates for placing quadrats are obtained as sequences of random numbers, using computer software, or a calculator, or published tables.

3 Within each quadrat, the individual species are identified, and then the density, frequency, cover or abundance of each species is estimated Figure 18.6.

4 Density, frequency, cover, or abundance estimates are then quantified by measuring the total area of the habitat (the area occupied by the population) in square metres. The mean density, frequency, cover or abundance can be calculated, using the equation:

$$\text{population size} = \frac{\text{mean density (etc.) per quadrat} \times \text{total area}}{\text{area of each quadrat}}$$

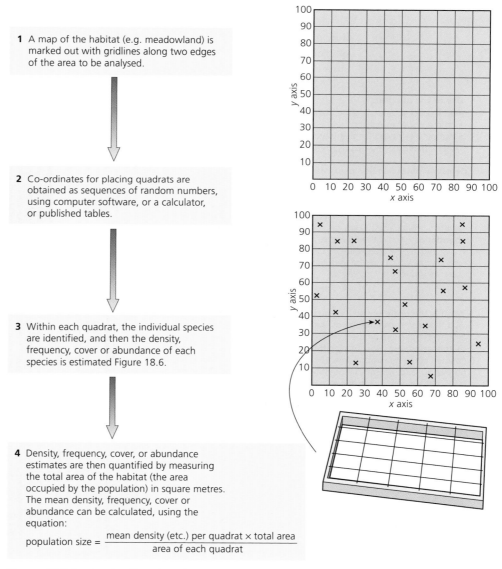

Figure 18.5 Random locating of quadrats

Random sampling of plant populations using quadrats

Quadrats are commonly used to estimate plant populations (Figure 18.6). A quadrat is a square frame that outlines a known area for the purpose of sampling. The choice of size of quadrat varies depending upon the size of the individuals of the population being analysed. For example, a 10 cm² quadrat is ideal for assessing epiphytic *Pleurococcus*, a single-celled alga, commonly found growing on damp walls and tree trunks. Alternatively, a 1 m² quadrat is far more useful for analysing the size of two herbaceous plant populations observed in grassland, or of the earthworms and the slugs that can be extracted from between the plants or from the soil below.

Quadrats are placed according to random numbers after the area has been divided into a grid of numbered sampling squares (Figure 18.5). The different plant species present in the quadrat may be identified. Then, using the quadrat an observer may estimate the density, frequency, abundance or cover of plant species in a habitat. The steps are outlined in Figure 18.6, including the issue of how many quadrats may need to be placed for an accurate assessment of population size. An area of chalk grassland is illustrated in which the use of quadrats would give estimates of the population sizes of two species of flowering plants.

Use of the quadrat:
- positioned at random within habitat being investigated
- different species present are then identified
- without destroying the plants and the microhabitats beneath them, plant species' density, frequency, abundance or cover can be estimated.

density = mean numbers of individuals of each species per unit area (time-consuming and may be hard to assess separate individuals)
frequency = number of quadrats in which a species occurs, expressed as % (rapid and useful for comparing two habitats)
cover = the % of ground covered by a species (useful where it is not possible to identify separate individuals)
abundance = subjective assessment of species present, using the DAFOR scale: D = dominant, A = abundant, F = frequent, O = occasional, R = rare (same observer must make abundance judgements, which may be useful as comparisons of two or more habitats, rather than objective scores)

What is the optimum size of quadrats? This varies with the habitat and the size of plants found. Look at the example here. In the 1 m quadrat there are six species present. How many different species are counted in the quadrat of sides 10, 20, 30, 40, 50, 60, 70, 80, and 90 cm? The optimum quadrat size is reached when a further increase in size adds no or very few further species present.
How many quadrats? When there is no further increase in the number of species found, sufficient quadrats have been analysed in that habitat.

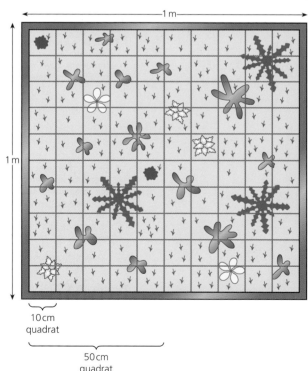

if more quadrats than about 20 are used, no additional species are found (in this habitat)

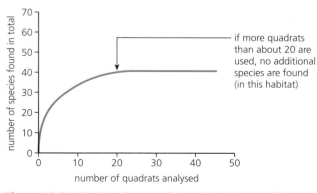

Figure 18.6 Estimating plant populations size using a quadrat

Figure 18.7 An area of chalk grassland with a rich flora

Estimating species distribution by means of a transect

Whereas some communities are relatively uniform over a given area and are suitable for random analysis (such as the grassland habitat shown in Figure 18.7), others show a trend in variation in a particular direction. Examples include seashore, pond or lake margin, saltmarsh or even an area where there is a change from dry soil to wetland. The appropriate technique to study such a trend of variation is the transect.

Transects are a means of sampling biotic (and abiotic) data at right angles to the impact of unidirectional physical forces. Although there may only be time to study one transect in detail (and this may be sufficient as a demonstration of the zonation of communities), a single transect may not provide an adequate sample or give an indication of differences from place to place. Transects should therefore preferably be replicated several times.

Where transects are carried out across a habitat where the land changes in height and where level is an important factor (such as a seashore, saltmarsh or pond margin), then the changes in level along the transect line can be measured and recorded as a **profile transect**. The surveying for this requires the use of survey poles and a field level device (Figure 18.8).

The community present along a transect can be analysed from a straight line such as a measuring tape, laid down across an apparently representative part of the habitat. The positions of every organism present that touches the line are recorded either all the way along the line or else at regular intervals. The result is a **line transect**.

A **belt transect** is a broad transect, usually half a metre wide. To produce it, a tape measure or rope is laid as for a line transect, but this time the organisms in a series of quadrats of half-metre width are sampled at (say) metre intervals. If the community changes little along the transect, then quadrats can be placed less frequently. Along with data on the biota, data on **abiotic** variables can be measured and recorded along the transect. For example, along a terrestrial transect the soil pH might be measured.

The results of a belt transect study of a seashore community are shown in Figure 18.8. The seashore (known as the littoral zone) is a part of the extreme margins of continents and marine islands periodically submerged below sea water, and so affected by tides. Tides are the periodic rise and fall of the sea level due to the attractions (gravitational pull) of Moon and Sun. The shore is an area rich in living things, and almost all are of marine origin.

The higher organisms occur on the shore, the longer the daily exposure to the air the organisms endure. Exposure brings the threat of desiccation and wider extremes of temperature than those experienced during submersion. 'Exposure' is an abiotic factor that influences distribution of organisms on the seashore.

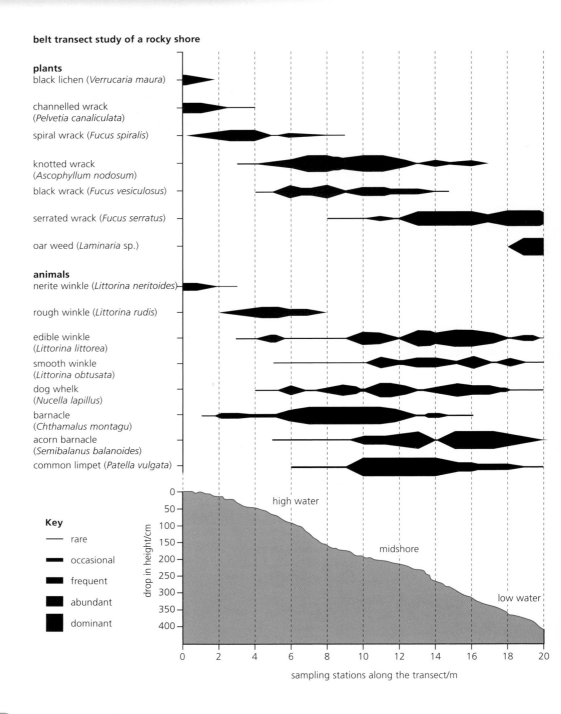

Figure 18.8 Profile and belt transect analysis of a rocky shore community

Mark, release and recapture (MRR) technique (the Lincoln Index)

We have already noted that an important step in determining the biodiversity of an area is the collection of accurate information on the size of populations present in a habitat.

A total count of all the members of a population is called a census, and would give us the most accurate data. However, the taking of a census is normally impractical. This may be because of the size of the habitat, or because of the quick movements of many animals, or because of species that are only active at dusk or after dark.

The capture, **mark, release and recapture (MRR)** technique, also known as the Lincoln Index, is a practical method of estimating population size of mobile animals (N), such as small mammals, woodlice or insects that can be captured and marked with a ring, tag, or dab of coloured paint or nail varnish.

Of the first sample caught for marking (n_1), it is essential to use a method of marking totally resistant to removal by moisture or by deliberate actions of the animals. It must be a method that does no harm to the captured animals, too. For example, these must not be more visible and therefore more vulnerable to predation than normal. It is also essential to trap relatively large samples if the results are to be significant.

The method then requires that marked individuals, on their release, are free to distribute themselves randomly in the whole population. After randomisation has occurred, a second sample is caught (n_2), some of which will be from the first sample (n_3), recognised by their marking. The size of the population is estimated by the formula:

$$N = \frac{n_1 \times n_2}{n_3}$$

MRR may be demonstrated on populations of woodlice discovered sheltering under stones/flower pots, etc. in an area of a garden or woodland. Alternatively, on populations of night-flying moths via light traps, or on populations of small mammals trapped in Longworth small mammal traps provided sufficient traps are available, for example.

Prior to experimenting on natural populations, the technique can be learnt and understood by application to non-living models. These may include beans dispersed in sawdust, or by using a 'population' of dried peas in a hessian bag. Here, samples are taken by hand, the sampler blindfolded, in effect. These trials enable the accuracy of the MRR method to be tested because the actual 'population' size (beans or dried peas, etc.) is easily known. It will quickly be evident that the size of samples captured determine the accuracy of the technique; small samples may give wildly inaccurate estimates. The steps to the MRR technique are outlined in Figure 18.10.

Questions

4 Using MRR, relatively large samples must be caught for significant results. Calculate the estimated size of the population in Figure 18.10 if the second sample had included two marked woodlice, not one.

5 As a new habitat was colonised, the size of a population of the common rough woodlouse *Porcellio scaber* was investigated by mark, release, recapture, with the following results:

Day	1/2	15/16	30/31	45/46	60/61	75/76
Captured, marked and released (n_1)	7	14	25	19	15	19
Captured in the second sampling (n_2)	6	12	24	16	9	14
Marked individuals recaptured (n_3)	2	2	3	2	1	2

a Calculate the size of the population at days 2, 16, 31, 46, 61 and 76.

b Plot a graph of estimated population size against time.

c Annotate your graph as fully as you are able.

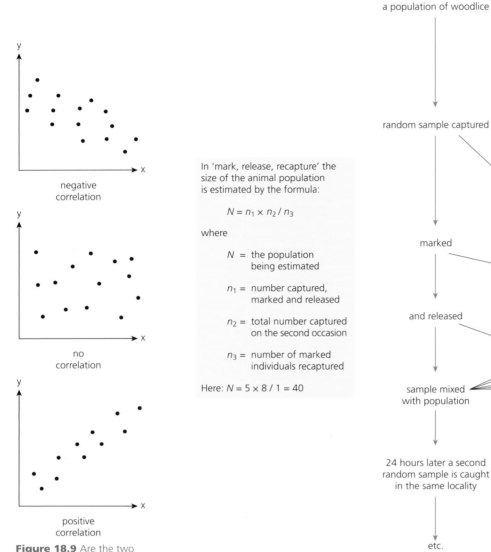

y

. . .
. . .
. . .
. . .
. .

→ x

negative
correlation

y

. .
. .
. .
. .
. .

→ x

no
correlation

y

. .
. .
. . .
. .
. .
.

→ x

positive
correlation

Figure 18.9 Are the two variables correlated?

In 'mark, release, recapture' the size of the animal population is estimated by the formula:

$$N = n_1 \times n_2 / n_3$$

where

N = the population being estimated

n_1 = number captured, marked and released

n_2 = total number captured on the second occasion

n_3 = number of marked individuals recaptured

Here: $N = 5 \times 8 / 1 = 40$

a population of woodlice

↓

random sample captured

↓

marked

↓

and released

↓

sample mixed with population

↓

24 hours later a second random sample is caught in the same locality

↓

etc.

Figure 18.10 Estimating animal populations using 'mark, release and recapture'

Analysing relationships between distribution and abundance of species and abiotic and biotic factors

There are different statistical tests for different kinds of quantitative data. In this case we are looking for the possibility of an association or correlation between two sets of data. When one factor is plotted against the other then we may find that:

● both factors increase together, indicating a **positive correlation**
● one factor increases whilst the other decreases, indicating a **negative correlation**
● the scatter diagram appears to have points randomly dispersed, indicating **no correlation**.

These differences are illustrated in Figure 18.9.

In investigating possible associations there are two issues:

● Is there an association between the two sets of variables?
● If so, what is the strength of the association?

To resolve these issues there are alternative statistical tests according to the type of data you have collected – either **Spearman's rank correlation** or **Pearson's linear correlation**. In Table 18.1 the differences between them are tabulated, and situations in which they would be chosen listed.

Table 18.1 Two tests of association

	Spearman's rank correlation	Pearson's linear correlation
When it is used	For ordinal (non-normal) data – values or observations that can be ranked, counted and ordered.	For normal data – values or observations that can be assigned a number (as a label) but data cannot be ranked.
Examples	How does the population size of grasshoppers vary in relation to when pesticides were last applied – are fewer insects found where pesticide application is recent? How do numbers of aquatic stone-fly nymphs relate to the hardness of the water – are more nymphs present in hard water? How does the depth of a stream relate to the proximity of the river bank – does depth of water increase with distance from the edge?	How does the height of fathers relate to height of their sons or daughters – do taller fathers have taller children? How does species diversity relate to latitude – does species diversity decrease with distance from the Equator? How does the size of land crabs relate to the density of their burrow – are the burrows of large crabs more dispersed?

Where your statistical test fits in

When you have decided on an ecological issue to investigate regarding some aspect of biodiversity in relation to an abiotic factor or another biotic factor, follow these steps:

1 State the null hypothesis (see page 381).
2 Conduct your investigation and obtain your data.
3 Present your results in a table and plot a scatter diagram.
4 Decide on the appropriate statistical test and carry this out (see below).
5 Draw a conclusion, stating whether you accept or reject the null hypothesis. Add a statement explaining the significance or implications of your conclusions.

Carrying out a statistical test

Use a statistical software package or programmed calculator to carry out the test. An appropriate source, accessed via the web is: Merlin: Statistical Software available to Biology Students. Merlin is a statistical package produced by Dr Neil Millar of Heckmondwike Grammar School, UK. This software is an add-in for Microsoft Excel. Merlin is available free of charge for educational and non-profit use. The package may be copied to laboratory and student computers from: **www.heckgrammar.kirklees.sch.uk/index.php?p=310**. Once you have data from laboratory or field investigations, Merlin can be used to carry out a statistical test. Merlin also includes a basic introduction to statistics for biology students, and a 'test chooser'. Data can be displayed in a range of graphs and charts, as appropriate.

Alternatively if you wish to analyse date by Pearson's linear correlation, you could use the following site: **www.socscistatistics.com/tests/pearson**.

Using Simpson's Index of Diversity to calculate biodiversity of a habitat

Early on in the development of a succession, for example at the pioneer stage when new land is being colonised, the number and diversity of species are low. At this stage the populations of organisms present are usually dominated by abiotic factors. For example, if extreme, unfavourable abiotic conditions occur (prolonged, very low temperatures, perhaps), the number of organisms may be severely reduced.

On the other hand, in a stable climax community many different species are present, many in quite large numbers. In this situation, adverse abiotic conditions are less likely to have a dramatic effect on the numbers of organisms present. In fact, in a well-established community the dominant plants set the way of life for many other inhabitants. They provide the nutrients, determine the habitats that exist, and influence the environmental conditions. These plants are likely to modify and reduce the effects of extreme abiotic conditions, too.

Thus the **diversity of species** present in a habitat is also an indicator of the **stability** of the community. Species diversity of a community may be measured by applying the formula, known as the **Simpson Diversity Index**:

$$\text{diversity} = \frac{N(N-1)}{\Sigma n\,(n-1)}$$

where N = the total number of organisms of all species found

and n = the number of individuals of each species

Questions

6 In a vegetable plot left fallow (left unsown, typically for a year) three weed species appeared. Individual plants were counted:

Species	n (no. of individuals)
Groundsel (*Senecio vulgaris*)	45
Shepherd's Purse (*Capsella bursa-pastoris*)	40
Dandelion (*Taraxacum officinale*)	10
Total (*N*)	**95**

Calculate the Simpson Diversity Index for this habitat.

7 Figure 18.11 is a study of the vegetation of a system of developing sand dunes. Note how both the species diversity and physical parameters change as the dunes age. **Calculate** the Diversity indices of the fore dune and semi-fixed dune (using a spreadsheet). **Comment** on the results.

Sand deposited by the sea and blown by the wind, builds into small heaps around pioneer xerophytic plants, such as marram and couch grass, at coasts where the prevailing wind is on-shore. The tufts of leaves growing through the sand accelerate deposition, and gradually drifting sands gather a dense cover of vegetation. Fixed sand dunes are formed.

point frame quadrat in use

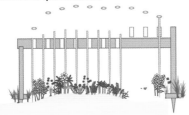

Behind the fore dune the fixed dunes can be seen.

The frame is randomly placed a large number of times, the 10 pins lowered in turn onto vegetation and the species (or bare ground) recorded.

Study of a sand dune succession from embryo state to fixed dune

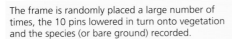

	Site 1	Site 2	Site 3	Site 4
Stage in succession	Embryo dune	Fore dune	Semi-fixed dune	Fixed dune
Number of pins dropped	220	220	220	220
sea couch grass	169	9	0	0
marram grass	1	123	19	0
fescue grass	0	0	126	182
spear-leaved orache	4	0	0	0
prickly saltwort	2	0	0	0
bindweed	1	44	0	0
bird's foot trefoil	0	0	0	6
biting stonecrop	0	0	0	1
buttercup	0	0	0	1
cat's ear	0	0	5	2
clover	0	0	0	68
common stork's bill	0	0	0	11
daisy	0	0	2	8
dandelion (common)	0	0	5	1
eyebright	0	0	0	4
hawkbit	0	0	20	1
ladies bedstraw	0	0	86	35
medick	0	0	0	62
mouse-ear chickweed	0	0	0	2
ragwort	0	0	8	1
restharrow	0	0	25	15
ribwort plantain	0	0	0	37
sand sedge	0	0	32	17
stagshorn plantain	0	0	0	36
tufted moss	0	0	0	119
yellow clover	0	0	0	5
Yorkshire fog	0	0	2	0
wild thyme	0	0	0	82
Total live hits	**177**	**176**	**330**	**706**
Bare ground hits	**48**	**68**	**6**	**0**
% bare ground	27	38	3	0
soil moisture (%)	5.0	8.5	16.0	14.3
soil density (g cm^{-3})	1.6	0.9	0.6	0.5
soil pH	8.0	7.5	7.0	7.0
wind speed (m s^{-1})	10.0	9.3	2.4	1.3

Figure 18.11 Sand-dune community development – a student field study exercise

18.2 Classification

Organisms studied locally may be used to show how hierarchical classification systems are organised.

By the end of this section you should be able to:

a) describe the classification of species into the taxonomic hierarchy of domain, kingdom, phylum, class, order, family, genus and species

b) outline the characteristic features of the three domains Archaea, Bacteria and Eukarya

c) outline the characteristic features of the kingdoms Protoctista, Fungi, Plantae and Animalia

d) explain why viruses are not included in the three domain classification and outline how they are classified, limited to type of nucleic acid (RNA or DNA) and whether these are single stranded or double stranded

Taxonomy – the classification of diversity

Classification is essential to biology because there are too many different organisms to sort out and compare unless they are organised into manageable categories. Biological classification schemes are the invention of biologists, based upon the best evidence available at the time. With an effective classification system in use, it is easier to organise our ideas about organisms and make generalisations.

The science of classification is called **taxonomy**. The word comes from *taxa* (singular: *taxon*), which is the general name for groups or categories within a classification system. The scheme of classification has to be flexible, allowing newly discovered living organisms to be added where they fit best. It should also include fossils, since we believe living and extinct species are related.

The process of classification involves:
- giving every organism an agreed name
- imposing a scheme upon the diversity of living things.

The binomial system of naming

Many organisms have local names, but these often differ from locality to locality around the world, so they do not allow observers to be confident they are talking about the same thing. For example, the name *magpie* represents entirely different birds commonly seen in Europe, in Asia, and in Sri Lanka (Figure 18.12). Instead, scientists use an international approach called the **binomial system** (meaning 'a two-part name'). By this system everyone, everywhere knows exactly which organism is being referred to.

Figure 18.12 'Magpie' species of the world

European magpie (*Pica pica*) **Asian magpie (*Platysmurus leucopterus*)** **Sri Lankan magpie (*Urocissa ornate*)**

So, each organism is given a scientific name consisting of two words in Latin. The first (a noun) designates the **genus**, the second (an adjective) the **species**. The generic name comes first, and begins with a capital letter, followed by the specific name. By convention, this name is written in *italics* (or is underlined). As shown in Figure 18.13, closely related organisms have the same generic name; only their species names differ. You will see that when organisms are frequently referred to the full name is given initially, but thereafter the generic name is shortened to the first (capital) letter. Thus, in continuing references to humans in an article or scientific paper, *Homo sapiens* would become *H. sapiens*.

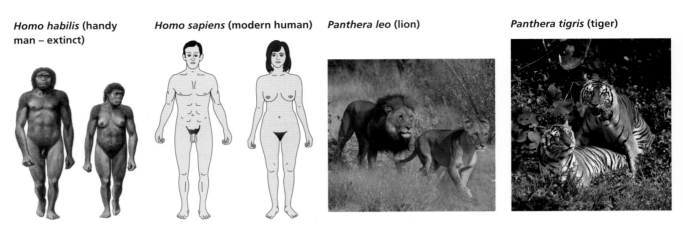

Homo habilis (handy man – extinct)   **Homo sapiens** (modern human)   **Panthera leo** (lion)   **Panthera tigris** (tiger)

Figure 18.13 Naming organisms using the binomial system – generic name (noun) + specific name (adjective) + common name

The scheme of classification

In classification, the aim is to use as many characteristics as possible in placing similar organisms together and dissimilar ones apart. Just as similar **species** are grouped together into the same **genus** (plural: *genera*), so too, similar genera are grouped together into **families**. This approach is extended from families to **orders**, then **classes**, **phyla** and **kingdoms**. This is the hierarchical scheme of classification; each successive group containing more and more different kinds of organism. The taxa used in taxonomy are given in Figure 18.14.

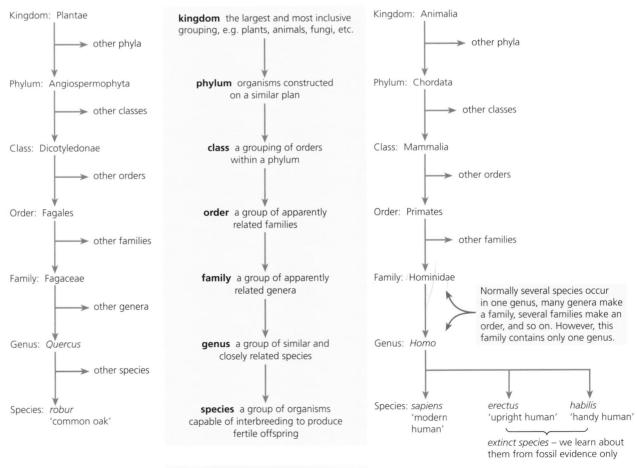

Kingdom: Plantae → other phyla
Phylum: Angiospermophyta → other classes
Class: Dicotyledonae → other orders
Order: Fagales → other families
Family: Fagaceae → other genera
Genus: *Quercus* → other species
Species: *robur* 'common oak'

kingdom the largest and most inclusive grouping, e.g. plants, animals, fungi, etc.

phylum organisms constructed on a similar plan

class a grouping of orders within a phylum

order a group of apparently related families

family a group of apparently related genera

genus a group of similar and closely related species

species a group of organisms capable of interbreeding to produce fertile offspring

Kingdom: Animalia → other phyla
Phylum: Chordata → other classes
Class: Mammalia → other orders
Order: Primates → other families
Family: Hominidae

Normally several species occur in one genus, many genera make a family, several families make an order, and so on. However, this family contains only one genus.

Genus: *Homo*

Species: *sapiens* 'modern human'   *erectus* 'upright human'   *habilis* 'handy human'

extinct species – we learn about them from fossil evidence only

Figure 18.14 The taxa used in taxonomy and applied to two different kingdoms

(a mnemonic to remember the hierarchy of taxa: **K**ing **P**eter **C**alled **O**ut **F**or **G**enuine **S**cientists)

Domains and kingdoms

At one time the living world seemed to divide naturally into two kingdoms consisting of the **plants** (with **autotrophic** nutrition) and the **animals** (with **heterotrophic** nutrition). These two kingdoms grew from the original disciplines of biology, namely botany, the study of plants, and zoology, the study of animals. Fungi and microorganisms were conveniently 'added' to botany! Initially there was only one problem; fungi possessed the typically 'animal' heterotrophic nutrition but were 'plant-like' in structure.

The 'five kingdoms' concept

Then, with the use of the electron microscope came the discovery of the two types of cell structure, namely **prokaryotic** and **eukaryotic** (page 22). As a result, the bacteria with their prokaryotic cells could no longer be 'plants' since plants have eukaryotic cells. The divisions of living things into kingdoms needed overhauling. This led to the division of living things into five kingdoms.

Table 18.2 The five kingdoms

Prokaryotae	the prokaryote kingdom, the bacteria and cyanobacteria (photosynthetic bacteria), predominately unicellular organisms
Protoctista	the protoctistan kingdom, (eukaryotes), predominately unicellular, and seen as resembling the ancestors of the fungi, plants and animals
Fungi	the fungal kingdom, (eukaryotes), predominately multicellular organisms, non-motile, and with heterotrophic nutrition
Plantae	the plant kingdom, (eukaryotes), multicellular organisms, non-motile, with autotrophic nutrition
Animalia	the animal kingdom, (eukaryotes), multicellular organisms, motile, with heterotrophic nutrition
Note that viruses are not classified as living organisms – see below.	

Extremophiles and the recognition of 'domains'

Then came the discovery of the distinctive biochemistry of the bacteria found in extremely hostile environments (the **extremophiles** – initially, the 'heat-loving' bacteria found in hot-springs at about 70 °C).

Table 18.3 The range of extremophiles

Halophytes – 'salt-loving' bacteria	common in salt lakes and where sea water becomes trapped and concentrated by evaporation where salt has crystallised
Alkalinophiles – 'alkali-loving' bacteria	survive at above pH 10, conditions typical of soda lakes
Acidophiles – 'acid-loving' bacteria	bacteria of extremely acidic conditions
Thermophiles – 'heat-loving' bacteria	occur in hot-springs at about 70 °C. Some are adapted as to survive at temperatures of 100–115 °C (**hyperthermophilic** prokaryotes)
Cryophiles – bacteria of sub-zero temperatures	common at temperatures of −10 °C, as in the ice of the poles where salt depresses the freezing point of water
Since these names were first used, these organisms have been found in a broader range of habitats. For example, some occur in the oceans, and some in fossil-fuel deposits deep underground. Some species occur only in anaerobic environments such as the guts of termites and of cattle, and at the bottom of ponds among the rotting plant remains. Here they break down organic matter and release methane – with very important environmental consequences.	

Extremophiles are different

These microorganisms of extreme habitats have cells that we can identify as 'prokaryotic'. However, the larger **RNA** molecules present in the ribosomes of extremophiles are different from those of previously known bacteria. Further analyses of their biochemistry in comparison with that of other groups has suggested new evolutionary relationships. This discovery led to a new scheme of classification (Figure 18.15).

The classification of living organisms into three domains on the basis of their ribosomal RNA

These evolutionary relationships have been established by comparing the sequence of bases (nucleotides) in the ribosomal RNA (rRNA) present in species of each group.

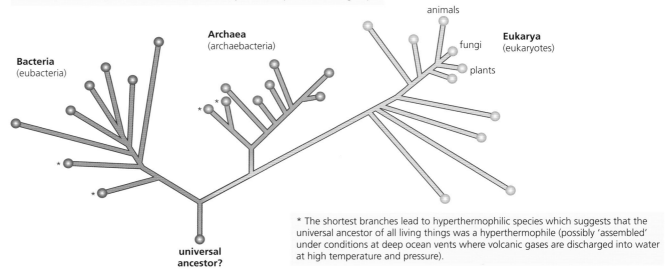

* The shortest branches lead to hyperthermophilic species which suggests that the universal ancestor of all living things was a hyperthermophile (possibly 'assembled' under conditions at deep ocean vents where volcanic gases are discharged into water at high temperature and pressure).

Archaea were discovered among prokaryotes of extreme and inaccessible habitats. Subsequently, other members of the Archaea were found more widely – in the gut of herbivores and at the bottom of lakes and mountain bogs, for example.

Figure 18.15 Ribosomal RNA and the discovery of archaea

All organisms are now classified into three domains

As a result, we now recognise three major forms of life, called **domains**. The organisms of each domain share a distinctive, unique pattern of ribosomal RNA, and there are other differences that establish their evolutionary relationships (Table 18.4). These domains are:

- the **Archaea** (the extremophile prokaryotes)
- the **Eubacteria** (the true bacteria)
- the **Eukaryota** (all eukaryotic cells – the protoctista, fungi, plants and animals).

This major advance, the recognition that the Archaea were a separate line of evolutionary descent from bacteria, was made by Carl Woese in 1977. You can read about him if you put his name into a search engine.

Table 18.4 Biochemical differences between the domains

Biochemical features	Domains		
	Archaea	Eubacteria	Eukaryota
DNA of chromosome(s)	circular genome	circular genome	chromosomes
1 bound protein (histone) present	present	absent	present
2 introns	typically absent	typically absent	frequent
Cell wall	present – not made of peptidoglycan	present – made of peptidoglycan	sometimes present – never made of peptidoglycan
Lipids of cell membrane bilayer	Archaeal membranes contain lipids that differ from those of eubacteria and eukaryotes (Figure 18.16).		

Question

9 In Figure 18.14 two animal species are classified from kingdom to species level. Suggest how this flow diagram needs to be modified to show its classification from 'domain' level.

Phospholipids of archaeal membranes

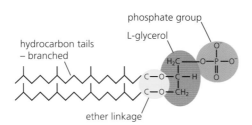

Phospholipids of eubacteria and eukaryote membranes

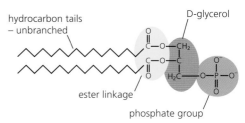

Figure 18.16 The lipid structure of cell membranes in the three domains

The characteristic features of the kingdoms Protoctista, Fungi, Plantae and Animalia

The Kingdom Protoctista

Includes the algae (including the multicellular seaweeds), the protozoa and the slime moulds.

Characteristics of the Protoctista:

- Single-celled organisms, or if multicellular, not differentiated into tissues.
- Cells are eukaryotic, with a distinct nucleus and membrane-bound organelles.
- Nucleus contains linear chromosomes of **DNA** with histone protein attached.
- Large ribosomes (80S) present.
- Some protoctista have cell walls, but in others these are absent.
- Nutrition is heterotrophic (in protozoa and slime moulds) or autotrophic (algae).

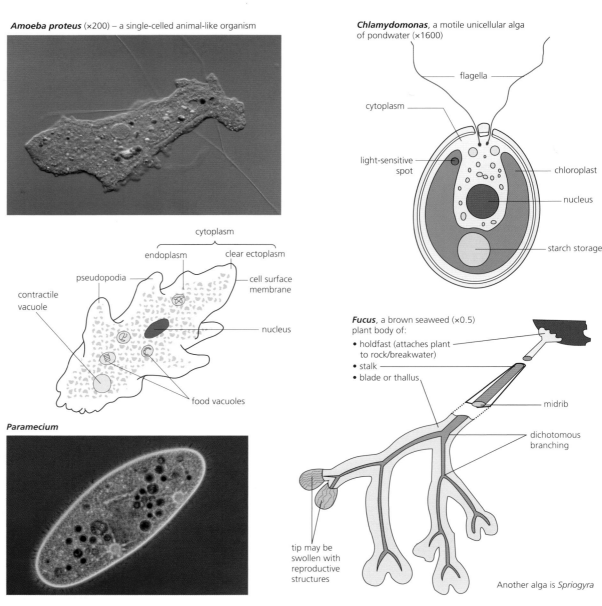

Amoeba proteus (×200) – a single-celled animal-like organism

Chlamydomonas, a motile unicellular alga of pondwater (×1600)

flagella

cytoplasm

light-sensitive spot

chloroplast

nucleus

starch storage

cytoplasm

endoplasm   clear ectoplasm

pseudopodia

cell surface membrane

contractile vacuole

nucleus

food vacuoles

Paramecium

Fucus, a brown seaweed (×0.5) plant body of:
- holdfast (attaches plant to rock/breakwater)
- stalk
- blade or thallus

midrib

dichotomous branching

tip may be swollen with reproductive structures

Another alga is *Spriogyra*

Figure 18.17 The Protoctista

The Kingdom Fungi

Includes the moulds, yeasts, mildews, mushrooms and bracket fungi.

The characteristics of the Fungi:

- Multicellular organisms (except the unicellular yeasts), not differentiated into tissues and non-motile.
- Cells are eukaryotic, with a distinct nucleus and membrane-bound organelles.
- Nucelus contains linear chromosomes of DNA with histone protein attached.
- Large ribosomes (80S) present.
- Cell walls are of chitin.
- Nutrition is hetertrophic.
- Body is made of branching, tube-like hyphae, often dividied by cross-walls into multinucleate sections.
- Reproduce by spores, produced by asexual and sexual processes.

False-colour SEM of yeast cell budding (x 4500)

Yeast feed on sugar solutions, wherever they occur.

Fungal spores 'germinate' quickly when they land on a suitable medium under favourable growing conditions.

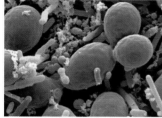

mycelium of hyphae

original spore

Mould fungi feed on exposed, moist organic matter.

hyphae and fungal fruiting bodies on the surface of the dung of herbivores

hyphal tip absorbs nutrients

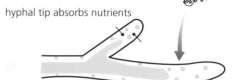

Coprinut (x 0.25) spores dispersed by flies

typical succession of fungi

Forest trees live in mutualistic association (both organisms benefit) with soil-inhabiting fungi. The fungal hyphae that permeate the soil for a wide area around the tree roots are continuous with those of the compact sheath around the roots. This sheath is the site of exchange of nutrients.

large bracket fungus releasing spores

a mushroom forms and releases spores

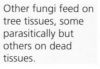

Other fungi feed on tree tissues, some parasitically but others on dead tissues.

Figure 18.18 The Fungi

The Kingdom Plantae

Includes the mosses, ferns, conifers and flowering plants (broad-leaved trees, herbaceous plants, leguminous plants and grasses).

   Characteristics of the Plantae:

- Multicellular organisms, differentiated into tissues; they are non-motile organisms (but the male gametes of mosses and ferns are motile).
- Cells are eukaryotic, with a distinct nucleus and membrane-bound organelles.
- Nucleus contains linear chromosomes of DNA with histone protein attached.
- Large ribosomes (80 S) present.
- Nutrition is autotrophic.
- Cell walls are largely of cellulose.

Mosses
– simple plants restricted to damp places
– without vascular tissue (xylem and phloem)
– without true roots
– spores produced in a capsule

Plants that reproduce by spores

Ferns
– with stem, leaves and roots
– adapted to terrestrial life
– vascular tissue present
– spores produced on the underside of leaves

Flowering plants
– reproduce by seeds
– may be herbaceous or woody
– with stem, leaves and roots with efficient vascular tissue
– stomata in the leaves
– have either broad leaves with net veins or narrow leaves with parallel veins

Broad-leaved plants

French bean plant
– plant of the pea and bean family
– found in Europe, Western Asia and North-West Africa
– now cultivated widely

Narrow-leaves plant (grass)

Sugar cane
– originally from India
– now cultivated widely
– commercial source of sucrose

Ebony tree
– a tropical tree of India

Figure 18.19 The Plantae

The Kingdom Animalia

Includes the non-vertebrates, such as worms and arthropods (which includes the insects), and the vertebrates, which are fish, amphibians, reptiles, birds and mammals.

Characteristics of the Animalia:

- Multicellular organisms, differentiated into tissues; many are motile organisms.
- Cells are eukaryotic, with a distinct nucleus and membrane-bound organelles.
- Nucleus contains linear chromosomes of DNA with histone protein attached.
- Large ribosomes (80 S) present.
- Cells sometimes have ciliar or flagella.
- Nutrition is heterotrophic.
- Cell walls are absent.

Non-vertebrate animals

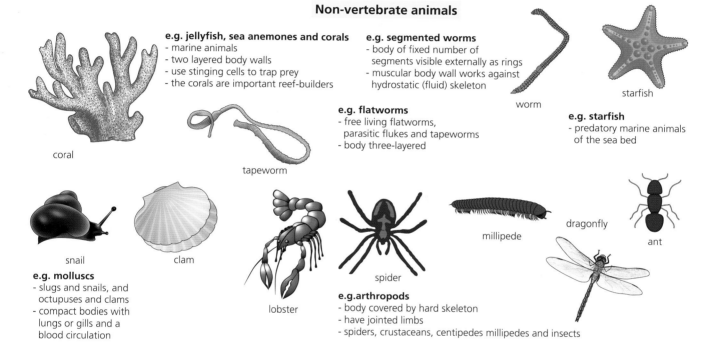

e.g. jellyfish, sea anemones and corals
- marine animals
- two layered body walls
- use stinging cells to trap prey
- the corals are important reef-builders

coral

tapeworm

e.g. flatworms
- free living flatworms, parasitic flukes and tapeworms
- body three-layered

e.g. segmented worms
- body of fixed number of segments visible externally as rings
- muscular body wall works against hydrostatic (fluid) skeleton

worm

starfish

e.g. starfish
- predatory marine animals of the sea bed

snail

clam

lobster

spider

millipede

dragonfly

ant

e.g. molluscs
- slugs and snails, and octupuses and clams
- compact bodies with lungs or gills and a blood circulation

e.g.arthropods
- body covered by hard skeleton
- have jointed limbs
- spiders, crustaceans, centipedes millipedes and insects

Vertebrate animals

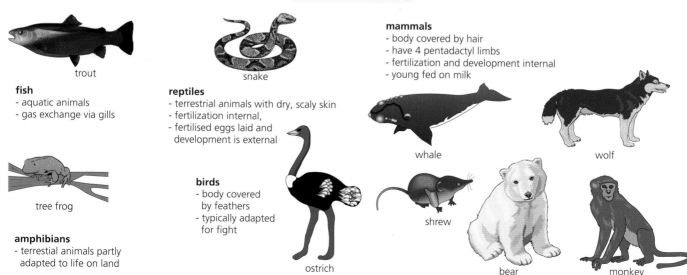

trout

fish
- aquatic animals
- gas exchange via gills

tree frog

amphibians
- terrestial animals partly adapted to life on land

snake

reptiles
- terrestrial animals with dry, scaly skin
- fertilization internal,
- fertilised eggs laid and development is external

birds
- body covered by feathers
- typically adapted for fight

ostrich

mammals
- body covered by hair
- have 4 pentadactyl limbs
- fertilization and development internal
- young fed on milk

whale

wolf

shrew

bear

monkey

Figure 18.20 The Animalia

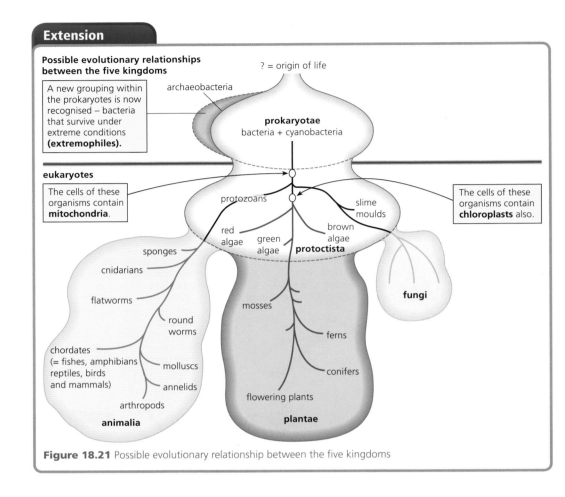

Extension

Possible evolutionary relationships between the five kingdoms

? = origin of life

A new grouping within the prokaryotes is now recognised – bacteria that survive under extreme conditions (**extremophiles**).

archaeobacteria

prokaryotae
bacteria + cyanobacteria

eukaryotes

The cells of these organisms contain **mitochondria**.

protozoans

slime moulds

The cells of these organisms contain **chloroplasts** also.

red algae   green algae   brown algae   **protoctista**

sponges

cnidarians

flatworms

round worms

chordates (= fishes, amphibians reptiles, birds and mammals)

molluscs

annelids

arthropods

animalia

mosses

ferns

conifers

flowering plants

plantae

fungi

Figure 18.21 Possible evolutionary relationship between the five kingdoms

Introducing the viruses

Viruses are disease-causing agents that are inactive outside a host cell. They are an assembly of complex molecules, rather than a form of life. Isolated from their host cell they are best described as 'crystalline'. Within susceptible host cells, however, they are highly active genetic programmes that take over the biochemical machinery of the host. Here, the component chemicals of the virus are synthesised, and then assembled to form new viruses. On breakdown (**lysis**) of the **host** cell, viruses are released, and may cause fresh infections. So, viruses are not living organisms, but may become active components of host cells. For these reasons, viruses are not included in the 'three domains' classification. Remember, antibiotics are ineffective against viruses.

Characteristics of viruses

Viruses are disease-causing agents, rather than 'organisms'. The distinctive features of viruses are:
● They are not cellular structures, but rather consist of a core of nucleic acid surrounded by a protein coat, called a capsid.
● In some viruses there is an additional external envelope of membrane made of lipids and proteins, as in HIV, Figure 10.15, page 204 and the influenza virus).

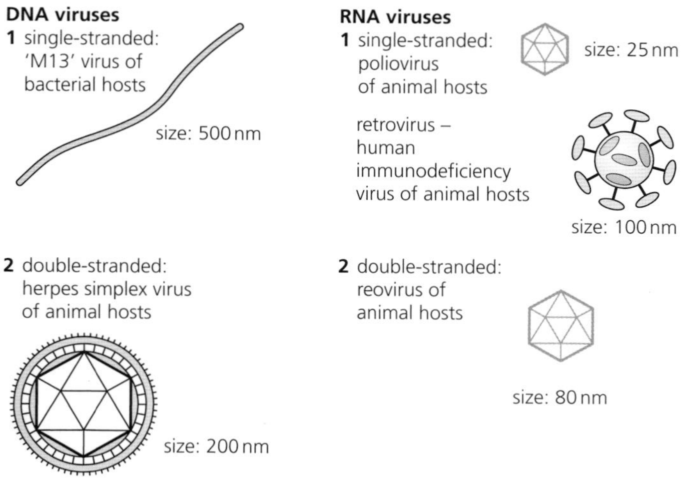

DNA viruses

1 single-stranded: 'M13' virus of bacterial hosts

size: 500 nm

2 double-stranded: herpes simplex virus of animal hosts

size: 200 nm

RNA viruses

1 single-stranded: poliovirus of animal hosts

size: 25 nm

retrovirus – human immunodeficiency virus of animal hosts

size: 100 nm

2 double-stranded: reovirus of animal hosts

size: 80 nm

- They are extremely small when compared with bacteria. Most viruses are in a size range of 20–400 nm (0.02–0.4 mm). They become visible only by means of the electron microscope.
- They can reproduce only inside specific living cells, so viruses function as endoparasites in their host organism.
- They have to be transported in some way between hosts.
- Viruses are highly specific to particular host species, some to plant species, some to animal species and some to bacteria.
- Viruses are classified by the type of nucleic acid they contain, either DNA or RNA, and whether they have a single or double strand of nucleic acid (Figure 18.22).

Figure 18.22 The classification of viruses

18.3 Conservation

Maintaining biodiversity is important for many reasons. Actions to maintain biodiversity must be taken at local, national and global levels. It is important to conserve ecosystems as well as individual species.

By the end of this section you should be able to:

a) discuss the threats to the biodiversity of aquatic and terrestrial ecosystems

b) discuss the reasons for the need to maintain biodiversity

c) discuss methods of protecting endangered species, including the roles of zoos, botanic gardens, conserved areas (national parks and marine parks), 'frozen zoos' and seed banks

d) discuss methods of assisted reproduction, including IVF, embryo transfer and surrogacy, used in the conservation of endangered mammals

e) discuss the use of culling and contraceptive methods to prevent overpopulation of protected and non-protected species

f) use examples to explain the reasons for controlling alien species

g) discuss the roles of non-governmental organisations, such as the World Wide Fund for Nature (WWF) and the Convention on International Trade in Endangered Species of Wild Fauna and Flora (CITES), in local and global conservation

h) outline how degraded habitats may be restored with reference to local or regional examples

The threats to biodiversity

Evolution, the development of life in geological time, is the process that has transformed life on Earth from its earliest beginnings to the organisms we know about today, living and extinct. Biodiversity is a result of 3.5 million years of evolution. Very many distinctly different organisms have evolved by natural selection and flourished at least for a while during this long time span. Consequently, today's biodiversity is only a tiny proportion of the diverse forms of life that have evolved in total. Most of these organisms and groups have become extinct, see Figure 17.25 on page 407.

It is fossils that provide the main source of information about these life forms that are now extinct – and give us a window on the true scale of diversity. Fossilisation is an extremely rare, chance event because scavengers and bacterial action normally dismember and decompose dead plant and animal structures before they can be fossilised. Of the fossils that do form, we can assume that most are never found or are overlooked or are accidentally destroyed before discovery. Nevertheless, numerous fossils *have* been found.

Question

10 Most fossils are preserved in sedimentary rocks. Why is this so and how are sedimentary rocks formed?

Fossil forms

petrification – organic matter of the dead organism is replaced by mineral ions mould – the organic matter decays, but the space left behind becomes a mould, filled by mineral matter

trace – an impression of a form, such as a leaf or a footprint, made in layers that then harden

preservation – of the intact whole organism; for example, in amber (resin exuded from a conifer, which then solidified) or in anaerobic, acidic peat

Steps of fossil formation by petrification

1. dead remains of organisms may fall into a lake or sea and become buried in silt or sand, in anaerobic, low- temperature conditions

2. hard parts of skeleton or lignified plant tissues may persist and become impregnated by silica or carbonate ions, hardening them

3. remains hardened in this way become compressed in layers, upthrust may bring rocks to the surface and erosion of these rocks commences

4. land movements may expose some fossils and a few are discovered by chance but, of the relatively few organisms fossilised, very few will ever be found by humans

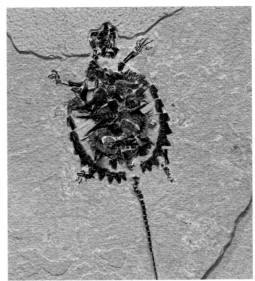

Sedimentary rock layers and the remains of extinct fossil species

Figure 18.23 Fossilisation

Biodiversity and habitat stability

Sometimes, entirely new land becomes available and is colonised by existing life forms. When this occurs, inevitably the initial number and diversity of species is low. At this stage the populations of organisms present are almost entirely dependent upon abiotic factors. For example, if extreme, unfavourable abiotic conditions occur (prolonged, very low temperatures, perhaps), the number of organisms may be severely reduced. On the other hand, in a well-established community, many different species are present, often in quite large numbers. In this situation, food webs are complex, with very many links and interconnections. Now, unfavourable abiotic conditions are much less likely to have a marked effect on the numbers of organisms present. We can see that the **diversity of species** present in a habitat is an indicator of the **stability** of the community. It is biodiversity that sustains ecosystems.

The associated genetic diversity sustains the stability of the ecosystem and the survival of species. We will return to this point shortly.

The current sharp decline in biodiversity

Today, the issue of human influence on biodiversity is an issue of major concern.

Why do you think this is?

In the long history of life on Earth, humans are very recent arrivals. Life originated about 3500 million years ago, but our own species arose only a little over 100 000 years ago. Initially, human activities had little impact on the environment. For one thing, during early human 'prehistory', population numbers were low. *Homo sapiens* were a rather 'struggling' species, living among many very successful ones. At this stage, survival must have been a very chancy affair.

Look back at Figure 10.2 on page 194.

The first significant increase in the human population occurred with the development of settled agriculture – a change that began in the fertile crescent in the Middle East about 10 000 years ago. The current human population explosion began at about the beginning of the **Industrial Revolution**, around 200 years ago, and it continues today. No one is certain when the rate of population growth will slow down, as it surely must.

When we plot world human population against time we see a steeply-rising J-shaped curve. Our population growth appears to be in an **'exponential** (log) **phase' of growth**. This is due to high birth rates and lower death rates, leading to rising life expectancy (people are living very much longer). Today the impact of humans on the environment is very great indeed – there is virtually no part of the biosphere which has not come under human influence and been changed to some extent. Many of these changes threaten the biodiversity in all ecosystems.

Only a part of planet Earth, its land, oceans and atmosphere are inhabited. Living things inhabit a narrow belt, from upper soil to the lower atmosphere, or, if marine, most occur near the ocean surface. This restricted zone is called the **biosphere**. The distribution patterns of living things within the biosphere show large, stable vegetation zones. These zones are called **ecosystems**. Examples include tropical rainforest and grassland. Here communities typically extend unbroken over thousands of square kilometres.

Question

11 What is the chief ecosystem in the country where you live?

The IUCN Red List of Threatened Species

Currently, the rate of extinctions is exceptionally high. Environmentalists seek the survival of endangered species by initiating and maintaining local, national and international action. For example, the **International Union for the Conservation of Nature (IUCN)**, working with appropriate local organisations, publishes a series of Regional Red Lists. These assess the risk of extinction to species within countries and regions. The lists are based upon criteria relevant to all species and all regions of the world (Figure 18.24). They convey the urgency of conservation issues to the international community and to policy makers. The aim is to stimulate action to combat loss of endangered species and of the habitats that support them. We look into this issue again later in this topic.

This bar chart is a summary of the percentage of organisms identified as 'Critically endangered', 'Endangered' and 'Vulnerable' within several groups. Data has been taken from the IUCN Red List website (**www.iucnredlist.org**) version 2014.2.

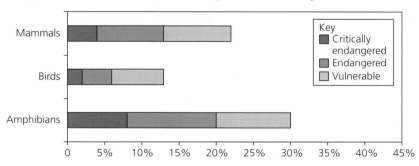

For all groups other than those above, the IUCN have only assessed a tiny proportion of the species in those groups, so comparisons cannot be made between and within these groups. The three groups opposite however have been comprehensively assessed.

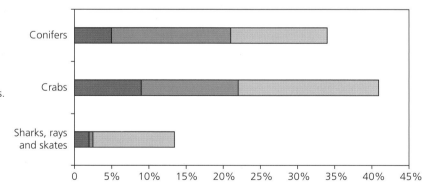

Figure 18.24 IUCN Red List data

The biodiversity crisis and the vanishing rainforests

Rainforests cover almost 2 per cent of the Earth's land surface, but they provide habitats to almost 50 per cent of all living species. It has been predicted that if all the non-vertebrates occurring in a single cubic metre of tropical rainforest soil were collected for identification, there would be at least one previously unknown species present. It is the case that tropical rainforests contain the greatest diversity of life of any of the world's terrestrial ecosystems.

Sadly, tropical rainforests are being rapidly destroyed. Satellite imaging of the Earth's surface provides the evidence this is so – if and where no other reliable sources of information are available. The world's three remaining tropical forests of real size are in **South America** (around the Amazon Basin), in **West Africa** (around the Congo Basin) and

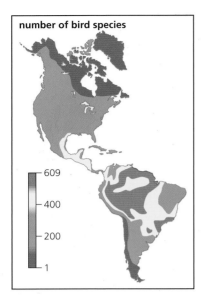

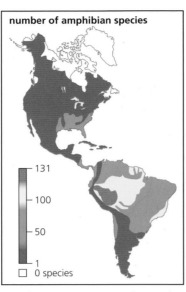

Figure 18.25 Most species have small ranges and these are concentrated unevenly

in the **Far East** (particularly but not exclusively on the islands of Indonesia). The current rate of destruction is estimated to be about one hectare (100 m × 100 m, a little larger than a football pitch) every second. This means that each year an area larger than 31 million hectares is cleared.

Why is this so devastating for biodiversity?

When the distribution of livings things across the Earth's surface is investigated we find that most species are not distributed widely at all (Figure 18.25). Instead, very many are restricted to a narrow range of the Earth's surface. Consequently, when tropical rainforests are destroyed, the only habitats of a huge range of plants is lost, and with them very many of the vertebrates and non-vertebrates dependent on them, too. This is one major reason why the fate of the remaining rainforest is such an urgent issue. We will return to deforestation and its consequences for the conservation of the African elephant and the orangutans of Asia, later.

The biodiversity crisis and coral ecosystems

Corals are colonies of small animals embedded in a calcium carbonate shell that they secrete around themselves. They form underwater structures known as coral reefs. These form mainly in warm, shallow water where sunlight penetrates and that is clear of sediments. The animal concerned is similar in structure to a sea anemone. However, within its bodies many tiny photosynthetic cells, called zooxanthellae, live, sheltered and protected. The relationship is one of mutual advantage (a form of symbiosis called mutualism), for the coral gets up to 90 per cent of its organic nutrients from these organisms.

Coral reefs are the 'rainforests of the oceans', because they are the most diverse of the ecosystems known. Although they occupy less than 0.1 per cent of the area of the oceans, coral reefs are home to about 25 per cent of all marine species. Today, coral reefs are dying all around the world. Their existence is threatened by coral mining, and pollution from run-off water from agricultural land (rich in organic and inorganic molecules). Rising sea temperatures and rising sea levels (due to global warming), and acidification of the oceans due to rising levels of atmospheric carbon dioxide are also major threats. About 60 per cent of the world's reefs are at risk due to human-related activities. The threats are greatest in Southeast Asia.

The importance of biodiversity

The ecological importance of diversity has been discussed above. The need to maintain biodiversity is also based upon genetic, economic and ecological principles, as well as upon ethical issues.

Genetic diversity and the evolution of new species

How diversity within a species arises via mutations (page 124) and during meiosis and fertilisation (pages 348–9) has already been discussed.

In a population, a group of individuals of a species living close together and able to interbreed, the alleles of the genes located in the reproductive cells of those individuals make up a **gene pool**. A sample of the alleles of the gene pool will contribute to form the genomes (gene sets of individuals) of the next generation, and so on, from generation to generation.

The size of an interbreeding population has a direct impact on the genetic diversity of the individuals. A very small population can be described as an **inbreeding** group – the individual are closely related. In fact, the smaller the population the more closely related the offspring will be. The important genetic consequence of inbreeding is that it leads to **homozygosity** – there is progressively less variation in the population as the number of loci at which there are homozygous alleles increases. Whilst the individuals of that population may initially be well adapted, in the face of environmental changes they are less able to adapt. We can say that the 'genetic fitness' of the population is compromised.

In the case of small, isolated populations, genetic variability is most critical. For example, in populations of species that become endangered, it is a question whether they posses sufficient genetic diversity to be able to adapt to future changes in the environment. If they do not, their survival is unlikely. This is a practical problem facing modern zoos attempting captive breeding programmes (page 445). It is also an issue for other attempts to protect endangered organisms in the wild where population numbers have been reduced to small, isolated groups in former strongholds.

Genetic diversity as the reservoir of essential genes for agriculture

Our domesticated animals and cultivated plants were derived originally from wild species. This is the origin of many of the organisms that provide our food, fuel, timber and pharmaceutical chemicals. Consequently, the wild relatives of domesticated species are a continuing source of genes for resistance to pests and diseases, and also to overcome stresses due to drought, rising salinity and extreme temperatures. For example, some modern crops have had resistance to pathogens added by the introduction of genes from 'wild' relatives. The loss of primitive varieties means the loss of the genetic variations essential for sustained crop and herd improvement. If genetic diversity of a species becomes low, the species is at risk, because once genetic variants are lost they cannot be recovered.

Biodiversity and the origin of new drugs

Many new drugs and other natural products, in some form or another, are extracted from substances manufactured by plants. The discovery of new, useful substances often starts with rare, exotic or recently discovered species. The whole range of living things is functionally a gene pool resource, and when a species becomes extinct its genes are permanently lost.

Biodiversity and ecotourism

Responsible travel to natural areas which conserves the environment and improves the welfare of the local people is referred to as **ecotourism**. The ecotourist seeks to observe wildlife and habitats that may be unfamiliar to them as well as being fragile and (hopefully) pristine, for example. Usually, this involves visits to national parks or special reserves where the diverse, local plants and animals are already conserved in as natural an environment as possible. The aim is to learn about ecology and diversity with minimum disruption or destruction, and to empower local and national communities. This may be by aiding local development funds and resources for their fight against poverty and disease. Ecotourism aims to support sustainable development by empowering the local community economically and socially.

The climatic and scientific importance of biodiversity

As the human population expands and extends its impact on the natural world, the threat imposed on diverse habitats increases. Yet natural habitats are critically important 'outdoor laboratories' where we learn about the range of life that has evolved and how it has evolved. In some habitats, including in tropical rainforests, the majority of organisms present are as yet unknown. Furthermore, autotrophic organisms everywhere (algae in the seas, grasses on the plains and prairies and forest and woodland trees) are all carbon dioxide sinks that help reduce climate change, to varying extents. Change in global patterns of rainfall result from deforestation, just as changes in wind and sea current patterns lead to extreme or unusual weather.

The ethical and aesthetic basis for maintaining biodiversity

A diversity of habitats and the wildlife they sustain are more than just exhilarating venues to visit and be inspired by. They are part of the inheritance of future generations which should be secured for future people's enjoyment, too. There is a moral obligation on humans to pass on to future generations the diversity we have inherited and enjoyed.

Some habitats are also the home of peoples who have avoided 'development'. These places and the people who inhabit them have a right to their traditional ways of life. Their livelihoods and homes are dependent upon the biodiversity they have evolved among and with. Similarly, many higher mammals, including relatively close relatives of *Homo*, live exclusively in some of these habitats. The needs of all primates must be respected, as indeed should the needs of the plants and animals around them. Damage to marine environments, especially to coral reefs, has resulted from overfishing and pollution. Marine food chains sustain biodiversity, as well as the fishing industry and tourism, on which many human communities depend.

Methods of protecting endangered species

Conservation by promotion of nature reserves

Nature reserves are carefully selected land areas set aside for restricted access and controlled use, to allow the maintenance of biodiversity locally. This is not a new idea; the New Forest in southern Britain was set aside for hunting by royalty over 900 years ago. An incidental effect was that it functioned as a sanctuary for wildlife!

Today, this solution to extinction pressures on wildlife includes the setting up and maintenance of areas of special scientific interest as nature reserves, of national parks (the first national park was Yellowstone, set up in North America in 1872), and of the African game parks which have been more recently established. In total, these sites represent habitats of many different descriptions, in many countries around the world. Some of the conservation work they achieve may be carried out by volunteers.

What biogeographical features of a reserve best promote conservation?

In a nature reserve the **area** enclosed is important – a tiny area may be too small to be effective. The actual dimensions of an effective reserve vary with species size and the lifestyle of the majority of the threatened species it is designed to protect.

Also, there is an **edge effect**. A compact reserve with minimal perimeter is less effective than one with an extensive interface with its surroundings. The use of the surrounding area is important, too; if it is managed sympathetically it may indirectly support the reserve's wildlife.

Another feature is **geographical isolation** – reserves situated at a greater distance from other protected areas are less effective than reserves that are closer together. Also, it has been found that **connecting corridors** of land are advantageous. In agricultural areas these may simply take the form of hedgerows protected from contact with pesticide treatments that nearby crops receive.

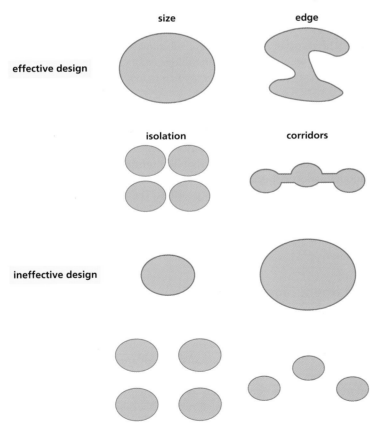

Figure 18.26 Features of effective nature reserves

The reserve or park is best managed to the advantage of wildlife by:

- the employment of local people as trained wardens and rangers, who patrol the area to ensure people comply with local regulations. They note the condition of the wildlife, record the changing population sizes and protect the wildlife where the need arises
- the supervision of agricultural and hunting activities, maintaining the emphasis on traditional practices applied to modest or minimal levels
- the control of permitted industrial activities, typically mining or quarrying, logging, and fishing in marine reserves, and of the associated transport infrastructures needed.

For visitors and ecotourists:

- visitor centres are provided and staffed to educate and inform on the local issues of conservation
- access is controlled and restricted to specific footpaths, tracks and roads to avoid critical wildlife habitats.

Table 18.5 What active management of nature reserves involves – a summary

Continuous monitoring of the reserve so that causes of change are understood, change may be anticipated and measures taken early enough to adjust conditions without disruption, should this be necessary. For example, coral reefs require monitoring for the impact of ecotourism.
Maintenance of effective boundaries and the limitation of unhelpful human interference. The enthusiastic involvement of the local human community communicates the messages of the purposes of conservation (a local 'education' programme, in effect) and that everyone has a part to play in conservation. Also, **protection of major migration routes** of large herds of herbivores and the big predators that follow them, as in Africa and North America.
Measures to facilitate the successful completion of lifecycles of any endangered species for which the reserve is home, together with supportive conditions for vulnerable and rare species.
Restocking and reintroductions of once common species from stocks produced by captive breeding programmes at zoological and botanical gardens. This makes possible reintroduction of species (usually predators) hunted almost to extinction in farming areas, including wolves in Europe and in the northern national parks in the USA.

Question

14 What are the particular challenges in the management of a nature reserve in your locality or country?

Zoological gardens and their captive breeding programmes

Endangered species typically have very low population numbers and are in serious danger of becoming extinct. For some species whose numbers have declined drastically, captive breeding may be their last hope of survival. Today, many zoos cooperate to manage individuals of the same species held in different zoos as a single population. A 'stud book' – a computerised database of genetic and demographic data – has been compiled for many of these species. This provides the basis for the recommendation and conduct of crosses designed to preserve the gene pool and avoid the problems of inbreeding. Animals may be shipped between zoos or the technique of artificial insemination may be used instead. Some species can be very hard to breed in captivity, for example, due to the stresses of captivity which may disrupt their breeding cycles. With others there has been a high success rate. Captive breeding permits the monitoring of the mother and her fetus during known pregnancies. To bring about pregnancies that are otherwise occurring at low frequencies, banks of eggs and sperm can be stored frozen. This allows the use of artificial insemination and even *in-vitro* fertilisation techniques (Figure 18.28) and the implantation of embryos in surrogate mothers.

A successful outcome is when individuals bred in captivity are released and have survived in the wild. Examples include red wolves, Andean condors, Bald Eagles and Golden Lion Tamarins. However, the release of captive-bred organisms is not always a success. The destruction of natural habitats may be a problem – for example, natural food sources may be limited. Animals reared in captivity may not adapt to unfamiliar surroundings and may easily fall prey to predators. They may experience difficulty in integrating with remaining wild individuals through failure to communicate. Another criticism of the process includes the fear that genetic diversity may have so declined already that a species cannot be regenerated. Also, because the work concentrates on a few, highly attractive species and the cost is high, funds are diverted from more effective habitat conservation and it gives a false sense that extinction problems are being solved.

Questions

15 What is meant by *in vitro* and *in vivo*?
16 Outline two ways in which captive breeding has made an effective contribution to the survival of endangered animal species.

Questions

17 The survival of endangered species is supported by nature reserves and by captive breeding programmes at zoos and seed banks at botanical gardens. By means of a table, contrast these approaches, identifying particular strengths and possible drawbacks.

18 Why do the seeds in long-term storage in seed banks have to be germinated and new ones stored?

The maintenance of botanic gardens and of seed banks

A seed is the product of sexual reproduction in flowering plants. It contains an embryonic plant and a food store within a protective structure. Many seeds naturally enter a period of dormancy on release from the parent plant. The dormancy periods are variable, but in many plants, seeds are capable of remaining dormant but viable for many years. This latter feature of the seed is exploited in seed banks where conditions that will extend dormancy are created and maintained.

Storing seeds in seed banks is a relatively inexpensive and space-efficient method of conservation. Here, natural dormant seeds are kept viable for long periods, typically in conditions of low humidity and low temperature. The steps, following collection and preparation are:

- seed drying (to below 7 per cent water)
- packaging (in moisture-proof containers)
- storage (at a temperature of −18°C)
- periodic germination tests
- re-storage or replacement.

Samples of individual species of seeds typically take up very little space, so each sample (consisting of many seeds) can represent a large gene pool that is being preserved. At regular intervals samples are germinated and the plants grown in the facilities of a botanic garden so that fresh seeds are available to replenish the stores. Seeds are easily exchanged between botanic gardens internationally, too. The possibility of the reintroduction of endangered species of plants to their natural habitats is a possible outcome.

Seeds with very limited longevity are not stored. Typically such species must be maintained as growing plants in botanic gardens. Cocoa, coconut and rubber are examples.

Figure 18.27 Wellcome Trust Millennium Building, Wakehurst Place – home to the Millennium Seed Bank (an initiative of the Royal Botanic Gardens, Kew, in the UK)

New developments in conservation

Using assisted reproduction

The introduction of captive breeding programmes was discussed on page 445. Techniques developed in modern reproductive technology have been adapted for use in these programmes (Figure 18.28).

Representative endangered species of mammals, held in zoos, may be effectively conserved by the process of fertilisation of eggs outside the body (**in-vitro fertilisation – IVF**). The key step in IVF is the successful removal of sufficient eggs from the ovaries. To achieve this, normal menstrual activity is temporarily suspended with hormone-based drugs.

Then the ovaries are induced to produce a large number of eggs simultaneously, at a time controlled by the team of zoo vets. In this way the correct moment to collect the eggs can be known accurately.

Egg cells are then isolated from surrounding follicle cells and mixed with sperm. Viable sperms may be held in sperm banks in a frozen state, and therefore can be from male animals held in zoos elsewhere in the world. In this way, the gene pool of endangered species is significantly less restricted.

If fertilisation occurs, the fertilised egg cells are incubated so that embryos at the 8-cell stage may be placed in the uterus. If one (or more) imbed there, then a normal pregnancy may follow. Spare embryos can also be frozen, and they too can be shared with other zoos.

The young mammals created in this way may be exchanged too, and surrogate mothers may be used. The steps to IVF as illustrated are routine in principle, but in many endangered species the success rate has been limited, so far.

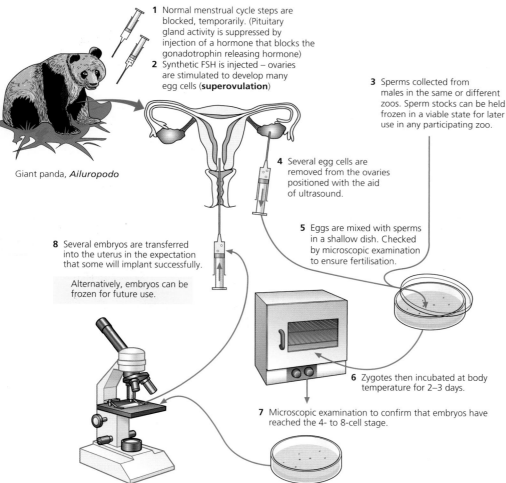

The process is illustrated here in the Giant panda, but has been applied to several other species – in some cases with marked success.

Giant panda, *Ailuropodo*

1 Normal menstrual cycle steps are blocked, temporarily. (Pituitary gland activity is suppressed by injection of a hormone that blocks the gonadotrophin releasing hormone)

2 Synthetic FSH is injected – ovaries are stimulated to develop many egg cells (**superovulation**)

3 Sperms collected from males in the same or different zoos. Sperm stocks can be held frozen in a viable state for later use in any participating zoo.

4 Several egg cells are removed from the ovaries positioned with the aid of ultrasound.

5 Eggs are mixed with sperms in a shallow dish. Checked by microscopic examination to ensure fertilisation.

6 Zygotes then incubated at body temperature for 2–3 days.

7 Microscopic examination to confirm that embryos have reached the 4- to 8-cell stage.

8 Several embryos are transferred into the uterus in the expectation that some will implant successfully.

Alternatively, embryos can be frozen for future use.

Figure 18.28 The process of *in-vitro* fertilisation, adapted for animal conservation

Alien and invasive species – culling and control

Alien species are introduced plants and animals that have been accidentally or deliberately transferred from habitats where they live, to new environments where the abiotic conditions are also suitable for them. If the alien takes over local ecosystems in an aggressive way, detrimental to the food chains of their new habitat, its presence leads to the decline of native species. The alien is then described as an **invasive species**.

The success of the **American Grey squirrel** (*Sciurus carolinensis*) in Britain at the expense of the red squirrel (Figure 18.29) is a good example. The grey squirrel was accidentally introduced here in the nineteenth century. The red squirrel is now restricted to just two locations in the UK, having once been widespread over much of the country. The grey squirrel has spread rapidly, chiefly because:

● It is able to consume a wider range of locally growing nuts
   (including the very common woodland oak tree's acorn that the grey can digest, but which the red squirrel cannot), together with the hazel nuts and pine cones, the limited diet of the red squirrel.
● The aggressive and nimble grey squirrel is a fair match for local predators and lacks some natural predators – alien species are frequently introduced with few if any of their natural predators.
● The grey squirrels carry the 'squirrelpox' virus which kills many red squirrels without causing any symptoms in the grey squirrel (and the niches of the two species overlap).

Figure 18.29 Red and grey squirrels

Grey squirrels – a control measure

In woodlands where the grey squirrel has become such an aggressive pest that it seriously damages young trees, foresters have deployed a chemical method of control. Bait, consisting of maize seeds impregnated with a high dose of oestrogenic chemicals, has been introduced into woodland during the squirrel's breeding season. The bait is protected in dispensers that only the squirrel has easy access to (Figure 18.30). The outcome is that the fertility of the squirrels is compromised and the local populations are not maintained – at least for a number of years. This is, in effect, an example of 'culling' by contraception.

Figure 18.30 Contraceptive bait deployed for the grey squirrel population

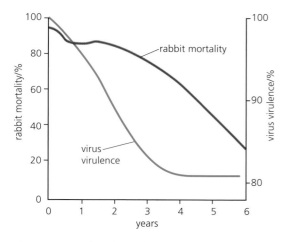

Figure 18.31 Changing rabbit mortality and virus virulence, following introduction of the myxoma virus as a biological control measure

The rabbit – a biological control measure

Another example of an alien species that has become 'invasive' is the **rabbit** (*Oryctolagus* sp.), which was deliberately introduced from Europe into Australia in 1859. This voracious herbivore spread rapidly. No natural predators present were an effective threat (**interspecific competition** failed), and rabbits quickly overran large parts of the continent. Grassland, available to herds of herbivores, principally cows and sheep, was seriously damaged. Agricultural production was plunged into a crisis. This continues, even now, with varying degrees of severity.

In desperation, and as a **biological control measure**, myxoma virus that had been discovered in South American rabbits (a different species from the European rabbit), was introduced in 1950. Myxoma virus causes a parasitic disease of rabbits, **myxamatosis**, but on its 'home' continent its effects were mild. The changing effects of this virus on rabbit mortality (1950–1956) are shown in Figure 18.31, as is the resulting change in virulence of the virus. We may assume that initially, a small number of the huge population of rabbits had immunity. As their vulnerable relatives were killed off the immune rabbits prospered from the diminished competition for grass. Rapidly, the bulk of the population were immune, and any that failed to develop immunity were quickly taken out.

In their original habitat an 'alien' species will have evolved in the company of **natural parasites** and **predators** – and come to exist in balance with other organisms of the community. In a new habitat the alien's natural enemies may be absent, so they grow at the expense of native species, crowding them out.

Japanese knotweed

A most unfortunate example of this is **Japanese knotweed** (*Fallopia japonica*), deliberately introduced into northern Europe in the early nineteenth century as an ornamental plant for garden ponds and lakes.

The advance of Japanese knotweed in the UK is an example of the spread of an invasive species devoid of the usual limiting factors that keep it under control in its native Japan (Figure 18.32). Here, knotweed plants grow into dense, submerged thickets that overshadow and crowd out native water plants, block waterways and public access to stream banks. In Japan, the plant grows in a balanced relationship with other water plants and causes no significant problems. Meanwhile, in Britain, the chemicals in the plant's leaves and roots discourage predation by local leaf browsers or root parasites. In Japan, predators have evolved that can cope with these defence mechanisms. Farmers, gardeners and local councils face huge and increasing bills to suppress this rampant invasive species.

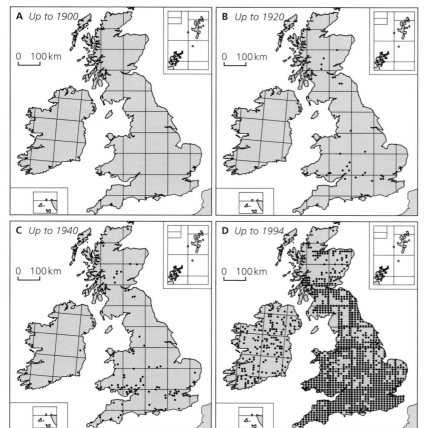

Figure 18.32 Distribution of knotweed in the UK, 1900–94

The roles of non-governmental organisations in conservation

World Wide Fund for Nature (WWF)

The WWF is an international organisation with over 5 million supporters worldwide, located in more than 100 countries. Conservation and environmental projects are supported, with the largest share of its funding coming from individuals. The aim is to '*build a future in which humans live in harmony with nature*'.

In the areas of both global and local conservation the focus is on maintaining biodiversity, especially in oceans and coast, forests, and fresh water ecosystems, by supporting particular initiatives.

You can read more about the work of the WWF at: **www.worldwildlife.org**.

The Convention on International Trade in Endangered Species of Wild Fauna and Flora (CITES)

The Convention on International Trade in Endangered Species (CITES), was set up in March 1973 to regulate worldwide commercial trade in wild animal and plant species. It is one of the largest and oldest conservation agreements in existence. The aim and goals of CITES are to ensure that international trade does not threaten the survival of any species. Since its inception, the number of nation states that have declared their commitment to the convention has reached more than 170. Those that participate in CITES are obliged to implement decisions in their domestic legislation programmes.

CITES classifies plants and animals according to three categories, based on how threatened they are:
1 Species that are in danger of extinction – the commercial trade of these plants and animals is prohibited outright.
2 Species that are not threatened with extinction but that might suffer a serious decline in number if trade is not restricted, so their trade is regulated by permit.
3 Species that are protected in at least one country that is a CITES member and that has petitioned others for help in controlling international trade in that species.

Categories 1 and 2 are effective in global conservation, whereas the impact of Category 3 is in local conservation. In the final analysis, CITES is only effective as long as the member countries genuinely abide by collective decisions.

You can read more about the work of CITES at: **www.cites.org**.

Restoration of degraded habitats

Habitats become degraded by industrial and related pollutants in many different forms. Soil and aquatic habitats are especially vulnerable. Ultimately, degraded habitats require restoration for the mutual benefit of humans and wildlife.

Bacteria and the bioremediation of the environment

Bioremediation is the process of exploiting microorganisms in the removal of pollutants from the environment. For example, we see this activity in the spontaneous removal of engine oils that drip from machines and engines, cars and other vehicles. These commonly seen deposits accumulate on nearby surfaces, but are normally removed (almost unnoticed by us), due to their degradation by naturally occurring (chemoheterotrophic) bacteria, provided the deposits are kept moist by occasional rain. A wide range of naturally occurring, ubiquitous prokaryotes and unicellular eukaryotes are able to hydrolyse and oxidise mineral oil to carbon dioxide.

When a major spill from an oil tanker occurs, leading to massive oil slicks that eventually reach sea shore habitats, it has been found that about 80 per cent of the non-volatile components of the oil may be oxidised and removed within a year of the pollution event. Whilst vast numbers of marine and littoral species of non-vertebrates are destroyed by the hydrocarbon pollutants in the initial disaster, removal of oil slicks may be speeded by spraying the oil with essential inorganic nutrients, chiefly phosphates and nitrates, which aid the saprotrophic bacteria and thus speed up bacteriological oxidation.

Figure 18.33 Investigating the degradation of mineral oil 'spills'

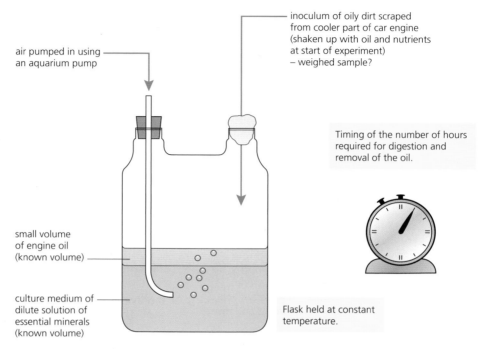

air pumped in using an aquarium pump

inoculum of oily dirt scraped from cooler part of car engine (shaken up with oil and nutrients at start of experiment) – weighed sample?

Timing of the number of hours required for digestion and removal of the oil.

small volume of engine oil (known volume)

culture medium of dilute solution of essential minerals (known volume)

Flask held at constant temperature.

Bioremediation of soils

Pesticides are products of the agrochemical industry that are widely used to control harmful organisms that are a danger to crops or herds. Their use extends to the horticultural industries that are not 'organically' orientated, of course, and the use of pesticides has been enthusiastically adopted by many people who maintain gardens in developed countries.

Modern pesticides are substances designed to kill specific types of pest, including plant weeds (by herbicides), insects (by insecticides), fungi (by fungicides), and slugs and snails (molluscicides). Pesticides have enormously improved productivity in agriculture, but their use has generated problems in the environment. Herbicides are the most widely used group of pesticides, but insecticides are also widely used and they may cause the greater problems for humans.

Pesticides are a wide range of different compounds. Many of them are organic molecules suitable as carbon sources and electron donors for various soil microorganisms, but others are not. Those that can be biodegraded are eventually removed from soils (Table 18.6), but their breakdown may not be entirely due to the activity of microorganisms. Some substances leach away into underground water, and others may spontaneously degrade with time.

Table 18.6 Persistence of a selection of pesticides

Substance	Time for disappearance from soil
Herbicides	
2,4-D	4 weeks
2,4-T	20 weeks
Insecticide (chlorinated compound)	
DDT	4 years
Aldrin	3 years
Insecticide (organophosphate compound)	
Malathion	1 week
Parathion	1 week

Industrial solvents, now widely used in developed countries at least, are organic compounds such as dichloroethylene, trichloroethylene, chloroform and some brominated and fluorinated related compounds, to name but a few. These toxic substances (some are also suspected of being carcinogens) are frequently detected as persistent contaminants of ground waters. This is a well-documented problem in parts of the USA, for example.

A variety of different bacteria are able to break down these molecules, including the removal of the chlorine component. For example, the bacterium *Dehalobacterium* uses the compound dichloromethane in its metabolism. Substrate-level phosphorylation (ADP + P_i → ATP) occurs in the process.

In areas where the underground waters are seriously polluted, the activities of natural populations of bioremedial microorganisms may be enhanced *in situ* by the drilling of wells and vents so that micronutrients and sources of oxygen (as compressed air or peroxide solutions) can be added for the benefit of the microorganisms. The natural clean-up reactions are accelerated in this way, although the process may take months and years to complete. No doubt the dispersal of the pollutants and the depth of the geological strata they occupy is a factor.

Summary

- The process of classification involves naming organisms using the **binomial system** so that each has a generic and a specific name. Genera are arranged in a hierarchical classification, of which the **kingdom** is the largest and most inclusive category. There are five kingdoms.

- The **Prokaryotae** (prokaryotes) include the bacteria, cyanobacteria (photosynthetic forms) and archaebacteria (extremophiles). They are unicellular or filamentous prokaryotic organisms consisting of very small cells, 1–5 μm. Their nutrition is heterotrophic or autotrophic.

- The **Protoctista** (protoctists) include the algae (including the multicellular seaweeds), the protozoa and the slime moulds. They are single-celled eukaryotic organisms or, if multicellular, not differentiated into tissues. Their nutrition is heterotrophic or autotrophic.

- The **Fungi** include the moulds, yeasts, mushrooms and bracket fungi. They are multicellular eukaryotic organisms (except the unicellular yeasts), not differentiated into tissues and non-motile. Their nutrition is heterotrophic.

- The **Plantae** (the green plants) includes the mosses, ferns, conifers and flowering plants (broad-leaved and the grasses). They are multicellular eukaryotic organisms, differentiated into tissues. The plants are non-motile organisms, but the male gametes of mosses and ferns are motile. Their nutrition is autotrophic.

- The **Animalia** (Animals) includes the non-vertebrates, such as worms and arthropods (which includes the insects), and the vertebrates, which include the fish, amphibians, reptiles, birds and mammals. They are multicellular eukaryotic organisms, differentiated into tissues. Many are motile organisms. Their nutrition is heterotrophic.

- **Biodiversity** refers to the vast number of living things that exist. About 1.7 million different species are known, but a much larger number of organisms may exist, undiscovered. The **need to maintain diversity** is based on ecological arguments, but includes diversity as the source of genes for improvement of agricultural crops, as a source of new drugs, to resist climate change, to support ecotourism and for ethical and aesthetic reasons.

- A **species** is a group of individuals of common ancestry that closely resemble each other and which are normally capable of interbreeding to produce fertile offspring.

- The exponential expansion of the human population is a relatively recent phenomenon, but the **current size of the human population** results in huge demands on the Earth's living space and resources. Many **environments are being degraded** and biodiversity is being reduced. Today, species are in danger of **extinction** due to the loss of their natural habitats, changes in the environment or due to over-exploitation. These causes are all attributable to human activity.

- **Conservation** is the management of resources and habitats so as maintain the existing range of habitats and to attempt to prevent future reductions in biodiversity. This includes **captive breeding programmes** and the establishment and maintenance of **reserves and national parks**, where there is supervised protection and an opportunity for people to learn about biodiversity and its conservation.

Examination style questions

1 Fig. 1.1 shows part of a tropical rainforest. Tropical rainforests have a high biodiversity.

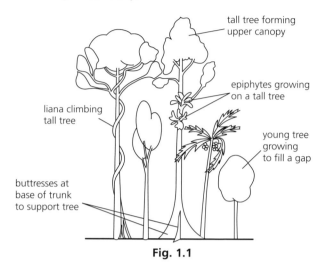

Fig. 1.1

a) Explain what is meant by *biodiversity*. [3]

b) Suggest why tropical rainforests have a high biodiversity of animal species. [2]

c) Discuss why it is important to maintain biodiversity. [4]

[Total: 9]

(Cambridge International AS and A Level Biology 9700, Paper 04 Q2 November 2007)

2 The African elephant once roamed most of the continent of Africa. The table below gives data on more recent estimates of the population size of this giant herbivore.

Date of estimate	Population size
1930s	5–10 million
By 1979	1.3 million
By 1989	600 000 (< 1% of original numbers)

a) Outline the diverse range of factors that are most likely to have directly contributed to this decline. [12]

b) If herds of the African elephant continue to decline sharply, explain what impact this is likely to have on heterozygosity in these animals and the impact this change is likely to have. [8]

[Total 20]

3 a) The squirrel monkey, *Saimiri sciureus*, of Costa Rica has become an endangered species.

The picture below shows a squirrel monkey.

Explain what is meant by the term *endangered species*. [2]

b) Discuss possible ways in which the squirrel monkey could be protected. [4]

[Total: 6]

(Cambridge International AS and A Level Biology 9700, Paper 41 Q1 November 2009)

19 Gene technology

The discovery of the structure of DNA by Watson and Crick in the early 1950s and discoveries since have led to many applications of gene technology in areas of medicine, agriculture and forensic science. This topic relies heavily on prior knowledge of DNA structure and protein synthesis studied in the topic on Nucleic acids and protein synthesis. Where possible, students should carry out practical work using electrophoresis, either with DNA or specially prepared dyes used to represent DNA or proteins.

19.1 Principles of genetic technology

Genetic engineering involves the manipulation of naturally occurring processes and enzymes.

Genome sequencing gives information about the location of genes and provides evidence for the evolutionary links between organisms.

By the end of this section you should be able to:

a) define the term recombinant DNA
b) explain that genetic engineering involves the extraction of genes from one organism, or the synthesis of genes, in order to place them in another organism (of the same or another species) such that the receiving organism expresses the gene product
c) describe the principles of the polymerase chain reaction (PCR) to clone and amplify DNA (the role of *Taq* polymerase should be emphasised)
d) describe and explain how gel electrophoresis is used to analyse proteins and nucleic acids, and to distinguish between the alleles of a gene
e) describe the properties of plasmids that allow them to be used in gene cloning
f) explain why promoters and other control sequences may have to be transferred as well as the desired gene
g) explain the use of genes for fluorescent or easily stained substances as markers in gene technology
h) explain the roles of restriction endonucleases, reverse transcriptase and ligases in genetic engineering
i) explain, in outline, how microarrays are used in the analysis of genomes and in detecting mRNA in studies of gene expression

Introducing genetic engineering

Genetic engineering is the technique of changing the genetic constitution of an organism, brought about by means other than conventional breeding and which usually would not occur in nature. It involves the transfer of a gene or genes and other sequences from one species to another – often an unrelated species, so that the host organism is able to express a new gene product. The outcome is new varieties of organisms, often (but not exclusively) of microorganisms. These are described as genetically modified (GM) or **transgenic** organisms containing **recombinant DNA**. This type of DNA is formed by laboratory techniques, bringing together genetic material from different sources. Transgenic organisms produce, in their cells, proteins that were not previously part of their species **proteome**. The proteome is the complete set of proteins that a cell or organism can make. These additional proteins, many of them enzymes, have been engineered with the intention of bringing about significant change in the host organism, usually with a specific purpose in mind.

The techniques of the genetic engineer offer numerous practical benefits in biotechnology, medicine and agriculture, and also to science and the study of how genes operate in the control of cells of organisms. At the same time, outcomes of genetic engineering may generate environmental and **ethical issues** of concern to society. We will return to this issue later.

Other applications

Today, in addition to the creation of GM organisms, gene technologies are applied to:

- **DNA sequencing** – the creation of genomic libraries of the precise sequence of nucleotides in samples of DNA of individual organisms. The nucleotide sequence in the whole human genome was the product of the Human Genome Project. Today, other genomes have been completely sequenced, including those of some viruses, bacteria, fungi and many other eukaryotes.
- **Genetic fingerprinting** – in which DNA is analysed in order to identify the individual from which the DNA was taken to establish identity and the genetic relatedness of individuals. It is now commonly used in forensic science (for example to identify someone from a sample of blood or other body fluid). It is also used to determine whether individuals of endangered species in captivity have been bred or captured in the wild.

The gene technologist's tool kit

It is the discovery and isolation of four naturally-occurring enzymes that make possible the manipulation of individual genes. We have already met these enzymes in other contexts, but it is significant that, in this context, some of the enzymes used are now obtained from bacteria we call extremophiles (page 431). In these organisms, cells (and therefore their enzymes) are adapted to function at optimum rates under some extreme conditions, including, for example, very high or low temperatures. We shall discuss how these enzymes work shortly. They are introduced in Table 19.1.

Table 19.1 The genetic technologist's toolkit of enzymes

Question

1 You have studied how the virus called HIV infects a human cell (page 205). What is the role of reverse transcriptase in this process?

These enzymes are extracted mainly from microorganisms or viruses and are used to manipulate nucleic acids in very precise ways		
Enzyme	**Natural source**	**Application in genetic engineering**
restriction enzyme (restriction endonuclease)	cytoplasm of bacteria (combats viral infection by breaking up viral DNA)	breaks DNA molecules into shorter lengths, at specific nucleotide sequences
DNA ligase	with nucleic acid in the nucleus of *all* organisms	joins together DNA molecules during replication of DNA
DNA polymerase	with nucleic acid in the nucleus of *all* organisms	synthesises nucleic acid strands, guided by a template strand of nucleic acid
reverse transcriptase	in retroviruses only	synthesises a DNA strand (cDNA) complementary to an existing RNA strand

Gene technology–an early break through

One of the earliest successful applications of these techniques was when the human genes for **insulin** production were transferred to a strain of the bacterium *Escherichia coli*. Insulin consists of two short polypeptides linked together by sulfide bonds. Once the hormone is assembled from its component polypeptides, it enables body cells to regulate blood sugar levels (page 293). Regular supplies of insulin are required to treat insulin-dependent diabetes. Cultures of *E. coli* have been 'engineered' to manufacture and secrete human insulin when cultured in a bulk fermenter with appropriate nutrients. The insulin is extracted and made available for clinical use.

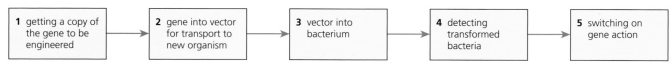

| 1 getting a copy of the gene to be engineered | → | 2 gene into vector for transport to new organism | → | 3 vector into bacterium | → | 4 detecting transformed bacteria | → | 5 switching on gene action |

Figure 19.1 The steps in the genetic engineering of *E. coli* for insulin production

The steps in the genetic engineering of *E. coli* for insulin production

Step 1: Obtaining a copy of the human insulin gene by isolating mRNA

One way to obtain a copy of the gene for insulin involves starting with messenger RNA, rather than searching for the gene itself among the chromosomes. The human pancreas contains patches of cells (the islets of Langerhans, Figure 14.7, page 293) where insulin is produced. Here the relevant gene in the nuclei of the cells is transcribed to produce messenger RNA. This passes out of the nucleus to the ribosomes in the cytoplasm. Here the base sequence of the RNA is translated into the linear sequence of amino acids of the insulin protein.

There is a particular advantage in making a copy of a gene from the messenger RNA it codes for. This is explained later in this topic. First, however, the process. This involves:
- messenger RNA for insulin being isolated from a sample of tissue from a human pancreas
- the use of the enzyme **reverse transcriptase** (obtained from a retrovirus other than HIV), alongside the isolated messenger RNA to form a single strand of DNA
- the conversion of this DNA into double-stranded DNA using **DNA polymerase** (Figure 19.2). In this way the gene is manufactured. This form of an isolated gene is known as **complementary DNA (cDNA)**.

Question

2 Outline
  a why we describe DNA as double stranded
  b where DNA polymerase occurs and its role in the cell.

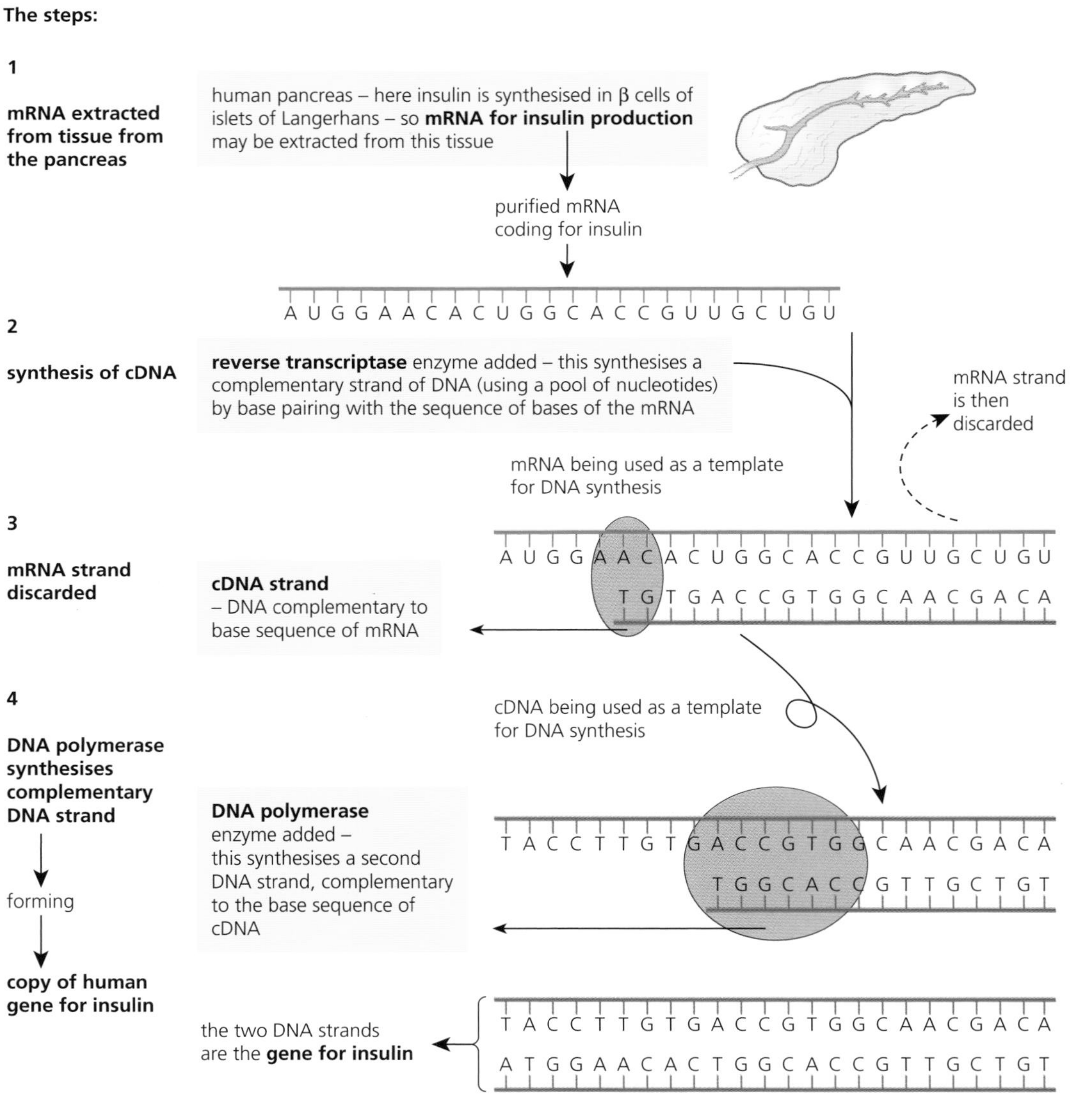

Figure 19.2 Using reverse transcriptase to build the gene for human insulin

The steps:

1

mRNA extracted from tissue from the pancreas

human pancreas – here insulin is synthesised in β cells of islets of Langerhans – so **mRNA for insulin production** may be extracted from this tissue

purified mRNA coding for insulin

A U G G A A C A C U G G C A C C G U U G C U G U

2

synthesis of cDNA

reverse transcriptase enzyme added – this synthesises a complementary strand of DNA (using a pool of nucleotides) by base pairing with the sequence of bases of the mRNA

mRNA strand is then discarded

mRNA being used as a template for DNA synthesis

3

mRNA strand discarded

cDNA strand – DNA complementary to base sequence of mRNA

A U G G A A C A C U G G C A C C G U U G C U G U
T G T G A C C G T G G C A A C G A C A

4

DNA polymerase synthesises complementary DNA strand

forming

copy of human gene for insulin

cDNA being used as a template for DNA synthesis

DNA polymerase enzyme added – this synthesises a second DNA strand, complementary to the base sequence of cDNA

T A C C T T G T G A C C G T G G C A A C G A C A
T G G C A C C G T T G C T G T

the two DNA strands are the **gene for insulin**

T A C C T T G T G A C C G T G G C A A C G A C A
A T G G A A C A C T G G C A C C G T T G C T G T

Once the gene is isolated, the double-stranded DNA undergoes amplification; many copies are made. There are various situations in which a genetic engineer is able to produce or recover only a very small amount of DNA (such as at a crime scene). It is now possible to submit such minute DNA samples to a process known as the **polymerase chain reaction (PCR)**. In the polymerase chain reaction, DNA is replicated in an entirely automated process, *in vitro,* to produce a large number of copies. A single molecule is sufficient as the starting material, should this be all that is available. The products are always exact copies. The heat-resistant polymerase enzyme used in this process is known as ***Taq* polymerase** after the thermophilic bacterium, *Thermus aquaticus,* found in hot springs and hydrothermal vents, from which it was originally isolated. This protein is able to remain an effective catalyst at a temperature of 97.5 °C for the length of time required in the PCR.

The polymerase chain reaction

The steps to the polymerase chain reaction are of special interest because they show us how important to genetic engineering was the discovery of the extremophiles with their uniquely adapted enzymes.

> The polymerase chain reaction involves a series of steps, each taking a matter of minutes. The process involves a heating and cooling cycle and is automated.
> Each time it is repeated in the presence of excess nucleotides, the number of copies of the original DNA strand is doubled.

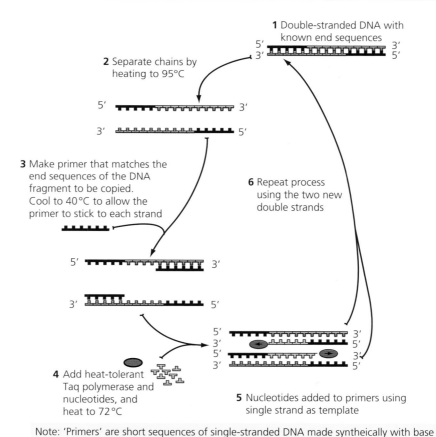

1 Double-stranded DNA with known end sequences

2 Separate chains by heating to 95°C

3 Make primer that matches the end sequences of the DNA fragment to be copied. Cool to 40°C to allow the primer to stick to each strand

6 Repeat process using the two new double strands

4 Add heat-tolerant Taq polymerase and nucleotides, and heat to 72°C

5 Nucleotides added to primers using single strand as template

Note: 'Primers' are short sequences of single-stranded DNA made syntheically with base sequences complimentary to one end (the 3' end) of DNA.
Remember: DNA polymerase synthesises a DNA strand in the 3' to 5' direction.

Figure 19.3 The polymerase chain reaction

Question

3 a What is meant by
 sticky ends?
 b How does a sticky
 end attach to a
 complementary sticky
 end?

Step 2: Inserting the DNA into a plasmid vector

In genetic engineering it is the **plasmids** of bacteria that are commonly used as the vector for the transference of amplified genes. Many bacterial cells have plasmids in addition to their chromosome. These are small, circular, double-stranded DNA molecules that are passed on to daughter cells when the bacterium divides. They were seen in Figure 1.23 (page 23) and are shown in Figure 19.5. Plasmids are isolated from the cytoplasm of a sample of the strain of bacteria being used for the amplification process.

The DNA of the plasmid is cut open using a **restriction enzyme** (restriction endonuclease). The restriction enzyme chosen cuts the DNA at a specific sequence of bases, known as the restriction site. The restriction enzyme selected leaves exposed specific DNA sequences, referred to as sticky ends. These are short lengths of unpaired bases and are formed at each cut end (Figure 19.4).

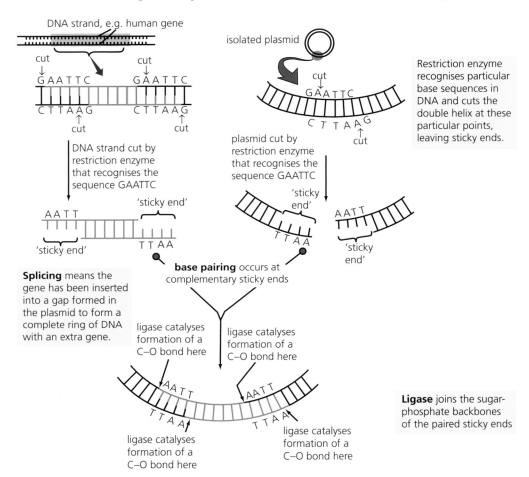

Figure 19.4 Gene splicing – the role of the restriction enzyme and ligase

Sticky ends with the same sequence as those on the plasmid are also created on the isolated insulin gene. This is done by adding short lengths of DNA that are then 'trimmed' with the same restriction enzyme. In this way complementary 'sticky ends' now exist at the free ends of the cut plasmids and the cDNA of the insulin gene, making it possible for them to be 'spliced' together into one continuous ring of DNA. The enzyme **ligase** catalyses the formation of a new C–O bond between ribose and phosphate of the DNA backbones being joined together. Energy from ATP is needed to bring the reaction about. The steps are summarised in Figure 19.5.

A note on restriction enzymes

Restriction enzymes occur naturally in bacteria. They are believed to have evolved as a defence mechanism against invading viruses because inside the bacterial cell they selectively cut up foreign DNA. Many different restriction enzymes have been identified as useful by genetic engineers. They are named after the bacterium from which they are isolated (genus, species and strain).

Table 19.2 Some commonly employed restriction enzymes

Name	Source organism	Recognition sequence
*Eco*RI	*Escherichia coli*	GAATTC
*Bam*HI	*Bacillus amyloliquefaciens*	GGATCC
*Hind*III	*Haemophilus influenza*	AAGCTT

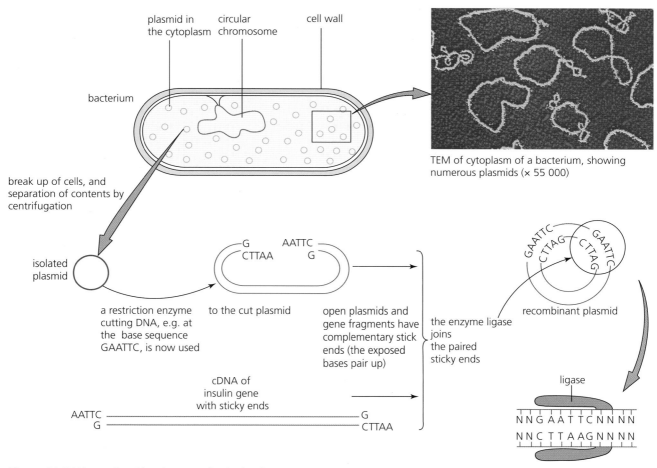

TEM of cytoplasm of a bacterium, showing numerous plasmids (× 55 000)

Figure 19.5 Using a plasmid as the vector for the insulin gene

Step 3: Inserting the plasmid vector into the host bacterium

In the next step, recombinant plasmids have to be returned to bacterial cells. This is a challenge, since the cell wall is a barrier to entry. The ways this may be done are shown in Figure 19.6. Once this has been brought about, the bacteria are described as **transformed**.

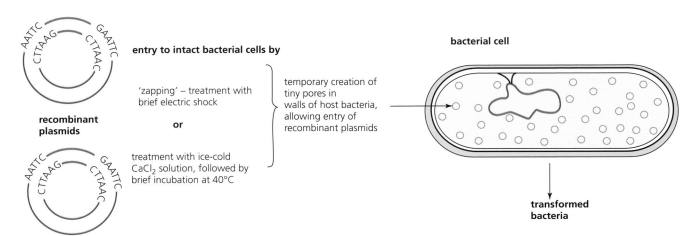

Figure 19.6 The return of recombinant plasmids to the bacterium

Step 4: Identifying transformed bacteria prior to cloning

Unfortunately, only a limited number of the cells of a culture of bacteria, treated as shown in Figure 19.6, will have successfully taken up recombinant plasmids and thus be able to synthesise insulin. The proportion of transformed bacteria may be as low as 1 per cent. It is only these bacteria that should be cloned and then cultured in a fermenter if a significant amount of product (insulin) is to be produced. One way they have been selected is by the use of **plasmids with antibiotic resistance genes (R-plasmids)**. The process is summarised in Figure 19.7.

Note that whilst the antibiotic resistance genes of R-plasmids were the first used markers, because of specific long-term safety issues that their use raises, other markers have been developed. We return to this issue later.

Step 5: Switching on gene action – why promoters are needed

The next stage is for the transformed bacteria to be cloned and then cultured in a fermenter. At this point the transferred gene needs to be 'switched on' so that the required protein – in this case, insulin – is produced in significant quantities. This product may then be extracted from the medium and purified.

In the living cell, some genes are expressed throughout the life of the cells, normally at quite low concentrations. We can assume in these cases that the proteins coded for are required constantly.

Other genes have to be activated, or switched on, first. One very efficient switch mechanism is the lac promoter, a promoter found in prokaryotic cells activated by the presence of lactose in the growth medium. A human gene such as the insulin gene, which is from a eukaryotic cell, has to be transferred into an **expression vector**, such as a plasmid so that a lac promoter occurs alongside it. The result is that, when lactose is present in the medium, recombinant bacteria containing this expression vector will be activated to produce large quantities of the desired polypeptides, insulin in this case, which can then be extracted.

Quaternary insulin production

We have already noted that although insulin is a relatively small protein, it has quite a complex quaternary structure (page 49). Insulin actually consists of two polypeptide chains, A and B, joined together by covalent disulfide bonds ($-CH_2-S-S-CH_2-$). Two separate genes code for the two polypeptides that make up insulin.

As a consequence, separate plasmids have to be engineered for the two chains of insulin, A and B. Genetically engineered *E coli* containing both types of plasmid are then selected.

A 'start' codon (ATG, also the code for methionine – see Figure 6.10, page 119) has to be inserted appropriately to the cDNA for both chain A and B. To stop the transcription process at the correct place, the 'stop' triplet codes (TAA followed by TAG) are added at the end of the cDNA, too. In this adjusted condition, both genes were added into plasmids with lac promoters.

Finally, when the two polypeptide chains are separated from the bacteria they are treated first with cyanogen bromide to cut them free from the longer polypeptide chain at the methionine amino acid. Then they are further treated to promote disulfide bond formation and development of a quaternary structure.

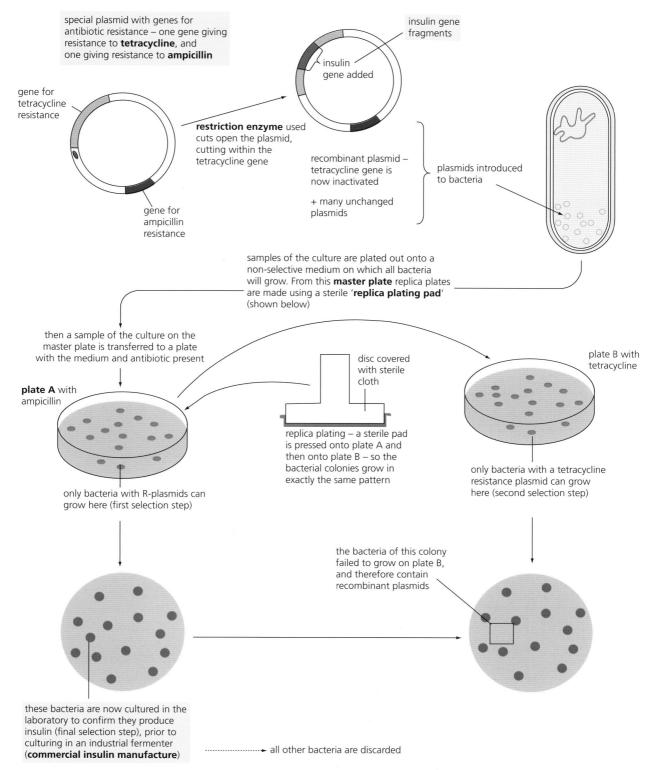

Figure 19.7 The use of R-plasmids for selecting genetically modified bacteria

461

A new development in insulin production

All this was part of the *original* process of producing human insulin from *E. coli*. The complications of producing human insulin from transformed bacteria have now been overcome by genetically engineering yeast cells for insulin production, rather than *E. coli*. Remember, yeast is a eukaryotic cell with a range of organelles, but it also contains plasmids. In eukaryotic cells, proteins for secretion from the cell are manufactured by ribosomes of the rough endoplasmic reticulum and transferred to the Golgi apparatus (Figure 1.16, page 18). Here, insulin is converted into its normal quaternary structure prior to discharge from the transformed yeast cells. However, since yeast is a eukaryote, the *lac* operon must be replaced by genes for promoters and other control sequences, in addition to the insulin genes in the 'engineered' plasmids. *Remind yourself of the control of gene expression in eukaryotes, on page 374.*

Alternative sources of insulin for clinic use

Before genetically engineered human insulin was available for clinic use, insulin was extracted and purified from the pancreases of pigs or cattle obtained from abattoirs after their slaughter. Insulin from both animal sources was used to treat insulin-dependent diabetes, although neither was exactly identical to human insulin.

Today, it is argued that genetically engineered human insulin has the following advantages:

- it is chemically identical to the body's own insulin and triggers no immune response as 'foreign' proteins tend to do when introduced into a patient's body
- it is acceptable to patients who previously declined 'pig' or 'cow' insulin for religious regions or because, as vegetarians, the use of animal products was unacceptable
- an immediate response is generated by the genetically engineered insulin because it exactly 'locks on to' the insulin receptors in the relevant cell surface membranes; the response brought about by insulin from other species is slower.

However, animal sources remain available and may be used in preference, if required.

Question

4 Distinguish between
  a genotype and genome
  b restriction endonuclease and ligase
  c a bacterial chromosome and a plasmid.

Genetic modification of eukaryotes

Manipulating genes in eukaryotes is a more difficult process than it is in prokaryotes. The reasons for this include:

- plasmids, the most useful vehicle for moving genes, do not occur in eukaryotes (except in yeasts) and, if introduced, may not survive and be replicated there
- eukaryotes are diploid organisms so each gene occurs twice in the nucleus of a cell, once on each of the homologous chromosomes, and may occur in different forms (alleles). Prokaryotes have a single, circular 'chromosome', so only one gene has to be engineered into their chromosome
- the transcription of eukaryotic DNA to messenger RNA is more complex than in prokaryotes; it involves the removal of short lengths of 'non-informative' DNA sequences, called 'introns'
- the mechanism for triggering gene expression in eukaryotes is more complex than in prokaryotes. Genes for the promoter and other control sequences have to be transferred, as well as the desired gene.

Despite these difficulties, several varieties of transgenic plants and animals have been produced.

Question

5 Why are the genes of prokaryotes generally easier to modify than those of eukaryotes?

Alternative markers for genetic engineering

A marker is a gene that is deliberately transferred along with the required gene during the process of genetic engineering. It is easily recognised and used to identify those cells to which the gene has been successfully transferred. In the production of insulin in *E. coli* the original markers were antibiotic resistance genes (Figure 19.7).

Now alternative markers are preferred. This is because of the potential danger of the antibiotic resistance genes being accidentally transferred to other bacteria, including eventually, pathogenic strains of *E. coli* or even pathogens causing other diseases. Whilst this has not been reported, the possibility must exist. The outcome would be an 'engineered' disease-causing microorganism that was resistant to one or more antibiotic.

Genes for fluorescent (or easily stained) substances are now used as markers (Figure 19.8) instead of antibiotic resistant genes, alongside the cDNA of the desired gene. These genes are then linked to a special promoter. The marker gene is expressed only when the desired gene has been successfully inserted into the genome of the host. Organisms that have been transformed in this way will glow under UV light (or can be identified by staining).

The crystal jellyfish (*Aequorea victoria*) occurs in the plankton off the west coast of North America. It shows bright green fluorescence when exposed to light in the blue to ultraviolet range.

Fluorescence is due to the presence of a green fluorescent protein (GFP). This protein and the gene that codes for it have been extracted.

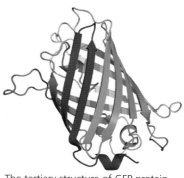

The tertiary structure of GFP protein

The GFP gene is now used as a marker gene in genetic engineering. For example, it may be inserted into plasmids alongside a cloned gene and its promoter. When these plasmids are returned to bacteria and the promoter is activated, the transformed bacteria will be detected by exposure to blue light.

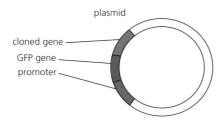

The tertiary structure of GFP protein

The GFP gene has been introduced and maintained in the genome, and then expressed in species of bacteria, yeast, fish, plants, insects, and mammals. For example, these GM mice glow green under blue light because the GFP gene has been introduced into their DNA.

Figure 19.8 The use of fluorescent markers

Other methods of isolating genes for gene technology

Converting messenger RNA that codes for the insulin genes into cDNA, as described above, has the advantage of yielding the gene *without* the short lengths of 'nonsense' DNA (known as **introns**). After transcription of the coding strand in the nucleus this editing is a natural and essential stage. It is not easily for a genetic engineer to carry this out *in vitro*.

There are alternative methods. For example, it is possible to synthesise a copy of the gene in a laboratory if the linear sequence of amino acids of the polypeptide or protein is known. The data in the DNA genetic dictionary (Figure 6.10, page 119) gives the sequence of nucleotides from which to construct a copy of the gene.

As a result of the **Human Genome Project** (HGP), the location of each human gene and the base sequence within its DNA structure are known. Now, where a single-stranded DNA probe, complementary to a particular gene can be produced and made radioactive or fluorescent in ultraviolet light, original genes can be located and isolated. To do so, the DNA has first to be cut into fragments. Fragments are then denatured into single strands by heating, separated by means of electrophoresis (Figure 19.9), and then treated with the probe. The fragments of DNA containing the gene are located.

Both of these methods are currently applied to isolate genes required for other gene technology projects.

Electrophoresis

Electrophoresis is a process used to separate molecules such as proteins and fragments of nucleic acids. It is widely applied in many studies of DNA. For example, it is central to the investigation of the sequence of bases in particular lengths of DNA, known as **DNA sequencing**. It is also used in the identification of individual organisms and species known as **genetic fingerprinting**. We will review these applications shortly.

The principle and practice of electrophoresis

In **electrophoresis**, proteins or nucleic acid fragments (either DNA or RNA) are separated on the basis of their net charge and mass. Separation is achieved by differential migration of these molecules through a supporting medium – typically either agarose gel (a very pure form of agar) or polyacrylamide gel (PAG). In these substances the tiny pores present act as a **molecular sieve**. Through these gels, small particles can move quite quickly, whereas larger molecules move much more slowly.

Molecules separated by electrophoresis also carry an **electrical charge**. In the case of DNA, it is the phosphate groups in DNA fragments that give them a net negative charge. Consequently when these molecules are placed in an electric field they migrate towards the positive pole (anode). The distance moved in a given time will depend on the mass of the molecule or fragment – the smaller fragments moving further in a given time than the larger fragments. So, in electrophoresis, separation occurs by the **size** and by the **charge carried.** This is the double principle of electrophoretic separations.

The DNA fragments are produced by the action of a restriction enzyme (Figure 19.4). We have seen that each type of restriction enzyme cuts DNA at a particular base sequence, as and when these occur along the length of the DNA molecules. Consequently, the fragments are of different lengths. A series of wells are cut close to one end of the gel and the gel is submersed in a salt solution that conducts electricity. Then a small quantity of a mixture to be separated is placed in a well. Several different mixtures can be separated in a single gel at the same time (Figure 19.9). An animation of electrophoresis is available at **www.dnalc.org/resources/animations/gelelectrophoresis.html**.

After separation the fragments are not immediately visible because they are tiny and transparent. Their presence is identified either by **staining** or by the use of a **gene probe**.

Gene probes are single-stranded DNA with a base sequence that is complementary to that of a particular fragment or gene whose position or presence is sought. The probe must be made radioactive so that when the treated gel is exposed to X-ray film the presence of that particular probe and the fragment it attached to will be disclosed. Alternatively, the probe must have a fluorescent stain attached. It will then fluoresce distinctively in ultraviolet light, thereby disclosing the presence of the particular fragment or gene sought.

The alternative approach, demonstrated in Figure 19.9 involves the use of a stain to locate the position of all DNA fragments.

Table 19.3 Stains used in electrophoresis

Ethidium bromide	DNA fragments fluoresce in short wave UV radiation
Methylene blue	Stains gel and DNA but colour fades quickly
Nile blue A	DNA visible in light and gel not stained

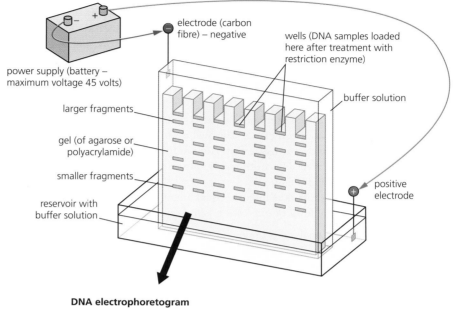

electrophoresis in progress

electrode (carbon fibre) – negative

power supply (battery – maximum voltage 45 volts)

wells (DNA samples loaded here after treatment with restriction enzyme)

buffer solution

larger fragments

gel (of agarose or polyacrylamide)

smaller fragments

reservoir with buffer solution

positive electrode

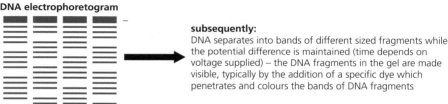

DNA electrophoretogram

–

+

subsequently:
DNA separates into bands of different sized fragments while the potential difference is maintained (time depends on voltage supplied) – the DNA fragments in the gel are made visible, typically by the addition of a specific dye which penetrates and colours the bands of DNA fragments

Figure 19.9 Electrophoresis – the separation of DNA fragments

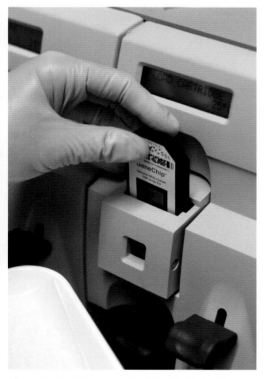

DNA microarrays

DNA microarrays are small, solid supports, typically made of glass, silicon or nylon, each about the size of a microscope slide. Onto this surface, DNA probes that correspond to thousands of different genes are attached in a regular pattern of rows and columns. Each dot on the array contains a DNA sequence that is unique to a given gene. The microarray is then used to measure changes in expression levels of individual genes from cells under investigation.

The DNA that is to be tested originates as mRNA being expressed by a particular cell. By the action of reverse transcriptase this mRNA is converted to cDNA. As the cDNA is being synthesised it is simultaneously linked to fluorescent dye. Then the DNA probes of the microarray are exposed to this cDNA. Any complementary sequences present will bind to the fixed probes. After this, the microarray is rinsed free of any cDNA that has not hybridised and become bound. Finally, the microarray is exposed to laser light that causes fluorescence where there has been hybridisation between cDNA and a DNA probe, and the results recorded. The brighter the light emitted the higher the level of expression of a particular gene in the cell from which the original mRNA was obtained.

A full human genome microarray would have as many as 30 000 spots. Others are made with clusters of genes that relate to particular conditions, such as cancers, for example.

Figure 19.10 Microarray analysis machine with microarray being inserted

Analysis of a simple microarray

A DNA microarray might be used to investigate the level of gene expression in a cancerous cell compared to that in a healthy (control) cell, for example. The steps to such an investigation are:

1 mRNA is extracted from the cancerous cells and the control cells.
2 cDNA is created from both samples of mRNA, that from the cancerous cell is tagged with a red fluorescent dye, whilst that from the control cell is tagged with green fluorescent dye.
3 The two samples are then mixed together and hybridised to the microarray.
4 At each probe location the red- and green-labelled cDNA compete to bind to the gene-specific probe.
5 When subsequently excited by laser light, each dot will fluoresce red if it has bound more red than green cDNA. This will occur if that particular gene is expressed more strongly in the cancerous tissue compared to the control tissue.
6 Conversely, a dot will fluoresce green if that particular gene is expressed more strongly in the control tissue.
7 Yellow dots indicate equal amounts of red and green cDNA have bound to that spot – meaning the level of expression is the same in both tissues.

The analysis is undertaken by a fluorescence detector (microscope and camera) to produce a digital image of the microarray. From this data a computer quantifies the red to green fluorescence ratio of each spot. Then the degree of difference in expression of each gene between the cancerous and the control tissues is calculated.

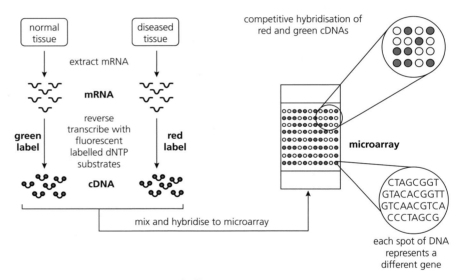

Figure 19.11 Microarray analysis of differential gene expression

19.2 Genetic technology applied to medicine

Bioinformatics combines knowledge from gene mapping, protein analysis and modern biotechnology in the service of science and society.

By the end of this section you should be able to:

a) define the term bioinformatics

b) outline the role of bioinformatics following the sequencing of genomes, such as those of humans and parasites, e.g. *Plasmodium*

c) explain the advantages of producing human proteins by recombinant DNA techniques

d) outline the advantages of screening for genetic conditions

e) outline how genetic diseases can be treated with gene therapy and discuss the challenges in choosing appropriate vectors, such as viruses, liposomes and naked DNA

f) discuss the social and ethical considerations of using gene testing and gene therapy in medicine

g) outline the use of PCR and DNA testing in forensic medicine and criminal investigations

Bioinformatics

Bioinformatics is a new discipline that combines three major components:

- **genomics** – the mapping of genomes
- **proteomics** – the analysis of the entire sets of proteins expressed by genomes
- **biotechnology** – the industrial and commercial application of biological science, particularly of microbiology, enzymology and genetics.

Developments in bioinformatics have created powerful techniques that allow scientists to find similarities in nucleotide sequences and discover the function of proteins among organisms of similar or different species. Nowadays, data is collected around the globe and from different research groups. The purpose of the resulting databases is to facilitate the sharing of this information, and to allow the widest possible exploitation of a huge and ever-growing body of knowledge and skills.

A flavour of the scale of developments can be gleamed from the following link to the 180 complete genomes sequenced:

www.genomenewsnetwork.org/resources/sequenced_genomes/genome_guide_p1.shtml

1 Knowledge of the genome of parasites and disease prevention

Malaria, the most significant insect-borne disease, poses a threat to 40 per cent of the world's population. The majority of the world's malaria cases are found in Africa, south of the Sahara, and it is here that 90 per cent of the fatalities due to this disease occur. Malaria is caused by *Plasmodium*, a protoctistan, which is transmitted from an infected person to another by blood-sucking mosquitoes of the genus *Anopheles*. Of the four species of *Plasmodium*, only one (*P. falciparum*) causes severe illness.

The search for a vaccine, begun in the 1980s, was initially unsuccessful. Now the situation is more hopeful. For example, the entire genome sequence of *P. falciparum* has been analysed. It consists of 14 chromosomes encoding approximately 5300 genes. This development is a major step in the attempts to design a successful vaccine, and one candidate is undergoing advanced trials. However, the complex parasitic life-style of *Plasmodium* (page 197) makes it a wily foe. *Plasmodium* has quickly developed drug resistance against each currently effective drug. The development of an effective malaria vaccine remains a major international challenge.

2 Human proteins and the treatment of disease

The production of a human protein by recombinant DNA techniques for the treatment of disease was an early success in the development of genetic engineering. The human genes for **insulin** were initially transferred to the bacterium, *E. coli* (and later more usefully to yeast) and human insulin was made available to treat Type I diabetes (page 295).

Haemophilia, a rare genetically determined condition in which the blood fails to clot normally, results from a failure to produce adequate amounts of blood proteins known as **clotting factors**. These are glycoproteins normally released from cells at the site of a haemorrhage (Figure 19.12). Today, haemophilia may be effectively treated by the administration of the specific clotting factor a patient lacks. For example, haemophilia A is due to a deficiency of functionally active **factor VIII**, normally the product of a gene on the X chromosome. The human gene for factor VIII has been transferred to the genome of cells obtained from hamster kidney tissue. These genetically engineered cells are then grown *in vitro*, using an animal cell culture medium. The factor VIII protein is then isolated, purified and administered to patients.

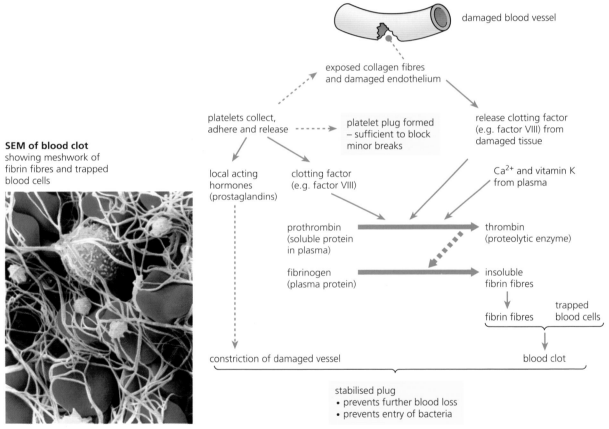

SEM of blood clot
showing meshwork of fibrin fibres and trapped blood cells

Figure 19.12 The blood-clotting mechanism

Severe combined immunodeficiency (**SCID**) is an inherited disorder due to absent or malfunctioning T- and B-lymphocytes of the immune system. The result is a defective antibody response. In SCID patients the immune system is so compromised it is effectively absent, and victims are extremely vulnerable to infectious disease from a very early age.

One form of SCID is due to a lack of the enzyme **adenosine deaminase** (**AD**), coded for by a gene on chromosome 20. In the presence of a defective form of the gene, the substrate of the enzyme this gene codes for accumulates in cells. Immature cells of the immune system are sensitive to the toxic effects of this substrate molecule, and they fail to mature.

Now stem cells, harvested from the umbilical cord blood at birth, have been genetically modified with the normal human AD gene (cDNA), introduced by retovirus. These GM stem cells are then transfused into the neonate for successful gene therapy.

So, in summary, gene therapy is an application of genetic engineering with the aim of supplying a missing gene to body cells in such a way that it remains permanently functional, when this is thought safe and ethically sound. Gene therapy is a very recent, and highly experimental, science. The steps to the process of gene therapy are discussed below.

3 Screening for genetic diseases

Genetic disorders are heritable conditions that are caused by a specific defect in a gene or genes. Most arise from a mutation involving a single gene. The mutant alleles that cause these conditions are commonly **recessive**. In these cases a person must be homozygous for the mutant gene for the condition to be expressed. However, people with a single mutant allele are 'carriers' of that genetic disorder. Quite surprising numbers of us are carriers for one or more such conditions.

An exception is Huntington's disorder, due to a **dominant** allele on chromosome 4. In this case, an individual will be affected even if they have a single allele (they will be heterozygous for the defective gene). The disease is extremely rare (about 1 case per 20 000 live births).

A tendency to suffer from certain cancers also runs in families. An example is breast cancer. In this case, there are two genes (known as BRCA1 and BRCA2), each with one mutant allele, which can be inherited. There is also a 60 per cent probability that the healthy allele will mutate at some stage, before the person reaches the age of 50 years, thereby triggering breast cancer. Both BRCA1 and BRCA2 are natural tumour-suppressing genes when present in the homozygous state.

Genetic disorders generally afflict about 1–2 per cent of the human population. These include sickle-cell disease, Duchene muscular dystrophy, haemophilia (page 371), severe combined immunodeficiency (SCID), and **cystic fibrosis** (**CF**) – on which we focus here.

The incidence of cystic fibrosis varies around the world. The available evidence suggests that in Asia it is quite rare, but in Europe and the USA it occurs in roughly one newborn person in every 3000–3500 births. In the UK cystic fibrosis is the most common genetic disorder. Here, one person in 25 is a carrier.

Causes of cystic fibrosis

Cystic fibrosis is due to a mutation of a single gene on chromosome 7, and it affects the epithelial cells of the body. The CF gene is 180 000 base pairs long. This codes for a protein known as CFTR, which functions as an ion pump (Figure 19.13). The pump transports chloride ions across membranes and water follows the ions, so epithelia are kept smooth and moist when this protein is present and working.

In cystic fibrosis the most common mutation involves a deletion of just three of the gene's nucleotides and results in the loss of an amino acid (phenylalanine) at one location along a protein built from almost 1500 amino acids in total. The mutated gene codes for no protein or for a faulty protein. The result is epithelium cells that remain dry and a build-up of thick, sticky mucus.

Question

6 What do you understand by the term *mutant allele*?

CFTR protein *in situ* in plasma membrane

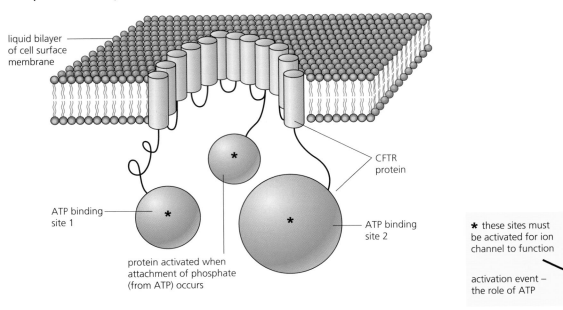

liquid bilayer of cell surface membrane

ATP binding site 1

protein activated when attachment of phosphate (from ATP) occurs

CFTR protein

ATP binding site 2

* these sites must be activated for ion channel to function

activation event – the role of ATP

ATP

ADP + P$_i$

ions

How CFTR regulates water content in mucus

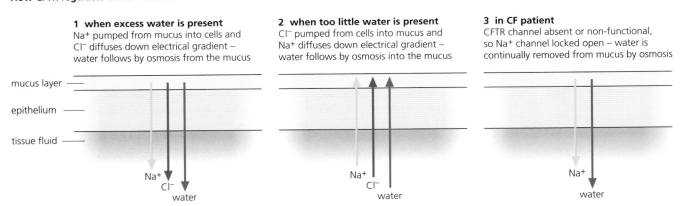

1 when excess water is present
Na$^+$ pumped from mucus into cells and Cl$^-$ diffuses down electrical gradient – water follows by osmosis from the mucus

2 when too little water is present
Cl$^-$ pumped from cells into mucus and Na$^+$ diffuses down electrical gradient – water follows by osmosis into the mucus

3 in CF patient
CFTR channel absent or non-functional, so Na$^+$ channel locked open – water is continually removed from mucus by osmosis

mucus layer

epithelium

tissue fluid

Na$^+$ Cl$^-$ water

Na$^+$ Cl$^-$ water

Na$^+$ water

Figure 19.13 The CFTR protein – a channel protein

The signs and symptoms of cystic fibrosis

The effects of faulty CFTR ion pumps are felt:

- in the pancreas – secretion of digestive juices by the gland cells in the pancreas is interrupted by blocked ducts. Digestion is hindered.
- in the sweat glands – salty sweat is formed. This is used in the diagnosis of cystic fibrosis.
- in the lungs – these become blocked by mucus and are prone to infection. This effect is most quickly life-threatening if not treated promptly.

In adult patients, the membranes of epithelium cells in their reproductive organs are affected too. Infertility frequently results. In women with cystic fibrosis, the mucus naturally present in the cervix becomes a dense plug in the vagina, restricting entry of sperm into the uterus. In men with cystic fibrosis, dense mucus in the sperm ducts blocks these tubes, significantly reducing the sperm count when semen is ejaculated.

The process of genetic screening

We can now locate genes responsible for human genetic diseases, if we choose. This is one outcome of the Human Genome Project. For example, the gene that codes for cystic fibrosis (CFTR) was sequenced in 1989. This made possible the screening of human cells for the presence of the mutated *CFTR* gene (Figure 19.14). So it is possible for couples to be genetically screened to assess how likely they are to have children who will have cystic fibrosis (because both are 'carriers'), for example. Alternatively, it is possible for all members of a family group be offered screening where there is a history of a particular condition.

Prenatal screening is used to test for the presence of a number of genetic conditions in the unborn child. This allows the detection of:

- chromosomes abnormalities such as Down syndrome (trisomy 21), trisomy 13 and trisomy 18. In the latter two abnormalities, only very rarely do affected infants survive their first year.
- single gene disorders, such as haemophilia, sickle cell anaemia and cystic fibrosis.
- neural tube defects, such as spina bifida and anencephaly.

Prenatal screening may be carried out by:

- **chorionic villus sampling** – undertaken at weeks 11–13 of the pregnancy, a sample from the placenta is taken.
- **amniocentesis** – undertaken at weeks 15–20 of the pregnancy, the fetal cells in the amniotic fluid are examined.

1 DNA extraction → cut with restriction enzyme

cells e.g. from skin tissue or white cells separated from a blood sample → mechanically broken up, then:
- cell debris removed by filtration and centrifugation
- cell membranes and nuclear envelopes broken down by incubation with detergent
- proteins removed by incubation with protease enzyme

so, DNA is freed from the nucleus, and from its histone 'scaffolding'

DNA precipitated in ice-cold ethanol

DNA

DNA resuspended in aqueous, pH-buffered medium, and incubated with restriction enzyme

restriction enzyme

restriction enzyme (restriction endonuclease) 'cuts' at a particular sequence of bases

2 DNA fragments separated by electrophoresis

Restriction endonuclease enzyme cuts DNA strands into fragments of differing length at the restriction sites, one of which will contain the required gene.

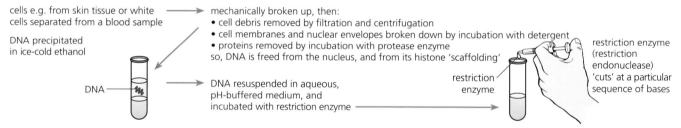

restriction sites

required gene

restriction fragments

DNA fragments are separated by gel electrophoresis, using an agarose gel:

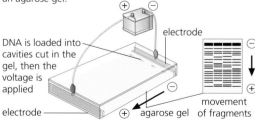

electrode

DNA is loaded into cavities cut in the gel, then the voltage is applied

electrode — agarose gel

movement of fragments

negative charge on DNA (due to phosphate groups) causes the fragments to move to the positive electrode (anode), but the gel has a 'sieving' effect – smaller fragments move more rapidly than larger ones

3 DNA transferred to nylon/nitrocellulose membrane (Southern blotting)

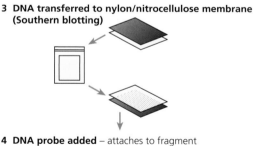

4 DNA probe added – attaches to fragment complementary to *CFTR* gene

5 DNA–probe complex detected – by X-ray film if radioactive probe or by fluorescence under UV light

Figure 19.14 Steps involved in genetic screening for cystic fibrosis

Newborn screening is undertaken in some countries, to detect conditions such as **phenylketonuria (PKU)**. Phenylketonuria, an error of protein metabolism, is caused by a mutation in a gene coding for the enzyme that converts the amino acid phenylalanine into tyrosine. In infants born with this mutation, the phenylalanine naturally obtained in the diet and not immediately used in the synthesis of new proteins builds up in the blood. An excess of phenylalanine in the body causes dangerous side effects, including growth deficiency and, eventually, severe mental retardation. Phenylketonuria can be detected at birth by means of a simple blood test and the symptoms can largely be avoided by adjusting the diet.

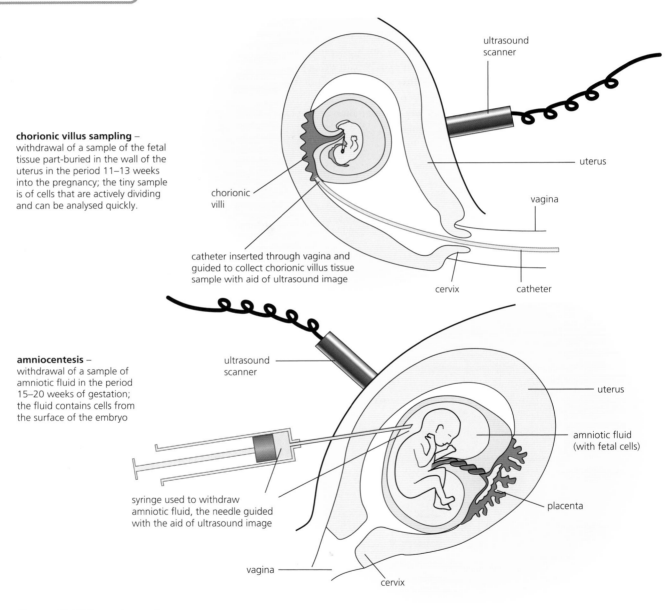

chorionic villus sampling – withdrawal of a sample of the fetal tissue part-buried in the wall of the uterus in the period 11–13 weeks into the pregnancy; the tiny sample is of cells that are actively dividing and can be analysed quickly.

amniocentesis – withdrawal of a sample of amniotic fluid in the period 15–20 weeks of gestation; the fluid contains cells from the surface of the embryo

Figure 19.15 Prenatal screening

The outcome of screening – genetic counselling

The outcome of genetic screening frequently does not necessarily allay a person's (or their partner's) fears. They are likely to be referred to a **genetic counsellor**. The role of the counsellor is to take a detailed case history of the person and their immediate family. A counsellor may then research the **family pedigree**.

Figure 19.16 A genetic counselling session

Reasons for referral to a genetic counsellor
- Screening has confirmed that one or both partners are carriers.
- One or both partners have a relative with a genetic disease.
- The partners are first cousins.
- One or both partners belong to an ethnic group with an above-average risk of a genetic disease.
- There is a history of earlier pregnancies ending in a miscarriage.
- The partners are aged 38 or over.

In these cases people need to know the risk of their offspring inheriting the defective allele or actually developing the disorder. One outcome may be the possibility of further treatment in cases where a defective allele has been detected. There may be consequences for future health, life expectancy, the appropriateness of future education and training and future employment. These are all issues where information and reassurance may be needed. Sometimes, genetic conditions affect the chance of obtaining life insurance. The role of the counsellor is to provide information in a way which allows their clients to make their own decisions.

A case study: investigating haemophilia by family pedigree

In the blood circulatory system of a mammal, if a break occurs there is a risk of uncontrolled bleeding. This is normally overcome by the blood clotting mechanism which causes any gap to be plugged. Haemophilia is a rare, genetically determined, condition in which the blood does not clot normally. The result is frequent, excessive bleeding.

There are two forms of haemophilia, known as haemophilia A and haemophilia B. They are due to a failure to produce adequate amounts of particular blood proteins (factors VIII and IX), both essential to the complex blood clotting mechanism. Today, haemophilia is effectively treated by the administration of the clotting factor the patient lacks.

Haemophilia is a sex-linked condition because the genes controlling production of the blood proteins concerned are located on the X chromosome. Haemophilia is caused by a recessive allele. As a result, haemophilia is largely a disease of the male since in him a single X chromosome carrying the defective allele (X^hY) will result in disease. For a female to have the disease, she must be homozygous for the recessive gene (X^hX^h), but this condition is usually fatal *in utero*, typically resulting in a miscarriage.

So, a family tree where haemophilia occurs typically shows no female haemophiliacs. However, there will also be female carriers (X^HX^h), with one X chromosome with the recessive allele. Female carriers have normal blood clotting, but for any children they have with a normal male, there is a 50 per cent chance of the daughters being carriers and a 50 per cent chance of the sons being haemophiliac.

The British queen, Queen Victoria (1819–1901) was a carrier of haemophilia. She transmitted the allele to three of her nine children and it was passed on to at least seven of her grandchildren. Look at the pattern of haemophiliac males and female carriers in this extensive royal family (Figure 19.17).

The transmission of this 'royal' haemophilia, which has had political and social consequences in the past 150 years, particularly in Russia and Spain, is well documented.

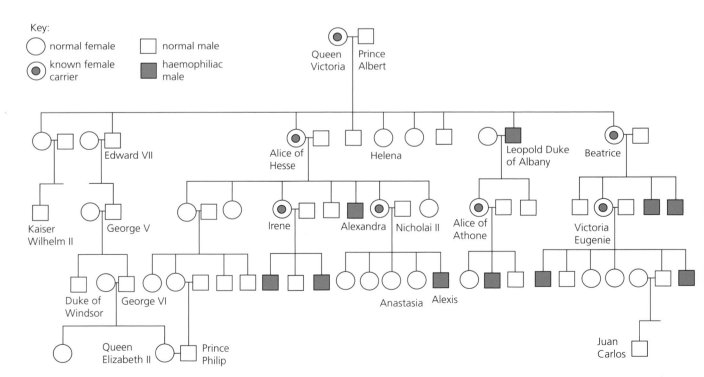

Figure 19.17 Queen Victoria and her heirs – a celebrated case!

4 Gene therapy

Somatic gene therapy applied to cystic fibrosis

Since the healthy *CFTR* gene codes for a protein that functions as an ion pump, the gene therapy initiative has involved getting copies of the healthy gene to the cells of the lung epithelium. This can be attempted in a number of ways.

The *CFTR* gene may be delivered in an aerosol spray. The spray contains tiny lipid bilayer droplets called liposomes to which copies of the healthy gene are attached. The liposomes fuse with cell membrane lipid and deliver the gene into the epithelium cell (Figure 15.16). In trials, the treatment is effective but the epithelium cells are routinely replaced so the beneficial effects only last a few days and the treatment has to be regularly repeated. The cure can only be more permanent when it is targeted on the cells that make epithelium cells and unfortunately this approach has failed to deliver the healthy gene effectively.

An alternative approach is based on a viral delivery of the healthy *CFTR* gene. Viruses such as the Adenoviruses have been used in place of liposomes. This virus is capable of entering and infecting lung epithelium cells. However, the virus is first genetically engineered to inactivate or remove from its genome the alleles that cause disease. Additionally, healthy *CFTR* alleles are added in. Unfortunately the treatment has not been successful. Viral vectors have triggered allergic or other immune responses and have had to be discontinued.

Other vectors for gene therapy

We saw that the vector for gene therapy in cystic fibrosis was a lipid envelope (a liposome). There are alternative vectors:

- **Viruses**, including retroviruses, have been used to transfer DNA into cells, as shown in Figure 19.18. Since a retrovirus is an RNA virus, the enzyme reverse transcriptase that the virus carries has a key role in the insertion of recombinant DNA in the host liver cells. Familial hypercholesterolaenia (FM) is a genetically inherited disorder in which low-density lipoproteins accumulate in the blood plasma, and instead of being absorbed into liver cells, contribute to the lipid being deposited as cholesterol in the arteries, leading to coronary heart disease. A retrovirus has also been used in the treatment of SCID. Adenovirus and vaccinia virus have also been successful vectors, and the conditions treated include inherited eye disease.

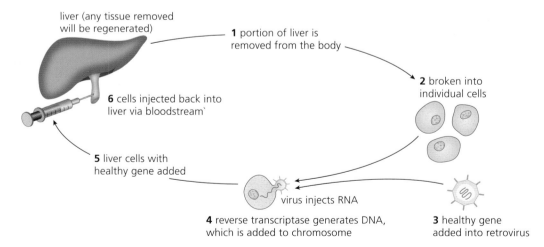

Figure 19.18 Gene therapy using a virus as vector

- **Naked DNA**. Genes have also been transferred by direct take-up of DNA from the environment. The host cells have to be capable of taking up the DNA – a condition described as 'competent'. Calcium chloride treatment has proved to be one of the best methods of preparing competent cells. The genetic diseases treated in this way include SCID.

The cystic fibrosis gene codes for a membrane protein that occurs widely in body cells, and pumps ions (e.g. Cl⁻) across cell membranes.

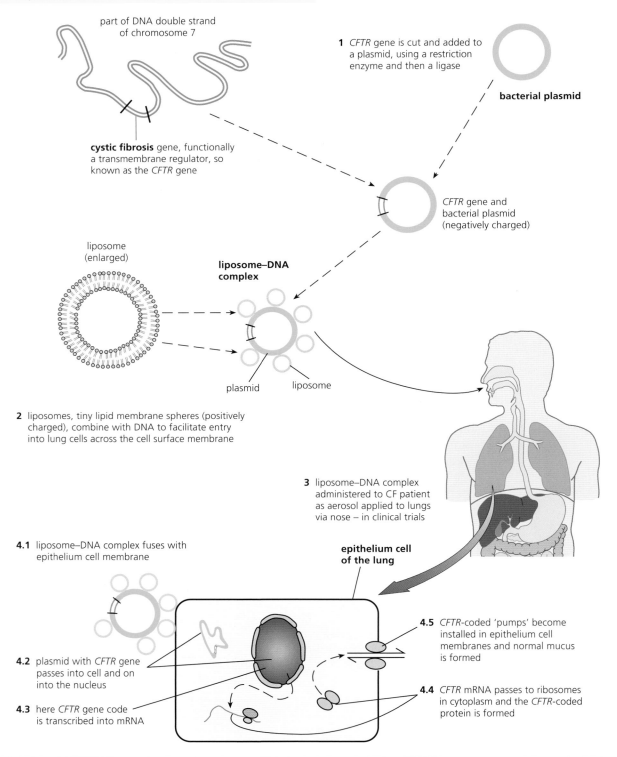

part of DNA double strand of chromosome 7

1 *CFTR* gene is cut and added to a plasmid, using a restriction enzyme and then a ligase

bacterial plasmid

cystic fibrosis gene, functionally a transmembrane regulator, so known as the *CFTR* gene

CFTR gene and bacterial plasmid (negatively charged)

liposome (enlarged)

liposome–DNA complex

plasmid   liposome

2 liposomes, tiny lipid membrane spheres (positively charged), combine with DNA to facilitate entry into lung cells across the cell surface membrane

3 liposome–DNA complex administered to CF patient as aerosol applied to lungs via nose – in clinical trials

4.1 liposome–DNA complex fuses with epithelium cell membrane

epithelium cell of the lung

4.5 *CFTR*-coded 'pumps' become installed in epithelium cell membranes and normal mucus is formed

4.2 plasmid with *CFTR* gene passes into cell and on into the nucleus

4.3 here *CFTR* gene code is transcribed into mRNA

4.4 *CFTR* mRNA passes to ribosomes in cytoplasm and the *CFTR*-coded protein is formed

In recent clinical trials some 20% of epithelium cells of CF patients were *temporarily* modified (i.e. accepted the *CFTR* gene), but the effects were relatively short-lived. This is because our epithelium cells are continually replaced at a steady rate, and in CF patients the genetically engineered cells are replaced with cells without *CFTR*-coded pumps. Patients would require periodic treatment with the liposome–DNA complex aerosol to maintain the effect permanently.

Figure 19.19 Somatic gene therapy: supplying the healthy *CFTR* gene to the lungs

Social and ethical aspects of genetic engineering and gene therapy

The benefits and hazards of gene technology

Genetically modified organisms have been produced for specific purposes, relatively quickly, where this proves possible. They offer enormous benefits However, geneticists are really producing new organisms when genes are transferred so there may be hazards as well as benefits.

Questions

8 Suggest what advantage would result from the transfer of genes for nitrogen fixation from nodules in leguminous plants to cereals such as wheat.

9 Occasionally newspaper articles make criticisms of genetic modification of food organisms, sometimes with alarmist headlines. Search the web for articles that give differing views on this. Prepare the following to use in a discussion with your peers.

a a concise summary of the criticisms frequently made, avoiding extremist language or unnecessary exaggeration

b a list of balancing arguments in favour of genetic modification of food organisms

c a concise statement of your own view on this issue

Table 19.4 The benefits and potential hazards of gene technology

The benefits so far	Hazards that have been anticipated
• In addition to human insulin production by modified *E. coli*, human growth hormone, and blood clotting factor VIII have been produced by genetically modified bacteria.	• Could a harmless organism such as the human gut bacterium *E. coli* be transformed into a harmful pathogen that escapes from the laboratory into the human population?
• Another application of genetically engineering is bioremediation. This is the removal of toxic compounds, carcinogens and pollutants, such as industrial solvents, contaminating ground waters and crude oil spills by genetically engineered bacteria which degrade these substances into safer molecules. The process is slow.	• Genes move between bacterial populations with time. Could the antibiotic-resistance genes in plasmid vectors become accidentally transferred into a pathogenic organism?
• Sugar beet plants have been genetically modified to be 'tolerant' of glyphosate herbicide applications. These plants are able to inactivate the herbicide when it is sprayed over them, but the surrounding weeds cannot and are killed. (This is discussed further in this topic.)	• The accidental transfer of herbicide-resistance from crop plants, such as sugar beet, to a related wild plant and then on to other plants is a possibility which could result in 'super weeds' whose spread would be hard to prevent.
• Similarly, natural resistance to attack by chewing insect has resulted from the addition of the genes for Bt toxin into crop plants (potato, cotton, and maize) by genetic engineering. This reduces the need for extensive aerial spraying of expensive insecticides which are harmful to wildlife.	• Natural insecticidal toxin engineered into crop plants, whilst an effective protection from browsing insects, might harm pollinating species such as bees and butterflies too.
	• The presence of the gene for the Bt toxin in the environment may lead to insects with resistance, too.
• There are on-going attempts to use gene technology to treat genetic diseases, such as cystic fibrosis (page 469) and severe combined immunodeficiency (SCID). It is hoped that some cancers may yield to this approach in the future.	• There are well publicised anxieties about food products from genetically-engineered species. Might the food become toxic or trigger some allergic reactions. So far there has been no evidence of these.

The social implications of gene technology

Some of the **social implications** of gene technology are apparent from the list of benefits and potential hazards in Table 19.4. We can summarise them as follows.

The **advantages** of gene technology for society may be:
- improved, cheaper medicines
- improved food supplies
- improved nutritional quality of foods
- a cleaner environment
- improved treatment of genetic diseases.

The **disadvantages** of gene technology for society may be:
- unexpected reductions in crop yields due to ecological disturbance
- farmers made dependent on specific varieties, needing fresh seed annually and expensive fertilisers
- reduced natural biodiversity resulting in a reduced possibility of new varieties arising
- a reduced effectiveness of antibiotics as more bacteria become resistant.

Ethical issues in gene technology

Ethical implications depend on wider values. Ethics are the moral principles that we feel ought to govern or influence the conduct of a society. Ethics are concerned with how we decide **what is right and what is wrong**. Today, developments in science and technology influence many aspects of people's lives. This certainly throws up increasingly complex ethical issues which we need to examine and respond to. Gene technology is just one case in point.

There are important forces in our lives concerned with right and wrong in various ways, but they are not themselves ethics (Table 19.5). After all, what is considered right and what is considered wrong varies across the world.

Table 19.5 Defining ethics by recognising what they are not

The law	Our laws are made by governments and may or may not be ethically-based. (In some cases, they may even only be a view held by a minority.)
Religion	In our societies there are followers of different religions and of none, yet ethics apply to us all.
Cultural norms	Many of these are little more than 'fashions' that seemed acceptable at one time and which may still be held uncritically.
Science	Whilst this may seek to give us an understanding of the origins of our world and life, and how these work, it does not suggest how we should act.
Our feelings or conscience	These are likely to be the product of our early environment, general outlook and temperament, and our individual experiences.

On the other hand, the principles we do use in coming to ethical decisions are identified in Table 19.6. You can apply them to the issues that gene technology raises, after you have read them, if you can accept them as comprehensive.

Regarding the **ethics of gene technology**, perhaps there are two broad issues that societies need to address in deciding what developments they wish to fund.
- Is there an important over-riding principle to be held to, that humans should not tamper with nature in a deliberate way? Are the changes in biodiversity and genetic diversity that inadvertently flow from it completely beyond our current knowledge? Alternatively, does all that we do influence and eventually change our environment so our role is to see that our activities are conducted responsibly?
- To what extent is gene technology a costly technology that is mostly beneficial to the health and life expectancy of wealthier people or perhaps to just those of developed nations? If the funds used for gene technology were made available for the more basic problems of housing, health and nutrition in less developed countries instead, could vastly more people benefit immediately? Alternatively, if we have the ability and resources to bring about beneficial development, is it our duty to do so?

What does your group think?

Table 19.6 Ethical standards to apply in decision making

Rights We have a duty to ensure these	Human rights have been recognised and articulated, for example: • The United Nations Universal Declaration of Human Rights: • 'All human beings are born free and equal in dignity and rights. They are endowed with reason and conscience and should act towards one another in a spirit of brotherhood'. • The European Convention on Human Rights, which includes issues of security, liberty, political freedom and welfare, and economic and group rights. • In countries with a Bill of Rights written into their constitution, as in the USA: 'We hold these truths to be self-evident, …'.
Justice A principle to guide actions	The principle that all should be treated equally; the idea of fairness in our actions to each individual is considered essential.
Utilitarianism A belief that the overall benefits should be greater than the costs	The principle of the greatest good for the greatest number and with actions that generate minimal harm to others' interests. In effect, this involves a cost–benefit analysis.
Common good The core conditions that are essential to the welfare of all	Where the life of a working, interacting community is inherently good, then all actions should prosper and support the common interest.
Virtue The placing of importance on a need to live a 'good life'	Ethical actions necessarily respect and embody values like truth, honesty, courage, compassion, generosity, tolerance, integrity, fairness, self-control and prudence.
Other criteria?	Do you or your peers have any other criteria about which the whole group can be convinced?

Question

10 Using a pedigree chart showing the incidence of haemophilia, what information is needed to identify a particular carrier in any generation?

Ethical issues raised by genetic screening

The nature and application of ethical principles in decision making has just been discussed.

It is possible to abort a fetus that has been identified as carrying a defective gene. Genetic counselling may result in ethical concerns for parents, medical personnel and for society as a whole. Other ethical issues that screening and counselling may present include:

● Who should decide who may be screened or tested?
● What known disorders should be considered by the medical profession or society? For example, should screening be available for disorders for which there is, as yet, no known cure?
● Who should meet the substantial costs of screening, given the escalating cost of public health provision, where this exists?
● Should the results be confidential? If not, who should have access to the information? Should it be available to potential employers or insurance providers?

What other issues do you think are relevant?

Specific issues to consider

The following issues are appropriate for discussion between individuals, and with peers in groups, as well as by patients and their medical teams, when they arise.

1 **Genetic disorders where treatment exists v. where none exist.**

For those in authority who supervise the allocation of public funds, and for the scientific and medical workers who are required to make professional decisions in the laboratory or clinic about the direction and emphasis on future developments, this issue will arise. *How should they respond?*

2 **Infertility and IVF.**

Fertilisation followed by implantation of an embryo does not always proceed in the normal way. Either the male or the female or both may be infertile, due to a number of different causes.

In **males** these may include:
● failure to achieve or to maintain an erect penis
● structurally abnormal sperm, sperm of poor mobility, short-lived sperm or too few sperm
● a blocked sperm duct, which prevents semen containing sperm from being ejaculated.

In **females** these may include:

- conditions in the cervix that cause death of sperm
- conditions in the uterus that prevent implantation of the blastocyst
- blocked or damaged oviducts, which prevent the egg from reaching sperm
- eggs that fail to mature or fail to be released.

In some cases, a couple's infertility may be treated by the process of fertilisation of eggs outside the body – **in-vitro fertilisation** (**IVF**). The key step in *in-vitro* fertilisation is the successful removal of sufficient egg cells from the ovaries. To achieve this, the normal menstrual cycle is temporarily suspended with hormone-based drugs. Then the ovaries are induced to produce a large number of eggs simultaneously, at a time controlled by the doctors. In this way the correct moment to collect the eggs can be known accurately. Eggs cells are then isolated from the surrounding follicle cells and mixed with sperm. If fertilisation occurs, the fertilised egg cells are incubated so that embryos at the eight-cell stage may be placed in the uterus. If one (or more) imbed there, then a normal pregnancy may follow.

The first baby created in this way, called a 'test tube baby', was born in 1978. The process of *in-vitro* fertilisation is illustrated in Figure 19.20. Today, the procedure is regarded as a routine one.

Figure 19.20 The process of *in-vitro* fertilisation

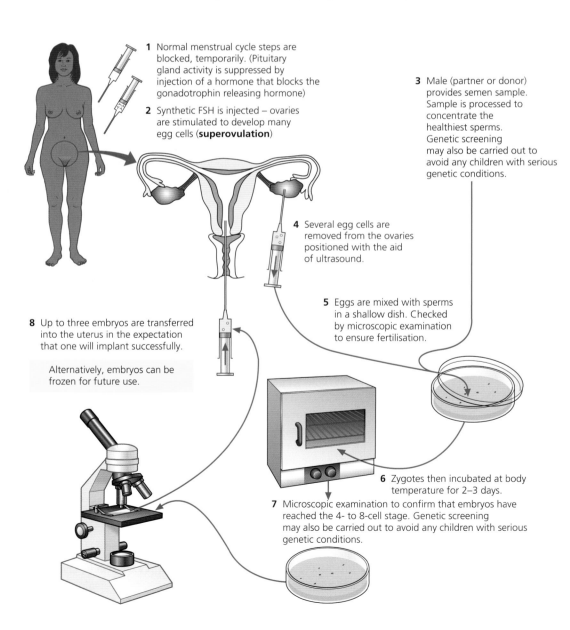

1 Normal menstrual cycle steps are blocked, temporarily. (Pituitary gland activity is suppressed by injection of a hormone that blocks the gonadotrophin releasing hormone)

2 Synthetic FSH is injected – ovaries are stimulated to develop many egg cells (**superovulation**)

3 Male (partner or donor) provides semen sample. Sample is processed to concentrate the healthiest sperms. Genetic screening may also be carried out to avoid any children with serious genetic conditions.

4 Several egg cells are removed from the ovaries positioned with the aid of ultrasound.

5 Eggs are mixed with sperms in a shallow dish. Checked by microscopic examination to ensure fertilisation.

8 Up to three embryos are transferred into the uterus in the expectation that one will implant successfully.

Alternatively, embryos can be frozen for future use.

6 Zygotes then incubated at body temperature for 2–3 days.

7 Microscopic examination to confirm that embryos have reached the 4- to 8-cell stage. Genetic screening may also be carried out to avoid any children with serious genetic conditions.

In-vitro fertilisation raises many issues for society. Infertility is a personal problem that generates stress and unhappiness in those affected. Infertile couples may seek assistance and a cure. The medical treatments involved are expensive and their deployment inevitably deflects finite resources away from other needs. Success rates are sometimes low. In the face of the world's booming human population, some argue against these sorts of development in reproduction technology.

Another controversial issue is that current fertility treatments tend to create several embryos, all with the potential to become new people. However, the embryos that are not selected for implantation ultimately must be destroyed – a stressful task in itself.

The ethical implications of *in-vitro* fertilisation are presented in Table 19.7.

Table 19.7 The ethical implications of *in-vitro* fertilisation

Arguments for IVF	Arguments against IVF
• For some otherwise childless couples, desired parenthood may be achieved. • Allows men and women surviving cancer treatments the possibility of having children later, using gametes harvested prior to radiation or chemotherapy treatments. • Permits screening and selection of embryos before implantation to avoid an inherited disease. • Offspring produced by IVF are much longed-for children who are more certain to be loved and cared for. • IVF can be used to treat the medical problem of infertility. Nobody should be allowed to deny treatment of a problem that can be cured.	• Allows infertility due to inherited defects to be passed on (unwittingly) to the offspring, who may then experience the same problem in adulthood. • Excess embryos are produced to ensure success and so an embryologist has to select some new embryo(s) to live and to allow the later destruction of other, potential human lives. • Multiple pregnancies have been a common outcome, sometimes producing triplets, quads or sextuplets, leading to increased risk to the mother's health, and also increased risks of premature birth and babies with cerebral palsy, for example. • It can be argued that infertility is not always strictly a health problem; it may have arisen in older parents who choose to delay having a family (a lifestyle issue). • There is an excess of unwanted children, cared for in orphanages or in foster homes. These children may have benefited from adoption by couples, childless or otherwise, keen to be caring 'parents'.

Question

11 Outline the key points that might be put by a genetic counsellor who felt that an alternative way to establish a family was more appropriate for a particular childless couple seeking *in-vitro* fertilisation treatment.

3 **Embryo biopsy, preselection and therapeutic abortions.**

Biopsy is the medical procedure that involves taking small samples of tissue for further analysis. It is also applied to the tissue concerned. You can see embryo biopsy in Figure 19.15 (page 472). We have noted how this procedure provides data that may need resolving by genetic counselling. It makes it possible to perform embryo selection (such as selecting the sex of a child) as well as therapeutic abortion. A situation where some people are able to make such decisions about the fate of an embryo or fetus, raises the most pressing ethical, social and cultural issues that need to be resolved where they arise. Nevertheless, these issues also arise for any wise and caring community or society. *Your discussions will contribute to this process.*

Bioinformatics in forensic medicine and criminal investigations

DNA profiling

DNA profiling exploits the techniques of genetic engineering to identify a person or organism from a sample of their DNA. The DNA of our chromosomes, which is unique, includes that of all our genes.

Southern blotting (named after the scientist who devised the routine):
- extracted DNA is cut into fragments with restriction enzyme
- the fragments are separated by electrophoresis
- fragments are made single-stranded by treatment of the gel with alkali.

1 Then a copy of the distributed DNA fragments is produced on nylon membrane:

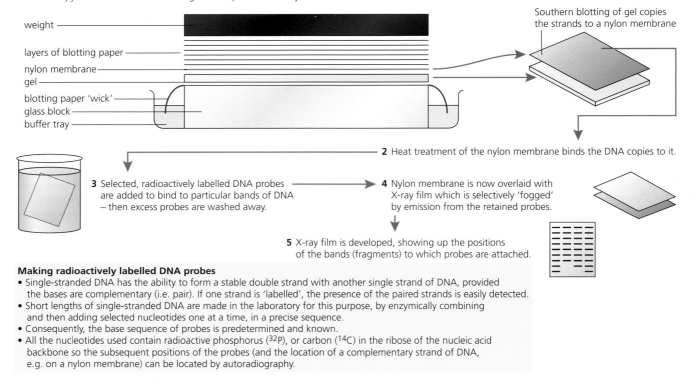

Southern blotting of gel copies the strands to a nylon membrane

weight
layers of blotting paper
nylon membrane
gel
blotting paper 'wick'
glass block
buffer tray

2 Heat treatment of the nylon membrane binds the DNA copies to it.

3 Selected, radioactively labelled DNA probes are added to bind to particular bands of DNA – then excess probes are washed away.

4 Nylon membrane is now overlaid with X-ray film which is selectively 'fogged' by emission from the retained probes.

5 X-ray film is developed, showing up the positions of the bands (fragments) to which probes are attached.

Making radioactively labelled DNA probes
- Single-stranded DNA has the ability to form a stable double strand with another single strand of DNA, provided the bases are complementary (i.e. pair). If one strand is 'labelled', the presence of the paired strands is easily detected.
- Short lengths of single-stranded DNA are made in the laboratory for this purpose, by enzymically combining and then adding selected nucleotides one at a time, in a precise sequence.
- Consequently, the base sequence of probes is predetermined and known.
- All the nucleotides used contain radioactive phosphorus (^{32}P), or carbon (^{14}C) in the ribose of the nucleic acid backbone so the subsequent positions of the probes (and the location of a complementary strand of DNA, e.g. on a nylon membrane) can be located by autoradiography.

What a probe is and how it works

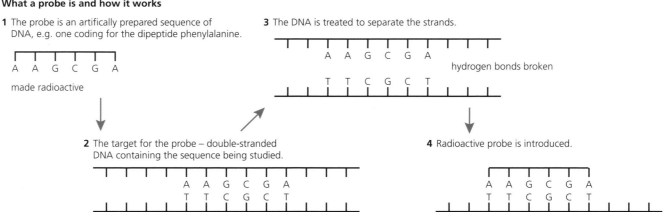

1 The probe is an artifically prepared sequence of DNA, e.g. one coding for the dipeptide phenylalanine.

A A G C G A
made radioactive

3 The DNA is treated to separate the strands.

A A G C G A
T T C G C T
hydrogen bonds broken

2 The target for the probe – double-stranded DNA containing the sequence being studied.

A A G C G A
T T C G C T

4 Radioactive probe is introduced.

A A G C G A
T T C G C T

Figure 19.21 Steps in genetic profiling

The bulk of our DNA is not genes and does not code for proteins. There are extensive 'non-gene' regions consisting of short sequences of bases repeated many times. Whilst some of these sequences are scattered throughout the length of the DNA molecule, many are joined together in major clusters. It is these long lengths of non-coding, 'nonsense' DNA that are used in genetic profiling. They may be referred to as **satellite DNA** or as **variable number tandems repeats (VNTRs)**. We inherit a distinctive combination of these apparently non-functional satellite DNA, half from our mother and half from our father. Consequently, each of us has a unique sequence of nucleotides in our satellite DNA (except for identical twins, who share the same pattern).

To produce a genetic 'fingerprint', a sample of DNA is cut with a restriction enzyme which acts close to the satellite DNA regions. Electrophoresis is then used to separate pieces according to length and size and the result is a pattern of bands.

To produce a DNA 'fingerprint' or profile:

1 A sample of cells is obtained from blood, semen, hair root or other body tissues, and the DNA is extracted. (Where a tiny quantity of DNA is all that can be recovered, this is precisely copied using the polymerase chain reaction (see Figure 19.3 on page 457) to obtain sufficient DNA to analyse.)

2 The DNA is cut into small, double-stranded fragments using a particular restriction enzyme chosen because it 'cuts' close to but not within the satellite DNA.

3 The resulting DNA fragments are of varying lengths, and are separated by **gel electrophoresis** into bands which are invisible, at this stage.

4 The gel is treated to split DNA into single strands and then a copy is transferred to a membrane.

5 Selected, radioactively labelled DNA probes are added to the membrane to bind to particular bands of DNA and then the excess is washed away. Alternatively, the probes may be labelled with a fluorescent stain which shows up under ultraviolet light.

6 The membrane is now overlaid with X-ray film which becomes selectively 'fogged' by radioactive emission from the retained probes.

7 The X-ray film is developed and shows the positions of the bands (fragments) to which probes have attached. The result is a profile with the appearance of a bar code.

Steps 4 to 7 make up the technique known as Southern blotting and are illustrated in Figure 19.21.

DNA profiling has applications in forensic investigations

DNA profiles are produced from samples taken from the scene of a serious crime, such as rape attacks, both from victims and suspects as well as from others who have certainly not been involved in the crime, as a control. The greatest care has to be taken to ensure the authenticity of the sample. There must be no possibility of contamination if the outcome of subsequent testing is to be meaningful. DNA profiling helps eliminate innocent suspects, and to identify a person or people who may be responsible or related. It cannot prove with absolute certainty, anyone's guilt or connection. An example is shown in Figure 19.22.

Other crimes such as suspected murder and burglaries may be solved in cases where biological specimens are left at the scene of the crime, such as a few hairs or a tiny drop of blood. Specimens may also be collected from people suspected of being present at the crime. DNA fingerprinting can help forensic scientists identify corpses otherwise too decomposed for recognition, or where only parts of the body remain, as may occur after bomb blasts or other violent incidents, including natural disasters.

Note that there are various circumstances where the amount of DNA available or which can be recovered (such as at a crime scene) is very small indeed – apparently too little for analysis, in fact. It is now possible to submit such minute samples to the process of polymerase chain reaction (PCR) in which the DNA is replicated in an entirely automated process, *in vitro*, to produce a large amount of the sequence. A single molecule is sufficient as the starting material, should this be all that is available. The product is exact copies in quantities sufficient for analysis.

Question

12 Explain why the composition of the DNA of identical twins challenges an underlying assumption of DNA fingerprinting, but that of non-identical twins does not?

DNA profiling in forensic investigation

Identification of criminals

At the scene of a crime (such as a murder), hairs – with hair root cells attached – or blood may be recovered. If so, the resulting DNA profiles may be compared with those of DNA obtained from suspects.

Examine the DNA profiles shown to the right, and suggest which suspect should be interviewed further.

Identification in a rape crime involves the taking of vaginal swabs. Here, DNA will be present from the victim and also from the rapist. The result of DNA analysis is a complex profile that requires careful comparison with the DNA profiles of the victim and of any suspects. A rapist can be identified with a high degree of certainty, and the innocence of others established.

Identification of a corpse which is otherwise unidentifiable is achieved by taking DNA samples from body tissues and comparing their profile with those of close relatives or with DNA obtained from cells recovered from personal effects, where these are available.

DNA profiles used to establish family relationships

Is the male (♂) the parent of both children?

Examine the DNA profiles shown to the right.

Look at the children's bands (C).

Discount all those bands that correspond to bands in the mother's profile (♀).

The remaining bands match those of the biological father.

DNA fingerprinting has also been widely applied in biology. In ornithology, for example, DNA profiling of nestlings has established a degree of 'promiscuity' in breeding pairs, the male of which was assumed to be the father of the whole brood. In birds, the production of a clutch of eggs is extended over a period of days, with copulation and fertilisation preceding the laying of each egg. This provides the opportunity for different males to fertilise the female.

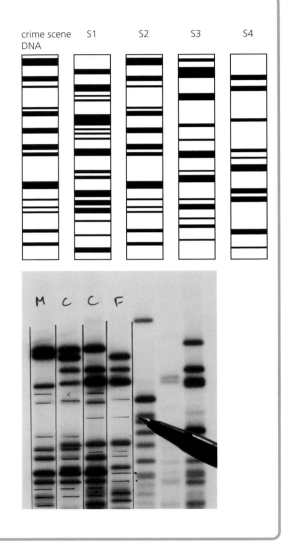

Figure 19.22 Applications of DNA profiling

DNA profiling has applications in determining paternity

Another very important application of DNA profiling is in issues of parentage. A range of samples of DNA of the people who are possibly related are analysed side by side. The banding patterns are then compared (Figure 19.22). Because a child inherits half its DNA from its mother and half from its father, the bands in a child's DNA fingerprint that do not match its mother's must come from the child's father.

DNA profiling also has wide applications in studies of wild animals, for example, concerning breeding behaviour, and in the identification of unrelated animals as mates in captive breeding programmes of animals in danger of extinction.

19.3 Genetically modified organisms in agriculture

The ability to manipulate genes has many potential benefits in agriculture, but the implications of releasing genetically modified organisms (GMOs) into the environment are subject to much public debate in some countries.

By the end of this section you should be able to:

a) explain the significance of genetic engineering in improving the quality and yield of crop plants and livestock in solving the demand for food in the world, e.g. Bt maize, vitamin A enhanced rice (Golden rice™) and GM salmon

b) outline the way in which the production of crops such as maize, cotton, tobacco and oil seed rape may be increased by using varieties that are genetically modified for herbicide resistance and insect resistance

c) discuss the ethical and social implications of using genetically modified organisms (GMOs) in food production

Genetic engineering and improvements in quality and yield

The World Health Organisation (WHO) has stated that hunger and related malnutrition are the greatest threat to the world's public health. Of the global population (currently 7.1 billion), some 870 million people suffer from chronic undernourishment. Most of these people live in developing countries.

The growing issue of food shortage is due to several factors:

- The size of human populations – this currently increases by 75 million people per year.
- The increasing wealth of populations in countries like China and Brazil. Growing numbers of people who once obtained the bulk of their nourishment from a diet high in grain (rice, wheat, maize) are increasing demand for meat and dairy products. Animals reared intensively require grain. It takes 5 kilos of grain to produce a single kilo of meat.
- Extreme weather conditions – climate change and water stress lessen agricultural production.
- Small farmers, whose production may sustain the needs of many of the poorest communities, have access to a shrinking land area. This is due to development of industrial farming, demands imposed by urbanisation, transport and the growth of cities, and the increasing use of land to grow alternative crops that supply raw materials for industry and urban populations. Crops grown for biofuels is just one example.

Genetically modified organisms have a part to play in worldwide improvement of agricultural production by increasing quantity and nutritional value, and by reducing loss due to pests and weed growth. However, the benefits arising from genetic modification may be at a price that the poorest (and most undernourished) are mostly unable to meet. There are urgent social and political factors to tackle too, in the battle against hunger. Nevertheless, there have been significant successes achieved by genetic modification.

Bt maize

The soil bacterium *Bacillus thuringiensis* is a natural source of an insecticide known as Cry toxin. Genes for the production of this toxin have been transferred from this bacterium to selected crops, making them toxic to certain insects if the animal should feed on the leaves, for example. The crops that have been genetically modified in this way are varieties of potato, corn and cotton. They are then referred to as Bt crops.

Bt maize has been grown in the USA since 1995. The resulting crops have been used as biofuels or in animal feed only. Subsequently, Bt potatoes and Bt cotton crops have been approved for use in USA. Bt cotton is the only genetically modified crop authorised for use in India and China. In Europe, Bt maize is grown mainly in Spain, where the crop is used as animal feed.

The advantages for Bt crops detected to date are:

- the level of toxin expression is sufficient to be toxic to pests
- expression of the Bt genes occurs within cells of the crops in such a way as to cause only insects that feed on the plant body to be killed
- toxin expression has been modulated by adding tissue-specific promoters that have limited their sites of production.

The use of these genetically modified varieties appears to have been advantageous, at least during the first 10 years. For example, the use of externally sprayed insecticides on corn and cotton crops has fallen substantially over this period.

The disadvantage has been a more recent, increasing resistance to Bt toxin among the insect pests of these crops. As a result, there has had to be a return to the use of externally-applied insecticides to maintain crop productivity.

Meanwhile, opponents of the use of Bt genetically modified crops continue to advance arguments and anxieties about their adoption (Table 19.8).

Question

13 Explain why any threat to honey bees and butterflies poses such a problem to agriculture and horticulture.

Table 19.8 Arguments advanced against the use of Bt crops and responses

Argument	Response
Pollen from Bt crops could kill butterflies, including the monarch butterfly.	This has been queried, since the pollen of Bt plants contains a very low level of toxin. Experimental evidence suggests this is not a realistic danger. Further, monarch butterflies do not consume maize pollen or even associate with the plants.
Bt genes may be transferred from Bt maize to natural maize in neighbouring plots.	A large-scale study initially failed to find evidence of contamination. Subsequent experiments have shown some small scale genetic contamination. Research continues.
Bt genes in crops may be linked to the recent disease of colony collapse in honey bee hives.	This is a recent problem and its cause, whether due to an unknown parasite or some toxin, is unknown. A study by an agricultural research group has found no evidence that Bt crops adversely affect bees, so far.

Vitamin A-enhanced rice

Vitamin A (retinol) is found only in foods of animal origin, but milk and many vegetables and fruits contain the deep orange pigment, **beta-carotene (β-carotene)**. This molecule (a photosynthetic pigment, page 264) is easily converted to retinol in the body. Beta-carotene is a good indirect source of vitamin A (Figure 19.23).

Vitamin A is essential for vision in dim light and a deficiency leads to night blindness. For children in many parts of the world a severe deficiency leads to total blindness. More recently, it has been shown that vitamin A is crucial to the functioning of our immune system. Because of this, a deficiency of vitamin A is the cause of very many childhood deaths from such common infections as diarrhoeal disease and measles.

We have already noted that rice is the staple for very many people. Whilst the green foliage of rice contains beta-carotene, the grain contains none. A genetically modified variety of rice has been produced which stores significant (and potentially life-saving) quantities of beta-carotene in the grain.

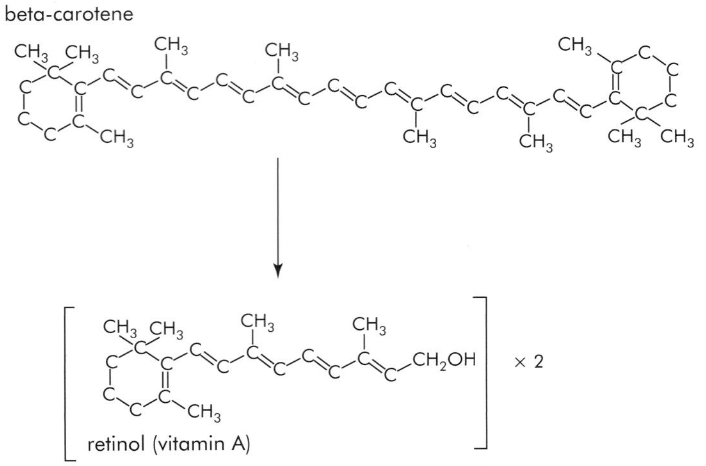

Figure 19.23 Beta-carotene as a source of vitamin A

We earlier saw that herbicide resistance had been introduced into some crops using the tumour-forming bacterium, *Agrobacterium* (page 395). This soil-inhabiting bacterium invades broad-leaved plants at the junction of stem and root, forming a huge growth called a tumour or crown gall. However, other plants, including the grasses are not attacked in this way. Consequently, a different method of delivering the genes for storage of beta-carotene in rice grain had to be used. In rice, the genes were delivered directly into the cells using extremely small gold or tungsten 'bullets' that were coated with the required DNA and then literally fired in with an appropriately sized gun. This method of genetic modification is used quite widely now.

A helping of about 300 g of these genetically modified rice grains, when cooked, contains sufficient precursor of vitamin A to meet a person's daily requirements. However, there remains a great deal of opposition to it. For example, field trials of this crop have not even been undertaken in some parts of the world, including in countries where vitamin A deficiency causes health problems. Opponents of the product argue that a more balanced diet that includes fresh green vegetables is a healthier solution. One might query whether this response is realistic, given the huge numbers of people that live in crowded urban environments where many items of a balanced diet are not available to most people. There is no doubt that genetic modification is a technology with many opponents.

GM salmon

Wild salmon (*Salmo salar*) have been genetically engineered to include growth-hormone regulating genes from Chinook salmon of the Pacific ocean. This has resulted in a much faster growth rate, and one that is maintained all the year round, rather than just in the spring and summer when growth is fastest in wild and fish-farmed salmon. These GM fish grow to market size in 18 months, rather than 3 years.

The role of crops with herbicide and disease resistance

The production of disease-resistant crops has been discussed previously (page 396).

Herbicide-resistant crops

Transgenic flowering plants may be formed using tumour-forming *Agrobacterium*. This soil-inhabiting bacterium sometimes invades broad-leaved plants at the junction of stem and root, forming a huge growth called a tumour or crown gall. The gene that induces tumour formation occurs naturally in a plasmid in the bacterium, known as a **Ti plasmid**. Useful genes may be added to the Ti plasmid (using restriction enzymes and ligase – see page 458), and the recombinant plasmid placed back into *Agrobacterium*. Then a host crop plant can be infected by the modified bacterium. The gall tissue that results may be cultured into independent plants, all of which also carry the useful gene. The plant will then make the gene product, which will be useful to humans. Herbicide-resistant sugar beet and other crop plants have been produced in this way (Figure 19.24).

Glyphosate is a powerful systemic herbicide. Once it has been absorbed by a green plant it is translocated to all tissues and organs and inhibits an enzyme essential for the production of amino acids. The outcome is that the whole plant dies. This enzyme is absent from animals, so glyphosate has very low toxicity for animal life. On reaching the soil, glyphosphate is inactivated and rendered harmless by soil bacteria.

The gene coding for the enzyme that inactivates glyphosphate has been identified in these soil bacteria, isolated and then transferred to the crop plants. These plants are now referred to as 'herbicide-resistant' because when the herbicide is applied the GM crop plant degrades the herbicide molecules and remains unharmed. Meanwhile, glyphosphate is also absorbed by weeds and kills even the largest weeds, including their extensive root systems.

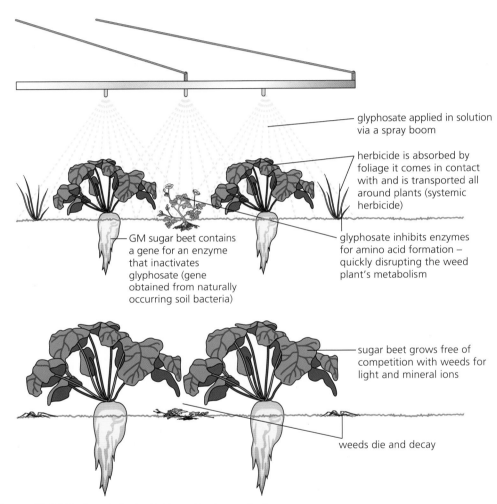

glyphosate applied in solution via a spray boom

herbicide is absorbed by foliage it comes in contact with and is transported all around plants (systemic herbicide)

GM sugar beet contains a gene for an enzyme that inactivates glyphosate (gene obtained from naturally occurring soil bacteria)

glyphosate inhibits enzymes for amino acid formation – quickly disrupting the weed plant's metabolism

sugar beet grows free of competition with weeds for light and mineral ions

weeds die and decay

Figure 19.24 Treatment of a herbicide-resistant sugar best crop

Table 19.9 Benefits and potential dangers of growing glyphosphate-resistant crops

Benefits	Dangers
• It eradicates weeds around sugar beet crops, where loss of yield due to competition with weeds is very high. • Glyphosate herbicide is transported all around the weed plant, killing even the largest weeds with extensive roots. • Glyphosate inhibits an enzyme for the production of essential amino acids in plants. This enzyme is absent from animals, so glyphosate has very low toxicity. • An average of five sprays of herbicide are applied to sugar beet in conventional cultivation, whereas with GM beet a maximum of three spays (and normally only one or two) are needed because of the greater activity of glyphosate. • Cereals normally precede spring-sown beet. Cereal stubble is an important source of food for bird life in autumn and winter. With glyphosate herbicide applied to GM beet being so effective, there is no need for weed control in stubble the preceding autumn – leaving wild birds' food sources available longer. • Glyphosate herbicide provides good weed control, allowing cultivations by discs or tines (conservation tillage) rather than by ploughing. These systems have fewer harmful effects on soil organisms, increase moisture, improve soil structure, and reduce tillage costs. • Glyphosate herbicide is applied as large droplets from coarse nozzles, rather than as a fine mist of tiny droplets that are prone to drift onto surrounding habitats.	• Fewer weed species on farmland and longer periods without weeds growing breaks wildlife food chains. • Selection of a glyphosate-resistant crop plant ties the grower to one particular herbicide product – choice is lost. • Will genetically modified plant material, when consumed by humans, release novel toxins or otherwise adversely affect existing enzyme systems in the human digestive system? • Genetically engineered genes may be vectored in plasmids containing an antibiotic-resistance gene to facilitate processes in the lab. This latter gene might be accidentally transferred to human gut bacteria via food eaten. • 'Superweeds' may develop by cross-pollination between herbicide-tolerant crop plants and compatible weed species (fat-hen, sea beet and weed beet are naturally occurring near relatives of sugar beet). Superweeds would be difficult to eradicate from crops. • There is a possibility of cross-pollination between GM crops and conventional and organic crops. The maintenance of sufficient distance or the devising of effective barriers to prevent or reduce pollen transfers between crops may be difficult to achieve. • If glyphosate herbicide droplets do reach hedgerow plants (or further) their size and chemical activity in plants means they are more likely to do damage.

Ethical and social implications of GM organisms in food production

In the description of the development of genetically-modified organisms (GMOs) presented in this topic, specific ethical, social and scientific issues have already been raised (see pages 477–81).

In preparing a paper for discussion with peers and your tutor, the issues can be researched and your own opinion developed within the following framework:

● What social and legal controls does society need, to influence these developments?

● What effects may GMOs have on the environment and on surrounding, unrelated species?

● What health risks are raised by GMOs, and what are their effects on human health?

● Who will have access to the technology, and how will scarce resources be allocated?

Because GMO technologies have been available for a relatively short while, some possible long-term effects may not yet have emerged. Remember, 'absence of evidence is not evidence of absence'. The greatest dangers may be one or more that have not yet emerged. On the other hand, if calculated risks are not taken, no progress is possible.

Summary

- **Gene technology** includes **genetic engineering, DNA sequencing** and **genetic fingerprinting**. Gene technology has applications in biotechnology, medicine and the pharmaceuticals industries, agriculture and forensic science. Gene technology generates many benefits for humans, but there are potential hazards, too, so gene technology raises **ethical issues**.

- **Genetic engineering** involves the transfer of genes from one species to another, possibly unrelated organism. Genes are transferred by inserting DNA into a vector, typically a **plasmid**, a tiny ring of double-stranded DNA obtained from a bacterium. The gene and the vector are cut by means of the same **restriction enzyme**, forming compatible **sticky ends** at the cuts. The gene and plasmid are then brought together and joined using the enzyme **ligase**. An alternative vector is the nucleic acid of a virus.

- Bacteria that have been genetically engineered to carry a new gene are **identified** by the use of plasmids that carry genes for resistance to two antibiotics, known as **R-plasmids**. Alternatively, a **fluorescent marker** is used.

- For a gene to be expressed (transcribed into messenger RNA) a **promoter** normally needs to be attached and activated. One example is the lac promoter mechanism present in certain bacteria.

- A gene may be **constructed from the messenger RNA** that it codes for, using the enzyme **reverse transcriptase**. Alternatively, copies of genes may be built up from nucleotides in the correct sequence. This is worked out from the **amino acid sequence** of the protein the gene codes for using the **genetic code**.

- **Genetically modified bacteria** are now used to make valuable products such as human **insulin** (for the treatment of diabetics), human **growth hormone** and several other hormones and enzymes of use in medicine, agriculture and other industries.

- **Genetically modified eukaryotes** are harder to produce. They carry two copies of a gene, unlike prokaryotes, so the engineering processes are often more difficult. However, food plants with **herbicide resistance**, **insect resistance** or improved food value have been produced.

- **DNA sequencing** involves the creation of genomic libraries of the precise **sequence of nucleotides** in samples of DNA of individual organisms. The nucleotide sequence in the whole human genome was the product of the **Human Genome Project**. Many other genomes have been completely sequenced, too.

- **Genetic profiling** involves the analysis of DNA to identify the individual from which the DNA was taken. It is used to establish identity and the genetic relatedness of individuals, for example in forensic science. The DNA samples are cut by **restriction enzymes** and the fragments separated. **Electrophoresis** is a process used to separate molecules such as proteins and nucleic acid fragments (of either DNA or RNA) on the basis of their net charge and size.

- **Genetic disorders** are heritable conditions that are caused by a specific defect in a gene or genes. Most arise from a **mutation** involving a **single gene**. Genetic disorders affect about 1–2 per cent of the human population and include sickle cell anaemia, haemophilia and **cystic fibrosis**. It is possible that the symptoms of a genetic disorder like cystic fibrosis may eventually be overcome by **gene therapy**. The Human Genome Project has made it possible to **screen** for the presence of the mutant cystic fibrosis allele. Consequently, affected people and their relatives may be referred to a **genetic counsellor** whose role is to take a detailed case history and to provide information from which their clients can make choices.

- Crop improvements are also attempted by genetic engineering. Projects have included herbicide-resistant oil seed rape and sugar beet, insect-resistant maize and cotton and vitamin A enhanced rice. The results have often been cautiously and even critically received, sometimes with possible good reasons. Genetic engineering remains a technology with many opponents. Very few countries have approved genetically modified crops for commercial production.

Examination style questions

1 a) A husband and wife who already have a child with cystic fibrosis (CF) elected to have their second child tested for the condition while still a fetus in very early pregnancy. The results of the test, a DNA banding pattern, were discussed with a genetic counsellor.
The relevant DNA banding pattern produced by electrophoresis is shown in Fig. 1.1.

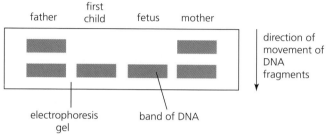

Fig. 1.1

With reference to Fig. 1.1, explain why,

ii) the fetus will develop CF [1]

ii) the positions of the bands of DNA of the first child and of the fetus indicate that the mutant allele for CF has a deletion in comparison with the normal allele. [2]

b) Explain briefly the need to discuss the result of the test with a genetic counsellor. [4]

[Total: 7]

(Cambridge International AS and A Level Biology 9700, Paper 41 Q6 November 2009)

2 In humans, the gene *RPE65* encodes a protein responsible for regenerating visual pigment in rod and cone cells after they have been exposed to light. A recessive allele of this gene causes impaired vision from birth, progressing to complete blindness in early adulthood. This condition is called LCA.
In 2008, trials were carried out into the possibility and safety of treating LCA using gene therapy.

a) Suggest and explain why LCA is suitable for treatment using gene therapy. [3]

b) Six adults with this condition were used in the study. Genetically modified adenoviruses (a type of virus that can cause respiratory infections) were used as vectors.
The vectors were injected beneath the retina of one eye of each of the participants.
Suggest two ways in which the genome of the adenoviruses used as vectors would differ from that of normal adenoviruses. [2]

c) Improvements were found in the vision of all the participants, but the small number in the trials made most of these improvements not statistically significant.
Suggest why these trials were designed to include such a small number of participants. [2]

[Total: 7]

(Cambridge International AS and A Level Biology 9700, Paper 41 Q5 November 2011)

3 a) Four naturally occurring enzymes make possible the genetic engineer's activities. List these enzymes, their sources, and the reaction they catalyse. [8]

b) In the *in-vitro* synthesis of human insulin, a copy of the gene has been made by first isolating mRNA from human pancreatic tissue. Outline the steps that this process involves. [6]

c) Explain the advantages gained by:

i) Making the insulin gene from mRNA compared to obtaining genes directly from relevant chromosomes. [3]

ii) The choice of insulin obtained by genetic engineering for the treatment of diabetics, compared to the use of insulin from animal sources. [3]

[Total: 20]

4 a) Give an illustrated account of the separation of DNA fragments by electrophoresis. [6]

b) Explain:

i) The nature and significance of 'variable number tandem repeat' sections of chromosomes in DNA fingerprinting.

ii) The steps in the production of a DNA profile. [10]

c) Outline the diverse applications of DNA profiling in science and society. [4]

[Total: 20]

5 a) Outline the steps to crop improvement by genetic modification with reference to canola sugar beet crops and herbicide resistance.

b) Identify both the potential benefits of genetic engineering of food crops for agricultural production and the possible hazards that may arise, illustrating your answer with reference to current examples.

Answers

1 Cell structure

1 Transfer of energy (respiration), feeding or nutrition, metabolism, excretion, movement and locomotion, responsiveness or sensitivity, reproduction, growth and development

2 a) 1 mm = 1000 μm.
Since 1000 ÷ 100 = 10, 10 cells of 100 μm will fit along 1 mm.

b) The drawing of *E. coli* is 64 mm in length.
Its actual length is 2.0 μm.
Therefore, magnification = 64 × (1000 ÷ 2) = ×32 000

3

Feature	Plant cells	Animal cells
Cell wall	Cellulose cell wall present	No cellulose cell walls
Chloroplasts	Many cells contain chloroplasts, site of photosynthesis	No chloroplasts; animal cells cannot photosynthesise
Permanent vacuole	Large, fluid-filled vacuole typically present	No large permanent vacuoles
Centrosome	No centrosome	A centrosome present outside the nucleus
Carbohydrate storage product	Starch	Glycogen

4 Cells vary greatly in size and in degree of specialisation of structure associated with their various functions.

5 Length of cell is 90 mm = 90 000 μm.
Magnification = 400
Actual length is 90 000 / 400 = 225 μm.

6 Magnification is ×60 (6 × 10)

7 See 'The magnification and resolution of an image', page 11.

9 The electron microscope has powers of magnification and resolution that are greater than those of an optical microscope. The wavelength of visible light is about 500 nm, whereas that of a beam of electrons used is 0.005 nm. At best the light microscope can distinguish two points which are 200 nm (0.2 μm) apart, whereas the transmission electron microscope can resolve points 5 nm apart when used on biological specimens. Given the sizes of cells and of the organelles they contain, it requires the magnification and resolution achieved in transmission electron microscopy to observe cell ultrastructure – in suitably prepared specimens.

10 Staining is necessary to provide contrast between the cytosol and the nucleus in an otherwise more or less transparent cell. However, the nucleus can then be seen. Ribosomes, however, are too small to be detected by light microscopy.

11

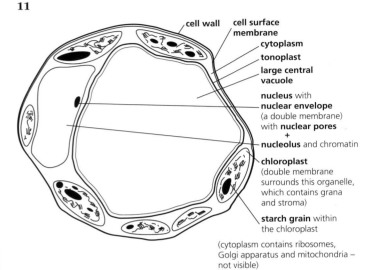

cell wall • cell surface membrane • cytoplasm • tonoplast • large central vacuole • nucleus with nuclear envelope (a double membrane) with nuclear pores + nucleolus and chromatin • chloroplast (double membrane surrounds this organelle, which contains grana and stroma) • starch grain within the chloroplast

(cytoplasm contains ribosomes, Golgi apparatus and mitochondria – not visible)

12 See 'ATP – the universal energy currency' pages 236–7.

13 a) A cell wall is characteristic of plant cells. It is entirely external to the cell, surrounding the cell surface membrane. The wall is not an organelle. Plant cell walls are primarily constructed from cellulose, an extremely strong material. When a growing plant cell divides, a cell wall is laid down across the old cell, dividing the contents. This primary cell wall has more layers of cellulose added, forming the secondary cell wall. In some plant cells the secondary layers of cellulose become very thick indeed. Cell walls are absent from animal cells. Prokaryotes have cell walls, but they are chemically different from those of plants.
The cell surface membrane is the membrane that surrounds and contains the cytoplasm of all cells. It is constructed almost entirely of protein and lipid. The lipid of membranes is phospholipid, arranged as a bilayer. The proteins of cell surface membranes are globular proteins which are buried in and across the lipid bilayer, with most protruding above the surfaces. The whole structure is described as a fluid mosaic. See Figure 4.3, page 76.

b) See page 16.

c) See page 16.

d) See Table 1.2, page 23.

e) See pages 1–20.

f) See Figure 1.14, page 15.

2 Biological molecules

1 Non-organic forms of carbon are found in the biosphere as carbon dioxide in the atmosphere; dissolved in fresh and sea water as hydrogencarbonate ions and as carbonates, e.g. calcium carbonate and magnesium carbonate, in the shells of non-vertebrate animals, mostly marine but some of fresh water and terrestrial habitats, and as chalk and limestone rocks, which are formed from fossilised shells.

2 See Background chemistry for biologists on the CD.

3 See Disaccharides, page 35.

4 See Polysaccharides, pages 36–9, particularly Figure 2.9.

5 See The functional groups of sugars, page 32.

6 A polymer is a large organic molecule made of repeating subunits known as monomers, chemically combined together. Among the carbohydrates, starch, glycogen and cellulose are all polymers built from a huge number of molecules of glucose (the monomer, combined together in different ways in the three, distinctive polymers).

7 See page 32 and Figure 2.18, page 45.

8 Given the naturally occurring pool of 20 different types of amino acids, in a polypeptide of only 5 amino acid residues there can be 20^5 different types of polypeptide, or 3 200 000. In a polypeptide of 25 amino acid residues there can be 20^{25} different types.

In a polypeptide of 50 amino acid residues there can be 20^{50} different types. A polypeptide of more than 50 amino acid residues is, by convention, called a protein.

So, there is virtually almost unlimited structural variation possible in the proteins of cells. Remember, as the sequence of amino acid residues changes, so does the structure and properties of the resulting protein.

9 See page 35 and Figure 2.19, page 46.

10 See Figure 2.22, page 47.

11 The term *macromolecule* means 'giant molecule'. We have seen that polysaccharides (e.g. cellulose) and proteins (e.g. lysozyme) are made up of vast numbers of repeating smaller molecules (glucose, in the case of cellulose; different amino acids in the case of proteins). Lipids, on the other hand, are composed of triglycerides – relatively much smaller molecules. However, as these are hydrophobic, they may clump together and give the appearance of being huge molecules.

12 See Background chemistry for biologists on the CD.

13 In gases, molecules are widely spaced and free to move about independently. In liquids, molecules are closer together. In the case of water, hydrogen bonds pull the molecules very close to each other, which is why water is a liquid at the temperatures and pressure that exists over much of the Earth's surface. As a result, we have a liquid medium which life exploits.

3 Enzymes

1 Enzymes work by binding to their substrate molecule at a specially formed pocket or crevice in the enzyme – the active site. Most enzymes are large molecules and the active site takes up a relatively small part of the total volume of the enzyme molecule. Nevertheless, the active site is a function of the overall shape of the globular protein and, if the shape of an enzyme changes for whatever reason, the catalytic properties may be lost.

2 a) A catalyst is a substance that alters the rate of a chemical reaction, but remains unchanged at the end.

b)

Inorganic catalysts	Enzymes
Typically a metal, such as platinum, in a finely-divided state, e.g. as platinised mineral wool	Typically made of protein
Able to withstand high temperatures, high pressures and extremes of pH, if necessary	Protein is easily denatured by high temperature, for example
	Extremely specific – most are specific to one type of substrate molecule

3 a)

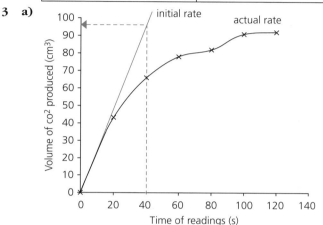

b) 96/40 = 2.4 cm^3 s^{-1}

4 Pre-incubation is required to ensure that when the reactants are mixed the reaction occurs at the known, pre-selected temperature.

5 a) pH is a measure of the acidity or alkalinity of a solution. (Strictly, pH is a measure of the hydrogen ion concentration.) A buffer solution acts to resist any change in pH when the solution is diluted or if a little acid or alkali is added. Many buffers used in laboratory experiments contain a weak acid, such as ethanoic acid (vinegar) and its soluble salt (e.g. sodium ethanoate). In this case, if acid is added, the excess hydrogen ions are immediately removed by being combined with ethanoate ions to form undissociated ethanoic acid. Alternatively, if alkali is added, the excess hydroxyl ions immediately combine with hydrogen ions forming water. At the same time, more of the ethanoic acid dissociates, adding more hydrogen ions to the solution. The pH does not change in either case.

b) In the body of the mammal, the blood is very powerfully buffered by the presence of a mixture of phosphate ions, hydrogencarbonate ions and blood proteins. The blood is held at a pH of between 7.35 and 7.45. pH is very important in living organisms because it affects the shape of enzymes, almost all of which are proteins. For the same reasons, buffers are often used in enzyme experiments. See also Background chemistry for biologists on the CD.

Answers

6 a) See Figure 3.13, page 66.
  b) See Enzyme concentration, page 67.

4 Cell membranes and transport

1 In periods of seasonal or prolonged low temperature, the proportion of unsaturated fatty acids in the plant cell membranes can be expected to increase.

2 A lipid bilayer is shown in Figure 4.2 (page 75). In the presence of sufficient lipid, molecules of lipid arrange themselves as a bilayer, with the hydrocarbon tails facing together. This latter is the situation in the cell surface membrane.

Several organelles of eukaryotic cells have a double membrane, including chloroplasts and mitochondria, in which there are present an outer and an inner membrane, both of which consist of lipid bilayers.

So an important difference between a lipid bilayer and a double membrane lies in the number of lipid bilayers present.

3 a)

Dimension (mm)	1 × 1 × 1	2 × 2 × 2	4 × 4 × 4	6 × 6 × 6
SA (mm²)	6	24	96	216
Volume (mm³)	1	8	64	216
SA/V ratio	6 / 1 = 6	24 / 8 = 3	96 / 64 = 1.5	216 / 216 = 1

  b) Plot the graph of SA/V using an Excel spreadsheet.
  c) As a cell increases in size the SA/V decreases.
  d) As a cell increases in size diffusion becomes progressively less efficient as a mechanism for removal waste, such as CO_2 from the interior of the cell.

4 a)

Dimensions/mm	SA / V ratio
10 × 10 × 10	600 / 1000 = 0.6
5 × 5 × 5	150 / 125 = 1.2
2.5 × 2.5 × 2.5	37.5 / 15.6 = 2.4

  b) Hydrogen ions enter the gelatin blocks by diffusion and there cause the observed colour change. Relatively large blocks (the 10 × 10 × 10 mm cubes) have a low *SA / V* ratio so little of the interior matter is close to the external environment and the diffusion path is a long one. Here, colour change is slowest. The reverse is true of the smallest gelatin blocks (the 2.5 × 2.5 × 2.5 mm cubes).

5 Compare Figure 4.8, page 81 and Figure 4.9, page 82. The difference is the mechanism of the permeability of the membrane transversed; in facilitated diffusion this is due to the properties of the substance that passes – it triggers the opening of pores through which diffusion occurs.

6 Water can diffuse cross a partially permeable membrane and so can some solutes, but not others.

7 The concentrated solution of glucose (where the concentration of free water molecules is low) will show a gain of water molecules at the expense of the dilute glucose solution.

A 'net' gain of water molecules refers to the fact that very many more water molecules diffuse into the concentrated solution than diffuse out.

8 The water potential of the jam is much lower than that of the fungal cytoplasm. A rapid net outflow of water occurs from the fungal hyphae, inactivating the fungal spore.

9 a) The dilute sucrose solution
  b) The concentrated sucrose solution
  c) The solution with the lower water potential (concentrated sucrose solution)

10 a) The dilute solution of glucose
  b) The dilute solution of glucose
  c) The concentrated solution of glucose

11 See Figure 4.16, page 88.

12 More water molecules will diffuse in than out. Ultimately the integrity of the cell surface membrane is threatened and survival of the animal cell is not ensured.

In a plant cell, however, as uptake of water continued the cell would become extremely turgid. The pressure potential would eventually offset the water potential of the cell and net water uptake would cease.

13 The ψ of cell M = -580 + 410 = -170 kPa.
The ψ of cell N = -640 + 420 = -220 kPa
Net water flow will be from cell M to cell N, because water flows from a less negative to a more negative water potential.

14 Uptake of ions is by active transport involving metabolic energy (ATP) and protein pumps located in the cell surface membranes of cells. As with all aspects of metabolism, this is a temperature-sensitive process and occurs more rapidly at 25 °C than it does at 5 °C.

Ions are transported across the membrane by specific, dedicated protein molecules. Because there are many more sodium ion pumps than chloride ions pumps, more of the former ion is absorbed.

15 Phagocytic cells of the lungs and of the airways serving the lungs, engulf bacteria, bacterial spores and dust particles – in fact, any small particles of foreign matter that reach those surfaces, rendering them less harmful.

16 a) See The cell surface membrane, pages 74–7.
  b) See Figure 4.5, page 78.
  c) See Figure 4.13, page 86.
  d) See Movement by bulk transport, page 93.

5 The mitotic cycle

1 See Chromosomes occur in pairs, page 98.

2 The result of the opposite charges is a strong attraction and high binding affinity between the DNA strand as it wraps around the histone-protein core of the nucleosome.

3 a) A haploid cell contains one set of chromosomes, the basic set.
  b) i) For example, in the layer of cells forming sperm, in the testes
     ii) For example, in the bone marrow where red blood cells are formed or in the generative layer of the skin at the base of the epidermis

4

	Meiosis	Mitosis
The products	2 identical cells, each with the diploid chromosome number.	4 non-identical cells, each with the haploid chromosome number.
The consequences and significance	Produces identical diploid cells. Permits growth within multicellular organisms, and asexual reproduction	Produces haploid cells. Contributes to genetic variability by: • reducing chromosome number by half, permitting fertilisation, and the combination of genes from two parents; • permitting random assortment of paternal and maternal homologous chromosomes; • recombination of segments of individual maternal and paternal homologous chromosomes during crossing over.

5 See Table 5.2 Common carcinogens known to increase mutation rates and the likelihood of cancer, page 105.

6 46 × 2 = 92

7 a) See table under part (b).
Your pie chart should comprise sectors with the following angles at the centre: Prophase 252°, Metaphase 36°, Anaphase 18°, Telophase 54°.

b)

Stage of mitosis	Percentage of dividing cells	Time taken by each step/minutes
Prophase	70	70 ÷ 100 × 60 = 42
Metaphase	10	10 ÷ 100 × 60 = 6
Anaphase	5	5 ÷ 100 × 60 = 3
Telophase	15	15 ÷ 100 × 60 = 9

6 Nucleic acids and protein synthesis

1 The nitrogenous bases are organic compounds. The carbon 'backbone' of nitrogenous bases is a ring molecule containing two or more nitrogen atoms in the ring, covalently bonded together. They are derived from one of two parent compounds, purine (a double ring compound) or pyrimidine (a single ring compound).
An inorganic base is a substance that can accept a hydrogen ion and so neutralise an acid. Strong bases are substances like sodium hydroxide. They are ionised compounds.

2 The most likely reason why many replication forks are required in the replication of a single strand of DNA may be that there are (typically) 80 million base pairs to be copied. A single replication fork would seem to be inadequate.

3 a) After three generations, 75 per cent of the DNA would be 'light'.

b)

The experimental results that would be expected if the Medelson-Stahl experiment were carried on for three generations.

4 See Figure 6.11, page 120.

5 See Figure 2.19, page 46.

6

RNA involved in transcription	
messenger RNA (mRNA)	See The stages of protein synthesis, Stage one: transcription, page 120.
RNA involved in translation	
messenger RNA (mRNA)	mRNA is now a linear molecule in the cytoplasm along which ribosomes move, 'reading' the genetic code and transcribing the information into a linear sequence of amino acid residues (= primary structure of the protein)
transfer RNA (tRNA)	See The stages of protein synthesis, Stage two: amino acid activation, page 121.

7 See also Figure 6.10, page 119 and Figure 6.16, page 124.
a) Gly (glycine), Asn (asparagine), Pro (proline), Phe (phenylalanine), Val (valine), Thr (threonine), His (histidine), Cys (cysteine)
b) CCT, TTA, GGA, AAA, CAA, TGA, GTA, ACA
c) See Triplet codes, codons and anticodons, page 122.

7 Transport in plants

1 The shape of the cellulose polymer allows close packing into long chains (see Figure 2.8, page 37) held together by hydrogen bonds. The cellulose fibres are laid down in porous sheets and have great tensile strength. The monomer from which cellulose is assembled, by a condensation reaction, is glucose, of which green plants typically have an excellent supply.

2 a)

	compact			flat and thin		
	small	medium	large	small	medium	large
dimensions/ mm	1 × 1 × 1	2 × 2 × 2	4 × 4 × 4	2 × 1 × 0.5	8 × 2 × 0.5	16 × 8 × 0.5
surface area / volume ratio	6.0	3.0	1.5	7	5.25	4.4

b) As the size of an object increases, so does the surface area. However, the increase is very much greater in a flat, thin object compared to in a compact object.
c) The surface area/volume ratio decreases as the size of an object increases but the decrease is very much greater in compact object than in flat, thin objects.
d) Movements of nutrients into an organism and of waste products out of an organism by diffusion alone is only likely to be possible in small organisms (effectively, unicellular organisms). However it may be a possible mechanism for transport in small multicellular organisms of a flat thin shape.

Answers

3　**TS of a sunflower stem (low power) – plan diagram**

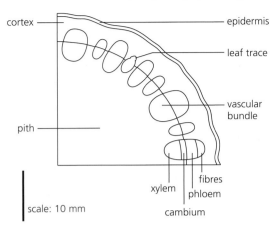

4 Your answer should refer to the features of their structure (see The movement of water through the plant, pages 135–8, particularly Figure 7.8), their intimate contact with the soil solution and to their position on the growing roots. The significance of a gradient in water potential between soil solution and the root cortex is explained in Uptake of water – the roles of the root hairs (page 136).

5 The casparian strip blocks the apoplast pathway of water movement across the cortex at the endodermis.

6 As temperature increases, evaporation of water from the surfaces of the cells in the leaf increases. At the same time, the movement of molecules of water is increased (their kinetic energy is raised), so diffusion occurs more quickly. Also, as temperature increases, the humidity of the air is decreased because warm air holds more water vapour. So the gradient in water vapour between the interior of the leaf and air outside is increased and diffusion of water vapour is enhanced.

7 See The transpiration stream, pages 140–1.

8 Transpiration is a direct consequence of plant structure, plant nutrition and the mechanism of gas exchange in leaves. In effect, the living green plant is a 'wick' that steadily dries the soil around it. Put like this, transpiration is an unfortunate consequence of plant structure and metabolism, rather than a valuable process. It means that a constant, adequate supply of water is critical to plant growth; if the water supply fails, plants cannot move elsewhere, as animals tend to do.

9 a) Plant growth is dependent on a supply of chemically combined nitrogen (for protein synthesis) of which nitrates are typically the most readily available. However, nitrates are also taken up by microorganisms and may be released in the soil at times other that when plant demand is at its peak. They are also very soluble and are easily leached away into ground water in heavy rain. By taking up nitrates whenever they are available (and storing them in cells), plant growth can be maintained.

b) Plant roots are metabolically very active and require oxygen for aerobic respiration and ATP formation (for ion uptake, for example). Waterlogged soil lacks soil air and the essential gas, oxygen.

10 Cultivation and cropping of plants to supply food for the market place involves the 'harvesting' of organic matter (and the nutrient ions built into it during growth). If the supply of essential minerals that plants require is not maintained in the soil, the productivity of the land will rapidly fall. Consequently, largely soluble nutrients are added as artificial fertilisers spread on the soil as granules at the time of peak crop growth. Alternatively, farmyard manure is applied to the soil at the time of ploughing. This decays slowly and releases nutrients more steadily.

11 Mitochondria produce ATP (the energy currency molecule). ATP transfers energy to enable the active transport of solutes in the sieve tubes. The presence of mitochondria confirms that phloem transport is an active process, using energy transferred during respiration.

12 a) The functioning of living cells (e.g. in the phloem tissue) would be impaired as proteins were denatured and cell surface membranes disrupted. Translocation would stop. Water flow in the xylem vessels would be unaffected.

b) The concentration of sugar might be lower in a sieve tube near the base of the stem, if starch is being stored in the cells at that point.

13 a) Sugar delivered by the phloem from the leaves is stored as starch (insoluble) on arrival. Many of the ions absorbed from the soil are used in the metabolism of the root cells or are carried to the stem and leaves in the xylem.

b) Sugars produced in the light in the chloroplasts accumulate in the cells of the leaf before being translocated away to 'sink' sites.

8　Transport in mammals

1 For example, it means that blood cannot be directed to a respiratory surface immediately before or after servicing tissues that are metabolically active (and so have a high rate of respiration). Rather, the blood circulates randomly around the blood spaces and blood vessels.

2 5×10^6 cells per mm^3 is equivalent to 5×10^9 cells per litre. So, in 5 litres (i.e. the whole body), there are $5 \times (5 \times 10^9) = 2.5 \times 10^{10}$ cells.

If a red blood cell lasts for 120 days, divide the total number of cells by 120 to find how many, on average, need to be replaced each day:

$2.5 \times 10^{10} \div 120 = 2 \times 10^8$ approximately

3

Pulmonary circulation	Systemic circulation
Carries deoxygenated blood under high pressure from the right side of the heart to the lungs	Carries blood under high pressure from the left side of the heart to the all the body organs
Returns oxygenated blood under lower pressure to the left side of the heart	Returns deoxygenated blood under lower pressure to the right side of the heart
Consists of pulmonary arteries delivering blood to lungs, capillary networks serving the air sacs of the lungs, and pulmonary veins carrying blood back to the left side of the heart	Consists of aorta and arteries delivering blood to the body organs, capillary networks serving all the tissues and cells of the organs of the body, and vena cava carrying blood back to the right side of the heart

4 a) The velocity and pressure of the blood as it enters the aorta are both high, but as it is about to re-enter the heart the pressure is very low. However, close to the heart, the velocity of the blood in the veins is similar to that of the main arteries leaving the heart. This is necessary to ensure adequate venous return to balance the cardiac output.

b) The blood speeds up on leaving the capillaries because the total cross-sectional area of the venules is much less than the capillaries. The total cross-sectional area of the venous system further decreases as venules join up to form the veins.

c) The velocity of the blood is at its lowest in the capillaries. Capillaries are the smallest blood vessels but their total cross-sectional area is typically 800 times greater than that of the aorta leaving the heart (diameter of about 25 mm). The slow blood flow in the capillaries allows for efficient exchange.

5 a) The components of the blood that are not found in tissue fluid are the blood proteins, mainly albumin, and red blood cells and platelets. (White blood cells are found in the tissue fluid – see page 153).

b)

	Tissue fluid	Lymph
Urea	Absent	Present
Respiratory gases	Contains oxygen but no carbon dioxide	Contains carbon dioxide but no oxygen
Products of digestion	Present in abundance	Reduced to products of lipid digestion – and with proteins in lymph draining from the liver
Lymphocytes	Very few or absent	Present

6 A fall of about 25 per cent

7 These non-elastic strands keep the heart valve flaps pointing in the direction of the blood flow. They stop the valves turning inside out when the pressure rises abruptly within the ventricles.

8 a) The pressure in the aorta is always significantly higher than that in the atria because blood is pumped under high pressure into the aorta and, during diastole and atrial systole, the semilunar valves prevent backflow from the aorta. Meanwhile, blood enters the atria under low pressure from the veins and the pumping action of the atria is slight compared to that of the ventricle, which generates our 'pulse'.

b) Pressure falls abruptly in the atrium once ventricular systole is underway as atrial diastole begins then.

c) The semilunar valve in the aorta only opens when the pressure in the ventricles exceeds that of the pressure in the aorta.

d) At the end of systole, pressure in the ventricles is high. When ventricular diastole commences the bicuspid valve will only open when pressure in the ventricles falls below that in the atria.

e) About 50 per cent of the cardiac cycle is given over to diastole – the resting phase in each heart beat. The heart beats throughout life and takes limited rest at these moments.

9 Gaseous exchange and smoking

1 Factors affecting the rate of diffusion and their consequences are identified and explained in Table 4.1.

2 Amount of oxygen in 1 litre of blood = 20 × 10 = 200 cm³
Amount of oxygen in 1 litre of water = 0.025 × 1000 = 25 cm³
Therefore, 175 cm³ more oxygen is carried by a litre of blood compared with the same quantity of water.

3

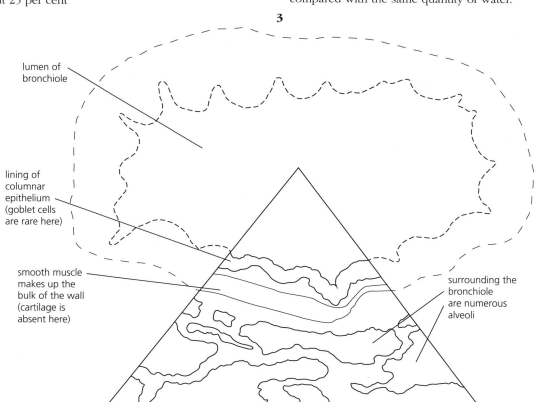

lumen of bronchiole

lining of columnar epithelium (goblet cells are rare here)

smooth muscle makes up the bulk of the wall (cartilage is absent here)

surrounding the bronchiole are numerous alveoli

Answers

4 About 10–15%. Land-living organisms obtain oxygen from the air, but diffusion within an organism occurs in solution, so the first step in oxygen uptake in the lungs is the dissolving of oxygen in the surface film of water at the respiratory surface. As a consequence, the respiratory surface is kept moist and water vapour will evaporate as gaseous exchange occurs. See Figure 9.6, page 176.

5 Gas exchange is the exchange of respiratory gases (oxygen and carbon dioxide) between cells of an organism and the environment; cellular respiration is the cellular process by which sugars and other substances are broken down to release chemical energy for other cellular processes. Gas exchange is a consequence of cellular respiration.

6 See Movement by bulk transport, page 93.

7

Feature	Effects and consequences
Surface area of alveoli	A huge surface area for gaseous exchange (Approximately 50 m^2, about the area of a doubles tennis court!)
Wall of alveoli	Very thin, flattened (squamous) epithelium (5 μm thick) so the diffusion pathway is short.
Capillary supply to alveoli	A network of capillaries around each alveolus supplied with deoxygenated blood from the pulmonary artery and draining into the pulmonary veins maintains the concentration gradients of oxygen and carbon dioxide.
Surface film of moisture	Oxygen dissolves in the water lining the alveoli and diffuses into the blood in solution.
Elastic fibres in walls	Stretched during inspiration, they then recoil, facilitating expiration.

8 Carbon dioxide is an acidic gas which, if it were to accumulate in the blood, would alter the pH of the plasma solution. The normal pH of the blood is 7.4. For life to be maintained it must remain within the range pH 7.0–7.8. This is largely because blood pH affects the balance of essential ions which are transported in the plasma solution. Efficient removal of respiratory carbon dioxide from the lungs is as important to life as efficient uptake of oxygen.

9 In 1950 the incidence of lung cancer in men was higher than in women, in keeping with the fact that the percentage of the adult female population that smoked cigarettes was lower, about 40 per cent compared with 60 per cent of males. In the period 1948–1975 the percentage of male cigarette smokers declined slowly whilst the percentage of female cigarette smokers increased slowly. Associated with these changing patterns in cigarette use, the incidence of lung cancer in males started to decline after 1985, whereas in females the incidence of lung cancer was slowly increasing throughout the period this statistic was recorded.

10 A control group in a medical investigation:
- comprises the same number of people as a patient group selected for an experimental treatment
- is as comparable with the patient group as possible, as regards age, gender, health, occupation and general life experiences of its members
- consists of people having the 'condition' for which treatment is being given to the experimental group.

The fate of the control group is followed in comparable detail to that of the treated patients in order to evaluate the efficacy of the experimental treatment. Note that the use of such a control group means that a treatment is withheld from people who might have benefited. This raises obvious ethical issues.

11
- Life expectancy has greatly increased – in the past, people died of other conditions first
- The development of antibiotics has led to a decrease in deaths from infectious diseases
- Improvements in knowledge and training of medics means that symptoms are more correctly attributed to actual causes
- Dietary changes
  People had less sedentary life styles

12
- Delivers carcinogens to the airways and air sacs
- Delivers carbon monoxide to red blood cells where it combines irreversibly with haemoglobin, preventing oxygen transport
- Raises blood pressure
- Triggers vasoconstriction of blood vessels
- Stimulates the secretion of viscous mucus by goblet cells of airways
- Inhibits the beating movements of cilia in the epithelium of the airways
- Leads to blocking of airways and the accumulation of dust and carcinogens in bronchioles
- Causes loss of natural elasticity of smallest bronchioles and air sacs
- Inflames the bronchi and causes persistent destructive coughing up of phlegm
- Leads to narrowing of airways and eventual destruction of air sacs
- Leads to the possible formation of a tumour – lung cancer

10 Infectious disease

1 Present in the wall of the stomach are millions of tiny pits called gastric glands. The cells in the gastric glands make and secrete the contents of the gastric juice. About 600 cm^3 of gastric juice is secreted per meal, secretion being coordinated with the presence of food. One of the components of gastric juice is hydrochloric acid (strength 0.15M) at about pH 1. This is sufficient to create an acid environment of pH 1.5–2.0 for the contents of the bolus, the optimum pH for the protein digesting enzymes of the gastric juice. The hydrochloric acid also kills many bacteria, among other useful effects.

2 In cholera, chloride ions move from epithelium cells into the lumen of the gut by active transport caused by protein pumps in the cell membrane, driven by ATP. Water moves by osmosis (a special case of diffusion), in which metabolic energy is not involved.

3 The insecticide DDT is harmless to humans at concentrations that are toxic to mosquitoes, but applied at these concentrations it rapidly collects in body fat and is transferred from prey to predator through the food chain, steadily increasing in concentration. It accumulates at the top of food chains at concentrations that may cause harm. The stability of the DDT molecule means that it only very slowly biodegrades. This persistence is why it is so effective, but is also why it is a health threat. By means of her book *Silent Spring*, it was Rachael Carson who publicised this problem.

4 These figures show us that, whatever the level of TB infection in a region, about 10 per cent of those infected die from the condition. Also, almost twice as many people in Africa are infected with TB as in Asia and the proportion of the population of Europe with TB is far lower than in either of the other two regions.

These differences in prevalence are most likely to be accounted for by three major factors:

- Housing, which is related to a general level of poverty and unemployment. Where the prevalence of TB is high the environmental and social conditions (particularly housing) are likely to favour transmission of the bacterium. Countries with high unemployment and low wages and countries experiencing local or civil wars are more likely to have a major part of the population living in crowded and unhealthy conditions.

- Access to medical care. The levels of diagnosis and treatment of illness vary substantially. More of the populations of African countries are without access to effective health infrastructures than in the other two regions, for example. Where there are fewer doctors or care centres per head of the population or in rural communities that are widely scattered, the chances of diagnosis, treatment and preventative measures are minimal.

- The cost of health provision. Some countries cannot afford to buy medicines or vaccines, or only the relatively well-off can afford them. Few countries have the equivalent of a health service so the poor do not get treatment.

5 HIV/AIDs causes particular problems for the people and the economies of less-developed countries such as Zimbabwe or Zambia because it almost exclusively handicaps the young adult population at their time of greatest economic activity and family building.

6 The pathogen is able to survive within only one specific host and there is no reservoir of the pathogen in other organisms (for example, any of the surrounding wildlife). By contrast, the TB bacillus exists as a reservoir in wild mammals (such as badgers in the UK).

7 When two events (A and B) regularly occur together, it may appear to us that event A causes event B. This is not necessarily the case; there may be a common event that causes both, for example, or it may an entirely false correlation. An example of the latter are those infants (very few) who develop the symptoms of autism shortly after the normal time in childhood when the MMR inoculations is administered. It was some time before detailed studies established that MMR inoculation and the onset of autism were not causally linked.

Because correlation does not prove cause was just one reason why Richard Doll's amassing of statistical evidence of a link between smoking and ill health was successfully resisted by the tobacco industry for an exceptionally long time. Now we know the various reasons why cigarette smoke triggers malfunctioning of body systems, ill health and diseases of various sorts. So, statistical confidence in the possibility of a causal link is a springboard to further investigation, not proof of a relationship.

8 The full range of reasons are contained in the text, How eradication came about, on pages 210. Individually they can be shown not to apply to one or more of the other diseases discussed in this topic.

9 a) Exposure of pathogenic bacteria to sub-lethal doses of antibiotic may increase the chances of resistance developing in that population of pathogens.

 b) By varying the antibiotics used there is increased likelihood of killing all the pathogens in a population, including any now resistant to the previous antibiotics used. This approach works until multiple-resistance strains have evolved, such as in strains of *Clostridium difficile* and *Staphylococcus aureus,* for example.

10 Antibiotics are added to the feed of intensively-reared livestock such as chickens and pigs. Here they reduce or prevent the incidence of many diseases, and the animals are found to grow better with them. Possible dangers arise from the over-exposure of bacteria to antibiotics – see Figure 10.29 (page 215).

11 Immunity

1 See What recognition of 'self' entails, page 220 and Figure 11.4, page 221.

2 Antigens may be present more or less anywhere in the body that can be contaminated from outside the body. Antibodies exist in the blood and lymph and may be carried in the blood plasma anywhere that blood 'leaks out' to, including sites of invasions. They also occur on B-lymphocytes.

3 See Figure 8.2, page 153.

4 Spleen cells and tumour cells cultured separately and then mixed together. Mixture is cultured to produced 'hybridoma' cells. Resulting cells are separated and tested for antibody required. Positive hybridoma cells are cultured further and produce antibody in significant quantities.

5 a) The existence of memory cells avoid the steps to the production of activated T-lymphocytes. The particular memory cell, once re-activated by the re-invading antigen, switches to production of an excess of the appropriate plasma cells and helper T-lymphocytes.

 b) The initial response to an antigen by the immune system is slow. For sufficient antibodies to overcome an infection takes weeks rather than days. When memory cells are present, many stages of the initial response are omitted and antibodies are quickly assembled.

6 Immunity is the resistance to the onset of disease after infection by harmful microorganisms or internal parasites. Long-lived specific immunity is a result of the action of the immune system and may be acquired naturally by previous infection, but can also be induced by vaccination. See A summary of the types of immunity, page 231.

Answers

7

	Natural	Artificial
Active immunity	antibodies received via the placenta	monoclonal antibodies to treat tetanus
Passive immunity	measles infection	polio vaccine

12 Energy and respiration

1

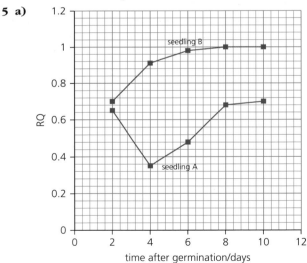

2 The important features of ATP are that it is:
- a substance that moves easily within cells and organisms – by facilitated diffusion
- formed in cellular respiration and takes part in many reactions of metabolism
- able to transfer energy in relatively small amounts, sufficient to drive individual reactions.

3 See Extension: The respiration of fats and proteins, page 239, and Excretion, blood water balance and metabolic waste, particularly Figure 14.14, page 297.

4 $C_{15}H_{31}COOH = C_{16}H_{32}O_2$

$C_{16}H_{32}O_2 + 23O_2 \rightarrow 16CO_2 + 16H_2O + Energy$

$$RQ = \frac{CO_2 \text{ produced}}{O_2 \text{ taken in}}$$

$$RQ = 16 \div 23 = 0.696$$

5 a)

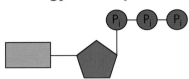

b) Seedling A has a pattern of RQ change in early germination typical of seeds with a food store consisting mainly of carbohydrate. (Actually, this data is from germinating wheat fruits which contain 70% carbohydrate, 2% lipid (oil) and 12% protein).
Seedling B has a pattern of RQ change in early germination typical of seeds with a food store consisting mainly of lipid. (Actually, this data is from germinating flax seeds which contain 2% carbohydrate, 55% lipid (oil) and 20% protein).

6 The following are produced during glycolysis: reduced NAD (c), ATP (e) and pyruvate (g).

7 a) In respiration, all the hydrogen atoms are gradually removed from glucose, catalysed by dehydrogenase enzymes. The hydrogen atoms are added to hydrogen acceptors, usually NAD (nicotinamide adenine dinucleotide, page 238), forming reduced NAD. NAD is a coenzyme that works with specific dehydrogenase enzymes in the oxidation of substrate molecules by the removal of hydrogen.

b) Decarboxylation is the removal of carbon from organic compounds by the formation of carbon dioxide. For example, glucose consists of six carbon atoms. All six carbon atoms are removed at different stages of respiration, one at a time, and given off as carbon dioxide. A specific decarboxylase enzyme is involved in each case. The first decarboxylation in aerobic respiration occurs in the reaction linking glycolysis with the Krebs cycle, when pyruvate is converted to a two-carbon molecule. The other decarboxylation reactions of aerobic respiration occur in steps in the Krebs cycle.

8 a) A substrate is a molecule that is the starting point for a biochemical reaction; it forms a complex with a specific enzyme. An intermediate is a molecule formed as a component of a metabolic pathway.

b) See Glycolysis, page 243, and The link reaction and the Krebs cycle, pages 243–5.

c) See Extension: Respiration as a series of redox reactions, page 238.

9 a) Glycogen

b) Starch

10 Endergonic reactions require energy input, because the products have more potential energy than the reactants – see Figure 3.1, page 57. Note: the alternative type of reaction is one in which the products have less potential energy than the reactants. These reactions transfer energy as heat and work and are called exergonic reactions.

11 Glycolysis (the conversion of glucose to pyruvate) which occurs in the cytosol (the aqueous part of the cytoplasm that surrounds organelles such as the mitochondria) does not require oxygen for completion.

12 Use of a single water bath at a favourable temperature – perhaps 25 °C; repetition of the experiment with a range of dilute glucose solutions (the substrate).

13 Lactate and a small amount of ATP are the final products of anaerobic respiration in muscle fibres.

14 In the absence of oxygen reduced NAD accumulates and oxidised NAD reserves are used up. In the absence of oxidised NAD, pyruvate production by glycolysis slows and stops, so subsequent steps in respiration must stop too.

15 Presence of aerenchyma tissue of stem, leaf and root; ability of roots to respire anaerobically – with relatively high tolerance of ethanol.

16 In the respirometer, the far side of the U-tube manometer is the control tube (A). Here, conditions are identical to those in the respirometer tube, but in the former, no living material is present. However, any change in external temperature or pressure is equally experienced by both tubes, and their effects on the level of manometric fluid are equal and opposite, and they cancel out.

13 Photosynthesis

1 See Figure 3.1, page 57.

2 Chlorophyll is a mixture of pigments that absorbs many of the component wavelengths of white light except green, which it transmits or reflects (hence the green colour of chlorophyll).

3 Knowing the R_f value of a metabolite or pigment helps in the identification of that substance when the composition of a different mixture is investigated. However, R_f values depend upon the same composition and concentration of solvent being used and also the same type and grade of chromatography paper.

4 The hydrophobic 'tail' of the chlorophyll molecule is unable to form hydrogen bonds with water molecules, making chlorophyll insoluble in water.

5 Chromatography allows the complete separation of components of mixtures. It is an ideal technique for separating biologically active molecules obtained from samples of tissue since biochemists are often able to isolate only very small samples.

6 Reduced NADP and ATP are the products of the light-dependent stage of photosynthesis required (and promptly used) in the light-independent stage. The light-independent stage occurs in the stroma, so the formation of reduced NADP and ATP on the side of thylakoid membranes that face the stroma makes them instantly available.

7 There are three ways in which the gradient in protons between the thylakoid space and the stroma is generated:
- the splitting of water on the thylakoid membrane on the thylakoid space side is a source of hydrogen ions there
- the energy made available as excited electrons from photosystem II are passed along the electron carrier chain to photosystem I is used to pump hydrogen ions from the stroma to the thylakoid space
- the formation of reduced NADP on the stroma side of the thylakoid membrane removes hydrogen ions there (reducing their concentration in the stroma).

8 a) The ultimate fate of electrons displaced from the reaction centre of photosystem I in non-cyclic photophosphorylation is to be retained in reduced NADP.

b) The ultimate fate of electrons displaced from the reaction centre of photosystem I in cyclic photophosphorylation is to be returned to the place in photosystem I from which they came.

9 In this demonstration of the Hill reaction, the isolated chloroplasts are provided with an electron acceptor dye (DCPIP). The chlorophyll at the reaction centre is oxidised by light energy and in this state is able to split water by removing electrons. So DCPIP is the electron acceptor in this demonstration of the Hill reaction and the electrons come from water.

10 Autotrophic nutrition of plants supports not only the plant, but also (indirectly) the primary consumers that feed on it and any higher-level consumers that are in the same food chain or web. Consequently, the carbon atoms making up almost all living things have been in a carbon dioxide molecule that has been catalysed by rubisco at some stage.

11 A thermometer is required in the glass tube with the pond weed (in a position that does not interfere with the supply of light to the plant), so that the temperature at which photosynthesis is measured is known. Also, it is advisable to check that the pond weed sample is not subjected to an unplanned rise in temperature as the intensity of light is increased, for example.

12 a) The biochemical steps of photosynthesis are dependent on the action and activity of numerous enzymes. As these are made of protein, they are progressively denatured and rendered inactive at high temperatures.

b) As water evaporates from the surfaces of mesophyll cells during transpiration, cooling occurs.

13 a) Prevention of autolysis of cell organelles (particularly the chloroplasts) as hydrolytic enzymes, held in previously intact lysosomes, are released into the homogenised cytoplasm.

b) Room temperature was selected as the working temperature at which the Hill reaction was investigated at.

c) To maintain a constant pH favourable for the enzymes of the chloroplast.

d) Tube 1 – colour change from blue to colourless
Tube 2 – colour remains blue

e) Tube 3 – confirms that chloroplasts are required for the Hill reaction.
Tube 4 – confirms the role of DCPIP as the hydrogen acceptor.

14 Adapt part of the diagram of the chloroplast (diagrammatic view) in Figure 13.22. Select your annotations from Table 13.2 (page 280).

15 See What happens in photosynthesis and Figure 13.6 (page 266).

16 The starting materials (carbon dioxide and water) are in an oxidised (low energy) state. The end product of photosynthesis – glucose, is in a reduced (higher energy) state.

17 See Photosynthesis in some tropical plants (pages 280–1) and Figure 13.24 (page 282).

14 Homeostasis

1 a) Carbon dioxide is an acidic gas which, if it were to accumulate in the blood, would alter the pH of the plasma solution. The normal pH of the blood is 7.4. For life to be maintained it must remain within the range pH 7.0–7.8. This is largely because blood pH affects the balance of essential ions which are transported in the plasma solution. Efficient removal of respiratory carbon dioxide from the lungs is as important to life as efficient uptake of oxygen.

b) Receptor: chemoreceptors in the medulla, carotid artery and aorta
Co-ordinator: inspiratory and expiratory centres of the respiratory centre in the medulla
Effectors: muscles of the ribs and the diaphragm that bring about ventilation of the thorax.

2 Blood from the gut via the hepatic portal vein, as well as containing the absorbed products of digestion, will be low in O_2 and relatively high in CO_2. So it is the supply of oxygen that the highly metabolic liver cells constantly require that is delivered by the hepatic artery.

Answers

3 The following organelles would be directly involved in hormone production and discharge:
- the nucleus – the site of the gene for the hormone (a protein) concerned, together with the enzyme machinery for the production of relevant messenger RNA
- pores in the nuclear envelope – exit points for the messenger RNA on its way to the cytosol
- ribosomes attached to the endoplasmic reticulum (RER) – the site of the formation of the hormone
- vesicles cut off from the RER – transport of the hormone across the cytosol to the cell surface membrane
- the cell surface membrane – to which the vesicles containing the hormone fuse allowing discharge of the hormone.

4 When a diabetic patient assesses their blood glucose by means of a biosensor, the measurement is of the amount of glucose present at the moment the reading is taken. When readings are taken of glucose levels in the urine, there has been an inevitable lag time before blood glucose level is reflected in the urine. Colorimetric readings involve a level of subjective judgment that the biosensor avoids, too.

5 a) The force that drives ultrafiltration in the glomerulus is the blood pressure generated by the muscles of the ventricles. This pressure is heightened by the efferent arteriole of the glomerulus being of a smaller diameter than the afferent arteriole.

   b) Water, useful ions, glucose and amino acids, along with urea.

6 a) Brush border is made up of microvilli. Microvilli provide a vastly increased surface area of cell surface membrane in contact with the filtrate. In the membrane there are protein pumps which actively and selectively absorb useful metabolites.

   b) See Facilitated diffusion, pages 81–2, and Figure 14.19, page 301.

7 In a counter-current flow system fluids flow in opposite directions in parallel and adjacent tubes. So the limbs of the loop of Henle and the vasa recta as it occurs besides the loop of Henle are both counter-current systems. Exchange occurs between these systems and a gradient is maintained along the entire exchange surface.

8 a) A negative feedback system consists of a receptor, a coordinator and an effector. In osmoregulation we have:
Receptor: osmoreceptors in the hypothalamus
Co-ordinator: hypothalamus and posterior pituitary gland from where ADH is released
Effector: collecting ducts of the nephrons in the medulla of the kidneys.

   b) The osmoreceptors in the hypothalamus, together with the receptors in the aorta and carotid arteries detect the change in the water content of the blood. Nerve impulses from the receptors are transmitted to the hypothalamus and pituitary gland.

9 See Figure 4.17 and Figure 4.18 (pages 88 and 89).

10 Stomata open and close in response to changes in turgor in the guard cells. See The opening and closing of stomata, pages 308–9 and Abscisic acid and water stress in the leaf page 309.

15 Control and co-ordination

1

Motor neuron	Sensory neuron	Relay neuron
Receives impulses via many dendrites	Receives impulses via many (typically long) dendrons, each protected by a myelin sheath	Receives impulses via many (typically short) nerve fibres, which are not myelinated
Passes impulse to a muscle fibre or gland via a single axon, which is protected by a myelin sheath	Passes impulse to dendrites of a relay neuron or its cell body via a single axon, which is protected by a myelin sheath	Passes impulse to dendrites or the cell body of motor neurons via many short nerve fibres, which are not myelinated

2

Stimulus detected	Receptor	Location of receptor
Mechanoreceptors		
Movement and position	Stretch receptors, e.g. muscle spindles, proprioceptors	Skeletal muscle
Blood pressure	Baroreceptors	Aorta and carotid artery
Thermoreceptors		
Internal temperature	Cells of the hypothalamus	Brain
Chemoreceptors		
Blood O_2, CO_2, H^+	Carotid body	Carotid artery
Osmotic concentration of the blood	Osmoregulatory centre in the hypothalamus	Brain

3 In positive feedback the effect of a deviation from the normal or set condition is to create a tendency to reinforce the deviation. Positive feedback intensifies the corrective action taken by a control system, leading to a 'vicious circle' situation. Imagine a car in which the driver's seat was set on rollers (rather than being secured to the floor) being driven at speed. The slightest application of the foot brake causes the driver to slide and to press harder on the brake as the car starts to slow with an extreme outcome.
Biological examples of positive feedback are rare, but one can be identified at the synapse. When a wave of depolarisation (a nerve impulse) takes effect in the postsynaptic membrane, the entry of sodium ions triggers the entry of further sodium ions at a greater rate. This is a case of positive feedback. The depolarised state is established and the impulse moves along the postsynaptic membrane.
In negative feedback the effect of a deviation from the normal or set condition is to create a tendency to eliminate the deviation. Negative feedback is a part of almost all control systems in living things. The effect of negative feedback is to reduce further corrective action of the control system once the set-point value is reached.

4 a) See The resting potential, page 316.
   b) See The action potential, page 317.

5 See The refractory period, page 318.

6 See Speed of conduction of the action potential, page 319.

7 **a)** Transmitter substances are produced in the Golgi apparatus in the synaptic knob and held in tiny vesicles prior to use.

b) See Figure 15.10, page 321.

8 See The ultrastructure of skeletal muscle, and Figures 15.11, 15.12, and Fig 15.13 pages 322–324.

9 The effects of hormones are restricted to cells with specific receptors on the cell surface membrane to which they attach or with which they react, so they affect only the target cells or tissues.

10 In negative feedback the effect of a deviation from the normal or set condition is to create a tendency to eliminate the deviation. Negative feedback is a part of almost all control systems in living things. The effect of negative feedback is to reduce further corrective action of the control system once the set-point value is reached.

In positive feedback the effect of a deviation from the normal or set condition is to create a tendency to reinforce the deviation. Positive feedback intensifies the corrective action taken by a control system, leading to a 'vicious circle' situation. Biological examples of positive feedback are rare, but one can be identified at the synapse. When a wave of depolarisation (a nerve impulse) takes effect in the postsynaptic membrane, the entry of sodium ions triggers the entry of further sodium ions at a greater rate.

11 **a)** At about the mid-point in the menstrual cycle (day 14) the high and rising level of oestrogen suddenly stimulates the secretion of LH and, to a slightly lesser extent, FSH, by the pituitary gland. LH stimulates ovulation (the shedding of the mature secondary oocyte from the Graafian follicle) and the secondary oocyte is released from the ovary. See also Figure 15.19, page 331.

b) The absence of fertilisation, together with the falling levels of FSH and LH in the blood (from about day 25 onwards) allows the corpus luteum to degenerate.

12 A placebo is a pill (or medicine) which is given for psychological reasons rather than for any physiological effect. It may be used in the place of a specific medication or drug.

13 Growth consists of a permanent, irreversible increase in size; development is a change in shape, form and complexity of an organism.

16 Inherited change

1 **a)** See Chromosomes occur in pairs, page 98.

b) Mitosis and meiosis are divisions of the nucleus by very precise processes; they ensure the correct distribution of chromosomes between the daughter cells.

The daughter cells produced by mitosis have a set of chromosomes identical to each other and to the parent cell from which they were formed. In growth and development it is essential that all cells carry the same information as the existing cells. Similarly when repair of damaged or worn out cells occurs, the new cells are exact copies of what they replace.

Meiosis occurs in the life cycle of all organisms that reproduce sexually. In meiosis four daughter cells are produced, each with one member of each homologous pair of the chromosomes of the parent cell, known as the haploid (*n*) state Halving of the chromosome number is essential since at fertilisation the number is doubled.

2 A major event of interphase is the replication of the chromosomes. By the omission of this stage between meiosis I and meiosis II, the chromosome number is halved in the four cells produced following meiotic cell division.

3 **a)** Metaphase I and II

b) Mid prophase I

c) Late prophase I

d) Anaphase II

e) Anaphase I

f) Telophase II

4

	Mitosis – a replicative division	Meiosis – a reductive division
Consequences	All cells carry the same genetic information as the existing cells from which they are formed, and which they share with surrounding cells	• Ensures maintenance of the chromosome number in body cells from generation to generation • Is a source of variation (important for the survival of the species in a changing world)
Product	Two identical cells, each with the diploid number of chromosomes	Four, non-identical cells, each with the haploid number of chromosomes
Significance	• Permits growth and repair within multicellular organisms • Also, for asexual reproduction	Contributes to genetic variability by: • reducing the chromosome number by half, permitting subsequent fertilisation and the combination of the genes of two parents • permitting the random assortment of maternal and paternal chromosomes • allowing recombination of segments of individual maternal and paternal homologous chromosomes during crossing over

5 The idea that the characteristics of parents were blended in their offspring was disproved by investigating the progeny of crosses of organisms with contrasting characteristics (such as 'tall' and 'dwarf' pea plants) through a second and sometimes subsequent generations. Large samples were used and the experiments were repeated, confirming that the observed ratios were statistically significant.

6 We can expect a ratio of 3:1 among the progeny only if three conditions are met:

- fertilisation is random – each fertilisation is independent of all the others
- there are equal opportunities for survival among the offspring
- large numbers of offspring are produced.

Answers

In breeding experiments with plants such as the pea plant, exact ratios may not be obtained because of parasite damage, due to the action of browsing predators on the anthers or ovaries in some flowers or because some pollen types fail to be transported by pollinating insects as successfully as others.

7 The layout of your monohybrid cross will be as in Figure 16.14 (page 355), but the parental generation (P) will have genotypes (if you have chosen C to represent the gene for coat colour):

$$C^R C^R \times C^W C^W$$

where C^R represents the allele for red coat and C^W represents the allele for white coat. The gametes the parental generation produce will be:

$$C^R \text{ and } C^W$$

The offspring (F_1) will have the genotype $C^R C^W$ and the phenotype will be 'roan'.

In a sibling cross of the F_2 generation the gametes of both siblings will be:

$$\tfrac{1}{2} C^R \quad \text{and} \quad \tfrac{1}{2} C^W$$

	C^R	C^W
C^R	$C^R C^R$	$C^R C^W$
C^W	$C^R C^W$	$C^W C^W$

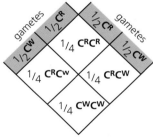

From the genetic diagram, the offspring (F_2) to be expected and the proportions are:

Genotype	$C^R C^R$	$C^R C^W$	$C^W C^W$
Ratio	1	2	1
Phenotype	red	roan	white

8 • Mr and Mrs Jones are blood groups A and B, which might produce any one of:
AA × BB → blood group AB
AO × BO → blood groups A, B, AB or O
AA × BO → blood groups AB or A
AO × BB → blood groups AB or B
So Mr and Mrs Jones could be the parents of *any* of the four children.

• Mr and Mrs Lee are blood groups B and O, which might produce either of:
BB × OO → blood group B
BO × OO → blood group B or O
So Mr and Mrs Lee might be the parents of *either* the blood group B or the blood group O child.

• Mr and Mrs Gerber are both blood group O, which can produce only:
OO × OO → blood group O
So Mr and Mrs Gerber must be the parents of the blood group O child.

• This also means that, by elimination, Mr and Mrs Lee must be the parents of the blood group B child.

• Mr and Mrs Santiago are blood groups AB and O, which might produce:
AB × OO → blood groups A or B
Since Mr and Mrs Lee are the parents of the blood group B, Mr and Mrs Gerber must be the parents of the blood group A child.

• By elimination, Mr and Mrs Jones must be the parents of the blood group AB child.

9 a) See Red–green colour blindness, page 359.
   b) See Sex linkage, page 358, and Figure 16.17, page 359.

10 The recombinants among the F_2 progeny are those with round seeds and green cotyledons and those with wrinkled seeds and yellow cotyledons.

11

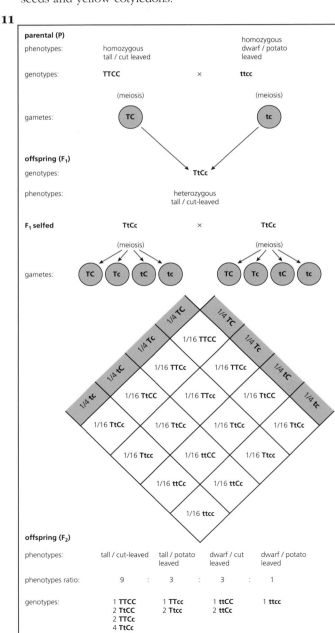

12 a) BBSS, BBSs, BbSS or BbSs

b) From a test cross, using a spaniel that has a red spotted coat. Among the puppies produced, the presence of red coat or of spotted pattern indicates that the other parent spaniel is heterozygous for that character.

13 A mutant organism (or cell) is one carrying altered genetic material which makes it different from its parent (or from its precursor cell).

14 a)

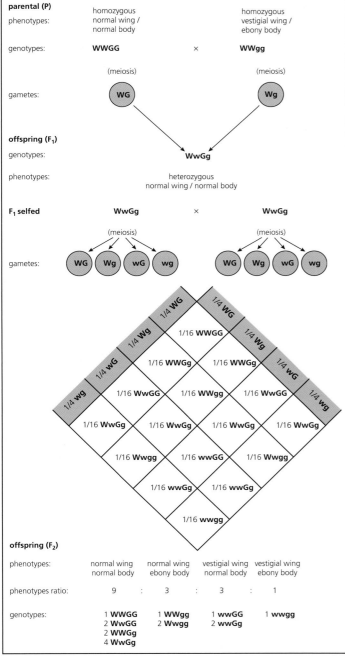

parental (P)

phenotypes: homozygous normal wing / normal body   homozygous vestigial wing / ebony body

genotypes: **WWGG** × **WWgg**

(meiosis)   (meiosis)

gametes: **WG**   **Wg**

offspring (F₁)

genotypes: **WwGg**

phenotypes: heterozygous normal wing / normal body

F₁ selfed **WwGg** × **WwGg**

(meiosis)   (meiosis)

gametes: **WG Wg wG wg**   **WG Wg wG wg**

offspring (F₂)

phenotypes: normal wing normal body | normal wing ebony body | vestigial wing normal body | vestigial wing ebony body

phenotypes ratio: 9 : 3 : 3 : 1

genotypes:
1 WWGG   1 WWgg   1 wwGG   1 wwgg
2 WwGG   2 Wwgg   2 wwGg
2 WWGg
4 WwGg

b) See diagram above.

c) The recombinants are those with normal wings and ebony bodies and those with vestigial wings and normal bodies.

15 First calculate the χ^2 statistic.

Category	Predicted ratio	Observed number, O	Expected number, E	O − E	(O − E)²	(O − E)²/E
tall, axial flowers	1	55	52	3	9	0.173
tall, terminal flowers	1	51	52	−1	1	0.019
dwarf, axial flowers	1	49	52	−3	9	0.173
dwarf, terminal flowers	1	53	52	1	1	0.019
Total		208	208			0.384

So $\chi^2 = 0.384$ and there are three degrees of freedom. From a table of the distribution of χ^2 (see page 364), the probability (P) of obtaining a deviation as large as (or larger than) the one we have by chance alone falls between the probability levels of P = 0.90 and P = 0.95.
This is not significant – there is no departure of the observed from the expected values.

16 Gene mutations are due to changes in the sequence of bases in the DNA of a gene, such as when one base is replaced by another in the coding strand of DNA that makes up the linear sequence of bases.

17 a) Liz and Diana

b) i) David and Anne

ii) James, William, Arthur and Fredrick

c) 8 (Richard and Judith, Anne, Charles, Sophie, Chris, Sarah, and Gail)

d) James and William, Arthur and Diana, etc.

18 A dominant allele is expressed in all individuals that inherit it (even if only heterozygous for that allele), whereas a recessive condition appears only in offspring that are homozygous for that allele (and therefore with parents who were both at least carriers of the allele – a relatively rare event).

19 The genes controlling the production of the blood proteins concerned in haemophilia are located on the X chromosome. Haemophilia is caused by a recessive allele. As a result, haemophilia is largely a condition of the male since in him a single X chromosome carrying the defective allele ($X^h Y$) will result in disease. For a female to have the haemophilia, she must be homozygous for the recessive gene ($X^h X^h$), a condition that is frequently fatal *in utero*.

17 Selection and evolution

1 a) Your answer could include two of these soluble nutrients essential to phytoplankton growth: nitrate; calcium or magnesium ions;

b) The numbers of phytoplankton decrease rapidly in May and June due to the level of predation by zooplankton and possibly due a shortage of essential nutrients.

c) The increasing quantity of nutrients in October came from the decay of the dead phytoplankton and zooplankton.

Answers

2 a) A non-native species may rapidly become the most common species in a new environment due to an absence of established predators and parasites, and possibly due to limited competition for resources.

b) Competition for food among the non-native population and the rapid spread of parasites among them, particularly if their large numbers cause them to live in close contact.

3 a) Peaks in predator populations occur after peaks in the prey population. A rise in the number of predators occurs because of the large number of prey available to be caught and because fewer than the maximum number of predators the habitat can support are competing for the prey.

b) An increases in the number of predators increases the competition for food and predator numbers exceed the supply of prey. Predators die so their population falls.

4 See But gene pools do change!, page 392.

5 The breeding of domesticated plants and animals has created varieties with little external resemblances to their wild ancestors. Darwin bred pigeons and noted there were more than a dozen, distinctive varieties of pigeon, all of which were descendents from the rock dove (Figure 13.13). Darwin argued that if so much change can be brought about in so few generations, then species must be able to evolve into other species by the gradual accumulation of minute changes as environmental conditions alter and natural selection operates.

6 In natural selection, environmental factors act on phenotypes so that organisms with the best suited phenotypes survive and breed, passing on the alleles to their offspring. This leads to improved survival or reproductive ability of these phenotypes and results in changes in the gene pool. Artificial selection (selective breeding) is the intentional breeding by humans for certain phenotypes or combinations of phenotypes. This can lead to reduced ability to survive in the wild and reduces the size of the gene pool.

7 Your ideas are likely to include:
- the identification of genes and alleles as Mendel's factors
- chromosomes as vehicles for carrying genes
- the behaviour of chromosomes in cell divisions
- meiosis as the mechanism of separating alleles – so that only one from each parent is passed on to the next generation.

8 By evolution we mean the gradual development of life in geological time. For example, we know that life appeared on Earth about 3500 million years ago and that most of the great diversity of living forms has appeared subsequently, with time. Before early geologist realised the Earth is extremely old, biblical calculations suggested that life had been 'created' in 4004 BC, or about 6000 years ago! If this were still believed, then there would be insufficient time for evolution by natural selection.

9 See Neo-Darwinism, page 400.

10 A variety is a grouping within a species whose members differ in some significant respect from other members of the species; a species is a group of organisms sharing morphological similarities that are reproductively isolated and can interbreed to produce fertile offspring. Different varieties of a species can interbreed.

The examples you propose are most likely to be ones familiar in habitats near where you live.

11 The factors you are likely to suggest include:
- Mutation – random, rare, spontaneous change in the genes in the gonads that lead to the possibility of new characteristics in the offspring, that come to confer an advantage in the struggle for existence
- Selective predation of members of the population with certain characteristics that are genetically controlled
- Emigrations of members of the population or immigrations of members of other populations
- Sudden hostile physical condition, for example, cold, flooding or drought, may quickly reduce a natural population to a very few survivors. On the return of a favourable environment, the numbers of the affected species may return to normal, for example, due to reduced competition for food sources. However, the new population has been built from a very small sample of the original population, so there are likely to be more crosses between closely related individuals (resulting in fewer heterozygotes and more homozygotes) and some alleles may be lost altogether. (This is called random genetic drift.)
- A barrier may develop within a population, instantly isolating a small sample of the original population, which may carry an unrepresentative selection of the gene pool, yet be the basis of a new population. (This is another form of genetic drift. It is called the founder effect.)

18 Biodiversity, classification and conservation

1 Possible abiotic factors include light, humidity, topography, soil (pH, moisture content, organic matter, etc.)

2 a) ecosystem
b) population
c) abiotic factor
d) community
e) habitat
f) abiotic factor
g) biomass

3

	plant	animal
a high water location	spiral wrack	rough winkle
a low water location	oar weed	acorn barnacle

4 The estimated population size would be halved, i.e. 40/2 = 20.

5 a) The size of the population at days 2, 16, 31 and 46 and 76:

Day	$n_1 \times n_2 / n_3$	N
2	$7 \times 6 / 2$	21
16	$14 \times 12 / 2$	84
31	$25 \times 24 / 3$	200
46	$19 \times 16 / 2$	152
76	$19 \times 14 / 2$	133

b) Plot the graph using an Excel spreadsheet – concerning 'time in days' and 'N', which will you plot on the x and the y axis?

Answers

c) The annotations to the curve are an opportunity to speculate on:

- on the initial rapid rise in numbers (continuing im-migrations, initial absence of competition for abundant resources, lack of presence of predators when population new/low in numbers)
- on subsequent stabilisation of population numbers (balance between reproduction/predation pressures, the idea of carrying capacity for that habitat).

6 The Simpson Diversity Index for this habitat:

Species (no of individuals)	n	n − 1	n(n − 1)
Groundsel	45	44	1980
Shepherd's purse	40	39	1560
Dandelion	10	9	90
Total (N)	95		

$\sum n\,(n-1) = 1980 + 1560 + 90 = 3630.$

$D = 95 \times \dfrac{94}{3630} = \dfrac{8930}{3630} = 2.46$

7 The original data in Figure 18.11 were obtained by a team of pre-University students in an activity they designed and carried out (independently of teachers) as the culmination of a field course. Respond to this question in like-manner. Select two small teams of students, each to tackle the calculation of one of the diversity indices. Then, by group discussion, evaluate the effect of age/stage of dune development on diversity. Present your data and conclusions as a poster or display newspaper item.

8 Precisely defined and internationally agreed scientific names facilitate cooperation between observers by identifying the exact species that is being investigated and reported on.

9 The central column needs to start with 'Domain: one of the three major forms of life'. The left- and right-hand columns need to parallel this addition with 'Domain: Eukaryota'.

10 Sedimentary rock layers are usually laid down under water, often under anaerobic conditions, and formed from sediments washed in from the land. They are added to, year after year. Objects that fall to the bottom (for example, dead organisms) are covered, compressed and eventually their molecules react with or are replaced by mineral ions. Fossils laid down in the deeper strata of sedimentary rocks are older than the fossils of the higher (more recent) strata.

Dead organism that fall elsewhere are much more likely to be dismembered by scavengers and/or decayed by microorganisms before any fossilisation is possible.

11 The chief ecosystems are:

- Tundra, e.g. alpine tundra, which occurs on the highest mountains, well above the tree-line
- Grassland, e.g. steppe, prairie and pampas, or savannah, which is tropical grassland
- Desert
- Forest or scrubland

12 Human actions and activities that represent a threat to the Earth's wildlife include:

- Deforestation – Destruction of trees destroys the habitats of numerous species of animals, plants, fungi and bacteria which therefore decline in numbers.
- Desertification – This is speeded up by:
  - overgrazing of land,
  - deforestation, or
  - climate change.
- Excessive application of pesticides in modern industrial agriculture. Pesticides have improved productivity in agriculture, but their use has reduced biodiversity.
- Air pollution – The levels of carbon dioxide in the atmosphere have been rising since the Industrial Revolution in the developed countries of the world. This gas is a major contributor to global warming and destructive climate change.
- Water pollution.

13

14 Information may be available from local wardens or rangers via their head office.

Answers

15 *in vitro* refers to biological processes occurring in cell extracts (literally 'in glass'); *in vivo* refers to biological processes occurring in living organisms (literally 'in life').

16 Possible answers:
- The possibility of artificial insemination using sperm obtained from captive individuals from a different gene pool. Similarly, *in-vitro* fertilisation and the implantation of embryos in surrogate mothers
- Maintenance of accurate breeding records so that the genetic relatedness of progeny is known for future crosses and prior to exchanges of gametes or individuals between different zoos
- Monitoring of the health and development of progeny and on-going supervision of the success of methods of release back into the wild to minimise immediate predation of released stock.

17

Terrestrial and aquatic nature reserves	Captive breeding programmes of zoos and seed banks at botanical gardens
Habitats that are already rare are especially vulnerable to natural disaster – rare habitats themselves are easily lost, if a range of examples are not preserved as nature reserves	Originally, zoos were collections of largely unfamiliar animals kept for curiosity, with little concern for any stress caused, but now captive breeding programmes make good use of these resources
When a habitat disappears the whole community is lost, threatening to increase the total number of endangered species	Captive breeding maintains the genetic stock of rare and endangered species
A refuge for endangered wildlife allows these species to lead natural lives in a familiar environment for which they are adapted and to be a part of their normal food chains	The genetic problems arising from individual zoos having very limited numbers to act as parents is overcome by inter-zoo co-operation (and artificial insemination in some cases)
The wildlife of a reserve may be monitored for early warning of any further deterioration in numbers of a threatened species so that remedial steps can be taken	Animals in zoos tend to have significantly longer life expectancies and are available to participate in breeding programmes for much longer than wild animals do
The offspring of endangered species are nurtured in their natural environment and gain all the experiences this normally brings, including the acquisition of skills from parents and peers around them	Captive breeding problems, for most species it is applied to, have been highly successful, although the young do not grow up in the 'wild', so there is less opportunity to observe and learn from parents and peers
There is an established tradition of maintaining reserves and protected areas in various parts of the world, so there is much experience to share on how to manage them successfully	Breeding programmes generate healthy individuals in good numbers for attempts at re-introduction of endangered species to natural habitats, a particularly challenging process given that natural predators abound in these locations

Nature reserves are popular sites for the public to visit (in approved ways), thereby maintaining public awareness of the environmental crisis due to extinctions and individual responsibilities that arise from it	Zoos and botanical gardens are accessible sites for the public to visit (as they are often situated in urban settings where many may have access), so they contribute effectively to public education on the environmental crisis
Reserves are ideal venues to which to return endangered individuals, the products of captive breeding programmes, providing realistic conditions for re-adaptation to the habitat but facilitating the monitoring of the re-introduction	Seed banks are a convenient and efficient way of maintaining genetic material of endangered plants, as they may exploit the ways seeds survive long periods in nature

18 The stocks in seed banks must be viable. Over very long periods, viability of seeds decreases, so fresh formed seeds are needed regularly. Also, genetic viability is maintained by this process because new seeds are the products of sexual reproduction and so maintain the diversity of the gene pool.

19 Gene technology

1 Viral DNA takes over the ribosomal machinery of the host cell, causing the production of viral proteins that are enzymes. Viral enzymes cause the replication of viral DNA and the production of other molecules that make up new viruses (see Figure 10.19, page 206).

2 a) See Nucleic acids – the information molecules, pages 110–13, and Figure 6.3, page 113.
b) See Table 19.1, page 455.

3 a) A sticky end consists of a short sequence of exposed bases of unpaired nucleotides forming a single-stranded extension at the end of a length of DNA. It is created by the action of a particular restriction enzyme that recognises a specific base sequence in a length of DNA and cuts the DNA in this distinctive way. See also Figure 19.4, page 458.
b) Sticky ends attach to each other by complementary base pairing. They are held together by the formation of hydrogen bonds between complementary bases. The enzyme ligase then catalyses the formation of C–O bonds between the sugar–phosphate backbones of the two DNA strands.

4 a) See Table 16.2, page 357.
b) See Table 19.1, page 455.
c) The bacterial chromosome is a single circular molecule of DNA within the cytoplasm, but attached to the cell surface membrane at one point. Plasmids are tiny circular molecules of DNA, free-floating in the cytoplasm of some bacteria. They replicate independently of the bacterial chromosome.

5 The genes of prokaryotes are generally easier to modify than those of eukaryotes because:
- prokaryotes have a single, circular chromosome, so only one copy of a gene has to be engineered into their DNA, whereas eukaryotes have two alleles of every gene

- plasmids, the most useful vehicle for moving genes, occur in prokaryotes but mostly do not in eukaryotes – the exceptions are some fungi, including yeast, and a few plants
- transcription of DNA into RNA in prokaryotes does not require the removal of 'non-informative DNA' (introns), but in eukaryotes it does
- bacterial walls can be crossed by plasmids (after suitable treatments).

6 A mutant allele is one of a pair of alleles in which (exceptionally) there has been a change in the structure of the DNA. Typically this involves a single base substitution, a translocation, a transposition or an insertion or deletion of a base or short sequences of bases.

7 A person with a single allele for cystic fibrosis is a 'carrier' and does not express the disease; whereas someone with a single allele for Huntington's disorder will become affected by the disease eventually.

8 The fruits of the cereals (which include wheat and rice) are the staples of human diets all over the world. They provide the bulk of essential energy- rich foods. The ability of leguminous plants to fix atmospheric nitrogen for amino acid and protein production enable these plants to grow well without the addition of (expensive) nitrogen-based fertilisers. It also makes food products from them (peas and beans, for example) relatively rich in proteins. Cereals that are genetically modified so that they could also fix nitrogen (if they were created) would grow well without nitrogen-based fertilisers. They may also add even more to human nutrition by helping to the overcome protein deficiency as well as meeting calorie needs.

9 Many newspaper reports are often rather superficial. You may have to amplify the criticisms slightly in order to be able to introduce balancing arguments. Your own opinion needs stating simply, with one or two clearly explained reasons.

10 The presence of one or more sons with haemophilia among the progeny of the members of any generation in a family tree is the way that carriers can be identified among the parents.

11 This is a highly personal issue which needs handling with great sensitivity. This is certainly the case for any counselling team whenever this situation arises. They will need to ensure the couples' preferences emerge and take precedence in ensuing discussions.

As you tackle this question in group discussion with peers, it is best if each individual explains their own view and is listened to carefully. It may be that a consensus concerning a suitable response can be arrived at – but you may find that you cannot all agree. In this situation, you should try to understand the view of those you do not agree with, rather than get into conflict with them.

12 DNA fingerprinting relies on the fact that each individual has unique DNA. The exception is identical twins; their DNA is identical. The DNA of non-identical twins is not.

13 Flowering plants are a major component of human diets and very many are insect-pollinated. Honey bees and butterflies are some of these insect pollinators. A reduction in their numbers would lead to fewer seeds being formed. The supply of fruits and seeds for cultivation and as food would be threatened.

Index

Index

Index

Index

Index

Acknowledgements

The publishers would like to thank the following for permission to reproduce copyright material:

Photo credits:

p.5 *l* © J.C. REVY, ISM/SCIENCE PHOTO LIBRARY, *r* © Gene Cox; **p.8** © Gene Cox; **p.10** © Power and Syred/Science Photo Library; **p.11** *l* © Biophoto Associates/Science Photo Library, *r* © J.C. Revy, ISM/Science Photo Library; **p.12** © Sinclair Stammers/Science Photo Library; **p.13** *t* © Dr Kevin S. Mackenzie, School of Medical Science, Aberdeen University, *m* © Eye Of Science/Science Photo Library, *b* © Power and Syred/Science Photo Library; **p.14** *l* © Phototake Inc./Alamy, *r* © Biophoto Associates/Science Photo Library; **p.17** *t* © Medimage/Science Photo Library, *b* © Omikron/Photo Researchers, Inc./Science Photo Library; **p.18** *t* © Carolina Biological Supply, Co/ Visuals Unlimited, Inc./Science Photo Library, *b* © CNRI/Science Photo Library; **p.19** © Don W. Fawcett/Science Photo Library; **p.20** *t* © Dr Kari Lounatmaa/Science Photo Library/Getty Images, *b* © Dr Jeremy Burgess/Science Photo Library; **p.23** © Kwangshin Kim/ Science Photo Library; **p.26** *tl* © A. Dowsett, Health Protection Agency/Science Photo Library, *bl* © Nigel Cattlin/Alamy, *br* © Nigel Cattlin/Alamy; **p.33** *both* © Andrew Lambert Photography/Science Photo Library; **p.37** © Biophoto Associates/Science Photo Library; **p.38** *both* © Andrew Lambert Photography/Science Photo Library; **p.39** *t* © Phototake Inc./Alamy, *bl* © Bon Appetit/Alamy, *bm* © quayside – Fotolia, *br* © Profotokris – Fotolia; **p.41** © Trevor Clifford Photography/Science Photo Library; **p.42** © Science Photo Library/ Alamy; **p.48** © Steve Gschmeissner/Science Photo Library; **p.71** © Cordelia Molloy/Science Photo Library; **p.77** *both* © Don W. Fawcett/ Science Photo Library; **p.88** *both* © Biophoto Associates/Science Photo Library; **p.100** © Biophoto Associates/Science Photo Library; **p.107** *all* © Michael Abbey/Science Photo Library; **p.108** © Wim van Egmond/Visuals Unlimited, Inc./Science Photo Library; **p.114** *l* © A. Barrington Brown/Science Photo Library, *m* © Science Photo Library, *r* © Science Photo Library; **p.129** *l* © Ed Reschke/Peter Arnold/Getty Images, *r* © Dr Keith Wheeler/Science Photo Library; **p.130** © Gene Cox ; **p.131** *both* © Gene Cox; **p.132** © Gene Cox; **p.133** © Gene Cox; **p.136** *both* © Gene Cox; **p.139** © Gene Cox; **p.141** *t* © Biophoto Associates/Science Photo Library, *b* © Richard Johnson; **p.144** *l* © CJ Clegg, *r* © Gene Cox; **p.147** © Biophoto Associates/Photo Researchers, Inc./Science Photo Library; **p.153** © Gene Cox; **p.155** © Phototake Inc./Alamy; **p.164** © Gene Cox; **p.169** © doc-stock/ Alamy; **p.174** © Steve Gschmeissner/Science Photo Library; **p.175** *t* © Dr Gladden Willis, Visuals Unlimited/Science Photo Library, *b* © Biophoto Associates/Science Photo Library; **p.176** *t* © Science VU/ Visuals Unlimited, Inc./Science Photo Library, *b* © Alaska Stock/ Alamy; **p.177** © Andrzej Tokarski – Fotolia; **p.179** © Phototake Inc./ Alamy; **p.180** © CMEABG/UCBL1, ISM/Science Photo Library; **p.182** *l* © Science Photo Library/Alamy, *r* © Dr Tony Brain/ Science Photo Library; **p.183** © Moredun Animal Health Ltd/Science Photo Library; **p.185** © AP/Press Association Images; **p.188** © Ulet Ifansasti/Getty Images; **p.194** © Dr. Hans Ackermann/Visuals Unlimited, Inc./ Science Photo Library; **p.196** *l* © Science & Society/SuperStock, *r* © Paula Bronstein/Getty Images; **p.200** *t* © James Cavallini/Photo Researchers, Inc./Science Photo Library, *m* © Adam Hart-Davis/ Science Photo Library, *b* © BSIP SA/Alamy; **p.201** *l* © Biophoto Associates/Science Photo Library, *m* © Dr P. Marazzi/Science Photo Library, *r* © Voisin/Phanie/Rex Features; **p.206** © Thomas Deerinck, Ncmir/Science Photo Library; **p.210** © Medical-on-Line/Alamy; **p.214** *l* © RGB Ventures LLC dba SuperStock/Alamy, *r* © Paul Gunning/Science Photo Library; **p.220** © Gene Cox; **p.227** © Science Photo Library/Alamy; **p.235** *l* © Martyn F. Chillmaid/Science Photo Library, *r* © LIU JIN/AFP/Getty Images; **p.249** © Medical-on-Line/ Alamy; **p.251** © Biophoto Associates/Science Photo Library; **p.256** *t* © jjayo – Fotolia, *m* © Dr Keith Wheeler/Science Photo Library, *b* © C J Clegg; **p.279** © Dr Kenneth r. Miller/Science Photo Library; **p.282** © Dr. Ken Wagner/Visuals Unlimited, Inc./Science Photo Library; **p.283** © June Green/Alamy; **p.287** *tl* © Silver – Fotolia, *ml* © Kevin Schafer/Corbis, *mr* © Janthonjackson – Fotolia, *bl* © Jan-Dirk Hansen – Fotolia, *br* © Juniors Bildarchiv GmbH/Alamy; **p.288** © Nickel ElectroLtd; p.293 © Astrid & Hanns-Frieder Michler/Science Photo Library; **p.295** © dalaprod – Fotolia; **p.296** *t* © Cordelia Molloy/Science Photo Library, *b* © Garo/Phanie/Rex Features; **p.299** © age fotostock/SuperStock; **p.300** © Steve Gschmeissner/Science Photo Library/Getty Images; **p.320** © Prof S. Cinti/Science Photo Library; **p.323** *t* © Gene Cox, *b* © Mediscan, University of Aberdeen; **p.324** © Mark Rothery, A Level Biology, www.mrothery.co.uk; **p. 326** © Mark Rothery, A Level Biology, www.mrothery.co.uk; **p.332** © cristi180884 – Fotolia; **p.335** *l* © Martin Shields/Alamy, *r* © FLPA/Alamy; **p.337** © J Garden/Alamy; **p.345** © Science VU/B. John, Visuals Unlimited/Science Photo Library; **p.351** © Mary Evans Picture Library/ Alamy; **p.359** © Sally and Richard Greenhill/Alamy; **p.363** © Graphic Science/Alamy; **p.366** *l* © Frank Greenaway/Dorling Kindersley/Getty Images, *m* © alexfiodorov – Fotolia, *r* © Frank Greenaway/Dorling Kindersley/Getty Images; **p.370** © Friedrich Stark/Alamy; **p.380** *l* © Gene Cox, *m* © tina7si – Fotolia, *r* © Martin Fowler/Alamy; **p.385** *l* © Minden Pictures/SuperStock, *r* © Adrian Sherratt/Alamy; **p.388** © greenwales/Alamy; **p.392** *both* © CJ Clegg; **p.394** *l* © Dave Watts/Alamy, *m* © Jon Durrant/Alamy, *r* © Nigel Cattlin/Alamy; **p.395** © wolfavni – Fotolia; **p.398** *l* © Craig Aurness/Corbis, *m* © inga spence/Alamy, *r* © Tanguy LeNeel/Alamy; **p.406** © tbkmedia.de/Alamy; p.407 *l* © Danita Delimont/Alamy, *r* © All Canada Photos/Alamy; **p.410** *l* © crisod – Fotolia, *r* © Suzanne Long/Alamy; **p.413** © Picture Hooked/ Malcolm Schuyl/Alamy; **p.414** © The Natural History Museum/Alamy; **p.418** *l* © Paul Heinrich/Alamy, *r* © Isarpix/Alamy; **p.422** © Martyn F. Chillmaid/Science Photo Library; **p.423** © Paul Glendell/Alamy; **p.428** CJ Clegg; **p.429** *l* © idp wildlife collection/Alamy, *m* © Lip

Kee / http://www.flickr.com/photos/lipkee/5657636385/sizes/o/in/pool-42637302@N00/ http://creativecommons.org/licenses/by-sa/2.0, *r* © FLPA / Alamy; **p.430** *l* © Mauricio Anton/Science Photo Library, *m* © DDniki – Fotolia, *r* © Gerry Ellis/Minden Pictures/FLPA; **p.434** *t* © Power And Syred/Science Photo Library, *b* © Dr David Patterson/Science Photo Library; **p.435** *tl* © Medical-on-Line/Alamy, *ml* © Sabena Jane Blackbird/Alamy, *mr* © Antony Ratcliffe/Alamy, *bl* © blickwinkel/Alamy, *br* © blickwinkel/Alamy; **p.436** *t* © Craig Lovell/Eagle Visions Photography/Alamy, *bl* © Images of Africa Photobank/Alamy, *bm* © Bjorn Svensson/Science Photo Library, *br* © blickwinkel/Alamy; **p.440** © John Cancalosi/Alamy; **p.446** © Robert Bird/Alamy; **p.448** *tl* © John Devries/Science Photo Library, *tr* © protopic – Fotolia, *b* CJ Clegg; **p.453** © Elena Schweitzer – Fotolia; **p.459** © Dr Gopal Murti/Science Photo Library; **p.463** *tl* © Hiroya Minakuchi/Minden Pictures/FLPA, *br* © Makoto Iwafuji/Eurelios/Science Photo Library; **p.465** © Jan Van De Vel/Reporters/Science Photo Library; **p.468** © CNRI/Science Photo Library; **p.473** © Will & Deni Mcintyre/Science Photo Library; **p.484** © David Parker/Science Photo Library.

Photos in CD material:

Topic 6 Do and understand mini – molecular DNA model kit © CJ Clegg; **Topic 9** Do and understand – TS of human trachea © Dr Gladden Willis, Visuals Unlimited/Science Photo Library, TS of lung tissue © Science Photo Library/Alamy, TS of alveoli © Phototake Inc./Alamy; **Topic 17** Data handling – yeast cells © Biophoto Associates/Science Photo Library.

t = top, *b* = bottom, *l* = left, *r* = right, *m* = middle

Text credits:

Past examination questions reproduced by permission of University of Cambridge International Examinations.

Artwork and data credits:

pp.145-5 Table 7.3 adapted from J.Bonner and A.W. Galston, Principles of Plant Physiology (W.H.Freeman, 1958); **p.157** Figure 8.6 from Colin Clegg (revised by S.Ingham) from Exercise Physiology and Functional Anatomy (Feltham Press, 1995), reproduced by permission of the author; **p.166** Figure 8.15 H.G.Q. Rowett, adapted from Basic Anatomy & Physiology, 3/e (John Murray, 1987); **p.184** Table 9.2 from Raymond Pearl, 'Tobacco Smoking and Longevity' from Science #2253 (4 March 1938); **p.184** Table 9.3: Data on smoking, American Cancer Society, from http://txtwriter.com/onscience/Ospictures/smokingcancer2.2.jpg; **p.185** Figure 9.15 from Cancer Research UK, http://www.cancerresearchuk.org/cancer-info/cancerstats/types/lungs/incidence; **p.188** Figure 9.19 World Health Organisation, from http://www.who.int/whr/1999/en/whr99_ch5_en.pdf, Figure 9.20a National Institutes of Health, from http://en.wikipedia.org/wiki/File:Cancer_smoking_lung_cancer_correlation_from_NIH.svg (15 August, 2007), Figure 9.20b, World Health Organisation, from www.who.int/whr/1999/en/whr99_ch5_en.pfd (15 August, 2007); **p.194** Figure 10.2 re-used from Alastair Gray, World Health and Disease (Open University Press, 1985), reproduced by permission of the author; **p.197** Figure 10.7 World Health Organisation, from Weekly Epidemiological Record, 71, (1996); **p.202** Figure 10.14 adapted from S.J.G. Kavanagh and D.W.Denning, 'Tuberculosis: the global challenge' from Biological Sciences Review, 8; **p.206** Figure 10.20 adapted from G.Pantaleo, C. Graziosi and A.S. Fauci, 'The immunopathogenesis of human immunodeficiency virus infection' from New England Journal of Medicine, 325(5); **p.208** Figure 10.21 from US Centers for Disease Control (23 April 2010), http://en.wikipedia.org/wiki/File:measles_US_1944-2007_inset.png; **p.213** Figure 10.26 The Health Protection Agency; **p.214** Chris Smyth, 'Antibiotics crisis means scratch could kill', from The Times (May 1, 2014); **p.214** Figure 10.28 US Department of Health & Human Services, 2013 from http://www.cdc.gov/drugresistance/threat-report-2013/pdf/ar-threats-2013-508.pdf; **p.215** Figure 10.29 US Department of Health & Human Services, 2013 from http://www.cdc.gov/drugresistance/threat-report-2013/pdf/ar-threats-2013-508.pdf; **p.257** Figure 12.18 M. Ingrouille, Azolla plants, adapted from Diversity and Evolution of Land Plants, p.238 (Figure 7.6) (Chapman & Hall, 1992); **p.257** Figure 12.18 R.C. Moran, Azolla, graph adapted from 'The little nitrogen factories' from Biological Sciences Review, 10(2): 2-6, p.5 (Philip Allan Updates); **p.276** Figure 13.20, from http://www.omafra.gov.ca/english/crops/facts/00-077.htm; **p.373** Figure 16.27, from http://upload.wikimedia.org/wikipedia/commons/thumb/2/22/Lac_Operon.svg/512px-; **p.378** Figure 17.1 adapted from 'The Heights and Weights of Adults in Great Britain' (1999), Office for National Statistics © Crown Copyright; **p.386** Figure 17.8, data from R.J. Berry, Inheritance and Natural History (Collins New Naturalist Series), (HarperCollins, 1977); **p.396** Figure 17.17, drawn using data from J.G.Vaughan and C.A. Geissler, The New Oxford Book of Food Plants (Oxford University Press, 1997); **p.398** Figure 17.20 adapted from Julius Janick et.al., Plant Science: An Introduction to World Crops, 2nd edition (W.H.Freeman, 1974); **p.404** Figure 17.23 adapted from A.D.Bradshaw and T. McNeilly, Evolution and Pollution (Cambridge University Press, 1981), reproduced by permission of Cambridge University Press; **p.441** Figure 18.24 IUCN Red List data (www.iucnredlist.org), reproduced by permission of the International Union for Conservation of Nature (IUCN); **p.457** Figure 19.3 adapted from P. Moore, Recombinant DNA Technology (The Biochemical Society, 1994) © the Biochemical Society, reproduced by permission of Portland Press; **p.466** Figure 19.11 from A.McLennan et.al., Molecular Biology, 4th Edition (Garland Science, 2013).

Permission for re-use of all © Crown Copyright information is granted under the terms of the Open Government Licence (OGL).

Thanks are due to the following sources of inspiration and data for illustrative figures:

p.59 Fig 3.3 *Random collision possibilities* adapted from Dianne Gull and Bernard Brown, 'Enzymes: fast and flexible', *Biological Sciences Review* 6.2, November 1997; **p.442** Fig 18.25 Adapted from Stuart L. Pimm and Clinton Jenkins (2005) 'Sustaining the Variety of Life' © *Scientific American* Inc. 293:3.

Every effort has been made to trace all copyright holders, but if any have been inadvertently overlooked the Publishers will be pleased to make the necessary arrangements at the first opportunity.